D1739946

Molecular Nanomagnets

MESOSCOPIC PHYSICS AND NANOTECHNOLOGY

MOLECULAR NANOMAGNETS

Dante Gatteschi
Department of Chemistry
University of Florence, Italy

Roberta Sessoli
Department of Chemistry
University of Florence, Italy

Jacques Villain
Départment de Recherche Fondamentale sur la Matière Condensée,
DSM, C.E.A. Grenoble, France

OXFORD

UNIVERSITY PRESS

Great Clarendon Street, Oxford OX2 6DP

Oxford University Press is a department of the University of Oxford.
It furthers the University's objective of excellence in research, scholarship,
and education by publishing worldwide in

Oxford New York

Auckland Cape Town Dar es Salaam Hong Kong Karachi
Kuala Lumpur Madrid Melbourne Mexico City Nairobi
New Delhi Shanghai Taipei Toronto

With offices in

Argentina Austria Brazil Chile Czech Republic France Greece
Guatemala Hungary Italy Japan Poland Portugal Singapore
South Korea Switzerland Thailand Turkey Ukraine Vietnam

Oxford is a registered trade mark of Oxford University Press
in the UK and in certain other countries

Published in the United States
by Oxford University Press Inc., New York

British Library Cataloguing in Publication Data

Data available

Library of Congress Cataloging in Publication Data

Gatteschi, D. (Dante)
Molecular nanomagnets / Dante Gatteschi, Roberta Sessoli,
Jacques Villain.
p. cm.
Includes bibliographical references.
ISBN-13: 978–0–19–856753–0 (acid-free paper)
ISBN-10: 0–19–856753–7 (acid-free paper)
1. Magnetism. 2. Nanostructured materials—Magnetic properties.
I. Sessoli, Roberta. II. Villain, Jacques. III. Title.
QC753.2.G38 2006
538—dc22   2005030492

Typeset by Newgen Imaging Systems (P) Ltd., Chennai, India
Printed in Great Britain
on acid-free paper by
Biddles Ltd., King's Lynn

ISBN 0–19–856753–7   978–0–19–856753–0

3 5 7 9 10 8 6 4 2

PREFACE

Magnetism has been known to humans for millennia, and for millennia interpretations of the nature of this elusive force capable of moving inert bodies have been produced. An early example is provided by Plinius, who in *Naturalis Historia* wrote: 'Quid ferri duritia pugnacius? Pedes ei importuit et mores. Trahitur namque magnete lapide, domitrixque illa rerum omnia materia ad inane nescio quid currit atque, ut propius venit, adsilit, tenetur amplexuque haeret.' Plinius expresses his surprise for the fact that iron, a typical example of hard matter, is irresistibly attracted by lodestone until they embrace. One and a half millennium later, G.B. Porta in his book *Natural Magick* in 1589 similarly wrote 'iron is drawn by the Loadstone, as a bride after the bridegroom, to be embraced; and the iron is so desirous to join with it as her husband, ...'. Magnetism was understood as a soul of inert matter which transformed it into something like a living organism and expressions like animal magnetism and organic magnetism gradually became popular, especially among charlatans. It was in the nineteenth and twentieth centuries that the nature of magnetism was finally understood, but the magnetic materials were still structurally based on metals or oxides. Finally towards the end of the twentieth century the first examples of magnets based on organic matter were discovered and a new research field, which is commonly defined as molecular magnetism, was opened.

A particularly appealing area in molecular magnetism is that of molecules which show a slow relaxation of the magnetization at low temperature, behaving as tiny magnets and thus known also as single-molecule magnets. These were discovered at the end of the twentieth century and immediately attracted much interest for their relevance to fundamental phenomena, like the coexistence of quantum and classical phenomena and for the opportunities of developing new types of magnetic materials.

The present book is particularly devoted to these single-molecule magnets although more general aspects of molecular nanomagnetism are also addressed. This research field is rapidly expanding and requires the cooperation of chemists, for the challenge of designing and synthesizing new examples of magnetic molecules with tailor-made properties, and of physicists, who can experimentally measure the properties and work out the theoretical models required for their interpretation.

Many research articles, reviews, and book chapters dealing with molecular nanomagnetism have recently appeared but a book covering the different aspects of this new domain was lacking. It was also felt that a field where the chemical and physical expertise is so intimately mixed could be tackled only by a joint effort of people with different backgrounds. The book is in fact written by the chemists pioneers in this field and by a theorist who has been one of the protagonists of its development. The book is explicitly addressed to an audience of chemists and physicists aiming to use a language suitable for the two communities.

Establishing a common language is certainly a very difficult task. The authors tried to be helpful to the other researchers, especially new-comers, by taking advantage of their reciprocal ignorance in the complementary field. The chapters of the book have been tested in this way, starting with a draft version which was returned full of question marks which showed that what is obvious for one person may be completely obscure for another. A trial and error approach progressively diminished the number of question marks in subsequent versions of the manuscript. The present text has been released when the question marks were acceptably few in the various chapters. The appendix section is fairly large because it was felt appropriate to leave the more demanding mathematical passages available for the interested reader, keeping technicalities to a minimum in the main text.

A further important improvement of the text has been achieved by the careful reading of the different sections by some patient and friendly colleagues who accepted of being the first test of the approach of the book. We thank Giuseppe Amoretti, Pierre Averbuch, Steve Blundell, Andrea Cornia, Pierre Dalmas de Reautier, Julio Fernández, Anna Fort, Andrew Kent, Alessandro Lascialfari, Achim Müller, Wolfgang Wernsdorfer, and Richard Winpenny for the many improvements they have provided to us. Of course all the errors and obscure passages that remain have to be attributed to the authors. We are indebted to our closer collaborators and our families who demonstrated great patience when finding us often occupied with the writing of this manuscript. We also wish to thank the many colleagues who kindly permitted the reproduction of their graphic material. Paolo Parri is gratefully thanked for the graphic elaboration of the front cover.

Dante Gatteschi
Roberta Sessoli
Firenze, July 2005

Jacques Villain
Grenoble, July 2005

CONTENTS

CONTENTS

INTRODUCTION

Molecular magnetic materials have been added to the library of magnetism only at the end of the twentieth century through the concerted action of chemists and physicists. Before this, all the known magnets were based on metallic and ionic lattices, ranging from magnetite, the first magnet discovered by man, to iron. The interest in functional molecular materials was not limited to magnetism but rather arose from the discovery that purely organic compounds could be electrical conductors, and even superconductors (Jérome and Schulz 2002). This prompted much research, because it was immediately clear that organic conductors could open up new technological applications, taking advantage of low cost and the possibility of tuning the properties using chemical techniques. With an obvious extension, the possibility of organic magnets was taken into consideration. After some false starts at the beginning of the 1990s, Kinoshita and co-workers (Tamura *et al.* 1991) in Japan reported the first evidence of a purely organic ferromagnet, based on a nitronyl nitroxide, whose structure is sketched in Fig. 1.1.

Organic radicals, i.e. systems with at least one unpaired electron, are in general unstable, but the nitroxides, which have an unpaired electron essentially localized in a NO group, are relatively stable, and have been widely used as spin probes and spin labels (Berliner and Reuben 1981). An early example of the investigation of ferromagnetic interactions involving organic radicals was provided by Veyret and Blaise (1973). Magnetic ordering was observed (Saint Paul and Veyret 1973)

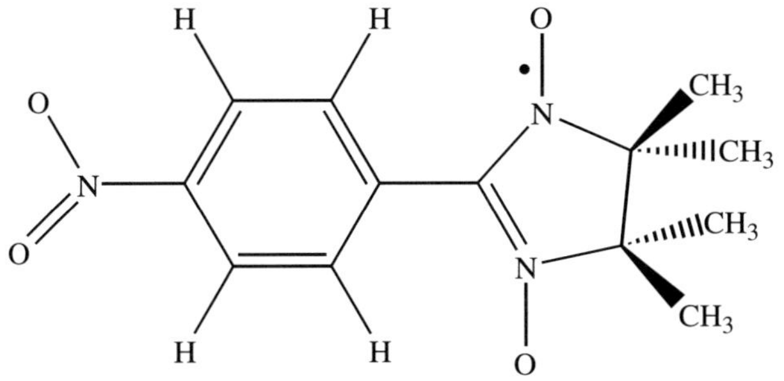

FIG. 1.1. Sketch of the molecular structure of the para-nitrophenyl nitronyl nitroxide, NITpNO$_2$Ph.

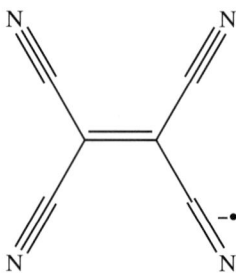

FIG. 1.2. Sketch of the structure of the TCNE.$^-$ radical. All the CN groups are equivalent.

but the material at a deeper investigation revealed to undergo a metamagnetic transition (Chouteau and Veyret-Jeandey 1981). Nitronyl nitroxides, like NITpNO$_2$Ph shown in Fig. 1.1, are a variation on the nitroxide theme, containing two equivalent NO groups in a five-membered ring with one unpaired electron delocalized on the two NO groups. The equivalence of the NO groups in Fig. 1.1 can be verified by writing the symmetric formula in which the double bond and the unpaired electron in the five-membered ring are moved to the symmetric counterpart.

NITpNO$_2$Ph is ferromagnetically ordered only below 0.6 K; nevertheless it was important because it showed that it is indeed possible to have a permanent magnet in which the magnetic orbitals, i.e. those containing the unpaired electrons, are s and p in nature, rather than the d and f orbitals involved in classical magnets. Currently the purely organic magnet with the highest critical temperature is a sulphur-based radical which orders as a weak ferromagnet below 35 K (Palacio et al. 1997).

Before NITpNO$_2$Ph some other examples of molecular ferro- and ferrimagnets had been reported, based on molecular lattices comprising various transition metal ions and also transition metal ion-organic radicals pairs (Miller et al. 1987; Kahn et al. 1988; Caneschi et al. 1989). In this way, a high-temperature ferrimagnet was obtained, using vanadium ions attached to the radical anions of tetracyanoethylene, TCNE$^-$ sketched in Fig. 1.2 (Manriquez et al. 1991). The structure is not known because V(TCNE)$_2$ is highly insoluble and no single crystals suitable for crystallographic analysis were obtained. However the compound orders above room temperature. The ferrimagnetic order arises from the antiferromagnetic coupling between the $S = 3/2$ of V^{2+} and the $S = 1/2$ of the TCNE.$^-$ radicals.

Another room-temperature ferrimagnet is a Prussian blue type compound comprising chromium(III)[1] and vanadium(II) and vanadium(III) ions, of formula [V(II)$_{0.42}$V(III)$_{0.58}$(Cr(CN)$_6$)$_{0.86}$]2H$_2$O (Ferlay et al. 1995).[1]

---

[1] We will use in the following different notations for the formal charge of the metal ions, namely: chromium(III), Cr^{3+}, Cr^{III}. The three must be considered to be equivalent.

Beyond providing some new magnetically ordered systems, molecular magnetism provided several new types of low-dimensional magnetic materials, which attracted the interest of a growing number of physicists, looking for new types of magnetic materials. For instance, materials which provided evidence for the so-called Haldane conjecture (Haldane 1983) are molecular in nature; Kagome-type lattices were obtained (Awaga *et al.* 1994; Wada *et al.* 1997); various types of one-dimensional ferro-, antiferro- and ferrimagnets were obtained, also using unusual constituent spins (Lascialfari *et al.* 2003).

In fact starting from the 1980s there was a marked shift of interest in the field of magnetism of molecular systems, which can be summarized as the transition from magnetochemistry to molecular magnetism. Magnetochemistry is essentially the use of magnetic techniques for obtaining structural information on simple paramagnetic systems, and it is a branch of chemistry which uses physical measurements (Carlin 1986). Molecular magnetism, on the other hand, is an interdisciplinary field, where chemists design and synthesize materials of increasing complexity based on a feedback interaction with physicists who develop sophisticated experimental measurements to model the novel properties associated with molecular materials (Kahn 1993). If one wants to fix a starting date for molecular magnetism the best candidate seems to be the NATO Advanced Study Institute, ASI, which was held in Castiglion della Pescaia in Italy in 1983 (Willet *et al.* 1983). The title of the ASI, 'Structural–magnetic correlations in exchange coupled systems', reflects the interest of the chemist organizers for understanding the conditions under which pairs of transition metal ions could give rise to ferromagnetic interactions. Looking at the list of participants it is clear that there was a blend of chemists and physicists, many of whom met for the first time. A common language started to be developed and useful collaborations were established for the first time. The proceedings of that ASI have been intensively referenced, and have been the textbook for the first generation of scientists active in molecular magnetism.

Important as they have been, the efforts in designing and synthesizing bulk magnets starting from molecules always meet the difficulty that molecules are not easy to organize in a three-dimensional net of strong magnetic interactions. This can be rather easily done with ions or metals, where the building blocks are spherical, while it is often far from being obvious with molecular building blocks, which are in general of low symmetry. This is one of the main reasons why a comparatively large number of low-dimensional materials have been obtained by using molecular building blocks.

However, this difficulty may turn out to be an advantage if the target is changed from three-dimensional magnets to low-dimensional and, in particular, zero-dimensional magnets. Indeed the interest in finite-size magnetic particles had developed in the 1980s as a consequence of the growing interest in the so-called nanoscience. It was realized that nanosize objects can be particularly interesting because matter organized on this scale has enough complexity to give rise to new types of properties, and yet it is not too complex and can be investigated in depth in much detail. The interest in nanoscience (and, in perspective, for nanotechnology) spans all the traditional disciplines. In condensed matter physics the first steps were perhaps made in the field of conductors and semiconductors, as a result of the impetus on the miniaturization processes associated with more efficient computers. One of the challenges is the realization of objects of size so small that they save rise to the coexistence of classical and quantum properties. The most interesting results were in the field of quantum dots and quantum wires (Bimberg *et al.* 1999), which correspond to objects whose size is in the nanometre range in three or two directions, respectively. Progress was made possible by the development of experimental techniques, which allowed 'seeing' and investigating the properties of particles of a few nanometres. Among them a particular relevant place was kept by scanning probe microscopy techniques, like atomic force microscopy, scanning tunnel microscopy, etc. (Bai 2000).

Magnetism could not be an exception, and one of the relevant themes was the possibility of observing quantum tunnelling effects in mesoscopic matter. A scheme, showing the size effects in the magnetization dynamics and hysteresis loop going from multidomain magnetic particles to molecular clusters, has been extracted from an interesting review (Wernsdorfer 2001) and is given in Fig. 1.3.

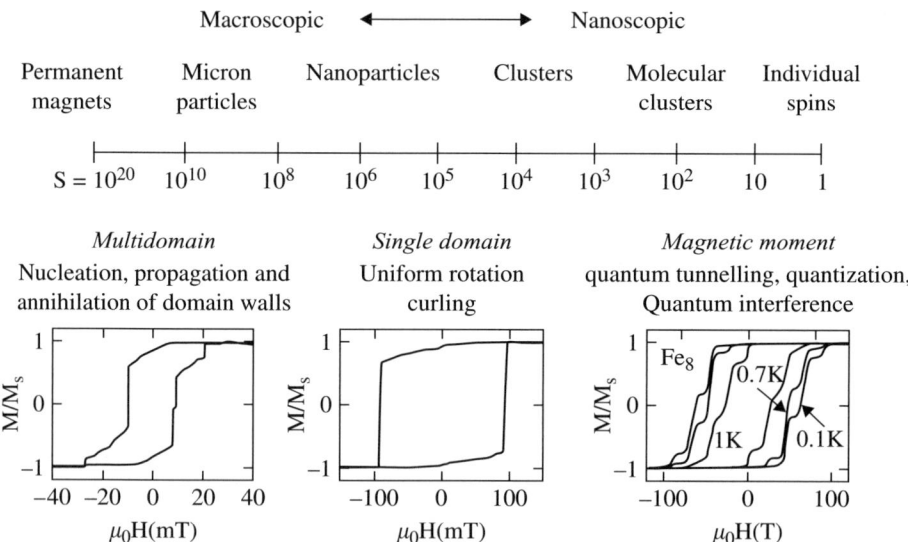

FIG. 1.3. The transition from macroscopic to nanoscopic magnets. From Wernsdorfer (2001). The hysteresis loops are typical examples of magnetization reversal via nucleation, propagation and annihilation of domain walls (left), via uniform rotation (middle), and quantum tunnelling (right). Reprinted with permission of John Wiley & Sons.

At the macroscopic limit the particles contain at least billions of individual spins, which are coupled in such a way that the individual moments will respond all together to external stimuli. The magnetic energy is minimized by forming domains, regions in space within which all the individual moments are parallel (antiparallel) to each other. The orientation of the moments of the domains will be random in such a way that in the absence of an external magnetic field the magnetization of the sample is zero. The transition from a domain to the neighbouring one will occur through a region where the local magnetic moments are rapidly varying, called the Bloch walls (Morrish 1966). The width of the Bloch walls, d, depends on the exchange coupling constant J, which tends to keep the spins ordered and to make the walls as large as possible, in order to minimize the effort needed to change the orientation of the moments, and on the magnetic anisotropy, which tends to minimize the Bloch walls to reduce the probability of high-energy orientations. Obviously the width of the domain walls depends on the nature of the magnetic material.

When the sample is magnetized all the individual moments will eventually be parallel to each other and the magnetization reaches its saturation value. If the field is decreased the formation of domains will not be reversible in such a way that the magnetization at zero field will not be zero, like in the non-magnetized case. The finite value of the magnetization in zero field is called the remnant magnetization. In order to demagnetize the sample it is necessary to go to a negative field, which is called the coercive field. This value is used in order to classify the bulk magnets: a small value of the coercive field is typical of soft magnets, while in hard magnets the coercive field is large. The M/H plot, shown in Fig. 1.3 on the left, shows a hysteresis loop, which tells us that the value of the magnetization of the sample depends on its history. This is the basis of the use of magnets for storing information.

On reducing the size of the magnetic particles a limit is reached when the radius of the particle is small compared to the Bloch wall depth. Energetically the process of domain wall formation is no longer economical and the particle goes single domain.

By further reducing the size of the particles, another effect sets in (Néel 1949). The magnetic anisotropy of the sample, A, depends on the size of the particle:

$$A = KV \tag{1.1}$$

where V is the volume of the particle and K is the anisotropy constant of the material. Let us suppose that the anisotropy of the magnetization is of the Ising type, i.e. the stable orientation of the magnetic moment of the particle is parallel to a given direction z. The energy of the system as a function of the orientation of the magnetic moment is pictorially shown in Fig. 1.4.

The bottom of the left well corresponds to magnetization down, the bottom of the right well to magnetization up, and the top to the magnetization at $90°$ from the easy axis. On reducing the size of the sample eventually the barrier for the reorientation of the magnetization will become comparable to the thermal

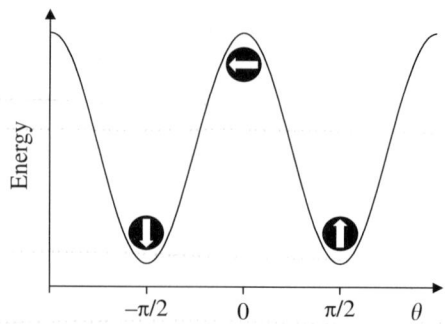

FIG. 1.4. Energy of an Ising (easy axis) type magnet as a function of the angle of the magnetization from the easy axis.

energy. If the sample is prepared with the magnetization up (right well) some of the particles will have enough energy to jump over the barrier and reverse their magnetization. If the particles are given enough time, half of them will be in the left and half in the right well at equilibrium because the two minima have the same energy. The system will no longer be magnetized in zero magnetic field, like a paramagnet. If an external field is applied then one of the two wells will lower its energy and the other will increase it. The two wells will have different populations and the system behaves like a paramagnet, but since the response to the external perturbation comes from all the individual magnetic centres, it will be large. These kinds of particles are called *superparamagnets*, and they find some interesting application, like in magnetic drug delivery, in magnetic separation of cells, and as a contrast agent for magnetic resonance imaging (Pankhurst *et al.* 2003).

An important feature of the superparamagnet is that the observation of either static or dynamic magnetic behaviour depends on the time-scale of the experiment used for investigating it. For instance, using an ac magnetic susceptibility measurement with a field oscillating at $\nu = 100$ Hz, static behaviour, with a blocked magnetization, will be observed if the characteristic time required for the particles to go over the barrier is longer than $\tau = (2\pi\nu)^{-1}$, while dynamic behaviour is observed for shorter τ. The so-called blocking temperature corresponds to the temperature at which the relaxation time of the magnetization equals the characteristic time of the experiment.

The characteristic time for the reorientation of the magnetization can be easily calculated assuming that it occurs through a thermally activated process. This gives rise to an exponential dependence on the energy barrier with so-called Arrhenius behaviour, as observed in many other classes of thermally activated physical and chemical processes:

$$\tau = \tau_0 \exp \frac{KV}{k_B T}. \tag{1.2}$$

This behaviour is typical of a classical system. In principle, when the size of the magnetic particles reduces, it may be possible to invert the magnetization also through the quantum tunnel effect (Leggett 1995). This effect should show up at low temperature, where it should provide the most efficient path for magnetic relaxation only if the wavefunctions of the left and of the right well have some overlap. The quest for quantum effects in magnetic nanoparticles is certainly one of the goals of this book.

The size of the particles needed to observe superparamagnetic behaviour ranges from 2–3 to 20–30 nm, depending on the nature of the material. Magnetic nanoparticles are obtained in many different ways, ranging from mechanical grinding to sol–gel techniques (Sugimoto 2000). An original procedure uses naturally occurring materials like ferritin, the ubiquitous iron storage protein. Iron is needed in the metabolism of living organisms, and it must be stored in some place in order to use when it is needed. Nature chose ferritin to do this job in animals, plants, fungi, and bacteria. Man has an average of 3–4 g iron and ca. 30 mg per day are exchanged in plasma. Structurally ferritin comprises a proteic shell, apoferritin, and a mineral core, of approximate composition FeOOH. The size of the internal core is ca. 7 nm, giving rise to superparamagnetic behaviour in the iron oxide particles, which can contain up to ca. 4000 metal ions (St. Pierre et al. 1989).

An interesting feature is that it is possible to substitute the iron oxide core with other magnetic oxides, like magnetite, taking advantage of the proteic shell for limiting the size of the magnetic particles (Wong et al. 1998). Indeed ferritin was used in one of the early attempts to observe quantum phenomena in mesoscopic magnets, but only conflicting evidence was obtained (Gider et al. 1995; Tejada et al. 1997). The problem is that the observation of quantum phenomena is made difficult by the fact that either the experiments are performed on individual particles, or, if an assembly of them is used, they must be absolutely monodisperse. Monodisperse means a collection of identical particles, because quantum phenomena scale exponentially with the size of the particles, and it would be impossible to unequivocally observe quantum phenomena in polydisperse assemblies.

Definite improvements have been made recently in the techniques to obtain monodisperse assemblies of magnetic particles. In some cases it has been possible to obtain identical particles that have been 'crystallized' (Redl et al. 2003; Sun and Murray 1999). In fact if spherical particles all identical to each other are put together they will try to occupy space in the most efficient way, giving rise to a close packed array exactly as atoms do in crystals.

An alternative to using magnetic nanoparticles, i.e. of reducing the size of bulk magnets in a sort of top-down approach, is that of using a molecular approach in a bottom-up approach (Gatteschi et al. 1994). The idea is that of synthesizing molecules containing an increasing number of magnetic centres. In the ideal process one would like to be able to add one magnetic centre at a time, starting from one and going up to say a few thousand magnetic centres. The theoretical advantage of the molecular approach is that molecules are all identical to each other, therefore allowing the performance of relatively easy experiments on large assemblies of identical particles, and still being able to monitor elusive quantum effects. Molecules can be easily organized into single crystals, therefore allowing the performance of accurate measurements. Further, they can be investigated in solutions, thus destroying all the intermolecular magnetic interactions that might give rise to spurious effects. As an alternative to single crystals it is possible to organize them in self-assembled monolayers and address them with microscopic techniques like STM. Therefore molecular nanomagnets have great promise and they well deserve the effort needed to design and synthesize them.

The idea of making molecules of increasing size by adding the magnetic centres one at a time is certainly appealing, but unfortunately it is not like that that chemistry goes. However, some successful strategies have led to noticeable results such as the spectacular increase in the size of manganese molecular clusters achieved by Christou and co-workers and schematized in Fig. 1.5 (Tasiopoulos *et al.* 2004).

Manganese-containing molecules have been intensively investigated, as will become apparent in the rest of the book. It must be recalled that the interest in manganese clusters is not only for the magnetic properties but also for mimicking the centres present in Photosystem II, the system responsible for water oxidation in photosynthesis (Christou 1989).

The interest in magnetic molecules with large spin was first related to the possibility of using them as building blocks to obtain bulk ferromagnets. Perhaps

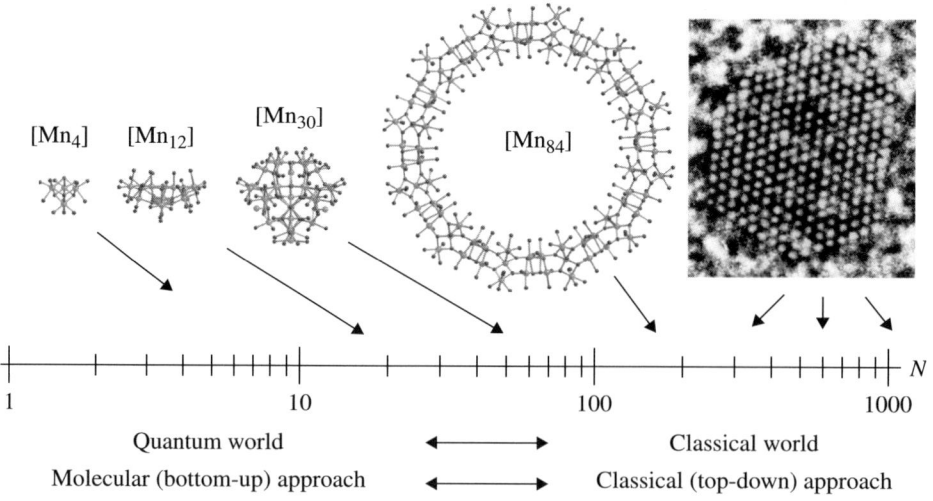

FIG. 1.5. Increasing size and nuclearity of molecular clusters containing manganese ions that approach the size of nanosized magnetic particles. From Tasiopoulos *et al.* (2004). Reprinted with permission of Wiley-VCH.

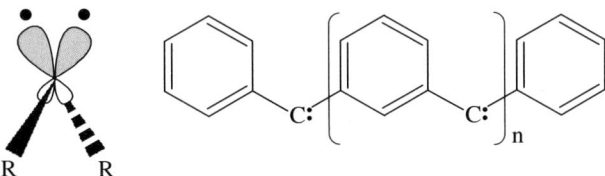

FIG. 1.6. Left: sketch of a carbene centre. Right: structure of linear aromatic polycarbene radicals.

the first interesting systems were unstable organic radicals, based on carbene groups, which actually were designed to show the possibility of yielding strong ferromagnetic coupling in organic matter (Itoh 1978). A carbene is a very reactive carbon centre with two unpaired electron, as shown in Fig. 1.6.

The valence orbitals of the carbene centre can be considered as sp^3 hybrid orbitals: two are used for normal two-electron bonds with R substituents, while the other two remain as non-bonding orbitals with *one* electron each. Since the two magnetic orbitals are orthogonal to each other the carbene centre is in the ground $S = 1$ state. It has been found that connecting the carbene groups through benzene rings, strong ferromagnetic coupling is established, leading to ground states $S = n$, where n is the number of carbene groups, with no evidence of thermally populated states with a spin value smaller than S. A branched chain nonacarbene with an $S = 9$ ground state has been reported. The largest carbene so far reported has $S = 9$ (Nakamura *et al.* 1993; Lahti 1999).

In order to achieve large ground spin states it is easier to use an approach with building blocks containing transition metal or rare earth ions, which can have spin states as high as $S = 5/2$, and $S = 7/2$, respectively. In this way it is relatively easy to obtain spin clusters with a large S in the ground state. An early success was achieved with the ring $[Mn(hfac)_2(NITPh)]_6$, whose structure is shown in Fig. 1.7 (Caneschi *et al.* 1988).

It comprises six Mn^{2+} $(S = 5/2)$ ions coupled to organic radicals analogous to $NITpNO_2Ph$, with the NO_2 group substituted by a hydrogen atom. hfac is just the diamagnetic organic anion of hexafluoracetylacetone. The radical has $S = \frac{1}{2}$ and is strongly antiferromagnetically coupled to the metal ion in such a way that the cluster behaves as a ferrimagnetic ring, with $S = 6 \times 5/2 - 6 \times 1/2 = 12$ in the ground state. This ring has for some time been the cluster with the largest spin in the ground state and it showed how it is possible to obtain real rings on which to test theoretical models. It must be remembered that one ring had long been used, for instance, for modelling the properties of one-dimensional materials.

The breakthrough occurred when the magnetic properties of a compound which had been synthesized at the beginning of the 1980s (Lis 1980) were investigated in detail. $[Mn_{12}O_{12}(CH_3COO)_{16}(H_2O)_4]$, $Mn_{12}ac$, has the structure shown in Fig. 1.8. The analysis of the structure, which will be discussed in more detail in Section 4.7.1, shows that the cluster has crystal-imposed S_4 symmetry,

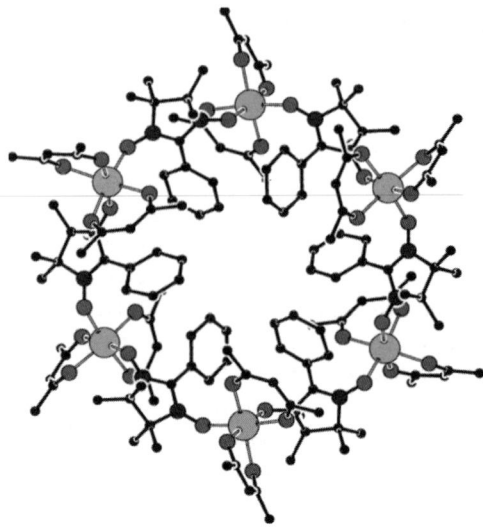

FIG. 1.7. Schematic view of the molecular cluster [Mn(hfac)$_2$(NITPh)]$_6$ along the trigonal axis. Manganese ions are the largest light-grey spheres.

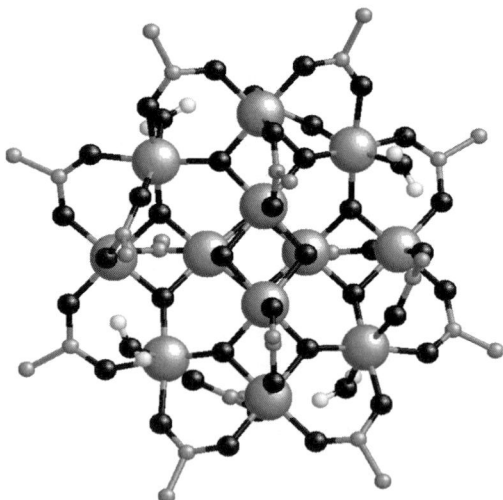

FIG. 1.8. Structure of the molecular cluster [Mn$_{12}$O$_{12}$(CH$_3$COO)$_{16}$(H$_2$O)$_4$] along the tetragonal axis. The manganese ions are reported as large grey spheres, oxygen in black, and carbon as small grey spheres. Only hydrogen atoms of water molecules have been drawn for sake of clarity.

with an external ring of eight manganese(III), $S = 2$, ions and an internal tetrahedron of four manganese(IV) ions. The temperature dependence of the magnetic susceptibility clearly indicates ferrimagnetic behaviour, and the low-temperature magnetization clearly indicates an $S = 10$ ground state. This is easily rationalized assuming that all the manganese(III) spins are up and the manganese(IV) spins are down. Magnetization data also showed strong magnetic anisotropy of the easy axis (Ising) type. The most exciting aspect was, however, that ac magnetic susceptibility measurements clearly indicated a slow magnetic relaxation below 10 K (Caneschi *et al.* 1991). The fact that the complex ac susceptibility strongly depended on the applied frequency of the oscillating magnetic field ruled out the possibility of a transition to bulk magnetic order. The system, measured as a polycrystalline powder, showed a magnetic hysteresis which is molecular in origin (Sessoli *et al.* 1993a), i.e. it is not associated with irreversibility effects in the domain-wall formation, like in classical magnets, but is bound to the slow relaxation of the magnetization of the individual molecules. In principle, it is possible to store information in one molecule of $Mn_{12}ac$, which has a diameter of ca. 1 nm. This observation produced some excitement and a proper name to the observed behaviour. After some variation, the term single-molecule magnet, SMM, was used, suggesting that the individual molecules behave as tiny magnets (Eppley *et al.* 1997; Aromi *et al.*1998; Christou *et al.* 2000). The name is certainly evocative, but in strict terms it is not correct. In order to have a magnet it is necessary that the spin correlation length diverges, and this is certainly impossible in a zero-dimensional material.

The conditions for the unusual behaviour of $Mn_{12}ac$, and of derivatives obtained by substituting the acetate groups with other carboxylates (Boyd *et al.* 1988; Aromi *et al.* 1998; Gatteschi and Sessoli 2003), soon appeared to be the large ground spin state and the large easy axis magnetic anisotropy. Using this point of view, it was soon realized that an octanuclear iron(III) cluster, Fe_8, a compound first reported by Wieghardt in the 1980s (Wieghardt *et al.* 1984) and investigated for its magnetic properties in the middle of the 1990s, indeed met the conditions for behaving as a SMM, showing an $S = 10$ ground state and an Ising type magnetic anisotropy (Barra *et al.* 1996). Ac magnetic susceptibility measurements showed that the magnetization of Fe_8 relaxes slowly at low temperature, but faster than in $Mn_{12}ac$, in agreement with its smaller magnetic anisotropy.

Even more exciting, these systems proved to be ideal testing grounds for theories of the coexistence of quantum and classical effects in magnets (Gunther and Barbara 1995); in particular they provided for the first time evidence of quantum tunnelling of the magnetization (Novak and Sessoli 1995; Friedman *et al.* 1996; Thomas *et al.* 1996; Sangregorio *et al.* 1997) and of oscillations of the tunnel splitting (Wernsdorfer and Sessoli 1999), an interference effect that is the magnetic analogue of the Berry phase. At this point SMMs attracted much attention from both chemists, who were trying to design new classes of SMMs with enhanced properties, and physicists, fascinated by the wealth of different new magnets on which to measure new properties and test theories.

The field of SMM, or more generally molecular nanomagnets, is coming of age and we felt it appropriate to try to provide a unified picture, trying to compensate the difficulties of chemists in following the great number of equations needed to understand the physics of the systems, and those of physicists in orienting themselves in the intricate forest of molecular compounds. Some important review articles are already available in the literature, and in particular a series of edited books is available in which all the major features of molecular nanomagnets are clearly outlined (Miller and Drillon 2001–2005). However a unified approach is still lacking and we are trying to produce it here. In general we will try to be as basic as possible in the treatment of the topics, leaving the more complex treatments to appendices. The general problem of the system of units will be introduced in Appendix A, together with physical constants and basic mathematical tools.

The organization of the book is the following. In Chapter 2 we introduce the spin Hamiltonian approach, which is the background theory needed to provide a first-hand description of the magnetic properties of individual spins, of pairs and of more complex clusters. The basic theory needed for understanding the meaning of the spin Hamiltonian parameters will be worked out at the simplest possible level, but trying to clarify which are the factors responsible of the nature of the ground state of the individual magnetic centres, including the anisotropy, and of the magnetic interactions between different centres.

Chapter 3 is devoted to the observation of microscopic magnetism, working out in some detail the most commonly used magnetic techniques. Also in this case the basic aspects of the techniques will only be briefly recalled, with an indication of relevant textbooks to be used for a sound background. However, it is the goal of the authors to allow the reader to be able to read the current literature with some acceptable understanding. The magnetic techniques that are presented include micro-SQUID and micro-Hall probe techniques and torque magnetometry; specific heat measurements, including equilibrium and out-of-equilibrium measurements; magnetic resonance techniques, including EPR, NMR, and muon spin resonance. Neutron techniques, in particular polarized neutron diffraction and inelastic neutron scattering, will conclude the section.

After the three introductory chapters, Chapter 4 will definitely introduce the reader to the field of single-molecule magnets. The first part is an essentially chemical one, aiming to familiarize the reader with the basic aspects of the art-and-science approach that chemists use to design and synthesize their compounds. In the following sections the three most investigated classes of compounds, namely Mn_{12}, Fe_8, and Mn_4, will be introduced, with some detailed description of their properties.

Chapters 5–12 are devoted to working out in some detail the theoretical background needed to understand the mechanism of relaxation of the magnetization in molecular nanomagnets, of tunnelling, and of quantum coherence. In particular Chapter 5 works out the thermally activated magnetic relaxation, which is responsible for the high-temperature behaviour of the SMM.

Chapter 6 introduces the magnetic tunnelling of an isolated spin, while Chapter 7 introduces the formalism of field theory applied to the tunnelling effect, including imaginary time and path integrals.

Time-dependent magnetic field at low temperature will be the theme of Chapter 8, in particular developing the Landau–Zener–Stückelberg formalism. The fundamental aspects of coherence, incoherence and relaxation, which characterize the interaction of a molecular spin with the external world, will be worked out in Chapter 9. Chapter 10 shows the tunnelling involving excited states, Chapter 11 coherence and decoherence, and Chapter 12 will tackle the basic problem of disorder.

Chapters 13 and 14 will lead the reader back to experiments. Chapter 13 will provide more insight into the subtleties of the magnetic properties of Mn_{12}, Fe_8 and Mn_4, while Chapter 14 will advocate the interest in other classes of magnetic molecules, which do not show SMM behaviour but are very interesting for their magnetic properties.

Chapter 15, comprising the conclusions, will also treat some emerging area, like that of single-chain magnets, SCM, i.e. one-dimensional magnetic materials whose magnetization relaxes slowly at low temperature, without cooperative phenomena.

2

MAGNETIC INTERACTIONS IN MOLECULAR SYSTEMS

Magnetic interactions in molecular systems are in principle the same as can be observed in continuous lattices (Herpin 1968). The relevance of pair interactions is, however, larger in molecular systems, because they are generally insulators and the magnetic interaction is strongly localized. The origin of the coupling between the two magnetic centres is twofold. One is purely magnetic and the other is electrostatic in nature. Alternatively, the two types of interaction may be described as through-space and through-bond, respectively. In principle we may consider that two magnetic centres interact via their magnetic fields. This is certainly possible, but elementary calculations using, for instance, point dipolar approximations, to be discussed below in Section 2.4, show that the interaction energies are a fraction of a kelvin, while values of 10–10^3 K are known to be operative between transition metal ions.

In fact it is well known (Kahn 1993) that the origin of the coupling is the electrostatic interaction responsible of the formation of the chemical bonds. In a very simple scheme the origin of the magnetic interaction is the formation of a weak chemical bond between the two magnetic centres. The simplest possible case is that of a system for which the ground state is orbitally non-degenerate on both centres. The states of the interacting centres are described by a set of orbitals, which in general can be considered as molecular orbitals, i.e. wavefunctions delocalized over all the atoms of the molecule. A rule of thumb for the magnetic interaction can be developed taking into consideration the magnetic orbitals. The magnetic orbitals are the singly occupied molecular orbitals (SOMO) of the magnetic centres (Slater 1968). If the magnetic orbitals are orthogonal to each other, the two spins of the electrons will be parallel to each other (ferromagnetic coupling), while if the magnetic orbitals have a non-zero overlap the spins will tend to orient antiparallel to each other.

The original description of the magnetic interaction (Anderson and Hasegawa 1955; Anderson 1959; Anderson 1963) was performed by using localized magnetic orbitals or a valence-bond approach. It is customary to use other approaches which have their justification in the molecular orbital (tight binding) approaches (Kahn and Briat 1976; Hay *et al.* 1975). Recently density functional theory (DFT) models proved to be extremely effective in calculating the magnetic interactions. They will be discussed in Section 2.3.5. Let us assume that a centre A interacts with a centre B through a bridging group L. Initially A and B are considered as isolated and each of them has some unpaired electrons. We assume that the group L has only paired electrons. The interaction is switched on by allowing the unpaired electrons of A to feel the electrons of B and vice versa. This requires that the total wavefunction is antisymmetric relative to the exchange of a pair of electrons. The exchange interaction, which is of paramount importance for magnetic phenomena, may occur either directly or through a formally diamagnetic ligand. Two different terms are used, namely exchange interaction for the former and super-exchange interaction for the latter. An early digest of the qualitative features responsible of the coupling between different centres is represented by the Goodenough–Kanamori rules (Goodenough 1958, 1963; Kanamori 1959, 1963).

These points will be made clearer in the following. For the moment we stop here, and before discussing in more detail the fundamental theory needed to understand in some detail the nature of the magnetic interactions, we will have to make a detour introducing the so-called spin Hamiltonian approach.

2.1 The spin Hamiltonian approach

The spin Hamiltonian, SH, approach is widely used in various spectroscopies in order to find a suitable short-cut which allows us to interpret and classify the obtained spectra without using fundamental theories. The SH approach eliminates all the orbital coordinates needed to define the system, and replaces them with spin coordinates, taking advantage of the symmetry properties of the system (Abragam and Bleaney 1986). Of course there are several approximations which are associated with this approach. A central one is that the orbital moment of the magnetic bricks is essentially quenched, as it often occurs in solids, and that it can be conveniently treated as a perturbation. We will see that this is often the case for many compounds, but we will also notice many cases where this approximation is far from being tenable. The systems with orbitally non-degenerate ground states are usually well treated with the SH approach, and we will focus on these for the moment.

2.1.1 *Zeeman and crystal field terms for isolated ions*

A magnetic centre with n unpaired electrons will have a ground state characterized by $S = n/2$. The $2S + 1$ spin levels associated to this multiplet will be split by low-symmetry components of the appropriate Hamiltonian, and by an applied magnetic field. We call the former a crystal field hamiltonian and the latter the Zeeman Hamiltonian. The notation 'crystal field hamiltonian' refers to a simplified treatment of the spin levels of the transition metal compounds, in which the effects of the atoms around the transition metal ions (the ligands or the donor atoms in chemical language) are considered as the only sources of ionic interactions.

The Zeeman Hamiltonian can be written as:

$$\mathcal{H}_Z = -\mathbf{H} \cdot \mathbf{m} = \mu_B \mathbf{H} \cdot \mathbf{g} \cdot \mathbf{S} \tag{2.1}$$

where μ_B is the Bohr magneton defined in Appendix A4, $\mathbf{H}$ is the applied magnetic field, $\mathbf{g}$ is a tensor[1] connecting the magnetic field and the spin vectors, and $\mathbf{S}$ is a spin operator. The corresponding magnetic moment $\mathbf{m}$ is given by:

$$\mathbf{m} = -\mu_B \mathbf{g} \cdot \mathbf{S}. \tag{2.2}$$

As will be seen, it is often a good approximation to assume the crystal field spin Hamiltonian to be a quadratic form of the spin operators, i.e.

$$\mathcal{H}_{CF} = \mathbf{S} \cdot \mathbf{D} \cdot \mathbf{S} \tag{2.3}$$

where $\mathbf{D}$ is a real, symmetric tensor. It therefore has three orthogonal eigenvectors. If the coordinate axes x, y, z are chosen parallel to these eigenvectors, $\mathbf{D}$ is diagonal and (2.3) takes the form

$$\mathcal{H}_{CF} = D_{xx} S_x^2 + D_{yy} S_y^2 + D_{zz} S_z^2 \tag{2.4}$$

where S_x, S_y, S_z are spin operators.

The physical properties are not changed if a constant is subtracted from a Hamiltonian. Subtracting $(1/2)(D_{xx} + D_{yy})(S_x^2 + S_y^2 + S_z^2) = (1/2)(D_{xx} + D_{yy}) S(S+1)$, one obtains

$$\mathcal{H}_{CF} = D S_z^2 + E(S_x^2 - S_y^2) \tag{2.5}$$

where

$$D = D_{zz} - \frac{1}{2} D_{xx} - \frac{1}{2} D_{yy}; E = \frac{1}{2}(D_{xx} - D_{yy}). \tag{2.6}$$

Subtracting the constant $DS(S+1)/3$ from (2.5), one obtains

$$\mathcal{H}_{CF} = D\left[S_z^2 - \frac{1}{3} S(S+1)\right] + E(S_x^2 - S_y^2). \tag{2.7}$$

With respect to (2.5), the advantage of (2.7) is that it satisfies Tr $\mathcal{H} = 0$, where Tr is the trace of the tensor. Another advantage of both (2.5) and (2.7) appears if one introduces the eigenvectors $|m\rangle$ of S_z, defined by $S_z|m\rangle = m|m\rangle$, where $m = -s, -(s-1), \ldots, (s-1), s$. If the vectors $|m\rangle$ are used as a basis, it is easily seen that the second term of (2.5) or (2.7) has no diagonal elements, i.e. $\langle m|S_x^2 - S_y^2|m'\rangle = 0$ if m' is equal to m. On the other hand, the first term of (2.5) or (2.7) has no off-diagonal elements.

Sometimes a parameter $B = 2E$ is used instead of E and (2.5) can be rewritten using the raising and lowering operators as

$$\mathcal{H}_{CF} = D S_z^2 + (B/2)\left(S_x^2 - S_y^2\right) = D S_z^2 + (B/4)\left(S_+^2 + S_-^2\right). \tag{2.8}$$

Ignoring a constant, one can use the alternative notation

$$\mathcal{H}_{CF} = -D' S_z^2 + B S_x^2 \tag{2.9}$$

where $D' = -(D + B/2)$.

---

[1] Formally only $\mathbf{g}^2$ has the properties of a second-rank tensor.

It follows from (2.6) that D is zero when $D_{zz} = D_{xx} = D_{yy}$, i.e. in cubic symmetry. In axial symmetry, $D_{xx} = D_{yy}$, and therefore $E = 0$, so that (2.5) reads

$$\mathcal{H}_{CF} = DS_z^2. \tag{2.10}$$

Thus, in axial symmetry, only the D parameter is needed to express the energies of the $(2S + 1)$ spin levels of the S multiplet at this level of approximation. The effect of the Hamiltonian (2.5), (2.7) or (2.9) is that of splitting the $(2S + 1)$ levels even in the absence of an applied magnetic field. Therefore this effect is often called zero-field splitting, ZFS.

These properties simplify the perturbation treatment of the second term of (2.9) if $|E|$ is sufficiently smaller than $|D|$. It is customary to limit its variation according to:

$$-1/3 \le E/D \le +1/3. \tag{2.11}$$

Letting E/D vary in a larger range is physically equivalent to renaming the reference axes. In fact from (2.6) one can derive:

$$D_{xx} = -D/3 + E; \quad D_{yy} = -D/3 - E; \quad D_{zz} = 2D/3. \tag{2.12}$$

For $E/D = 1/3$, $D_{xx} = 0$; $D_{yy} = -2D/3$; $D_{zz} = 2D/3$ and the splitting between the three components is maximum (maximum rhombic splitting). At $E/D = 1$, $D_{xx} = D_{zz} = 2D/3$; $D_{yy} = -4D/3$. The x and z components are identical to each other, meaning that the system is now axial with y as the unique axis.

The eigenvectors of (2.5) or (2.7) in axial symmetry (when $E = 0$) are the eigenvectors $|m\rangle$ of S_z, and the eigenvalues are

$$W(m) = Dm^2 - S(S+1)/3. \tag{2.13}$$

At this level of approximation $W(m) = W(-m)$. Including a non-zero E (i.e. reducing the symmetry below the axial one) the $|m\rangle$ and $|-m\rangle$ states remove their degeneracy if S is integer, while they remain degenerate in pairs if S is half-integer. This is due to time reversal symmetry, and the pairs of degenerate levels are called Kramers doublets (Kramers 1930) .

D can be positive or negative: in the former case the levels with lowest $|m|$ are the most stable, while for negative D the levels with highest $|m|$ lie lowest. Positive D corresponds to easy-plane magnetic anisotropy, negative D to easy-axis type magnetic anisotropy.

The calculated energies for $S = 2$ and $S = 5/2$, in the range $|E/D| = 0 - 1/3$ are shown in Fig. 2.1. The degeneracy of the levels is completely removed for the integer spin while the double degeneracy is retained in half-integer spin. The five levels of an $S = 2$ state are symmetrically split for the maximum rhombic splitting, $|E/D| = 1/3$, and the same is true for the three doublets of $S = 5/2$. It is also interesting to follow the variation of the eigenvectors. For $E = 0$ the states are pure eigenstates of S_z, each of them corresponding to a different m value. For the maximum rhombic splitting substantial admixing of the $|m\rangle$ levels is observed, as shown in Tables 2.1 and 2.2.

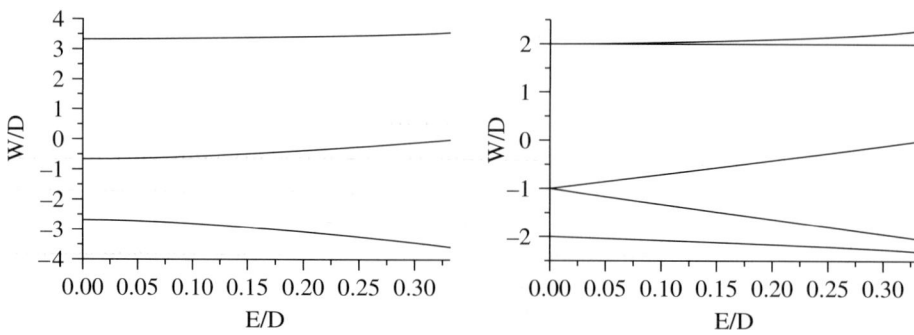

FIG. 2.1. Energy levels in zero field for an $S = 5/2$ (left) and $S = 2$ (right) multiplet as a function of the rhombic distortion factor E/D.

TABLE 2.1. Eigenvectors of an $S = 2$ spin for $E/D = 1/3$.

Eigenvalue	2	1	0	−1	−2
−2.3094	0.1830		−0.9659		0.1830
−2.0000		0.7071		−0.7071	
0.0000		−0.7071		−0.7071	
+2.0000	+0.7071				−0.7071
+2.3094	−0.6830		−0.2588		−0.6830

TABLE 2.2. Eigenvectors of an $S = 5/2$ spin state for $E/D = 1/3$.

Eigenvalue	5/2	3/2	1/2	−1/2	−3/2	−5/2
−3.52767	0.1364		−0.8881		+0.4390	
−3.52767		+0.4390		−0.8881		0.1364
0.0000	−0.1336		+0.4226		0.8964	
0.0000		0.8964		0.4226		−0.1336
3.52767	−0.9816		−0.1810		−0.0610	
3.52767		−0.0610		−0.1810		−0.9816

We will discuss in some detail the splitting of the $\pm m$ levels for $S = 2$, because this kind of discussion will be resumed later in conjunction with the tunnel splitting. The levels can be labelled as $\pm m$ when $E = 0$. $m = 0$ lies lowest when D is positive, then follow the $m = \pm 1$ levels, separated by D, and $m = \pm 2$, which lie $3D$ above $m = \pm 1$. When E becomes different from zero the splitting of the ± 1 levels is much larger than that of the ± 2 levels, because the E term mixes directly states which differ in m by ± 2. The second observation is that a non-zero E determines the admixture of odd and even m states only between themselves. If we imagine calculating the energies of the levels using the $E(S_x^2 - S_y^2)$ term as a perturbation on the energies determined by the D term, we see that the ± 1 levels are already split in first order, while the ± 2 levels are split only in second order. We will see in Chapter 6 that this is of paramount importance for the dynamics of the magnetization of systems with large S.

The Hamiltonian (2.3), which is quadratic in the S coordinates (spin components), is the simplest possible crystal field Hamiltonian. In principle, one has to include fourth, sixth, etc. order terms in the Hamiltonian in order to adequately reproduce the energy levels. Only even-order terms must be retained in zero field when calculating the effects on states belonging to the same configuration. In fact in this case the product of the bra and of the ket is even, and only an even operator can give non-zero matrix elements. Also it is not necessary to include all the terms up to infinity, it is sufficient to include only terms of order

$$N = 2, 4, 6, \ldots, 2S. \tag{2.14}$$

It is in general a good approximation to consider that higher order terms are comparatively smaller than the lower order terms. For spin $S = 2$ it is in principle required to include fourth-order terms in the zero-field splitting Hamiltonian. This can be done by introducing operators of the type S_z^4, S_x^4, etc., but a convenient way to exploit at best the point group symmetry is that of using the so-called Stevens operator equivalents. In a symbolic way they are indicated as:

$$\mathcal{H}_{CF} = \sum_{N,k} B_N^k \mathbf{O}_N^k \tag{2.15}$$

where the sum runs over all the N values defined in (2.14) and the integer k satisfies

$$-N \le k \le +N \tag{2.16}$$

B_N^k are parameters, and $\mathbf{O}_N^k$ are the so-called Stevens operators (Abragam and Bleaney 1986), whose explicit form is given in Appendix A5. The N numbers are limited according to (2.14) while the k values to be actually included in the sum depend on the point group symmetry. For instance, for tetragonal symmetry, and $N = 4$ only terms with

$$k = 0, \pm 4 \tag{2.17}$$

must be taken into consideration. The $k = 0$ term depends on operators of the type S_z^N, therefore it is diagonal in the S manifold. The second-order terms split the levels according to (2.13), and the inclusion of the $N = 4$, $k = 0$ terms changes the energy separations between levels with different $|m|$. The $k = 4$ term, on the other hand, couples states differing by ± 4 in m, therefore it can have important

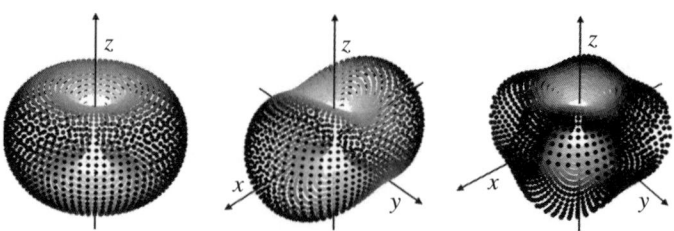

FIG. 2.2. The distance of the surface from the origin represents the classical potential energy of a spin experiencing a uniaxial crystal field with negative D (left), the same including a transverse second-order term (middle), or a transverse fourth-order term (right).

effects on the composition of the eigenstates. Physically it introduces an anisotropy in the xy plane according to which the x and y axes are equivalent between themselves, but different from the diagonals of the quadrants in the plane.

A convenient way to visualize the effects of the different crystal field terms on the magnetic anisotropy consists in plotting the potential energy of a classical spin as a function of Cartesian coordinates, as shown in Fig. 2.2 (Del Barco et $al.$ 2005).

Instead of using the Stevens operators simpler expressions have also been used, like the following Hamiltonian which is appropriate for tetragonal symmetry

$$\mathcal{H} = \alpha S_z^2 + \beta S_z^4 + \gamma(S_x^4 + S_y^4) \tag{2.18}$$

where $\alpha = D - B_4^0[30S(S+1) - 25]$, $\beta = 35B_4^0$, and $\gamma = \frac{1}{2}B_4^4$.

The energies associated to the levels obtained with the quadratic Hamiltonian (2.3) range from 100 mK for organic radicals to 10 K for some transition metal ions (Bencini and Gatteschi 1999). The fourth-order terms are typically one-hundredth of this. It must be remembered that the relative energies of the different order perturbations must be made comparing the $B_N^k \mathbf{S}^N$ values and not the parameters.

2.1.2 Electron nucleus (hyperfine) interaction terms

Another term in the spin Hamiltonian for individual magnetic bricks is that associated with the interaction with the magnetic nuclei which may be present. For historical reasons this is usually referred to as the hyperfine interaction (Abragam and Bleaney 1986). A convenient form for it is:

$$\mathcal{H}_{\mathrm{hf}} = \sum_i \mathbf{S} \cdot \mathbf{A}_i \cdot \mathbf{I}_i \tag{2.19}$$

where the sum is over all the magnetic nuclei of the brick, $\mathbf{I}_i$ is the angular momentum operator of nucleus i, and $\mathbf{A}_i$ is the tensor describing the electron–nucleus magnetic interaction. It is the sum of three contributions, namely the contact, the dipolar and the pseudocontact term. The contact term is given by the electron spin density on the magnetic nucleus, the dipolar term is given by the magnetic dipolar interaction between the electron spin and the nuclear spin, and the pseudocontact term is given by the magnetic dipolar interaction between the orbital moment of the electron and the nuclear spin. The energies involved are typically in the mK region, and may be much smaller than that.

2.1.3 *Spin Hamiltonian for pairs*

When two magnetic centres interact it is possible to extend the above formalism by summing the spin Hamiltonians for the two non-interacting centres, and then adding an interaction term, which we will call the spin–spin Hamiltonian, which at the simplest level can be written as (Heisenberg 1926; Dirac 1929; Van Vleck 1932):

$$\mathcal{H}_{SS} = \mathbf{S}_1 \cdot \mathbf{J}_{12} \cdot \mathbf{S}_2 \tag{2.20}$$

where $\mathbf{S}_1$ and $\mathbf{S}_2$ are the spin operators for brick 1 and 2, respectively, and $\mathbf{J}_{12}$ is a matrix describing the interaction, which is not necessarily symmetric and may have a non-zero trace. It is always possible to rewrite it in an equivalent way, breaking it into three contributions, corresponding to the scalar, the vector, and the tensor product of two vector operators:

$$\mathcal{H}_{SS} = -J_{12}\mathbf{S}_1 \cdot \mathbf{S}_2 + \mathbf{S}_1 \cdot \mathbf{D}_{12} \cdot \mathbf{S}_2 + \mathbf{d}_{12} \cdot (\mathbf{S}_1 \times \mathbf{S}_2) \tag{2.21}$$

where $J_{12} = -(1/3)\mathrm{Tr}\mathbf{J}_{12}$, $D_{12}^{\alpha\beta} = (1/2)(J_{12}^{\alpha\beta} + J_{12}^{\beta\alpha}) - \delta_{\alpha\beta}(1/3)\mathrm{Tr}(\mathbf{J}_{12})$; $d_{12} = (1/2)(J_{12}^{\beta\gamma} - J_{12}^{\gamma\beta})$ and α, β, γ are Cartesian components. The first term in (2.21) is referred to as the isotropic, the second as the anisotropic, and the third as the antisymmetric spin–spin contribution to the magnetic interaction. The isotropic term tends to keep the spins either parallel or antiparallel to each other, the third term to cant them by 90°. The second term tends to orient the spins along a given orientation in space.

In many cases the first term can be considered as dominant, introducing the other terms as perturbations. Under these conditions the total spin $\mathbf{S} = \mathbf{S}_1 + \mathbf{S}_2$ of an isolated pair of spins is a good quantum number and can be used to label the states of the pair. It is defined by the standard angular momentum addition rules and satisfies

$$|S_1 - S_2| \leq S \leq S_1 + S_2. \tag{2.22}$$

The energies of the states S are given by:

$$W(S) = -(J_{12}/2)[S(S + 1) - S_1(S_1 + 1) - S_2(S_2 + 1)]. \tag{2.23}$$

With the sign convention introduced in (2.21) for positive J_{12} the S with maximum multiplicity is the ground state (ferromagnetic coupling), while for antiferromagnetic coupling J_{12} is negative. Unfortunately (2.21) is not the only possibility of expressing the spin–spin Hamiltonian. The coupling constant is frequently indicated as either J_{12} or $-2J_{12}$ and in order to compare experimental results care must be taken to verify the type of Hamiltonian used by the authors.

The Hamiltonian (2.20) is the simplest possible correction to the energies of the pair. In fact they are bilinear in the spin coordinates, but it is also possible to include biquadratic terms, which have been occasionally used. However, in general the Hamiltonian (2.21) is believed to provide reasonable approximation to the energy of the coupled states.

An important limit is the case $|J| \gg |D_1|, |D_2|, |D_{12}|, |d_{12}|, |A_1^k|, |A_2^k|$, which is quite often realistic. This is often called the strong exchange limit. Then it is possible to relate the spin Hamiltonian parameters observed in the S state with those of the individual centres. In principle one has just to project the individual spins on the total spin, and this can be done using standard techniques.

The relations between the **g** tensor, the hyperfine coupling constants, and the zero-field splitting parameters of the coupled pair and those of the individual ions are given below (Bencini and Gatteschi 1990):

$$\mathbf{g}_S = c_1 \mathbf{g}_1 + c_2 \mathbf{g}_2 \tag{2.24}$$

$$\mathbf{A}_S^k = c_1 \mathbf{A}_1^k + c_2 \mathbf{A}_2^k \tag{2.25}$$

$$\mathbf{D}_S = d_1 \mathbf{D}_1 + d_2 \mathbf{D}_2 + d_{12} \mathbf{D}_{12} \tag{2.26}$$

where

$$c_1 = (1+c)/2; c_2 = (1-c)/2$$
$$d_1 = (c_+ + c_-)/2; d_2 = (c_+ - c_-)/2 \tag{2.27}$$
$$d_{12} = (1-c_+)/2$$

and

$$c = \frac{S_1(S_1+1) - S_2(S_2+1)}{S(S+1)}$$

$$c_+ = \frac{3[S_1(S_1+1) - S_2(S_2+1)]^2 + S(S+1)[3S(S+1) - 3 - 2S_1(S_1+1) - 2S_2(S_2+1)]}{(2S+3)(2S-1)S(S+1)}$$

$$c_- = \frac{4S(S+1)[S_1(S_1+1) - S_2(S_2+1)] - 3[S_1(S_1+1) - S_2(S_2+1)]}{(2S+3)(2S-1)S(S+1)}.$$

$$\tag{2.28}$$

For $S = \frac{1}{2}$, c_+ and c_- are zero. The reader can easily check these relations in the classical limit $S_1, S_2, S \gg 1$. The parallel case ($S = S_1 + S_2$) and the antiparallel case ($S = |S_1 - S_2|$) are of particular interest.

The use of these expressions has been verified in several cases (Bencini and Gatteschi 1990) and they provide an important insight into the properties of coupled states. We will see later how it is possible to extend the use of equations (2.24–2.28) to an arbitrary number of coupled spins, provided that the conditions for strong exchange are met. An important point, which needs to be well understood, is that equations (2.24–2.26) are tensorial relations, and one has to take into account not only the principal values of the various tensors, but also their relative orientations.

2.2   Single ion levels

Using the SH approach it is possible to express the energies of the spin levels in a very effective way, and obtain the values of the parameters from the comparison with experimental data. It is possible, for instance, to find that in a manganese(III) compound the ground state is characterized by an axial **g** tensor, with $g_\parallel = g_{zz} = 1.93$, $g_\perp = g_{xx} = g_{yy} = 1.96$, and a zero-field splitting parameter $D/k = -4.0$ K. These data are certainly useful if, for instance, one is interested in the interpretation of the temperature dependence of the magnetic susceptibility or of the EPR spectra but *per se* they do not provide any information on the electronic structure of the compound. In order to do this, it is necessary to resort to some fundamental theory which can explain why these values are obtained.

In the last few years there has been an explosion of possibilities of calculating the electronic structure of transition metal compounds, using density functional theory (DFT) techniques (Sen 2002; Noodleman *et al.* 1995). We will come back to this point later, treating the new possibility of calculating from first principles the magnetic coupling between molecular bricks. Now we prefer to introduce an empirical description of the low-lying energy levels using a relatively simple theory which has long been used for the properties of the transition metal and rare earth ions. We will neglect here the possible treatment for organic radicals. The origin of the treatment dates back to Bethe (1929) who coined the term Crystal Field, which we already used in Section 2.1. The physical assumption of a purely ionic interaction between the metal ion and the ligand was soon found to be unrealistic, and the basic crystal field theory evolved to ligand field theory.

Ligand field theory (Ballhausen 1962; Griffith 1961; Lever and Solomon 1999) is based on the assumption that the low-lying energy levels of a metal ion compound, like for instance the aquo ion depicted below, can be described without taking explicitly into account the ligand orbitals, performing a perturbation calculation on the configuration d^n that corresponds to the oxidation state of the metal ion. For instance Mn^{III} corresponds to the configuration d^4, while Mn^{IV} corresponds to d^3. The theory simply assumes that the effect of

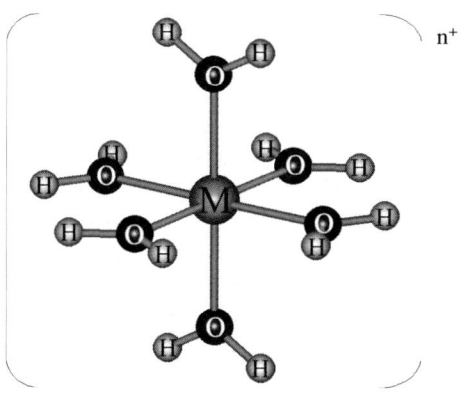

the ligands can be described as an electrostatic one, which can be represented by either a negative point charge or a point dipole. The electrons of the metal ions will try to avoid the region of space where the negatively charged ligands are. Therefore, in an octahedral complex, like the aquo ion depicted before, the orbitals dx^2-y^2 and dz^2, which point to the ligands, will have higher energies than the dxy, dxz, and dyz orbitals. The result of orbital splitting is also obtained on the basis of symmetry considerations. In octahedral symmetry, the dx^2-y^2 and dz^2 orbitals span the irreducible representation e_g of the group O_h, while the dxy, dxz, and dyz orbitals span t_{2g}, and therefore must have different energy. Similar qualitative conclusions are also reached developing an elementary molecular orbital (tight binding) treatment of the metal–ligand bond. The d orbitals of the metal ion are allowed interact with a linear combination of the ligand orbitals. The σ orbitals span the e_g irreducible representation of the group O_h, therefore they will couple with the corresponding orbitals of the metal, giving a bonding and an antibonding combination. Since the ligand orbitals lie lower in energy than the metal orbitals, the antibonding combination has a higher metal contribution. The π orbitals of the ligands span the t_{2g} representation, and similar considerations can be worked out. Since the π interaction is weaker than the σ one the antibonding $t_{2g}*$ orbitals lie lower than the e_g* ones, in qualitative agreement with the ligand field theory.

In order to describe the levels of the metal ions the following Hamiltonian must be taken into consideration:

$$\mathcal{H} = \mathcal{H}_{\text{el}-\text{el}} + \mathcal{H}_{\text{LF}} + \mathcal{H}_{\text{so}} \tag{2.29}$$

where $\mathcal{H}_{\text{el}-\text{el}}$ is the Hamiltonian relative to the electron–electron repulsion, $\mathcal{H}_{\text{LF}}$ the Hamiltonian relative to the ligand field interaction, and $\mathcal{H}_{\text{so}}$ the Hamiltonian relative to the spin–orbit coupling interaction. The magnetic consequences of the three terms are spin multiplicity, the quenching of the orbital angular moment, and the anisotropy, respectively.

The two Hamiltonians relative to electron–electron and ligand field interactions are of comparable energies for transition metal ions, while the spin–orbit coupling is substantially smaller, and often it is considered as a perturbation. Matters may be different for heavy transition metal and rare earth ions. In the former approximation the spin value S is a good quantum number and the states can be labelled as $|^{2S+1}\Gamma\gamma\rangle$, where S is the spin quantum number, and $\Gamma\gamma$ are the labels of the irreducible representation of the point group symmetry of the molecule induced by the orbital component. In fact it has been the extensive use of symmetry which has determined the great success of ligand field theory.

The matrix elements of the d basis set of the electron repulsion operator are often expressed as a function of the Racah parameters. Their values in free ions are known from spectroscopic analysis. In compounds they are usually reduced as a consequence of electron delocalization on the ligands. The energies of the d^n states are found in many standard textbooks (Griffith 1961; Lever and Solomon 1999). In spherical symmetry the electron states are labelled by using the total spin, S, and orbital, L, momentum using the so-called Russell–Saunders formalism, in which the states are labelled as ^{2S+1}L. For a given electron configuration the ground state is characterized by the highest value of S, according to Hund's first rule. If more than one state with the same S is present then the state with the highest L value is the ground state according to Hund's second rule.

The Hamiltonian relative to the ligand field interaction can be written as:

$$\mathcal{H}_{\mathrm{LF}} = \sum_i V_i. \tag{2.30}$$

The sum is over all the ligands (or better the donor atoms). Usually the potentials relative to the donor atoms are expressed as a sum of terms, which depend on the physical model used to justify the LF treatment. In the crystal field frame V_i is given by:

$$V_i = \frac{Z_i e^2}{\varepsilon_0 r_{ij}}. \tag{2.31}$$

Equation (2.31) describes the potential associated with an electron j of charge e at a distance r_{ij} from a negatively charged ligand of charge $Z_i e$. Usually the Hamiltonian (2.30) is expanded in spherical harmonics centred on the metal ion:

$$\mathcal{H}_{\mathrm{LF}} = \sum_{k=0}^{\infty} \sum_{q=-k}^{k} \frac{4\pi}{2k+1} \sum_i Z_i e^2 Y_k^{q*}(\theta_i, \varphi_i) \frac{r_<^k}{r_>^{k+1}} Y_k^q(\theta_j, \varphi_j) \tag{2.32}$$

where $r_<$ is the radius vector of the electron and $r_>$ is that of the ligand. The sum over k is actually limited to $k = 0, 2, 4$ for d electrons. The $k = 0$ term shifts all the levels by the same amount, therefore it may be omitted. The calculation of the matrix elements of (2.32) requires the calculation of radial integrals, which are used as parameters. In octahedral symmetry only the $k = 4$ and $q = 0, \pm 4$ terms are needed, leaving only one parameter which is often called Dq, not to be confused with the D parameter of (2.5). The d levels are split into two subsets of t_{2g} and e_g symmetry, respectively (Griffith 1961). The energies of the former are $-4Dq$ and of the latter $+6Dq$, giving an energy difference, which is accessible from spectroscopic measurements, of $10\ Dq$. It is found that the Dq values for different ligands are in the order:

$$\mathrm{I}^- < \mathrm{Br}^- < \mathrm{Cl}^- < \mathrm{S}^{2-} < \mathrm{N}_3^- < \mathrm{F}^- < \mathrm{OH}^- < \mathrm{OAc}^- < \text{oxalate}$$

$$\approx \mathrm{O}^{2-} < \mathrm{H}_2\mathrm{O} < \mathrm{NH}_3 < \text{bipy} < \mathrm{CN}^- < \mathrm{CO}. \tag{2.33}$$

This is called the *spectrochemical series* and is a useful tool for the qualitative analysis of the electronic structure of transition metal ions.

Another approach relates the parameters not to an ionic interaction but rather to a covalent interaction, and the parameters are called e_σ and e_π, respectively. This model is called the angular overlap model, AOM (Schäffer 1968). It has a very intuitive approach for the case of one ligand on the z axis. Its effect is to give the following energies for the d orbitals:

$$E(z^2) = e_\sigma; E(xz) = E(yz) = e_\pi; E(xy) = E(x^2 - y^2) = e_\delta \qquad (2.34)$$

where e_δ is usually taken as zero. The energies for the case of more ligands can be additively calculated by taking into account the required coordinate rotations through the so-called angular overlap matrix.

The ground states for octahedral transition metal ions can easily be determined by an aufbau approach to the various d^n configurations. They are shown in Fig. 2.3. It is seen that for the d^1, d^2, d^3, d^8, d^9 configurations there is only one way of putting the electrons in the two sets of levels. For the d^4 to d^7 configurations, depending on the energy separation between the two sets of levels, the electrons may prefer to occupy the d orbitals with their spin parallel, giving rise to high spin configurations, or to pair in the lowest-lying t_{2g} orbitals, giving

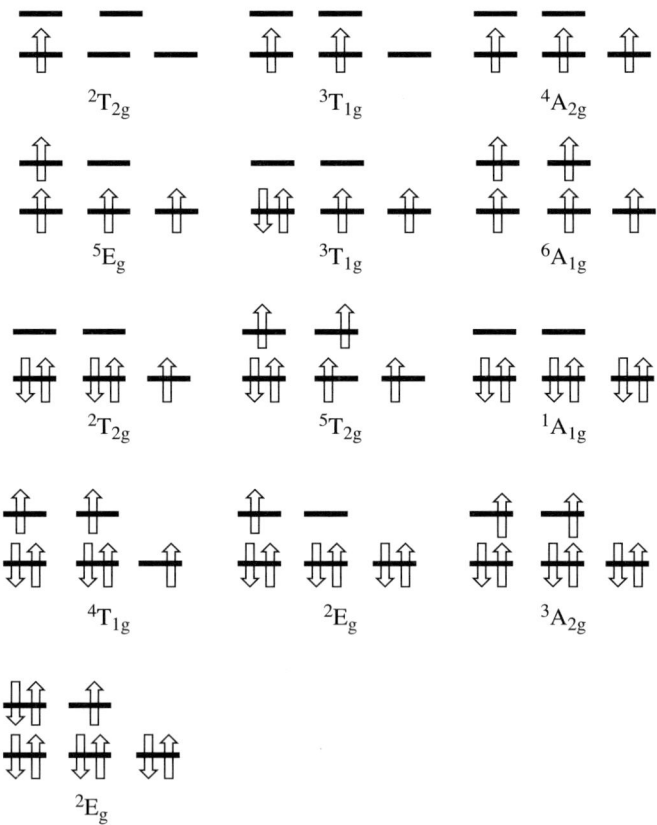

FIG. 2.3. Electronic configurations of d^n octahedral complexes. The number of d electrons increases from left to right and from up to down.

rise to low spin configurations. The choice of the configuration is dictated by the relative kinetic exchange energies, which favour the high-spin configuration, and the splitting energy, which favours the low-spin configuration.

For some compounds of iron(III), iron(II), and cobalt(II), the strength of the ligand field is comparable to the exchange energy and the low- and high-spin states can be thermally populated. Low temperatures favour the low-spin states, which are characterized by shorter bond lengths and lower entropy. The transition from low to high spin may show some irreversibility and a hysteretic behaviour, therefore these systems have been exploited to store information (Kahn and Martinez 1998).

Perusal of Fig. 2.3 shows that orbitally non-degenerate ground states occur for d^3 ($^4A_{2g}$), d^5 ($^6A_{1g}$), low spin d^6 ($^1A_{1g}$) and d^8($^3A_{2g}$), all the others have either E (doubly degenerate) or T (triply degenerate) ground states. Orbitally degenerate states are unstable, and two mechanisms may be operative for removing the degeneracy, namely the Jahn–Teller effect or spin–orbit coupling. The former is a consequence of phonon coupling (vibronic coupling for the chemists) which is enhanced by the breakdown of the Born–Oppenheimer approximation associated with orbital degeneracy. It can be shown that vibronic coupling is operative with the doubly degenerate E terms, therefore all the E terms drastically lower their symmetry due to phonon coupling. In general the deformation of the octahedron is of the tetragonally elongated type for d^4, d^7, and d^9. The orbital contribution is drastically quenched and as we will see the magnetic anisotropy is low for the **g** tensor, but relatively large for the zero-field splitting.

For T states matters are usually different. The states of lowest energy form a space inside which the spin–orbit coupling matrix elements are not zero, so that the spin–orbit coupling actively works to quench the orbital degeneracy. Low-symmetry components, however, can still be operative. The result is that in general the T states have a very complex magnetic behaviour, characterized by high magnetic anisotropy. A typical example is cobalt(II), which is characterized by a ground $^4T_{1g}$ state. In octahedral symmetry spin–orbit coupling removes the degeneracy of order twelve (three orbital and four spin components), giving rise to two quartets and two Kramers doublets. One Kramers doublet happens to be the ground state and it is well separated from the other multiplets by ca. 100 K. This can be treated as an effective $S = \frac{1}{2}$ state, characterized by a g value much different from the free-electron value $g_e \cong 2$ due to the strong orbital contribution. In fact for octahedral symmetry $g = 4.3$. If some low-symmetry component is present, it can produce strong deviations from this behaviour and very anisotropic g values. For instance, we may take into consideration the case of tetragonal elongation and compression, respectively. The order of the energies of the d orbitals in the two coordination environments is shown in Fig. 2.4.

Two unpaired electrons are in the two orbitals of the octahedral e_g manifold, which are no longer degenerate. In the tetragonally elongated case the unpaired electron in the t_{2g} subshell is in the xy orbital, and $g_\parallel \sim 2$, $g_\perp = 4$. For the tetragonally compressed case $g_\parallel \sim 9$, $g_\perp \sim 0$.

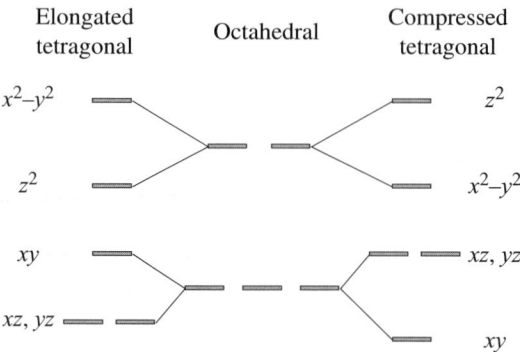

FIG. 2.4. Energies of the d levels in octahedral and tetragonal symmetry.

In order to understand the behaviour of a compound of a transition metal ion in a magnetic field it is necessary to explicitly introduce the spin–orbit coupling interaction (Carlin 1986). In a general way the corresponding Hamiltonian can be written as

$$\mathcal{H}_{so} = \sum_i \zeta_i \mathbf{l}_i \cdot \mathbf{s}_i \tag{2.35}$$

where ζ is the spin–orbit coupling constant of the ith electron, and $\mathbf{l}_i$ and $\mathbf{s}_i$ are the orbit and spin operators, respectively, for electron i. The spin–orbit coupling constant increases on passing from the light to the heavier elements. Therefore spin–orbit coupling effects are largely negligible in the magnetic properties of organic radicals, where only light elements are generally present, while it has an important role for transition metal ions and even larger for rare earth ions. The Hamiltonian (2.35) can be rewritten in a simplified form if the spin–orbit coupling contribution is calculated within a given Russell–Saunders term ^{2S+1}L:

$$\mathcal{H}_{so} = \lambda \mathbf{L} \cdot \mathbf{S} \tag{2.36}$$

where $\mathbf{L}$ and $\mathbf{S}$ are the total orbital and spin operators, respectively, and

$$\lambda = \pm \zeta/(2S) \tag{2.37}$$

where the plus sign applies for d^n configurations with $n < 5$, and the minus sign for $n > 5$. For $n = 5$, (2.37) gives zero.

In the perturbation treatment of the spin–orbit coupling interactions the spin Hamiltonian parameters for an orbitally non-degenerate ground state are given by:

$$\mathbf{g} = g_e \mathbf{I} - 2\lambda \mathbf{\Lambda} \tag{2.38}$$

TABLE 2.3. Calculated deviations of the g values from the free-electron value for some pseudo-octahedral transition metal ions.

Configuration	S	Ground state	Δg_x	Δg_y	Δg_z
d^1	1/2	$\mathrm{T_{2g}}$ [a]	$-2\lambda/\Delta_1$	$-2\lambda/\Delta_2$	$-8\lambda/\Delta_3$
d^3	3/2	$\mathrm{A_{2g}}$ [b]	$-8\lambda/\Delta_1$	$-8\lambda/\Delta_2$	$-8\lambda/\Delta_3$
d^4	2	$\mathrm{E_g}$ [c]	$-2\lambda/\Delta_1$	$-2\lambda/\Delta_2$	$-8\lambda/\Delta_3$
d^8	1	$\mathrm{A_{2g}}$ [d]	$-8\lambda/\Delta_1$	$-8\lambda/\Delta_2$	$-8\lambda/\Delta_3$
d^9	1/2	$\mathrm{E_g}$ [e]	$-2\lambda/\Delta_1$	$-2\lambda/\Delta_2$	$-8\lambda/\Delta_3$

[a] Ground state xy. Δ_1 and Δ_2 are the excitation energies to yz and xz, respectively. Δ_3 is the excitation energy to x^2-y^2.
[b] Δ_i are the energies of the split components of the excited $^4\mathrm{T_{2g}}$.
[c] Ground state configuration $(xy)(xz)(yz)(z^2)$. Δ_1, Δ_2, and Δ_3 are the energy excitations $yz \rightarrow x^2 - y^2$, $xz \rightarrow x^2 - y^2$, and $xy \rightarrow x^2 - y^2$, respectively.
[d] Δ_i are the energies of the split components of the excited $^3\mathrm{T_{2g}}$.
[e] Ground state x^2-y^2. Δ_1 and Δ_2 are the excitation energies to xz and yz, respectively. Δ_3 is the excitation energy to xy.

where g_e is the Landé factor of the free electron, $\mathbf{I}$ is the identity matrix, $I_{lm} = \delta_{lm}$, and $\mathbf{\Lambda}$ is given by

$$\mathbf{\Lambda} = \sum_n \frac{\langle g|\mathbf{L}|n\rangle\langle n|\mathbf{L}|g\rangle}{E_n - E_g} \qquad (2.39)$$

where $|g\rangle$ is the ground state function and the sum is extended over all the excited states, $|n\rangle$. E_n and E_g are the energies of the excited and of the ground state, respectively. $\mathbf{L}$ is the orbital angular momentum vector operator. Due to its definition the elements of $\mathbf{\Lambda}$ are positive, while λ is positive for the d^n configuration with $n < 5$ and negative for $n > 5$. Therefore the corrections to the free-electron g values are negative for the transition metal ions with $n < 5$ and positive for $n > 5$. The calculated expressions for some transition metal ions to be used in the of the book are shown in Table 2.3.

The orbital contribution to the single-ion zero-field splitting tensor is given by:

$$\mathbf{D} = -\lambda^2\mathbf{\Lambda}. \qquad (2.40)$$

The $\mathbf{D}$ tensor so calculated has a non-zero trace, therefore it must be modified using (2.7) to compare the calculated values with the ones obtained by using the Hamiltonian (2.5).

The above treatment is very useful for a semiquantitative rationalization of the single-ion properties. However, it must be recalled that a second-order perturbation treatment may be difficult to justify. A more correct approach is that of diagonalizing the Hamiltonian matrix, using the full matrix of the d^n configuration including the magnetic field, and then to compare the values obtained by the ligand field–spin–orbit coupling treatment with the values of the spin Hamiltonian approach. This has been advocated by Bencini et al. (1998). They calculate the energy levels of a given transition metal ion with a set of ligand field–magnetic field parameters, then they use a least squares fitting procedure in order to reproduce the same values of the energy levels, using the SH approach. Therefore the Ligand Field approach provides the SH parameters to be compared with those obtained through experiment.

An instructive example is provided by the manganese(III) ions which will be extensively taken into consideration in the following. Manganese(III) is a d^4 ion, therefore in octahedral symmetry it has a ground ^5E_g state, which is unstable due to Jahn–Teller effects. It is generally observed that the removal of the orbital degeneracy is performed through a tetragonal elongation, i.e. two donor atoms lengthen their bonds, and four shorten them. The split components of the ^5E_g, $^5A_{2g}$ and $^5B_{1g}$, the former lying lowest, are typically separated by 12–15 thousand K. In [Mn(dbm)$_3$], where Hdbm = 1,3-diphenyl-1,3-propanedione, the metal ion is in a roughly tetragonal coordination of six oxygen atoms. High-frequency EPR spectra show $g_x = g_y = 1.99$; $g_z = 1.97$; $D/k_B = -6.26$ K, $E/D = 0.06$. Using (2.37)–(2.39) the g values are correctly calculated but the calculated D value, -1.94 K, is much smaller than the experimental one (Barra *et al.* 1997a). This was taken as evidence that the simple second-order approximation does not hold. A full matrix diagonalization, which includes also the contribution of states with $S < 2$, gives much better agreement with experiment, $D/k_B = -6.55$ K, $E/D = 0.06$.

2.3 Exchange interaction

2.3.1 *Delocalization effects*

In a system in which the magnetic orbitals are essentially localized on two centres, as is often the case for metal ion derivatives, it is possible to use the so-called natural orbitals, i.e. assuming that in a first approximation the unpaired spin density is localized on the d orbitals and partially delocalized on the bridging ligands. These orbitals have a spin density on the bridging ligands, and the conditions for coupling are determined by the overlap density there. An example is shown in Fig. 2.5. It corresponds to two metal ions bridged by a cyanide, CN$^-$, group as it is found in the Prussian blue family of molecular magnets. The

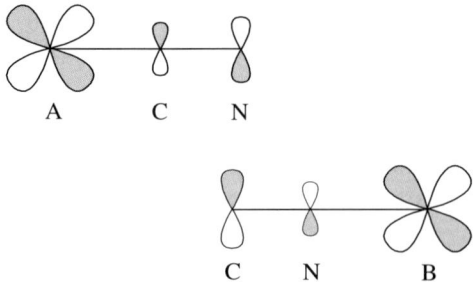

FIG. 2.5. Natural magnetic orbitals for two ions bridged by a CN$^-$ group.

convention of different shading to differentiate the signs of the wavefunction is used. Since the two natural orbitals have large spin densities on the bridging cyanide the two magnetic orbitals are not orthogonal to each other and the coupling is expected to be antiferromagnetic (Verdaguer *et al.* 1999a)

Orthogonalized localized orbitals are frequently used. They can be obtained from the molecular orbitals by a standard procedure. Let us suppose that a given pair of molecular orbitals is formed by linear combinations of the two atomic orbitals, centered on atom A and B, respectively:

$$
\begin{aligned}
\psi_g &= \frac{1}{\sqrt{1 + 2S^2}}(\phi_A + \phi_B) \\
\psi_u &= \frac{1}{\sqrt{1 - 2S^2}}(\phi_A - \phi_B).
\end{aligned}
\tag{2.41}
$$

where S is the overlap of the two orbitals and g and u stand for *gerade* and *ungerade*, respectively. The orthogonalized orbitals are given by:

$$
\begin{aligned}
\Phi_A &= \frac{1}{\sqrt{2}}(\psi_g + \psi_u) \\
\Phi_B &= \frac{1}{\sqrt{2}}(\psi_g - \psi_u).
\end{aligned}
\tag{2.42}
$$

Φ_A is essentially localized on A, but it also has a non-zero contribution on B, as shown in Fig. 2.6.

The mechanisms responsible of the coupling have been first introduced by Anderson (Anderson 1959), crystallized in a simple set of rules by Goodenough and Kanamori (Goodenough 1958, 1963; Kanamori 1959), and then extended and specialized by several authors, in order to take into account the large number of cases which have been reported in the last 20 years (Kahn and Briat 1976; Kahn 1993; Hay *et al.* 1975; Weihe and Güdel 1997a), and the advent of organic magnetism (Borden W. T. 1999; Lahti 1999).

In the following we work out essentially the Anderson model using orthogonalized localized orbitals. When the magnetic orbitals Φ_A and Φ_B are allowed to interact with each other there is some delocalization of the electron of A to B and vice versa. In order to predict the nature of the ground state, singlet or triplet, it is necessary to take into account excited states which can mix in the ground state. The first excited state corresponds to the transfer of one electron from Φ_A to Φ_B. This corresponds to a polarization A^+B^-, and necessarily the two spins

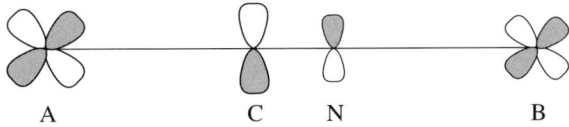

FIG. 2.6. Orthogonal localized magnetic orbital Φ_A for two ions bridged by one CN$^-$ group.

are antiparallel and thus in a singlet state. By second-order perturbation this singlet excited state depresses the energy of the ground singlet by

$$\Delta E_1 = -\frac{4t_{11}^2}{U} \qquad (2.43)$$

where U is the Coulomb repulsion energy for the two electrons occupying the same orbital and t_{11} is the so-called transfer integral, which corresponds to the one-electron matrix element connecting Φ_A and Φ_B. t_{11} depends on the overlap S between Φ_A and Φ_B. If $S = 0$, $t_{11} = 0$ and the stabilization of the singlet is zero. ΔE_1 can be considered as the antiferromagnetic contribution to the coupling. It was called *kinetic exchange* by Anderson.

A second term which must be taken into account is the exchange energy associated with the ground configuration. It stabilizes the triplet and is given by

$$\Delta E_2 = 2\,K \qquad (2.44)$$

where K is the exchange integral $\int \Phi_A^*(1)\Phi_B^*(2)(e^2/r_{12})\Phi_A(2)\Phi_B(1)\mathrm{d}\tau$. It was called *potential exchange* by Anderson. It can be considered as a ferromagnetic contribution to the coupling.

Up to now we have only considered singly occupied molecular orbitals (SOMO). If we consider also excitations to empty orbitals, or from completely filled orbitals, additional mechanisms of coupling become available. The first case to be considered is that in which the excitation involves two different magnetic centres. In this case the excitation also gives rise to a singlet and a triplet, but Hund's rule requires that the triplet lies lowest. These terms therefore stabilize the ground state triplet and correspond to ferromagnetic coupling. The energy correction ΔE_3 can be calculated using third-order perturbation theory as

$$\Delta E_3 = -4\frac{t_{12}^2}{U'}\frac{K_0}{U'} \qquad (2.45)$$

where K_0 is the one-centre exchange integral, U' is the energy difference between the ground state and the excited state, and t_{12} is the transfer integral between the two states. Comparing (2.45) and (2.43) we see that the former formally corresponds to the latter with a weighting factor K_0/U'. This is a number much smaller than 1, therefore this ferromagnetic correction is much smaller than the antiferromagnetic one.

2.3.2 *Spin polarization effects*

All these terms correspond to delocalization effects, i.e. to the formal transfer of one electron from one site to the other. There are additional terms which give a contribution to the coupling, which correspond to excitations on the same centre. These are spin polarization mechanisms. These mechanisms have long been known in the analysis of magnetic resonance experiments and in polarized neutron diffraction. The best way to understand their role is that of using an example.

We have already seen that nitronyl nitroxide radicals are very important building blocks for molecular magnets. They have the general formula sketched below:

The unpaired electron is in a π^* orbital, ψ_1, which in the Hartree–Fock approximation can be represented as shown in Fig. 2.7a. We remind that HOMO means highest occupied molecular orbital.

At this level of approximation, therefore, one should expect a positive spin density on the NO groups, and no density elsewhere. However, many experimental results show that this is not the case, and that, in particular, a relatively large negative spin density is present on the carbon atom between the two nitrogen atoms. The origin of this discrepancy is clearly the crudeness of the approximation inherent in the assumption that the ground state can be correctly represented by one configuration. Actually there are other states close to the ground one which can be admixed into it: at a more suitable level of approximation it is necessary to include a configuration interaction, CI. In particular the closest state is represented by the orbital depicted in Fig. 2.7b, ψ_2. A possible way to take into account electron correlation is that of using, in a given MO wavefunction ψ, a different orbital part for different spin components, $\pm\frac{1}{2}$. In fact, when electron exchange is taken into consideration, $\psi_2(\uparrow)$ is different from $\psi_2(\downarrow)$, because the former interacts with $\psi_1(\uparrow)$, and the latter does not. In other words the $\uparrow$ spin density at the NO groups interacts with the filled ψ_2 orbital stabilizing the $\uparrow$ spin density over the $\downarrow$ spin density in this orbital. Since on the whole there cannot be a finite spin density in ψ_2, because it is a filled orbital,

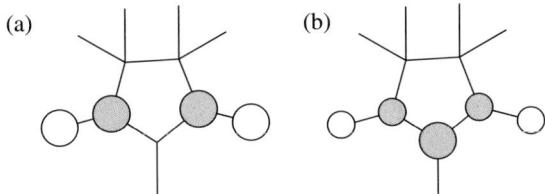

FIG. 2.7. Sketch of the π^* SOMO (a) and HOMO (b) of a nitronyl nitroxide radical. The surface of the circles is proportional to the amplitude of the wavefunction on the corresponding atom, and the shading distinguish positive and negative amplitudes.

the increase of ↑ spin density on the NO groups must be compensated by a corresponding increase of ↓ spin density on C, i.e. by a build-up of a negative spin density on that atom.

These spin polarization mechanisms are operative also in coupled systems, and they are responsible of some coupling mechanisms, which can be either ferro- or antiferromagnetic in nature.

2.3.3 Some examples

In general it is much easier to realize the conditions for antiferromagnetic coupling, therefore in the following we will show which are the conditions under which ferromagnetic coupling can be observed in coupled systems. The key is always that of realizing orthogonality conditions, and at the same time constraining the two magnetic orbitals to have a large overlap density. Examples of orthogonal orbitals are shown in Fig. 2.8. **1** is the case of a binuclear copper(II)–oxovanadium(IV) complex (Kahn 1993). Copper(II) has a dx^2–y^2 magnetic orbital, while oxovanadium(IV) has a dxy magnetic orbital. The two are obviously orthogonal, but they both have large overlap densities on the bridging L ligand, thus providing the required large overlap density for ferromagnetic coupling.

In all these cases the orthogonality between the orbitals is symmetry determined. It is possible, however, to realize orthogonality by accident. An example is shown in **2** of Fig. 2.8. The magnetic orbitals of the two copper ions can be considered to a good approximation to be essentially dx^2–y^2 plus a small amount of copper $4s$. For a general angle Cu–O–Cu the two orbitals have overlap $S \neq 0$, and antiferromagnetic coupling, but when the O–Cu–O angle is close to $96°$ the overlap accidentally goes to zero, and in complexes with angles close to this the coupling is ferromagnetic (Hatfield 1983).

Spin polarization mechanisms are responsible of ferromagnetic coupling in **3** of Fig. 2.8. It corresponds to the case in which a magnetic orbital, say dz^2 in a tetragonally elongated manganese(III) ion, has non-zero overlap with the empty dx^2–y^2 orbital of a neighbouring ion. We may imagine that a fraction of unpaired electrons is transferred into the dx^2–y^2 orbital, and this, according to Hund's rule, will polarize the unpaired electron in the dz^2 orbital, thus providing a ferromagnetic coupling.

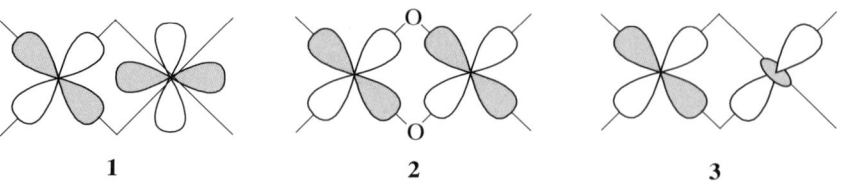

FIG. 2.8. Magnetic orbitals in Cu^{2+}–VO^{2+}, **1**, Cu^{2+}–Cu^{2+}, **2**, and Mn^{3+}–Mn^{3+}, **3**.

A very strict condition in order to have ferromagnetic coupling is to have a set of degenerate orbitals in a high symmetry, highly delocalized molecule. However, it is clear that strong limitations exist with this approach. In fact, the highest possible degeneracy is three for a molecule with cubic symmetry therefore it is not possible in this way to realize ground states with very high spin multiplicity. Further many of these systems are unstable towards Jahn–Teller distortions, which lower the symmetry removing the orbital degeneracy.

2.3.4 *Double exchange*

An additional type of magnetic interaction is observed in the case of mixed valence systems (Zener 1951; Anderson and Hasegawa 1955; Girerd 1983). In the case of inorganic materials this corresponds to the presence of metal ions in different oxidation states. A typical example is provided by the manganites, where manganese(III) and manganese(IV) ions are present. In mixed valence compounds the valences may be trapped, partially or totally delocalized. Generally these cases are referred to as Class I, II, and III according to the Robin and Day classification (Robin–Day 1967). Class I corresponds to the situation where a well-defined charge can be associated to each metal. In Class III, the complete delocalized case, one or more electrons hop rapidly from one centre to the other, while Class II is an intermediate case between Class I and Class III behaviour. Class I is not particularly appealing and the magnetic properties of the mixed valence pairs are qualitatively similar to the properties of similar pairs, and can be predicted using the technicalities described above. In Class II and Class III compounds it necessary to take into account the additional contribution coming from the delocalization of one unpaired electron, which is rapidly hopping from one centre to the other. The formal electron configurations of the Mn^{3+} and Mn^{4+} ions are shown in Fig. 2.9.

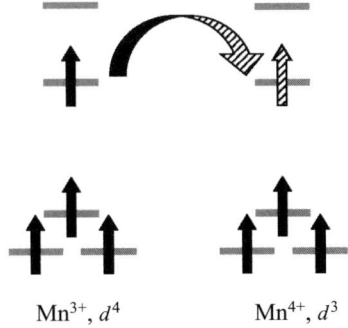

Mn^{3+}, d^4 Mn^{4+}, d^3

FIG. 2.9. Electron configuration of Mn^{3+} and Mn^{4+} and the electron transfer process (hatched symbols) responsible of double exchange. In a mixed valence species one electron can hop from the e_g orbital of Mn^{3+} to the empty orbital of Mn^{4+}.

The additional electron on Mn^{3+} can be delocalized on the other centre as schematized in Fig. 2.9. It is apparent that in the transfer it must keep its spin parallel to those of the Mn^{4+} electrons, according to the Pauli principle, therefore determining an effective ferromagnetic coupling between the two centres. This particular interaction has been called double exchange, or spin-dependent delocalization. The double exchange interaction doubles the number of states as compared to the trapped case. In fact one has to take into account the case where the hopping electron is on the right or on the left centre, respectively.

Double exchange can be introduced into the SH scheme giving an expression originally derived by Anderson and Hasegawa (1955):

$$E_{\pm}(S) = -\frac{J}{2}S(S+1) \pm B \left(S + \frac{1}{2} \right) \qquad (2.46)$$

where S is the total spin, comprised between $\frac{1}{2}$ and $2S_0 + 1/2$. S_0 is the spin of the configuration without hopping electrons, and $1/2$ gives the spin of the hopping electron. In the manganese(III)–manganese(IV) example, S_0 is the spin of the manganese(IV) centre, $S_0 = 3/2$. The $\pm$ refers to the symmetric and antisymmetric combination of states, respectively, and B is a parameter which can be expressed as:

$$B = \frac{t_{ab}}{2S_0 + 1} \qquad (2.47)$$

where t_{ab} is the transfer integral between the magnetic orbitals of centre a and b, respectively. The double exchange splits the S levels into two, stabilizing the ferromagnetic state. The energy separation between the pairs of levels is given by:

$$\Delta E_{\pm} = 2B(S + 1/2). \qquad (2.48)$$

An early review of the application of double exchange to transition metal binuclear complexes was reported by Girerd et al. (1983). More detailed applications have been reported by Tsukerblat et al. (Borras-Almenar et al. 1996; Borras-Almenar et al. 1998a; Borras-Almenar et al. 1999; Borras-Almenar et al. 2001a)

2.3.5   *Towards quantitative calculations of exchange interactions*

The calculation of the exchange coupling constants from first principles is a rather difficult problem, which is now finding encouraging results using the DFT approach. The main difficulty is associated with the fact that open shell systems must be considered as weakly interacting between themselves. In particular the electrons remain largely localized on the two centres as qualitatively shown above. This means that in principle it is necessary to do an open shell or unrestricted Hartree–Fock calculation with subsequent configuration interaction. *Ab initio* methods, which introduced the interaction perturbatively, were used by De Loth et al. (1985), but the method was far from being general. They considered the dimeric species copper acetate hydrate, $Cu_2(CH_3COO)_4(H_2O)_2$, and found a reasonable agreement with experiment by first constructing the magnetic orbitals and then perturbing them by adding excited configurations. The potential and kinetic exchanges have the largest contribution but they alone do not even reproduce the sign of the experimental coupling constant. In order to obtain better results they had to introduce spin polarization and high-order kinetic exchange.

A simpler model was introduced by Noodleman (Noodleman 1981; Noodleman and Davidson 1986; Noodleman et al. 1982; Noodleman et al. 1995). This model is widely used now, therefore we will briefly give its foundations in the following. In a coupled dimer there will be a number of doubly occupied canonical molecular orbitals describing paired electrons on the ligand and metal atoms, and a small number of weakly interacting magnetic orbitals. The model takes into consideration the two linked subunits and introduces a mixed or broken symmetry state, that is in each subunit the spins are parallel to each other and antiparallel to those of the other subunit. When the two subunits are allowed to interact a state is obtained which has $\langle S_z \rangle = 0$ but is not an eigenvector of S^2, therefore it is not a pure state. The Heisenberg exchange coupling constant can be calculated from the energies of the mixed spin state and the highest spin multiplet. The calculation is usually performed using DFT. The key equation for the broken symmetry approach is:

$$E_{\text{Smax}} - E_{\text{bs}} = -2S_{\text{max}}J. \tag{2.49}$$

The method was originally applied to binuclear iron–sulphur clusters, which attracted interest as models of the active centres in iron–sulphur proteins. Later it was applied to a large number of clusters (Kortus and Pederson 2000; Kortus et al. 2001, 2002), including also mixed valence species (Ciofini and Daul 2003). An interesting review on the subject is available (Ruiz E. et al. 2001).

2.4 Through-space and other interactions

The through-space interaction between the two magnetic centres can often be easily calculated through the point dipolar approximation. This assumes that the individual moments can be well represented by point dipoles, and that the separation between the dipoles is large compared to the spatial extension of the dipoles themselves. Under this approximation the dipolar component of $\boldsymbol{J}_{12}$ defined in (2.20) is given by

$$J_{12}^{\text{dip}} = \frac{\mu_B^2}{R^3} \left[\mathbf{g_1} \cdot \mathbf{g_2} - 3 \frac{(\mathbf{g_1} \cdot \mathbf{R})(\mathbf{R} \cdot \mathbf{g_2})}{|R|^2} \right] \tag{2.50}$$

or, in a more explicit style

$$\left(J_{12}^{\text{dip}} \right)_{\alpha\gamma} = \frac{\mu_B^2}{R^3} \left[g_1^{\alpha\xi} \cdot g_2^{\xi\gamma} - 3 \frac{(g_1^{\alpha\xi} R_\xi)(R_\varsigma \cdot g_2^{\varsigma\gamma})}{|R|^2} \right]$$

when $\mathbf{R}$ is the vector joining the two dipoles. Here repeated indices mean summation on those indices. α, γ, ξ, and ζ correspond to Cartesian coordinates $x, y,$

and z. It is important to notice that in the general case when $\mathbf{g}_1$ and $\mathbf{g}_2$ are two different tensors which are not multiples of the unit matrix J_{12}^{dip} is not symmetric. It can be decomposed as outlined in (2.20)–(2.21). In particular it has a non-zero trace, therefore it also contributes to the isotropic component J_{12}. For instance for a pair of interacting spins characterized by the following g tensors $\mathbf{g}_1 = g_e\mathbf{I}$, $g_{2,xx} = g_{2,zz} = 2.05$; $g_{2,yy} = 2.20$ separated by 0.3 nm, the D_{zz} component is calculated as -0.202 K. The dipolar contribution to the isotropic term is only -6.16×10^{-3} K. The anisotropic component has its maximum value parallel to the line connecting the two spins, and this is negative, indicating that the spins tend to be parallel to each other. In this example, where the direction connecting the two spins coincides with a principal axis of the anisotropic $\mathbf{g}$ tensor, all the components of the dipolar antisymmetric vector, $\mathbf{d}^{\text{dip}}$, are zero.

The isotropic through-bond interaction was described in Section 2.3. The anisotropic and antisymmetric terms arise from higher order perturbations. A relatively simple treatment was provided by Moriya (1960, 1963) who showed that the ground state function of an electron centred at the atom 1 must include excited states through spin–orbit coupling mixing:

$$|g_1\rangle = |g_1^0\rangle + \sum_{e_1^i} \frac{\langle e_1^i|\mathcal{H}_{\text{so}}|g_1^0\rangle}{\Delta_{e_1^i}} \tag{2.51}$$

where g_1^0 refers to the ground state in the absence of spin–orbit coupling, and e_1^i label the excited states of centre 1. Similar corrections are applied to the functions of centre 2. Spin–orbit mixing of excited states is responsible of the deviation of the g values of a paramagnet from the free-electron value $g_e = 2.0023$ (see equations 2.38–2.39). When the corrected wavefunctions are used to express the exchange interaction, they give rise to terms which depend on integrals involving ground and excited orbitals. The anisotropic parameters are proportional to $(\Delta g/g)^2$, while the antisymmetric terms correspond to $(\Delta g/g)$. Since $(\Delta g/g)$ is usually much smaller than 1 the antisymmetric term may be potentially larger than the anisotropic one, provided it is symmetry-allowed. However there are several difficulties to a quantitative estimation of these terms. In fact the integrals involving excited states are also needed. Often they are approximated with the isotropic term J, but they are in principle different, and there is good experimental evidence that they are in fact different. However the $(\Delta g/g)$ dependence is a relevant one, which allows us to neglect anisotropic exchange components in systems like high-spin iron(III), for which $g \sim g_e$.

All the above considerations hold on the assumption of orbitally non-degenerate ground states. While this is a good approximation for organic magnets, for transition metal ions it is not infrequent to encounter orbital degeneracy. When this situation is attained the isotropic term is no longer dominant in the bilinear expansion of (2.21), and the other components must be directly introduced. This makes the whole treatment very complicated. However, a simplified treatment is quite often used, with the following anisotropic Hamiltonian:

$$\mathcal{H}_{\text{ex}} = -J[\alpha S_{1z}S_{2z} + \beta S_{1x}S_{2x} + \gamma S_{1y}S_{2y}] \tag{2.52}$$

where, for $\alpha = \beta = \gamma = 1$, (2.52) becomes identical to the isotropic term of (2.21). This case is called Heisenberg exchange. For $\alpha = 1$, $\beta = \gamma = 0$, $\mathcal{H}_{ex}$ corresponds to the Ising model, and for $\alpha = 0$, $\beta = \gamma = 1$ to the XY model. Tsukerblat suggested a new model for taking into account orbital degeneracy (Borras–Almenar *et al.* 1998b, 1998c; 2001b).

2.5   From pairs to clusters and beyond

2.5.1   *Isotropic coupling*

The simplest extension to the case of more than two interacting centres is made by adding additional terms to the spin Hamiltonian (2.21). Let us suppose initially for the sake of simplicity that the isotropic term of the spin–spin interaction is the only one to be considered. Then (2.21) can be extended to:

$$\mathcal{H}_{ex} = -\sum_{i<j} J_{ij} \mathbf{S}_i \cdot \mathbf{S}_j \qquad (2.53)$$

where the sum is over all the pairs of the cluster.

At this level of approximation, i.e. considering isotropic exchange interactions only, the total spin is a good quantum number. Writing the eigenfunctions of the total spin can be a good strategy for reducing the size of the Hamiltonian matrices to be computed for calculating the energies of the spin levels. This is simple in principle but the problem rapidly becomes very complex, because the number of states increases very rapidly. In fact for a cluster of N identical spins S_i the states are $(2S_i + 1)^N$. If $S_i = 5/2$ for $N = 8$ the states are 1,679,616, and the total spin states range from $S = 20$ to $S = 0$. The number of states corresponding to each value of S is shown in Table 2.4.

In some highly symmetric cases it is possible to give closed formulas for the energy levels of the total spin states. The procedure relies on the so-called Kambe approach (Kambe 1950) which rewrites the spin Hamiltonian using the total spin. Let us take, for example, a cluster of six equivalent spins S_i on the vertices of an octahedron. The spin Hamiltonian can be written as:

$$\mathcal{H} = -J(\mathbf{S}_1 \cdot \mathbf{S}_2 + \mathbf{S}_1 \cdot \mathbf{S}_3 + \mathbf{S}_1 \cdot \mathbf{S}_4 + \mathbf{S}_1 \cdot \mathbf{S}_5 + \mathbf{S}_1 \cdot \mathbf{S}_6 + \mathbf{S}_2 \cdot \mathbf{S}_3$$
$$+ \mathbf{S}_2 \cdot \mathbf{S}_4 + \mathbf{S}_2 \cdot \mathbf{S}_5 + \mathbf{S}_2 \cdot \mathbf{S}_6 + \mathbf{S}_3 \cdot \mathbf{S}_4 + \mathbf{S}_3 \cdot \mathbf{S}_5 + \mathbf{S}_3 \cdot \mathbf{S}_6 \qquad (2.54)$$
$$+ \mathbf{S}_4 \cdot \mathbf{S}_5 + \mathbf{S}_4 \cdot \mathbf{S}_6 + \mathbf{S}_5 \cdot \mathbf{S}_6).$$

If we define a total spin operator:

$$\mathbf{S} = \mathbf{S}_1 + \mathbf{S}_2 + \mathbf{S}_3 + \mathbf{S}_4 + \mathbf{S}_5 + \mathbf{S}_6 \qquad (2.55)$$

it is easy to verify that (2.54) can be rewritten as:

$$\mathcal{H} = -\frac{J}{2}(\mathbf{S}^2 - \mathbf{S}_1^2 - \mathbf{S}_2^2 - \mathbf{S}_3^2 - \mathbf{S}_4^2 - \mathbf{S}_5^2 - \mathbf{S}_6^2). \qquad (2.56)$$

TABLE 2.4. Number of states for each
total spin value S in the case of 2,
4, 6, and 8 coupled spins 5/2.

S	$N = 2$	$N = 4$	$N = 6$	$N = 8$
0	1	6	111	2666
1	1	15	315	7700
2	1	21	475	11900
3	1	24	575	14875
4	1	24	609	16429
5	1	21	581	16576
6		15	505	15520
7		10	405	13600
8		6	300	11200
9		3	204	8680
10		1	126	6328
11			70	4333
12			35	2779
13			15	1660
14			5	916
15			1	462
16				210
17				84
18				28
19				7
20				1

The energies of the total spin levels become a function only of the total spin S, and they are given by

$$E(S) = -\frac{J}{2}[S(S + 1) - 6S_i(S_i + 1)]. \tag{2.57}$$

The systems for which the Kambe approach can be used have been reported (Belorizky and Fries 1993). In general it can be applied to highly symmetric spin arrangements.

For spin clusters which are not too large (typically up to say 8–10 $S = \frac{5}{2}$ spins) an elegant approach exploits the irreducible tensor operators ITOs (Silver 1976; Gatteschi and Pardi 1993). The formalism of ITOs is rather complex, but their use is after all very simple. The central idea is that of writing the operators in a standard way in order to exploit the symmetry of the full rotation group. This is a continuous group, i.e. it contains an infinite number of elements, and it is not generally taught in chemistry, where only the finite point groups are used. Nevertheless chemists are familiar with its irreducible representations, whose bases correspond to the eigenfunctions of the angular momentum operators. For instance a p function is the basis for the irreducible representation $J = 1$. When we combine two angular momenta S_1 and S_2 to give S values $|S_1 - S_2| \leq S \leq S_1 + S_2$, we exploit the rules for the reduction of a representation in the full rotation group. The basis functions which we use in the description of the coupled magnetic clusters are eigenfunctions of the spin momentum, therefore they are suitable for obtaining the irreducible representation in the coupled scheme. In order to be able to exploit symmetry to calculate the matrix elements it is necessary to write also the operators in such a way that they correspond to irreducible representations of the full rotation group.

Without giving any details of how this can be done, it is important to learn how to recognize the irreducible representations of the full rotation group appropriate to a given operator. An angular momentum operator like $\mathbf{S}_i$ is a vector operator, characterized by three components, x, y, and z. Therefore it can be considered as a basis for the $J = 1$ representation of the full rotation group. Formally we can replace it by a $\mathbf{T}_{J=1,M}(S_j)$ irreducible tensor operator of rank 1. The explicit form of the operators is the following: $T_{11}(S) = -(2)^{-1/2}S_+$; $T_{10}(S) = S_z$; $T_{1-1}(S) = (2)^{-1/2}S_-$, where S_+ and S_- are the well-known shift operators.

A scalar operator is defined by a $\mathbf{T}_{JM}$ ITO characterized by $J = 0$, $M = 0$. A second-rank tensor operator is characterized by $J = 2$ and $-2 \leq M \leq 2$. The Zeeman operator is a first-order tensor operator; the zero field splitting corresponds to a second-rank operator.

The matrix elements of the ITO can be formally expressed by using the Wigner–Eckart theorem:

$$\langle S_1 M_1 | \mathbf{T}_{JM}(S_1) | S_1' M_1' \rangle = (-1)^{S_1 - M_1} \langle S_1 \| \mathbf{T}_J(S_1) \| S_1' \rangle \begin{pmatrix} S_1 & J & S_1' \\ -M_1 & M & M_1' \end{pmatrix}$$
(2.58)

where $\langle S_1 \| \mathbf{T}_J(S_1) \| S_1' \rangle$ is called a reduced matrix element, which can be calculated once and for all for ITOs, as shown below:

$$\langle S_1 \| \mathbf{T}_0(S_1) \| S_1' \rangle = \delta_{S_1 S_1'} \sqrt{(2S_1 + 1)}$$

$$\langle S_1 \| \mathbf{T}_1(S_1) \| S_1' \rangle = \delta_{S_1 S_1'} \sqrt{S_1(S_1 + 1)(2S_1 + 1)}.$$
(2.59)

$\begin{pmatrix} S_1 & J & S_1' \\ -M_1 & M & M_1' \end{pmatrix}$ is a 3-j symbol, i.e. a number which can be easily calculated, as shown in Appendix A6, and it is also found in standard computer routines. In order to be different from zero the condition $M_1 + M + M_1' = 0$ must hold.

The importance of (2.58) is that in order to calculate the matrix elements of a given vector or scalar operator one needs only the symmetry properties of the functions. Since the spin eigenfunctions are indeed symmetry labelled, the calculation of the matrix elements becomes very simple. There is no real advantage in the use of ITOs for the calculation of the levels of a system comprising one spin, but matters become very different in the case of coupled spins. In fact in this case it is possible to define compound operators, which allow very fast calculations. If we are interested at the energy levels of the operator:

$$\mathcal{H} = -J\mathbf{S}_1 \cdot \mathbf{S}_2 \qquad (2.60)$$

we can define a compound ITO, $\mathbf{T}_0(S)$, starting from the operators $\mathbf{T}_{1M}(S_1)$ and $\mathbf{T}_{1M}(S_2)$. The direct product of two first-rank operators is a set of nine operators which can be reorganized as $\{T_1(S_1) \otimes T_1(S_2)\}_{JMj}$, where $J = 0, 1, 2$. $\{T_1(S_1) \otimes T_1(S_2)\}_{00} = T_0(S)$ must be scalar because in (2.60) we have the scalar product of the two operators.

The matrix elements of (2.60) in a given basis $|S_1 M_1 S_2 M_2 S M\rangle$ are given by:

$$\langle S_1 S_2 S M | \mathbf{T}_0(S) | S_1 S_2 S' M' \rangle$$

$$= \delta_{SS'} \delta_{MM'} (-1)^{S-M} \langle S_1 S_2 S | \mathbf{T}_0(S) | S_1 S_2 S \rangle \begin{pmatrix} S & 0 & S \\ -M & 0 & M \end{pmatrix}. \qquad (2.61)$$

The compound reduced matrix element can be written as:

$$\langle S_1 S_2 S | \mathbf{T}_0(S) | S_1 S_2 S \rangle$$

$$= (-1)^{S+S_1+S_2} \langle S_1 \| \mathbf{T}_1(S_1) \| S_1 \rangle \langle S_2 \| \mathbf{T}_1(S_2) \| S_2 \rangle \begin{Bmatrix} S_1 & 0 & S_1 \\ S_2 & S & S_2 \end{Bmatrix} \qquad (2.62)$$

where the symbol in curly brackets is a 6-j symbol, i.e. a number, akin to a 3-j symbol, which can be calculated given the values of the numbers indicated in it. Standard expressions for calculating 6-j symbols are given in Appendix A6.

The great advantage of using ITOs, although at the beginning the formalism looks awkward, is that the eigenfunctions of the total spin need to be written only symbolically, with a minimum need of memory storage on a computer program, and their matrix elements can be easily calculated in a standard way using equations (2.60)–(2.62). Let us make this point clear with an example, considering two spins $S_1 = S_2 = 5/2$. We know that the total spin values S are 0,1,2,3,4,5. When we use ITOs the eigenfunction is simply indicated by the set of four numbers $|S_1 S_2 S M\rangle$, $|5/2 \ 5/2 \ 0 \ 0\rangle$, and the matrix elements are simply computed using (2.60)–(2.62). The advantage is clear, even for two spins, but of course it is much larger when we increase the number of spins.

In order to manage to write a general expression for a matrix element for a cluster of n spins we still need some additional considerations. We must imagine the functions to be written in the order corresponding to the chosen coupling scheme. For instance, in the case of four spins for which we choose a coupling scheme in which we consider first S_1 and S_2 to give S_{12}, then S_{12} and S_3 to give S_{123}, and finally S_{123} and S_4 to give S, the functions should be written as: $|S_1 S_2 S_{12}, S_{12} S_3 S_{123}, S_{123} S_4 S M\rangle$. In this way all the spins are grouped into $n-1$ sets of three, corresponding to pairwise couplings, specified by the spins $S_\alpha, S_{\alpha+1}$, and $S_{\alpha+2}$ in the bra, and by the spins $S'_\alpha, S'_{\alpha+1}$, and $S'_{\alpha+2}$ in the ket.

If $\alpha = 1$ then $S_\alpha = S_1$, $S_{\alpha+1} = S_2$, $S_{\alpha+2} = S_{12}$; if $\alpha = 4$ then $S_\alpha = S_{12}$, $S_{\alpha+1} = S_3$, $S_{\alpha+2} = S_{123}$; If $\alpha = 7$ then $S_\alpha = S_{123}$, $S_{\alpha+1} = S_4$, $S_{\alpha+2} = S$.

The general matrix element for the term of the Hamiltonian (2.53) describing the interaction between the ith and lth spins of the cluster, is given by:

$$\left\langle S_1 S_2 \ldots S_n \ldots S_p \ldots SM | J_{i,l} \mathbf{S}_i \cdot \mathbf{S}_l | S_1 S_2 \ldots S_n \ldots S'_p \ldots S'M' \right\rangle$$

$$= \frac{\sqrt{3}}{\sqrt{2S+1}} J_{i,l} (-1)^{S-M}$$

$$\Pi_{m=1,n} \langle S_m \| T_k(S_m) \| S_m \rangle \Pi_{\alpha=1,4,\ldots 3n-2} \left\{ \begin{matrix} S_\alpha & S'_\alpha & k_\alpha \\ S_{\alpha+1} & S'_{\alpha+1} & k_{\alpha+1} \\ S_{\alpha+2} & S'_{\alpha+2} & k_{\alpha+2} \end{matrix} \right\}$$

$$[(2S_{\alpha+2} + 1)(2S'_{\alpha+2} + 1)(2k_{\alpha+2} + 1)]^{1/2} \tag{2.63}$$

where S_1 to S_n represent the individual n spins of the cluster and S_p stands for the intermediate coupled spins and k_α is the rank of the corresponding ITO. The symbol in curly brackets is a 9-j symbol, similar to the 3-j and 6-j symbols already introduced, only more complex. Standard formulae are available to calculate them. However, it can be easily shown that if we are interested in isotropic exchange only, then at least one of the k_ι indices is zero, and the 9-j symbols reduce to 6-j symbols, according to the relations shown in Table 2.5.

Although (2.63) looks formidable, it is not difficult at all to use it as can be shown by working out an example in detail (Gatteschi and Pardi 1993). Let us consider a cluster of four $S = 1/2$ spins, and let us assume the coupling scheme defined above, $|S_1 S_2 S_{12}, S_{12} S_3 S_{123}, S_{123} S_4 S\rangle$. The corresponding eigenfunctions of $\mathbf{S}^2$ are shown in Table 2.6

To calculate the matrix element (2.63), the values of the ranks k_ι of the operators are needed. Since we are considering a Hamiltonian which operates on the coordinates of spins 1 and 2, respectively, and since $\mathbf{S}_1$ and $\mathbf{S}_2$ are vector operators, then $k_1 = k_2 = 1$. The Hamiltonian requires the scalar product of the two vector operators, therefore the compound operator rank k_{12} must be a scalar, i.e. $k_{12} = 0$. In the Hamiltonian there is no operator corresponding to the coordinates of the spins 3 and 4, therefore $k_3 = k_4 = k_{123} = k = 0$. The diagonal

TABLE 2.5. Relations between 6-j and 9-j symbols in special cases. The first column gives the 9-j symbol, the second the equivalent 6-j product.

$\{S_a S_{a'} 1; S_b S_{b'} 1; S_c S_{c'} 0\}$	$\delta_{ScSc'}(-1)^{Sa'+1+Sb+Sc}[3(2S_c + 1)]^{-1/2}\{S_a S_{a'} 1; S_{b'} S_b S_c\}$
$\{S_a S_{a'} 1; S_b S_{b'} 0; S_c S_{c'} 1\}$	$\delta_{SbSb'}(-1)^{Sa+1+Sb+Sc'}[3(2S_b + 1)]^{-1/2}\{S_{a'} S_a 1; S_c S_{c'} S_b\}$
$\{S_a S_{a'} 0; S_b S_{b'} 1; S_c S_{c'} 1\}$	$\delta_{SaSa'}(-1)^{Sa+1+Sb'+Sc}[3(2S_a + 1)]^{-1/2}\{S_{c'} S_c 1; S_b S_{b'} S_a\}$
$\{S_a S_{a'} 0; S_b S_{b'} 0; S_c S_{c'} 0\}$	$\delta_{SaSa'} \delta_{SbSb'} \delta_{ScSc'}[3(2S_a + 1)(2S_b + 1)(2S_c + 1)]^{-1/2}$

TABLE 2.6. Eigenfunctions of
S^2 for four $S = \frac{1}{2}$ spins.

$\mathbf{S}_{12}$	$\mathbf{S}_{123}$	$\mathbf{S}$
1	$\frac{3}{2}$	2
1	$\frac{3}{2}$	1
1	$\frac{1}{2}$	1
0	$\frac{1}{2}$	1
1	$\frac{1}{2}$	0
0	$\frac{1}{2}$	0

matrix element of the function $|S_{12} = 1, S_{123} = 1/2, S = 0\rangle$ is given by:

$$\left\langle \frac{1}{2} 0 \,|-J_{12} S_1 \cdot S_2 | \, 1 \frac{1}{2} 0 \right\rangle$$

$$= J_{12} \left\langle \frac{1}{2} \left\| \frac{1}{2} \sqrt{3}\, \mathbf{T}_1 \right\| \frac{1}{2} \right\rangle \left\langle \frac{1}{2} \left\| \mathbf{T}_1 \right\| \frac{1}{2} \right\rangle \left\langle \frac{1}{2} \left\| \mathbf{T}_0 \right\| \frac{1}{2} \right\rangle \left\langle \frac{1}{2} \left\| \mathbf{T}_0 \right\| \frac{1}{2} \right\rangle$$

$$\times \left[(2 \times 1 + 1)(2 \times 1 + 1)(2 \times 0 + 1) \right]^{1/2} \left[\left(2 \times \frac{1}{2} + 1\right) \left(2 \times \frac{1}{2} + 1\right) (2 \times 0 + 1) \right]^{1/2}$$

$$\times \left[(2 \times 0 + 1)(2 \times 0 + 1)(2 \times 0 + 1) \right]^{1/2} \begin{Bmatrix} \frac{1}{2} & \frac{1}{2} & 1 \\ \frac{1}{2} & \frac{1}{2} & 1 \\ 1 & 1 & 0 \end{Bmatrix} \begin{Bmatrix} 1 & 1 & 0 \\ \frac{1}{2} & \frac{1}{2} & 0 \\ \frac{1}{2} & \frac{1}{2} & 0 \end{Bmatrix} \begin{Bmatrix} \frac{1}{2} & \frac{1}{2} & 0 \\ \frac{1}{2} & \frac{1}{2} & 0 \\ 0 & 0 & 0 \end{Bmatrix}.$$

$$(2.64)$$

For the sake of simplicity we omit the $S_i = 1/2$ spins and we give only the values
of S_{12}, S_{123}, S. It is apparent that all the 9-j symbols in (2.64) contain at least
one zero, therefore they can be reduced to 6-j symbols taking advantage of the
relations shown in Table 2.5. The first 9-j symbol is equal to $-1/18$, the second
to $1/\sqrt{12}$, and the third to $1/2$. The matrix element on the whole is calculated to
be $-1/4 J_{12}$. Proceeding in the same way it is possible to calculate all the matrix
elements.

Even with the simplifications allowed by the ITO approach, the number of
levels increases very rapidly, and if we look, for instance at Table 2.4 we immediately realize that it may be difficult to diagonalize so many large matrices. An
additional reduction in the size of the matrices can come from the introduction
of the point group symmetry appropriate to the system under consideration.
The procedure becomes even more complex. The techniques are described in the
literature (Raghu *et al.* 2001; Rudra *et al.* 2001).

2.5.2 *Magnetic anisotropy in clusters*

An important step must still be made in order to extend the treatment of
Section 2.1.3 to find correlations between the spin Hamiltonian parameters of
the individual centres and those of the clusters. We recall that it is possible to

find closed equations in the assumption of strong exchange. In order to do this we rewrite (2.24)–(2.26) in a formally slightly different but equivalent way:

$$\mathbf{g}(S_1 S_2 S) = c_\alpha(S_1 S_2 S)\mathbf{g}_1 + c_\beta(S_1 S_2 S)\mathbf{g}_2. \tag{2.65}$$

We label the coefficients defined in (2.24) relative to the first spin as α and to the second spin as β. When we pass to a cluster we must first of all specify the coupling scheme, i.e. decide the procedure to pass from the individual spins to the total spin. This is not unambiguous and the choice will be made according to the most convenient way of describing the total spin states. At any rate we will have to define $n-2$ intermediate spins. Let us work out as an example the case of a cluster of four spins, and let us decide as a coupling scheme:

$$\mathbf{S}_1 + \mathbf{S}_2 = \mathbf{S}_{12}; \mathbf{S}_3 + \mathbf{S}_4 = \mathbf{S}_{34}; \mathbf{S}_{12} + \mathbf{S}_{34} = \mathbf{S} \tag{2.66}$$

The required relations can be worked out by using (2.65) for the two intermediate spins S_{12} and S_{34} and for S. We may first calculate the coefficients for the intermediate state S_{12}: clearly (2.65) holds. For S_{34} we will write in analogy:

$$\mathbf{g}(S_3 S_4 S_{34}) = c_\alpha(S_3 S_4 S_{34})\mathbf{g}_3 + c_\beta(S_3 S_4 S_{34})\mathbf{g}_4. \tag{2.67}$$

Finally we proceed to calculate the relations for the total spin:

$$\mathbf{g}(S_{12} S_{34} S) = c_\alpha(S_{12} S_{34} S)\mathbf{g}(S_1 S_2 S_{12}) + c_\beta(S_{12} S_{34} S)\mathbf{g}(S_3 S_4 S_{34}). \tag{2.68}$$

Combining (2.66)–(2.68) we finally find:

$$\mathbf{g}(S_1 S_2 S_{12} S_3 S_4 S_{34} S) = c_\alpha(S_{12} S_{34} S)[c_\alpha(S_1 S_2 S_{12})\mathbf{g}_1 + c_\beta(S_1 S_2 S_{12})\mathbf{g}_2]$$
$$+ c_\beta(S_{12} S_{34} S)[c_\alpha(S_3 S_4 S_{34})\mathbf{g}_3 + c_\beta(S_3 S_4 S_{34})\mathbf{g}_4]. \tag{2.69}$$

Proceeding in a similar way we calculate the analogous relations for the $\mathbf{D}$ tensor generated by the local $\mathbf{D}_i$ tensors:

$$\mathbf{D}(S_1 S_2 S_{12} S_3 S_4 S_{34} S) = d_\alpha(S_{12} S_{34} S)[d_\alpha(S_1 S_2 S_{12})\mathbf{D}_1 + d_\beta(S_1 S_2 S_{12})\mathbf{D}_2]$$
$$+ d_\beta(S_{12} S_{34} S)[d_\alpha(S_3 S_4 S_{34})\mathbf{D}_3 + d_\beta(S_3 S_4 S_{34})\mathbf{D}_4]. \tag{2.70}$$

where the d_α and d_β can be calculated according to (2.27) and (2.28).

The $\mathbf{D}$ tensor generated by the spin–spin $\mathbf{D}_{ij}$ tensors are given by:

$$\mathbf{D}(S_1 S_2 S_{12} S_3 S_4 S_{34} S) = \sum_{i<j} d_{ij}(S_1 S_2 S_{12} S_3 S_4 S_{34} S)\mathbf{D}_{ij}. \tag{2.71}$$

Let us calculate the $d_{ij}(S_1 S_2 S_{12} S_3 S_4 S_{34} S)$ coefficients, starting from $d_{12}(S_1 S_2 S_{12} S_3 S_4 S_{34} S)$. This depends on the coordinates of spin 1 and spin 2, therefore it will contribute with a term $d_{\alpha\beta}(S_1 S_2 S_{12})$, the $d_{\alpha\beta}$ coefficients being defined in (2.27) and (2.28).

Further we must take into account also the single spin contribution of S_{12} introducing a term $d_\alpha(S_{12} S_{34} S)$:

$$d_{12}(S_1 S_2 S_{12} S_3 S_4 S_{34} S) = d_{\alpha\beta}(S_1 S_2 S_{12})d_\alpha(S_{12} S_{34} S). \tag{2.72}$$

The case is slightly different for $d_{13}(S_1S_2S_{12}S_3S_4S_{34}S)$, the spins 1 and 3 showing up in different intermediate spin operators. The $d_{\alpha\beta}(S_{12}S_{34}S)$ coefficient must be weighted according to the presence of spin 1 in S_{12} and spin 3 in S_{34}:

$$d_{13}(S_1S_2S_{12}S_3S_4S_{34}S) = d_{\alpha\beta}(S_{12}S_{34}S)c_\alpha(S_1S_2S_{12})c_\alpha(S_3S_4S_{34}). \qquad (2.73)$$

The other coefficients can be calculated by obvious extensions:

$$d_{14}(S_1S_2S_{12}S_3S_4S_{34}S) = d_{\alpha\beta}(S_{12}S_{34}S)c_\alpha(S_1S_2S_{12})c_\beta(S_3S_4S_{34})$$

$$d_{23}(S_1S_2S_{12}S_3S_4S_{34}S) = d_{\alpha\beta}(S_{12}S_{34}S)c_\beta(S_1S_2S_{12})c_\alpha(S_3S_4S_{34})$$

$$d_{24}(S_1S_2S_{12}S_3S_4S_{34}S) = d_{\alpha\beta}(S_{12}S_{34}S)c_\beta(S_1S_2S_{12})c_\beta(S_3S_4S_{34})$$

$$d_{34}(S_1S_2S_{12}S_3S_4S_{34}S) = d_{\alpha\beta}(S_3S_4S_{34})d_\beta(S_3S_4S_{34}). \qquad (2.74)$$

In order to gain a feeling of the formulae reported above it may be interesting to calculate the variation of the anisotropic interactions on increasing the size of the clusters (the number of the interacting spins). For a system of N spins S_i which are ferromagnetically coupled to give a total spin $S = NS_i$ it can be shown that the following relations are valid:

$$d_i = \frac{2S_i - 1}{N(2NS_i - 1)}; d_{ij} = \frac{2S_i}{N(2NS_i - 1)} \qquad (2.75)$$

where d_i and d_{ij} are defined by the relations

$$\boldsymbol{D}_S = \sum_i d_i \boldsymbol{D}_i + \sum_{i<j} d_{ij} \boldsymbol{D}_{ij} \qquad (2.76)$$

$$\mathcal{H} = \sum_i \mathbf{S}_i \cdot \mathbf{D}_i \cdot \mathbf{S}_i + \sum_{i<j} \mathbf{S}_i \cdot \mathbf{D}_{ij} \cdot \mathbf{S}_j. \qquad (2.77)$$

In the case of individual centres all parallel to each other the single-ion contribution to the D parameter is

$$D_S = \frac{2S_i - 1}{2S - 1} D_i. \qquad (2.78)$$

The energy difference between the states with highest and lowest $|M|$, value respectively, Δ, which is relevant for the low-temperature dynamics of the magnetization of single-molecule magnets is therefore given by

$$\Delta = |D|S^2 \qquad (2.79)$$

for integer spin and

$$\Delta = |D|S^2 - \tfrac{1}{4} \qquad (2.80)$$

for half-integer spin.

Let us consider integer spin $S_i = 2$. Then

$$\Delta(NS_i) = \frac{2S_i - 1}{2NS_i - 1} N^2 S_i^2 D_i. \qquad (2.81)$$

A ferromagnetic ring of eight manganese(III) ions, provided all the local aniso-tropy axes are parallel to each other, would have $\Delta(S = 16) = 768/31 D_i$ while for a single ion it would be $\Delta(S = 2) = 4D_i$, yielding an energy barrier ca. 6.2 times larger for the ring.

In the case of regular rings it is possible to evaluate the dipolar contribution for isotropic g in the ferromagnetic state. We define a local coordinate system where all the z axes are along the line connecting spin i and spin j and all the local y axes are parallel to each other and perpendicular to the ring plane and

$$D_S = -\frac{2S_i}{3(2S-1)}\left[D_{i,i+1} + D_{i,i+2} + \cdots + \frac{1}{2}D_{i,i+\frac{N}{2}}\right] \tag{2.82}$$

for even N, and

$$D_S = -\frac{2S_i}{3(2S-1)}\left[D_{i,i+1} + D_{i,i+2} + \cdots + \frac{1}{2}D_{i,i+\frac{N-1}{2}}\right] \tag{2.83}$$

for odd N.

A useful example is provided by antiferromagnetic rings of six $S= 5/2$ spins, which will be discussed in some detail in Section 14.1. We suppose that the ring lies in the xz plane. Therefore the ring has axial symmetry around y, and we need to calculate only the D_{yy} components of the individual dipolar tensors. The distances $r_{i,i+k}$ needed to calculate the dipolar tensors are given by

$$r_{i,i+k} = r_{i,i+1}\sqrt{\frac{1 - \cos\left(\dfrac{2k\pi}{N}\right)}{1 - \cos\left(\dfrac{2\pi}{N}\right)}}. \tag{2.84}$$

For a nearest-neighbour distance $r_{i,i+1} = 3.00$ Å we calculate the $D^{yy}_{i,i+k}$ components reported in Table 2.7.

TABLE 2.7. Components of the dipolar tensor for a hexagonal ring of six $S=5/2$ spins. The values of D are given in 10^{-3} K.

k	$r_{i,i+k}$	$D^{yy}_{i,i+k}$
1	3.00	0.9229
2	5.20	0.1772
3	6.00	0.1154
4	5.20	0.1772
5	3.00	0.9229

TABLE 2.8. Coupling coefficients for the lower
spin states originated by the coupling of two
intermediate spins $S_{135} = S_{246} = 15/2$ of a
hexagonal ring of six $S = 5/2$ spins.

S	c_1	d_1	d_{12}	d_{13}	d_{14}
1	0.1667	−2.4000	2.8556	−3.0000	2.8556
2	0.1667	−0.5442	0.6905	−0.6803	0.6905
3	0.1667	−0.2349	0.3296	−0.2937	0.3296
4	0.1667	−0.1224	0.1984	−0.1531	0.1984
5	0.1667	−0.0684	0.1353	−0.0855	0.1353
6	0.1667	−0.0381	0.1000	−0.0476	0.100
7	0.1667	−0.0194	0.07812	−0.0242	0.0781

The total spin states of the rings can be efficiently described by coupling first the odd site spins, then the even sites and finally summing the two intermediate spins together. Therefore the functions can be labelled as: $|S_1 S_3 S_{13} S_5 S_{135} S_2 S_4 S_{24} S_6 S_{246} S\rangle$. It can be shown that for antiferromagnetic coupling the lowest lying states are $|\frac{5}{2}\frac{5}{2}5\frac{5}{2}\frac{15}{2}\frac{5}{2}\frac{5}{2}5\frac{15}{2}S\rangle$. The coupling coefficients which are needed are: $c_1, d_1, d_{12}, d_{13}, d_{14}$, all the others being symmetry related. The coefficients for the lowest total spin S states originating by the coupling of the intermediate spins $S_{135} = 15/2$ and $S_{246} = 15/2$ are given in Table 2.8.

3

OBSERVATION OF MICROSCOPIC MAGNETISM

Experiments have played a seminal role in the field of molecular nanomagnetism. In fact, the experimental observation of an unusual imaginary component of the alternating current (ac) susceptibility in zero static field for $Mn_{12}ac$ could be considered as the starting point for this branch of molecular magnetism (Caneschi *et al.* 1991; Sessoli *et al.* 1993b). Since then many different experimental techniques have been used to characterize these classes of materials. This chapter focuses on highlighting what kind of information, relevant for molecular nanomagnetism, can be obtained from the different techniques commonly employed in magnetism. It is impossible to cover in depth the various topics, and we assume that the reader has a basic knowledge of experimental magnetism. We will provide, however, the relevant background references to which the interested readers may refer.

The information we are interested in can be grouped into three categories. The first one regards interactions among the different paramagnetic centres constituting the cluster and the energy spectrum of spin states that arises from them. Equally important is the information on the magnetic anisotropy of the system and the energy spectrum of the lowest spin multiplets. The last category concerns the dynamics of the magnetization. We will proceed by discussing the different techniques beginning from standard magnetometry and moving to more sophisticated techniques like magnetic torque measurements. Magnetic resonances, both electronic and nuclear, or even the more exotic muon resonance, are also widely employed. Neutron techniques, exploiting the magnetic moment of the neutron, are particularly useful to quantify energy splitting in zero field and the spin distribution on the molecule.

Most of the content of this chapter is obvious for experimentalists working in the field but can be of some use for scientists and students entering the field, as well as for theoreticians, who sometimes are not aware of the details that are behind the experimental data they are struggling with.

3.1 Magnetic techniques

3.1.1 *Standard magnetometry*

Most magnetometers are nowadays based on an inductive detection of the magnetic moment. According to Faraday's law a time-varying magnetic flux causes a current to flow in a closed circuit. The electromotive force is proportional to $-d\Phi/dt$. The pick-up coil is therefore sensitive to the flux generated by the

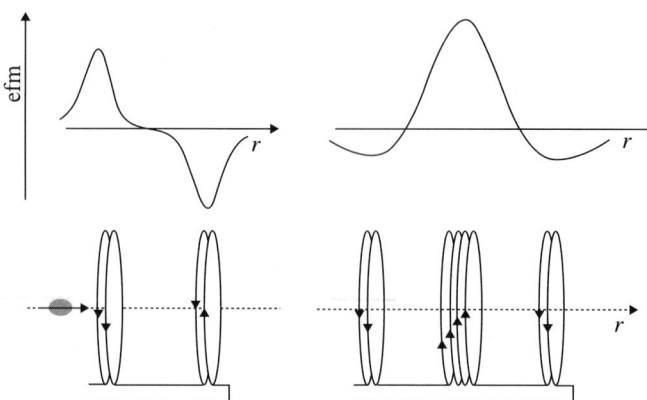

FIG. 3.1. Pick-up coils (gradiometers) to detect the magnetic flux of a magnetized sample that is moved inside the coils. Left: first-order gradiometer; right:
second-order gradiometer. On top of each coil the typical dependence of
the induced electromotive force on the position of the magnetized sample
is shown.

magnet itself and by the magnetized sample. The first contribution can be elimin-
ated by employing a special design for the pick-up coils, also called gradiometers
because they are sensitive only to the gradient of the magnetic field. The simplest
form is a first-order gradiometer constituted by two loops wound in opposite directions. The magnetized sample is moved from one loop to the other and the
difference between the induced voltages measured at the two positions is proportional to the magnetization of the sample. A further improvement is given by a
second-order gradiometer, where $2N$ windings are placed between two sets of N
windings, these last two wound in opposite directions to the central one. A schematic drawing of the two types of gradiometer and the induced electromotive force
is shown in Fig. 3.1.

In most high-sensitivity magnetometers the induced current is not directly
measured but the coils are inductively coupled to a Superconducting QUantum
Interference Device (SQUID). These devices combine two physical phenomena:
the quantization of the flux in a superconducting loop and Josephson tunnelling.
A SQUID is in fact a superconducting ring interrupted by a Josephson junction,
i.e. a non-superconducting barrier that can be tunnelled by the Cooper pairs.
Indeed there are two types of SQUIDs; the dc-SQUID consists of two Josephson
junctions connected in parallel, while the radiofrequency rf-SQUID has only one
junction, and to be operated must be biased with an alternating current. SQUIDs
are the most sensitive detectors of the magnetic flux. They are, in essence, a very
efficient flux—voltage transducer with a voltage output that is periodic in the
applied flux and the flux period is one flux quantum, $\Phi_0 = h/2e = 2.07 \cdot 10^{-15}$
weber.

It is not within our scope to cover here the theory of the operation and the practical aspects of SQUIDs and interested readers can refer to a specialized text (Clarke 1990) for general information. Companies commercializing SQUID magnetometers also offer interesting technical descriptions of their instruments.

The SQUID, being superconducting, is kept in liquid helium and is usually shielded from the magnetic field. Due to its sensitivity to the field, commercial SQUID-based magnetometers do not work at field higher than 80 kOe. For larger fields a different measuring technique is employed. The sample is moved at a frequency in the range 50–100 Hz up and down over a length of a few mm in the centre of one coil of a first-order gradiometer. The signal induced in the pick-up coils is read by a phase-sensitive detector (a lock-in amplifier). These very old types of magnetometers, developed by Foner at MIT (Foner 1959) are called vibrating-sample magnetometers, VSMs, and are now commercially available, usually equipped with a magnet producing fields over 100 kOe.

The measurement of the magnetization is generally done at variable temperature, and temperatures between 1.5 K and 300 K are currently available. The lower limit is reached by pumping over the liquid helium that enters the sample space through a needle valve. Variable temperatures are extremely important to get information on the energy spectrum of the spin levels. By decreasing the temperature the excited spin states are depopulated and the temperature dependence of the magnetization is thus related to the energy separation of the levels with different S value.

It is very common in the chemical literature to report the temperature variation of the magnetic susceptibility, χ, or its product with temperature, χT. The latter is particularly useful to highlight deviations from the paramagnetic behaviour of non-interacting spins, because in the last case χT assumes a constant value, as predicted by the Curie law:

$$\chi T = \frac{N g^2 \mu_B^2}{3 k_B} S(S+1) \tag{3.1}$$

for a mole of centres characterized by the spin value S. χ is field independent as long as one works at small fields and not too low temperatures, where the condition $g \mu_B H \ll k_B T$ is satisfied. It is important to stress here that what is usually measured is the magnetization of the sample, from which the M/H ratio is often reported. This ratio at moderate and strong applied fields does not always coincide with the magnetic susceptibility $\chi = dM/dH$, especially if the ground state has a large spin as in the materials of interest here.

When extracting the susceptibility from the magnetization we assume that the field experienced by the spin system has the same value of the applied field. This is not true, because a magnetized sample has poles of opposite signs on its surfaces, and these poles generate an additional field, which points in the opposite direction to the magnetizing one, and therefore is called the demagnetizing field:

$$\mathbf{H'} = \mathbf{H} - d\mathbf{M}. \tag{3.2}$$

$\mathbf{H}'$ is also known as the Maxwell field. The demagnetization factor, d, depends on the geometry of the sample. For a sphere, d is isotropic and equal to $\frac{4}{3}\pi$, in Gaussian (cgs) units. In the other cases, $\mathbf{d}$ is a tensor. For an infinitely long needle-shaped sample, the component of $\mathbf{d}$ along the axis of elongation approaches zero, and 2π perpendicular to it. Demagnetizing factors can be calculated for ellipsoids of revolution and are reported for instance in the textbook by Morrish (1966). A more general formulation is given in Appendix B.

We should take into account that the susceptibility relates M to H' and not to the applied field, and therefore differs from the M/H ratio, which we call here χ_{obs}, according to:

$$\chi = \frac{M}{H'} = \frac{M}{H - dM} = \frac{\chi_{\text{obs}}}{1 - d\chi_{\text{obs}}}$$

or

$$\chi_{\text{obs}} = \frac{\chi}{1 + d\chi} \tag{3.3}$$

where χ is the susceptibility per volume unit. A simple calculation for a spherical sample of $Mn_{12}ac$ can be performed assuming that at low temperature it behaves as a paramagnetic $S = 10$ spin. Taking into account the molecular weight, 2060.22 g/mol, and the density of the crystals, 1.895 g/cm^3, at 4 K we obtain $d\chi = 5.2 \times 10^{-2}$. Molecular nanomagnets are characterized by a low magnetic density and the effect of the demagnetizing field can be neglected especially if a moderate field is applied. It is for the same reason that the permeability can be assumed to be that of the vacuum, and in Gaussian units the magnetic field, H, and the magnetic induction, B, assume the same value.

The experimental temperature dependence of χ (or χT) can be used to extract information on the exchange magnetic interactions active in the system. These experimental data are usually sufficient to determine the exchange coupling constant in a pair of coupled spins or in clusters with small nuclearity and high symmetry. Example of pairs, triads, and tetramers can be easily found in any book on magnetochemistry (Carlin 1986; Kahn 1993). The experimental curve is simulated by weighting, according to the Boltzman population, the contributions to the susceptibility that come from the different spin states of the cluster according to equation (2.53).

A well-known example is the Bleaney–Bowers equation (Bleaney and Bowers 1952) for two isotropically coupled $s = \frac{1}{2}$ spins, where the total spin states $S = 1$ and $S = 0$ are separated by J and the magnetic susceptibility, in the low-field regime, is given by:

$$\chi = \frac{2Ng^2\mu_B^2}{k_BT} \frac{1}{3 + \exp(-J/k_BT)}. \tag{3.4}$$

In this case the only adjustable parameters are J and g.

As soon as the magnetic molecule increases in nuclearity, many independent parameters of the spin Hamiltonian need to be determined. An over-parametrization is often encountered in the best-fit procedure and the possibility of gaining additional information from different techniques becomes crucial. The use of a variable magnetic field, especially if a large field and a low temperature are available, often provides the lacking information. This is the case when the paramagnetic centres of the molecule are antiferromagnetically coupled. The state with the largest spin is necessarily not the ground one, but its energy decreases most rapidly by interacting with the field. A crossing of energy levels can be induced by the field and reveals itself as a sudden increase in the magnetization as schematized in Fig. 3.2.

The simplest molecular compounds to which this approach has been applied are antiferromagnetic dimers, such as $[Fe_2(salen)_2Cl_2]$(Shapira *et al.* 1999) and $[Fe_2(C_2O_4)(acac)_4]$(Shapira *et al.* 2001), where salen $= N, N'$-ethylene-bis-(salicylaldiminato), and acac $=$ acetylacetonate. Here, isotropic intradimer exchange interactions are dominant and the energy of spin levels $E(S, M)$ are well described by the formula

$$E(S, M) = -\frac{J}{2}S(S+1) + g\mu_B M H \qquad (3.5)$$

which follows directly from the spin Hamiltonian

$$H = -J\mathbf{S}_1 \cdot \mathbf{S}_2 + g\mu_B \mathbf{S} \cdot \mathbf{H} \qquad (3.6)$$

for two exchange-coupled spins $\mathbf{S}_1$ and $\mathbf{S}_2$ in a magnetic field $\mathbf{H}$ ($H = |\mathbf{H}|$). In (3.5), S is the total spin quantum number of the binuclear species while $M = -S, -S + 1, \ldots, S - 1, S$ labels the total spin component along $\mathbf{H}$. For two high-spin ferric ions $S_1 = S_2 = \frac{5}{2}$ and S thus ranges from 0 to 5. In the presence of antiferromagnetic interactions ($J < 0$) the ground state has $S = 0$ in zero applied field, as shown in Fig. 3.2(a). By contrast, in a strong magnetic field the $S = 5$ state must lie lowest, the high magnetic field limit being reached when the external field overcomes the antiferromagnetic interaction. By sweeping the magnetic field, it is thus possible to observe the cross-overs from $S = 0$ to 1, from 1 to 2, etc. as depicted in Fig. 3.2(b). These occur at evenly spaced magnetic field values H_n given by:

$$H_n = n\frac{|J|}{g\mu_B} \quad \text{with } n = 1, 2, \ldots 5. \qquad (3.7)$$

At each cross-over the value of $|M|$ in the ground state changes by one unit, so that the magnetization exhibits a sudden step-like increase at low temperature ($k_B T \ll |J|$). Alternatively, when dM/dH is measured (as in pulsed-field experiments) each magnetization step shows up as a peak in the dM/dH versus H plot, as shown in Fig. 3.2(c). Because the position of the steps is directly related to the magnitude of the exchange constant through equation (3.7), the method represents a useful alternative to the traditional approach based on the temperature

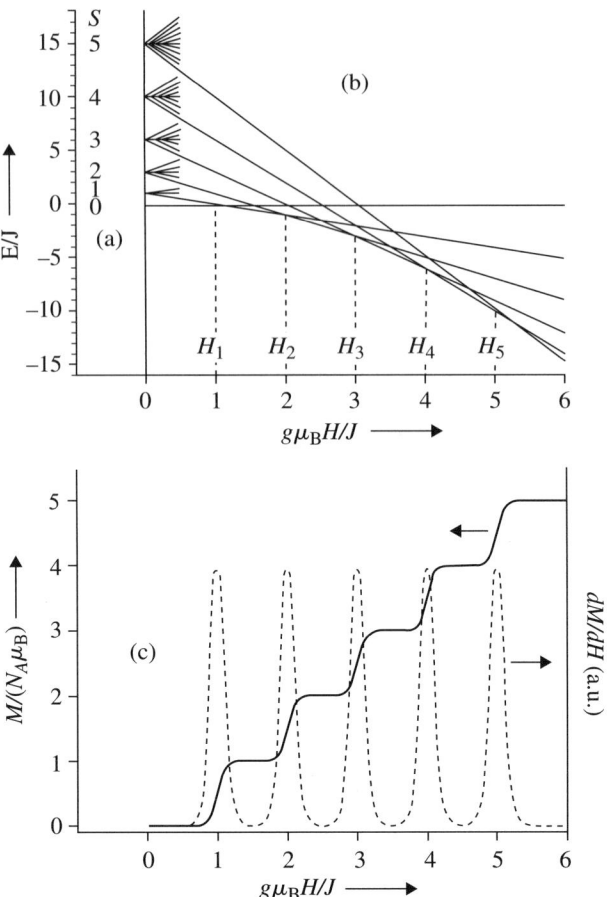

FIG. 3.2. (a) Calculated spin levels for a dimer of antiferromagnetically coupled
$S = 5/2$ spins in zero magnetic field. (b) Evolution of the spin levels in a
magnetic field H. The cross-over fields H_n are marked by dashed vertical
lines. (c) Molar magnetization (M) and differential susceptibility (dM/dH)
of the dimer at low temperature.

variation of magnetic susceptibility for the determination of J values, provided
that the experiment is performed at sufficiently low temperature (typically below
1 K). In Fig. 3.3 the data taken from Shapira (2001) in fields up to 45 T for
$[Fe_2(C_2O_4)(acac)_4]$ show all five predicted peaks, and $J/k_B = -10.4(3)$ K has
been extracted, assuming $g = 2.00$ for the high-spin ferric ion. This technique has
been widely used in the investigation of molecular rings of antiferromagnetically
coupled spins, as discussed in Chapter 14.

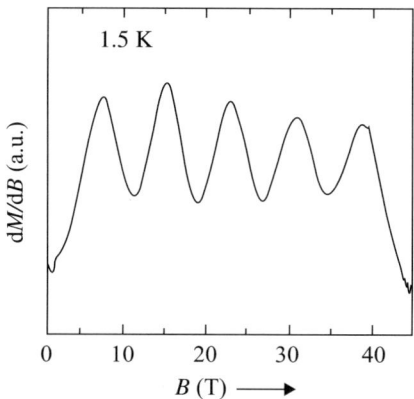

FIG. 3.3. Differential susceptibility of $[\mathrm{Fe_2(C_2O_4)(acac)_4}]$ measured in pulsed fields at 1.5 K. Reprinted with permission from Shapira *et al.* (2001). Copyright (2001) by the American Physical Society.

An important parameter is the magnetic anisotropy, which in many cases is associated to the axial zero-field splitting parameter D of equation (2.5). A rather precise estimation of D can be obtained also working on powder samples by analysing the field dependence of the magnetization at low temperature.

Some authors plot the magnetization versus the H/T ratio by varying the temperature and not the field. This procedure is correct for an isolated isotropic spin but is not adequate for polynuclear clusters, where excited spin states can be thermally populated. If the data are going to be analysed in terms of one spin state, i.e. the ground state split in zero field by the magnetic anisotropy, it is mandatory to stay at the lowest temperature and sweep the field.

When the data are obtained from a powder average it is important to avoid field-induced orientation of the crystallites due to their magnetic anisotropy. The degree of orientation is in fact hardly reproducible and difficult to quantify, but crucial in the quantitative analysis. Several techniques can be employed, like mixing the microcrystalline powder with grease or with wax. Eicosane, the aliphatic hydrocarbon $\mathrm{C_{20}H_{42}}$, has a very low melting point, 35 °C, and it is therefore suitable for compounds that decompose upon heating. This technique, however, fails with very anisotropic materials for which it can be necessary to press the ground powder in a pellet. In this last case it is important to take into account possible structural or chemical modification induced by the pressure.

When the magnetic data are obtained from a random orientation of aniso-tropic microcrystals the quantitative analysis requires us to calculate the magnetization by integrating over all the possible orientations of the external field. In the simplest case, when only the ground spin state is thermally popu-lated, the magnetization can be calculated starting from the diagonalization of

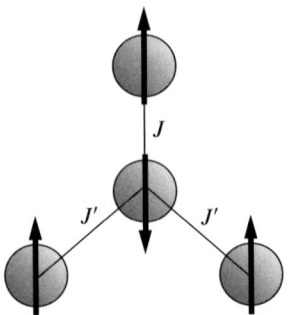

FIG. 3.4. Spin topology of 'iron stars' (four high-spin iron(III), spin $S = 5/2$) leading to a ground state with $S = 5$.

the $(2S + 1) \times (2S + 1)$ matrix describing the spin Hamiltonian:

$$H = \mathbf{S} \cdot \mathbf{D} \cdot \mathbf{S} + \mu_B \mathbf{H} \cdot \mathbf{g} \cdot \mathbf{S}. \tag{3.8}$$

In order to calculate the magnetization the energy of the $2S + 1$ states is evaluated at different but close field values and the derivative is evaluated numerically.

To work out a simple example we have considered the tetranuclear iron cluster of formula $[Fe_4(thme)_2(dpm)_6]$, where dpm- is the anion of dipivaloylmethane, and thme^{3-} stays for the anion of trishydroxymethylethane (Cornia *et al.* 2004). Several clusters belong to this class of molecule, also known as 'iron stars' (Barra *et al.* 1999; Saalfrank *et al.* 2001; Cornia *et al.* 2004) and schematized in Fig. 3.4. They are characterized by a high-spin iron(III) ion at the centre and three iron(III) on the vertex of a triangle that can be either equilateral or isosceles, depending on the crystal symmetry.

Antiferromagnetic interactions between the central and peripheral spins are mediated by the bridging ligands, usually alkoxides. The topology is, however, very favourable because the peripheral spins are antiparallel to the central one, but parallel to each other, resulting in a $S = 5$ ground state, usually well separated from the excited spin states. In Fig. 3.5 is reported the magnetization curve measured for this compound at $T = 1.9$ K. Superimposed on the experimental data are also reported the Brillouin function for $S = 5$ and the magnetization curve calculated for $D/k_B = -0.58$ K and $g = 1.93$, by simply averaging the magnetization calculated along the easy axis and in the hard plane according to $\langle M \rangle = 2/3 M_\perp + 1/3 M_{//}$. In both cases the agreement is very poor, while the curve obtained by integrating over all the possible orientations of the field, the solid line of Fig. 3.5, nicely fits the data.

Of course, more reliable information on the magnetic anisotropy can be obtained by the measurement of the magnetization of oriented single crystals, if large enough crystals are available. In this case other aspects, beyond the difficulty in getting large crystals, must be taken into account. First of all, not

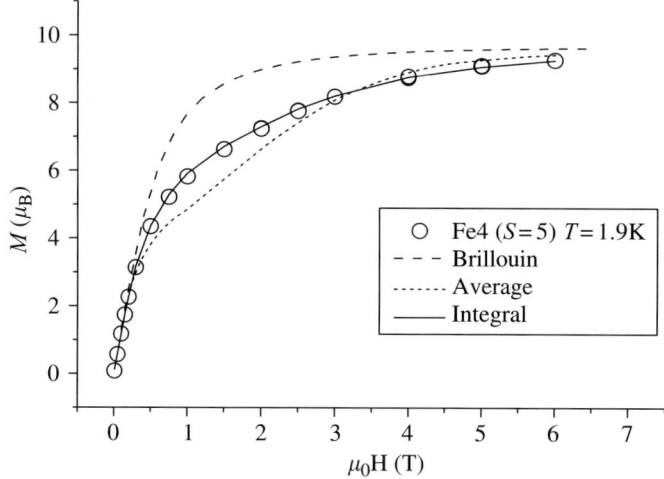

FIG. 3.5. Field dependence of the magnetization of a powder sample of
$[Fe_4(thme)_2(dpm)_6]$ and calculated magnetization curves with the Brillouin
function (–), including the magnetic anisotropy $D/k_B = -0.58$ K and averaging over the field orientations (solid line). The simple averaging formula
$\langle M \rangle = 2/3M_\perp + 1/3M_{//}$ (dotted line - -) does not fit the experimental data.

all crystalline materials are well suited to this kind of investigation. Even if the
molecules are characterized by a strong anisotropy they can be packed in the
crystal lattice in such a way that the anisotropy axes are not all parallel to each
other. This happens whenever a molecule has a symmetry that is lower than
that of its crystal space group, and in particular if a molecule in the crystal lattice is related to the equivalent one by a rotation axis or a mirror plane. These
symmetry elements are present in the most common crystal systems in coordination chemistry, i.e. the monoclinic and ortho-rhombic systems. An inversion
centre does not rise this problem. The triclinic system is therefore unaffected
by this problem but the experiments are complicated because it is not possible
to know *a priori* where the anisotropy axes are pointing. A particularly favourable case is that of high crystalline symmetry, with the molecules also sitting
on a high-symmetry site in the unit cell. This is a very special and favourable
case, because the principal direction of the magnetic anisotropy, i.e. the easy,
medium, and hard axes, are forced by crystal symmetry to coincide with some
special crystallographic directions, easily identifiable from the morphology of the
crystals, or, in the worst case, from single-crystal X-ray diffraction.

In the more general case the experimental procedure consists of measuring
the magnetic moment by rotating the sample along three orthogonal axes, as
commonly done in single-crystal electron paramagnetic resonance (EPR) spectroscopy discussed in Section 3.3.1. What seems a trivial procedure is, however,

very time-consuming, because in most magnetometers the magnetic field is generated by a vertical solenoid, and the simple rotation of the sample rod does not change the field orientation. Therefore it is necessary to remove the sample rod to change the orientation of the sample unless the system is equipped with a special sample holder that allows a horizontal rotation of the sample. Such a device, also commercially available, is a precious tool for the magnetic investigation of single crystals.

3.1.2 *Time-dependent measurements*

From the same instrumentation, a SQUID magnetometer or a VSM, we can get information also on the spin dynamics, provided it falls into the appropriate time-window. This window usually ranges from 100 s to days or even weeks or months, depending on the patience of the experimentalist. The simplest characterization of the dynamics of the magnetization consists in monitoring the time decay of the remnant magnetization.

This type of experiment is usually performed by applying a strong field at a temperature where the dynamics is still fast. The next step is to cool down the sample to the investigated temperature, then remove the field and start to measure the magnetization as a function of the elapsed time. This procedure is schematically shown in Fig. 3.6. The curve is then fitted starting from a single exponential law,

$$M(t) = M(0)\exp(-t/\tau) \tag{3.9}$$

where τ is the relaxation time. More complex behaviour is often encountered, like the stretched exponential decay first observed by Kohlraush in 1847 when

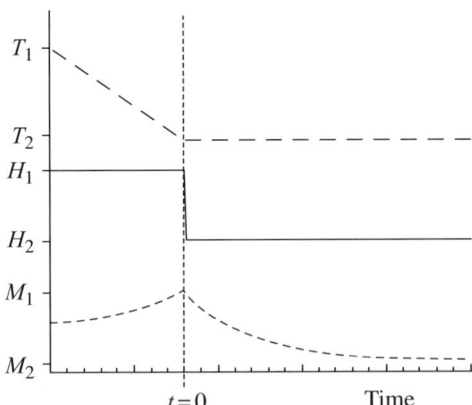

FIG. 3.6. Schematic view of the procedure to measure the time decay of remnant magnetization. The time dependence of the temperature (top), applied magnetic field (middle), and the measured magnetization (bottom) are reported.

he investigated the decay of residual charge in glasses:

$$M(t) = M(0) \exp[-(t/\tau)^\alpha] \tag{3.10}$$

where α is a number ranging from 0 to 1.

It can also be interesting to investigate the dynamics in going from an initial state in the magnetic field H_1, characterized by the equilibrium value M_1, to another field H_2 and magnetization M_2 with the time dependence of the magnetization given by

$$M(t) = M_2 + (M_1 - M_2) \exp(-t/\tau). \tag{3.11}$$

While M_1 is easy to detect it is important that the experiment lasts long enough to get a precise determination of M_2. Moreover, to avoid errors, the relaxation time must be much longer than the time needed to ramp the field. This instrumental parameter determines the shortest relaxation time that can be measured, which is in the range of a few minutes for standard instrumentation.

Other types of experiments are commonly performed to investigate the dynamics of the magnetization. Certainly the most common procedure is the measurement of the thermal irreversibility and thermo-remnant magnetization, summarized in Fig. 3.7. The sample is first cooled in zero applied field. At low temperature the relaxation time is much longer than the measuring time

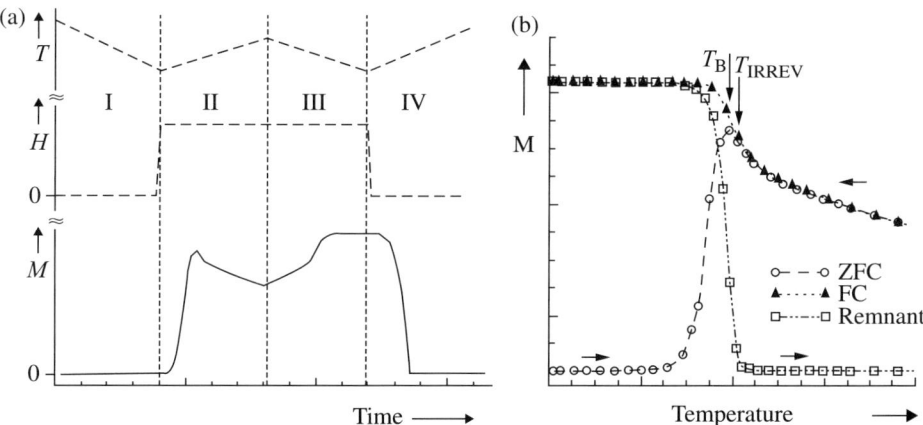

FIG. 3.7. (a) Schematic view of the procedure to measure the thermal irreversibility and thermo-remnant magnetization. The time dependence of the temperature (top), applied magnetic field (middle), and the measured magnetization (bottom) are reported. Step I corresponds to the cooling in zero field, step II to the measurement of ZFC magnetization, step III to FC magnetization, and step IV to the measurement of the thermo-remnant magnetization. (b) The magnetization measured in steps II, III, and IV is plotted as a function of the sample temperature.

and the application of a polarizing field, weak enough not to affect the relaxation time, does not produce a net magnetic moment because the magnetization is frozen. The magnetization that is measured as a function of increasing temperature after having cooled in zero field and having applied a weak field is called zero field cooled (ZFC). On increasing temperature usually the relaxation time diminishes and the magnetization is gradually unblocked. On increasing the temperature the magnetization increases because it approaches the equilibrium state that in an applied field is a magnetized one. However, the equilibrium value of M in a paramagnet decreases with temperature and this results in a maximum of the ZFC magnetization versus T curve. The system is at the equilibrium when the ZFC magnetization superposes with the magnetization measured when cooling down in the same field, the field cooled (FC) magnetization. The experimental procedure is usually completed with a fourth step that consists in removing the field at the lowest temperature and measuring again the magnetization on increasing the temperature. The equilibrium value in the paramagnetic phase has $M = 0$ and therefore the equilibrium is reached when the thermo-remnant magnetization disappears.

The time dependence of the magnetization is now not very easy to interpret but if the magnetization data of this experiment are plotted as a function of the temperature, as in Fig. 3.7b, the difference between the ZFC and FC magnetization, measured under the same conditions in H and T but with a different history of the sample, becomes evident.

The temperature at which the maximum in the ZFC magnetization occurs is known as the blocking or freezing temperature, T_B, which does not necessarily coincide with the temperature at which the ZFC and FC curves superimpose, which we can call the irreversibility temperature, T_{IRREV}. The larger the difference $(T_{IRREV} - T_B)$, the larger the distribution of relaxation times. In molecular nanomagnets the two points almost coincide. The procedure of Fig. 3.7a is commonly used in a preliminary screening to show the presence of irreversibility and to estimate T_B. It is, however, important to notice that T_B is strongly affected by the sweeping rate of the temperature, a parameter not always easy to control in most experimental set-ups. Therefore minor changes in T_B, especially if obtained from different instruments, should be critically evaluated. As these experimental procedures are also employed to characterize spin-glass materials the interested reader can easily find more information in the related literature (Mydosh 1993; Binder and Young 1986).

Slow relaxing magnetic materials are characterized by the opening of the magnetization versus field loop, known as the hysteresis loop. In molecular nanomagnets the appearance of magnetic hysteresis is not due to the irreversible growth of domains with the orientation of the magnetic moments parallel to the field, but rather to the fact that the magnetization of each molecule is relaxing too slowly compared to the time required to sweep the field and therefore the magnetization of the ensemble of the molecules in the sample does not reach the equilibrium value in the time-window of the experiment. The width of the hysteresis loop is therefore directly related to the dynamics of a single molecule. In molecular nanomagnets the magnetic field can have a very peculiar effect on the dynamics of the magnetization. The recording of the hysteresis loop has therefore played a much more important role in the characterization of the dynamics compared to conventional magnetic materials. We will discuss this aspect in more detail in the following chapters.

3.1.3 Micro-SQUID and micro-Hall probe techniques

If we were asked which experimental technique has given the most significant contribution to the investigation of molecular nanomagnets, the choice would surely fall on micro-SQUID magnetometry. This unconventional instrumentation was developed at the beginning of the 1990s in Grenoble, principally to investigate the magnetic reversal of a single magnetic nanoparticle (Chapelier et al. 1993; Wernsdorfer et al. 1995). The technique has been extended and upgraded by Wernsdorfer at the L. Néel Laboratory in Grenoble (Wernsdorfer 2001). It is based on coupling the micro-SQUID, a superconducting ring of micrometric dimensions, to the sample by placing the last one directly on the SQUID loop, as shown in Fig. 3.8.

The advantages of this technique go well beyond its very high sensitivity, which allows us to detect magnetic moments as small as 10^{-17} emu. The miniaturization of the device allows its insertion in a dilution refrigerator and in two small orthogonal superconducting solenoids that can be independently driven. We will see in the following how relevant is the possibility to sweep the magnetic field simultaneously but independently along both the easy and the hard axes of an anisotropic sample. Given the small size of the coils it is possible to ramp the field at very high speed and very short relaxation times can therefore be measured.

For this type of measurement single crystals of micrometric size are used, thus permitting single-crystal characterization also for those compounds that do not grow in large crystals. The small size of the sample improves the dissipation of the heat pulse that is generally produced when the magnetization reverses its

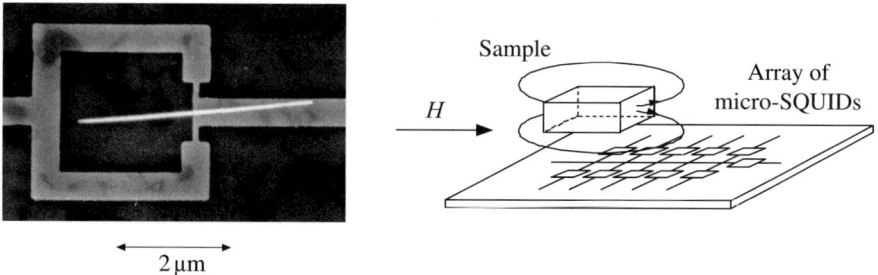

$2\,\mu m$

FIG. 3.8. Left: scanning electron micrograph of a Nb micro-SQUID fabricated by electron beam lithography. A Ni wire of diameter of about 90 nm was deposited on the SQUID. Right: an array of micro-SQUIDs used for a macroscopic crystal. From Wernsdorfer (2001). Reprinted with permission of John Wiley & sons.

orientation. In general samples of molecular nanomagnets exceed the dimension of a single micro-SQUID loop and are arranged on an array of these loops, as shown in Fig. 3.8. In the magnetometer developed at the Néel laboratory in Grenoble the signals of all the loops of the array are read simultaneously so that a sort of mapping of the magnetization of the macroscopic sample is possible. This procedure can show inhomogeneities of the sample, for instance due the presence of decomposition products, which can be non-homogenously distributed inside the crystal.

This technique, however, has some limitations. The highest operative temperature is limited by the critical temperature of the superconducting material employed for the SQUID loop (i.e. about 7 K for loops made of Nb), and the applied field is limited to $\mu_0 H \approx 2$ T. Moreover, absolute measurements of the magnetic moment cannot be done, because the output signal is strongly dependent on how the SQUID(s) are coupled to the sample, and therefore on its shape, positions, etc.

The field and temperature limitations can be overcome if the detection of the magnetic moment is done by exploiting the Hall effect on a micro-Hall probe. If a current is flowing through a conductor or a semiconductor in the form of a thin slab and a magnetic field is applied perpendicular to the current direction, the trajectory of the carriers is deflected. The result is a difference in the electric potential, V_H, at the extremes of the conductor perpendicular to the current according to the law

$$V_H = R_H I B. \qquad (3.12)$$

The Hall constant R_H is given in a first approximation by $R_H = 1/ned$, where n is the density of the carriers and e is the charge of the electron and d is the thickness of the slab.

The Hall probe, and in particular the active area of the cross, is sensitive to the magnetic induction produced by both the applied field and the magnetized sample. In order to get rid of the contribution of the field different set-ups can be exploited. In the original work by Kent *et al.* (1994), where for the first time micro-Hall probes where used to investigate arrays of nanoparticles, two identical Hall crosses were connected as shown in Fig. 3.9a. Opposite currents flow in the two crosses that are exposed to a field applied perpendicular to the large surface of the crosses. If the two crosses are identical and $I = I'$ no voltage difference is measured in the absence of a magnetized sample. The positioning of a magnetized sample on one of the crosses results in a voltage which is proportional to the magnetic induction of the sample. In practice, I and I' are chosen to be slightly different to compensate small differences in the contruction of the two crosses. AC currents are applied and the voltage is measured with a phase-sensitive detector.

An early application of micro-Hall probes to molecular nanomagnetism was the detailed investigation of the temperature dependence of tunnel resonance fields in a single crystal of $Mn_{12}ac$, discussed later in Chapter 13 (Bokacheva *et al.* 2000).

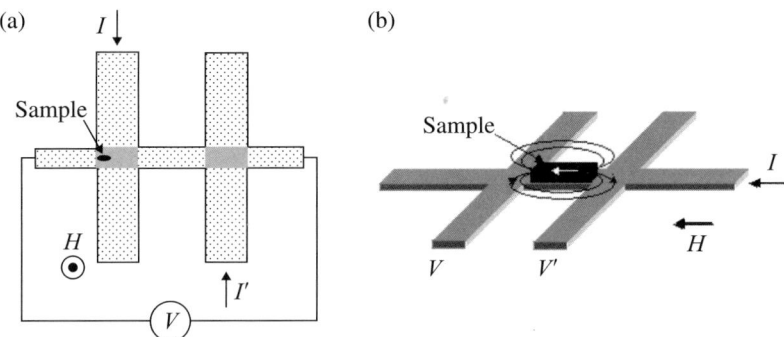

FIG. 3.9. Schematic view of two micro-Hall crosses with the applied field perpendicular to the probes (a) and in the plane of the probes and parallel to the current (b).

A different set-up, characterized by a lower parasitic signal, is shown in Fig. 3.9b. In this case the magnetic field is applied parallel to the current direction, and ideally it does not produce any Hall effect, even though in practice a misalignment of the field is always present. On the contrary, the magnetic flux of a magnetized sample, if placed close to the active area of the cross, has a component perpendicular to the cross and induces a voltage. If two crosses are used and the sample is large enough to be close to the active area of both surfaces the field generated by the magnetized sample has opposite direction in the two crosses. By measuring the difference $V - V'$ it is possible to eliminate the contribution due to the misalignment of the field. This design, which has been mainly used to investigate the molecular cluster Fe$_8$ (Sorace *et al.* 2003), has the disadvantage that the direction of the magnetic field is fixed.

The most frequently employed materials for these devices are two-dimensional heterostructure materials, such as GaAl/GaAlAs, known as a two-dimensional electron gas, or thin films of three-dimensional semiconductors, such as InGaAs. The ideal material has a low density of carriers but a high mobility. A typical value of R_H for 2D electron gas probes is 0.2 Ω/G, about two orders of magnitude larger than for 3D semiconductors. Hall probes can be used over a wide temperature range (1–50 K) without significant variation in sensitivity. At much lower temperatures ballistic transport reduces the sensitivity, and quantum Hall effects influence the measurements, but micro-Hall probes have nevertheless been employed even slightly below 100 mK. In this case 2D hole gas heterostructures have been employed because holes have a larger effective mass than electrons and therefore a lower mobility. The reduction of mobility of the carriers, on the contrary, limits the use of the device at higher temperature. The frequency range that can be spanned by the ac current is also quite large, from 1 mHz up to 100 kHz, with a sensitivity increase proportional to $\sqrt{v}$. The sensitivity at high frequency is, however, limited by Josephson noise. As in the case of the micro-SQUID, absolute values of the magnetization cannot be extracted from the measurements. However the high versatility, lower cost, and simpler operation of the micro-Hall probes have contributed to a much larger diffusion of this experimental technique compared to that of micro-SQUIDs, even if the last ones still retain a sensitivity about 3–4 orders of magnitude larger.

3.1.4 Torque magnetometry

Small single crystals can also be investigated with torque magnetometry, thanks to the high sensitivity that has been achieved with the introduction of cantilever devices. A cantilever is a flat member fixed at one end and hanging free at the other end. In the simplest realization a cantilever apparatus is a constructively simple device, as shown in Fig. 3.10. An elongated thin slab made of silicon or a non-magnetic metallic alloy (typically CuBe) is mounted, with one end hanging free, parallel to a fixed metal platform. This results in a parallel-plate capacitor, whose capacitance C (neglecting edge effects) is given by:

$$C = \varepsilon \frac{A}{d} \tag{3.13}$$

where A is the area of the plates, ε is the dielectric constant of the medium in which the cantilever is operated and d is the separation between the two electrodes (usually <100 µm). The sample is fixed to the cantilever surface close to the free end, and if a mechanical couple is present, this results in a flexion of the cantilever and in a change in the capacitance, C. For small deflections the device approaches a linear-response regime in which the capacitance variation is directly proportional to the torque t (Cornia et al. 2000):

$$\Delta C = C_{\mathrm{H}} - C_0 \propto -t_y \tag{3.14}$$

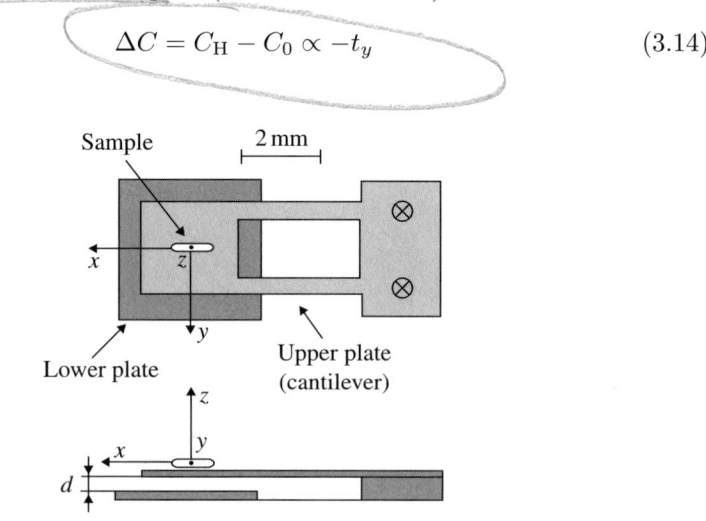

FIG. 3.10. Schematic view of a cantilever used for magnetic torque measurements. Top: front view; bottom: side view.

where C_H is the capacitance measured in the presence of an applied field and C_0 is that measured at rest, i.e. without an applied field.

Very high sensitivities can be achieved through a suitable design of the capacitor, for instance by increasing the flexibility of the cantilever itself and/or by reducing the separation between the electrodes.

With this technique it is possible to measure the force acting on an isotropic magnetic sample in an inhomogeneous magnetic field or the torque experienced by an anisotropic sample under the influence of a homogeneous magnetic field.

In the first case the force experienced by the magnetic sample is related to the gradient of the magnetic field $\mathbf{F} = \nabla(\mathbf{M} \cdot \mathbf{H})$. A non-homogenous field is usually achieved moving the sample far from the centre of the superconducting solenoid. If the profile of the magnetic field is known the magnetization can be easily extracted from the torque signal.

In the second case, most common in molecular nanomagnetism, the magnetic torque is given by

$$\mathbf{t} = \mathbf{M} \times \mathbf{H} \tag{3.15}$$

where $\mathbf{H} = (H_x, H_y, H_z)$ is the magnetic field and $\mathbf{M} = (M_x, M_y, M_z)$ is the bulk magnetization. If we choose a reference frame so that $\mathbf{M}$ and $\mathbf{H}$ lie in the xz plane, the magnetic torque vector $\mathbf{t} = (t_x, t_y, t_z)$ is necessarily parallel to y, as shown in Fig. 3.11. It follows that $t_x = t_z = 0$ while the y component is given by:

$$t_y = M_z H_x - M_x H_z = H^2 \left(\frac{M_z}{H_z} - \frac{M_x}{H_x} \right) \sin\theta \cos\theta \tag{3.16}$$

where θ is the angle between the magnetic field and the z axis. Hence, the experiment measures the anisotropy of the magnetization in the xz plane.

From equation (3.16) it follows that the origin of the magnetic torque is the fact that the magnetization and the applied field are non-collinear. The origin of

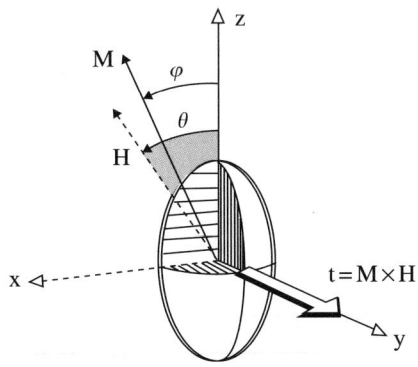

FIG. 3.11. Geometrical arrangement of the external field, the magnetization and the magnetic torque.

the non-collinearity is different in paramagnets and in permanent magnets. Since in the latter the direction of the magnetization is fixed, the torque reduces to

$$t_y = HM \sin \theta \tag{3.17}$$

where $H = |\mathbf{H}|$ and $M = |\mathbf{M}|$. Notice that the torque vanishes for $\theta = 0°$ or $180°$, and that the torque modulus is simply proportional to the bulk magnetization. Indeed, this approach has been used to measure magnetization of SMMs below the blocking temperature (Perenboom *et al.* 1998a,b).

More complex is the behaviour in paramagnetic materials, because both magnetization and magnetic anisotropy are field dependent. Nevertheless the technique represents a powerful tool for magnetic anisotropy investigations, as shown by the earlier applications of this technique to simple transition-metal complexes (De W. Horrocks and De W. Hall 1971; Mitra 1977).

To give an idea of the information that can be extracted from a magnetic torque measurement on paramagnetic samples we can work out a simple example for the same tetranuclear iron cluster whose structure is shown in Fig. 3.4, which at low temperature can be handled as an $S = 5$ characterized by $D/k_B = -0.58$ K and $E = 0$. The crystal-field induced anisotropy is of the easy-axis type, i.e. the magnetization at low temperature will preferentially orient along, say, z. The behaviour of this type of system is different in the weak-field ($g\mu_B H \ll k_B T$) and in the strong-field ($g\mu_B H \gg k_B T$) limits. In the former case, the magnetization is simply $\mathbf{M} = \chi \mathbf{H} = (\chi_{xx} H_x, \chi_{yy} H_y, \chi_{zz} H_z)$, so that if the field is rotated in the xz plane the axial anisotropy is just the difference between the z- and x-magnetic susceptibilities, $\chi_{zz} - \chi_{xx}$. Consequently, the torque t_y at a fixed θ-angle is simply proportional to the square of the magnetic field:

$$t_y = H^2(\chi_{zz} - \chi_{xx}) \sin \theta \cos \theta. \tag{3.18}$$

Notice that in contrast with the case of permanently magnetized samples (3.17) the torque is zero when the magnetic field is applied along a principal direction ($\theta = n\pi/2$ with n integer) (Cornia *et al.* 2000), and goes through extrema for $n\pi/4$.

In the high-field regime the magnetization reaches its saturation value. For a system with isotropic g, like the one we are considering, the axial anisotropy goes to zero. It has been shown (Cornia *et al.* 2000) that at high field the torque has a limiting value

$$\lim_{H \to \infty} t_y = -2DS \left(S - \frac{1}{2} \right) \sin \theta \cos \theta. \tag{3.19}$$

In principle the high-field limit of the torque could provide a direct measurement of the spin of the system and its magnetic anisotropy. In practice, however, absolute measurements of the torque are not performed, simply because the crystals employed are in the micrograms range and cannot be precisely weighted. Other features can be exploited to obtain quantitative information on the magnetic anisotropy, as shown in the following.

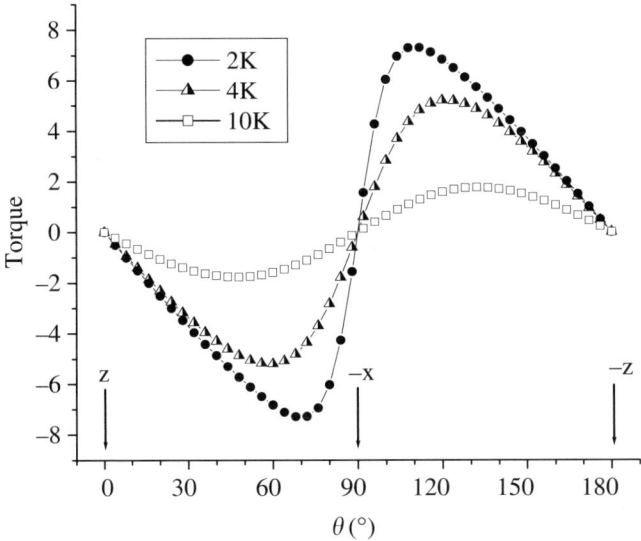

FIG. 3.12. Calculated angular dependence of the y component of the magnetic torque of a $S = 5$ spin state characterized by $D/k_B = -0.58$ K and $E = 0$. θ is the angle formed by the magnetic field with the molecular easy axis (z).

In Fig. 3.12 we show the angular field dependence of the torque calculated for the $S = 5$ system defined above.

The magnetic field, 2 T in strength, is rotated in the xz principal plane. At 10 K the curve has the typical sinusoidal shape given by the $\sin\theta \, \cos\theta$ term of (3.18), with zero torque when the field is parallel to the principal directions x and z. On lowering the temperature the curve becomes more and more asymmetric, the torque being larger close to the hard direction. The origin of this asymmetry lies in the departure from the weak-field limit. In Fig. 3.13 the field dependence of the torque is shown for two angles, $\theta = 5°$ and $\theta = 85°$, close to the easy and hard axis, respectively. The strong-field asymptotic limit is the same for these two angles, but close to the hard axis the torque has a non-monotonic behaviour, passing through a well-defined maximum. The position of the maximum is only slightly affected by small angle variations around the hard axis direction, and therefore it is a feature directly connected to the axial magnetic anisotropy. It is worth noticing, as shown in Fig. 3.13, that the field position of the maximum in the torque is strongly temperature dependent, raising the necessity of a good temperature calibration for the system.

One of the first examples of torque measurements on a SMM concerns the archetypal $Mn_{12}ac$ cluster (Cornia *et al.* 2000). The torque curves were recorded at 4.2 K by applying the magnetic field (0–30 T) close to the hard (xy) magnetic plane of a $Mn_{12}ac$ single crystal ($\theta = 90 \pm 1.1°$), see Fig. 3.14. In the weak-field

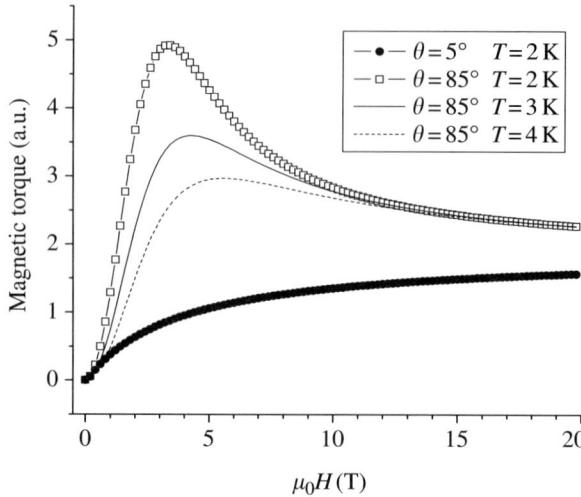

FIG. 3.13. Calculated field dependence of the magnetic torque of an $S = 5$ spin state characterized by $D/k_\mathrm{B} = -0.58$ K and $E = 0$. θ is the angle formed by the magnetic field with the molecular easy axis (z).

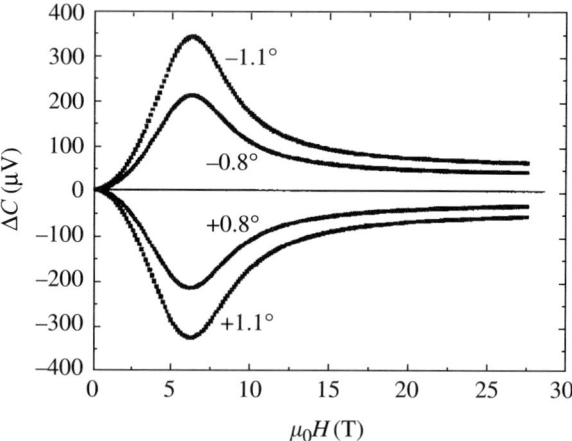

FIG. 3.14. Experimental torque data in μV for the detected capacitance change ΔC, recorded for a Mn$_{12}$ac single crystal by applying the magnetic field at different small angles from the hard axis. The temperature was fixed at 4.2 K. Reprinted from Cornia *et al.* (2000). Copyright (2000) Elsevier.

region the torque signal is proportional to H^2, as predicted by equation (3.18). A pronounced peak is detected at about 6.2 T, while in high fields an asymptotic field dependence is observed according to (3.19). The symmetry of the curves around the hard direction is well evident, and was indeed exploited to precisely align the crystal. The values $D/k_B = -0.67(1)$ K and $B_4^0/k_B = -3.4(1) \times 10^{-5}$ are found to provide the best fit with $g_{xx} = g_{yy} = 1.96$ and $g_{zz} = 1.93$. The agreement with the same parameters determined by HF-EPR (Barra *et al.* 1997b) and inelastic neutron scattering (Mirebeau *et al.* 1999) experiments is remarkable.

Magnetic torque measurements are useful not only to characterize the anisotropy of paramagnetic centres of the ground spin state of a cluster. In fact it can provide information on the exchange interactions active in the system. For instance, it has been recently exploited to determine the pattern of energy of the spin levels, as well as the D values in antiferromagnetic clusters (Cornia *et al.* 1999).

Before closing this section it is useful to remind the reader that the calculated curves of Figs 3.12 and 3.13 have been obtained by diagonalization of the spin Hamiltonian matrix in the presence of the applied field as already mentioned for Fig. 3.5. In the present case all the magnetization components, and not only that along the field direction, have to be computed to evaluate the torque.

Finally in the above treatment it has been assumed that the crystal magnetic axes, which are obtained from the experiment, are the molecular axes. This is only true when there is only one magnetically equivalent molecule in the unit cell. Associating the crystal and molecular axes in the general case requires some additional assumptions.

3.1.5 *Ac susceptometry*

This is a very simple technique that can give easy access to the susceptibility, dM/dH. The sample is inserted in a small coil, called the primary coil, most commonly made of copper wire. An alternating current flows in the coil thus generating a small oscillating magnetic field that usually does not exceed 10 Oe. A secondary coil is wound inside the primary one as schematized in Fig. 3.15.

The two coils can be designed so that the voltage induced in the secondary is zero in the absence of a sample. This is usually done by winding two secondary coils in opposite directions as for a first-order gradiometer. Once a sample is inserted its magnetic moment oscillates as an effect of the ac field and induces a voltage in the secondary coil that can be easily detected in amplitude and phase. To eliminate the signal induced by non-perfect balancing of the coils the sample is moved from the centre of a secondary coil to the centre of the other. The measurement can be done also in a static field H_0 parallel to the oscillating field h so that the applied magnetic field becomes

$$H = H_0 + h\cos\omega t \qquad (3.20)$$

where ω is the angular frequency of the ac current flowing in the primary.

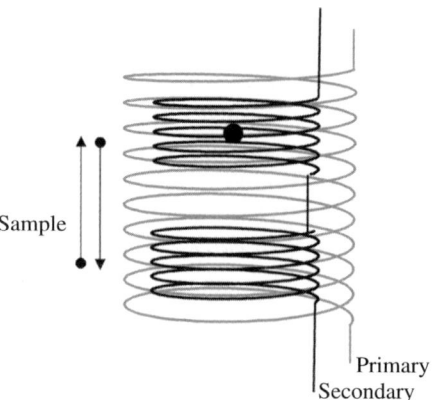

FIG. 3.15. A schematic view of the primary and secondary coil design usually employed in ac susceptometers.

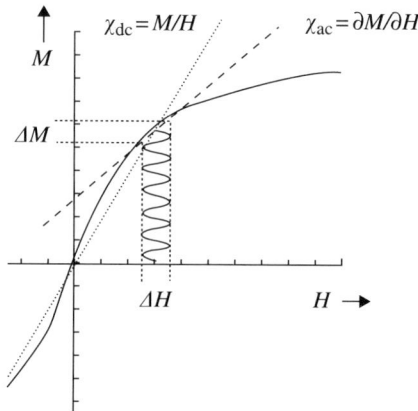

FIG. 3.16. A schematic magnetization curve where the differences between M/H and dM/dH are highlighted. The amplitude of the oscillating field is enlarged for clarity.

The advantage of measuring the ac susceptibility is clearer in Fig. 3.16. The magnetization curve can be a non-linear function of H but for the small field oscillations given by the primary coil the response is in general linear, and $\Delta M/\Delta H$ corresponds to the susceptibility. The signal amplitude is proportional to ΔM and therefore depends linearly on h but not directly on H_0. Usually it decreases with H_0 and therefore this type of measurement is particularly well suited to investigate the behaviour at low applied static field, without loss of sensitivity.

The major advantage of the technique resides, however, on the fact that the dynamics of the magnetization can be easily investigated by varying ω. Since the first experiments, done in the 1930s by C. J. Gorter (Gorter and Brons 1937) at the Kamerlingh Onnes Laboratory to investigate the magnetic relaxation of paramagnetic metal ions, the technique has seen a significant improvement of its sensitivity, thanks to the use of lock-in detectors or even SQUIDs and Hall probes. The principles of the measurement, however, remain the same.

The population of each state, for instance the m component of an s multiplet, according to the Boltzmann distribution is given by:

$$p_m = \frac{1}{Z} \exp(-E_m/k_{\rm B}T) = \frac{1}{Z} \exp(-\beta E_m) \tag{3.21}$$

where Z is the partition function and $\beta = 1/k_{\rm B}T$. The magnetization depends on p_m as

$$\langle M \rangle = \sum_m p_m M_m. \tag{3.22}$$

For simplicity we assume N spins $S = 1/2$ in the magnetic field given by (3.20). The equilibrium population of the two states $|1\rangle$ with $m = -1/2$ and $|2\rangle$ with $m = 1/2$ oscillates in time:

$$p_1/p_2 = \exp[-g\mu_{\rm B}(H_0 + h\cos\omega t)/k_{\rm B}T] \tag{3.23}$$

The establishment of thermal equilibrium requires a time τ. If the frequency ω of the ac field is low, $\omega\tau \ll 1$, the susceptibility which is measured is the isothermal one, $\chi = \chi_{\rm T}$. On the contrary, in the limit $\omega\tau \gg 1$, the system has no time to exchange energy with the external world and one measures the adiabatic susceptibility χ_S. In the intermediate regime, Casimir and Du Pré (1938) proposed the following interpolation formula for the measured susceptibility:

$$\chi(\omega) = \chi_S + \frac{\chi_T - \chi_S}{1 + i\omega\tau}. \tag{3.24}$$

If χ_S and χ_T are real, the real and imaginary components of the susceptibility are given by:

$$\chi' = \frac{\chi_T - \chi_S}{1 + \omega^2\tau^2} + \chi_S; \quad \chi'' = \frac{(\chi_T - \chi_S)\omega\tau}{1 + \omega^2\tau^2}. \tag{3.25}$$

Mathematically analogous relations have been determined for dielectrics by Debye (Mc Connell 1980).

It may be appropriate to recall that the complex susceptibility is defined by

$$M(t) = M_0 + Re[(\chi' - i\chi'')he^{i\omega t}] = M_0 + (\chi'\cos\omega t + \chi''\sin\omega t)h \tag{3.26}$$

The assumption of real χ_S and χ_T in (3.25) is justified if the frequency is so low that $\chi_S(\omega)$ and $\chi_T(\omega)$ are practically equal to the static values $\chi_S(0)$ and $\chi_T(0)$. At very high frequency, both $\chi_S(\omega)$ and $\chi_T(\omega)$ vanish. In practice, $\chi_S(\omega)$ may be interpreted as the susceptibility of an isolated magnetic molecule while $\chi_T(\omega)$ corresponds to equilibrium with phonons. An extension of the Casimir Du Pré theory, which takes thermal conductivity into account, has been given by Eisenstein (1951).

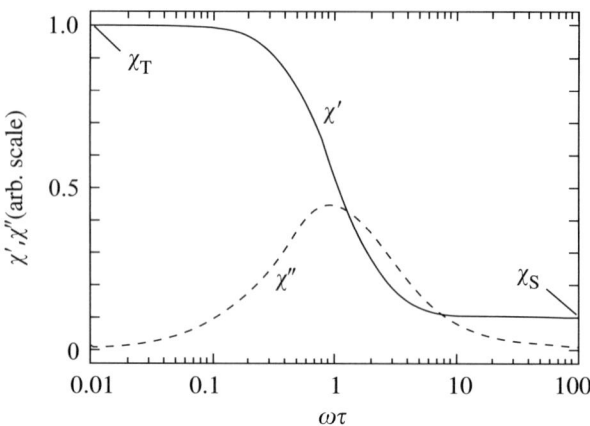

FIG. 3.17. Theoretical frequency dependence of the real and imaginary component of the magnetic susceptibility in a semi-log scale. χ_T and χ_S are the isothermal and adiabatic limit of the susceptibility, respectively.

Figure 3.17 reports the frequency dependence of χ' and χ''. χ'' goes through a maximum when $\omega\tau = 1$, while it goes to zero for $\omega \to 0$ and $\omega \to \infty$, contrary to χ' which has the limiting values χ_T and χ_S.

Another common way to report dispersion data is the Cole–Cole plot used for dielectrics (Cole and Cole 1941), known in magnetism as the Argand plot (Dekker *et al.* 1989), where χ'' versus χ' is used. Equation (3.25) transforms in a semicircle with its centre on the x axis, as shown in Fig. 3.18. At the top of the semicircle the frequency satisfies the relation $\omega^{-1} = \tau$, and thus the relaxation time can be easily extracted.

If the relaxation process is not characterized by a single τ but rather by a distribution of relaxation times a simple empirical law that can account for this is (Cole and Cole 1941):

$$\chi(\omega) = \chi_S + \frac{\chi_T - \chi_S}{1 + (i\omega\tau)^{1-\alpha}}$$

and

$$\chi'(\omega) = \chi_S + (\chi_T - \chi_S)\frac{1 + (\omega\tau)^{1-\alpha}\sin(\pi\alpha/2)}{1 + 2(\omega\tau)^{1-\alpha}\sin(\pi\alpha/2) + (\omega\tau)^{2-2\alpha}} \qquad (3.27)$$

$$\chi''(\omega) = (\chi_T - \chi_S)\frac{(\omega\tau)^{1-\alpha}\cos(\pi\alpha/2)}{1 + 2(\omega\tau)^{1-\alpha}\sin(\pi\alpha/2) + (\omega\tau)^{2-2\alpha}}.$$

The wider the distribution in relaxation times the larger is α. This parameter can be easily derived by the experimental ac data in the Argand plot because the semicircle becomes an arc of a circle with its centre translated in the fourth quadrant. The angle that subtends the arc is given by $\pi(1 - \alpha)$, as shown in

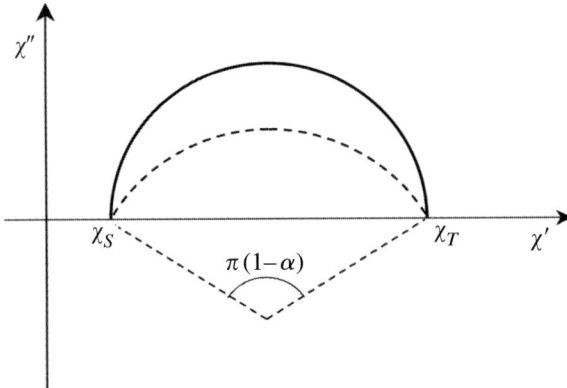

FIG. 3.18. Argand plot (or Cole–Cole plot for dielectrics) where at a given temperature χ'' is plotted versus χ', for each frequency. Solid line: no distribution in relaxation time; broken line, a distribution in τ according to equation (3.27).

Fig. 3.18. More complex behaviour can be also observed. For instance if different components of the magnetization relax with different processes, more semicircles, sometimes partially merged, can be obtained (Grahl *et al.* 1990).

In molecular magnetism the ac susceptibility is often measured scanning the temperature with a relatively small number of frequencies, usually in a limited range that does not exceed three decades. An Argand plot is therefore not always drawn with sufficient accuracy. $\chi(\omega)$ is then reported as a function of the temperature, rather than of ω. If τ varies with temperature, as in all mechanisms involving an energy exchange with the phonon bath, the condition $\omega\tau(T) = 1$ can be met by sweeping the temperature and this gives rise to a maximum in χ''. It is important to stress that the condition $\tau(T_{\max}) = \omega^{-1}$ is exact only if $(\chi_T - \chi_S)$ is temperature independent, which is not the case in the paramagnetic phase. However, the simple $\tau(T_{\max}) = \omega^{-1}$ relation is commonly used for paramagnetic materials because χ_T, being proportional to T^{-1}, can be considered as constant over the narrow temperature range of the maximum. An example of temperature and frequency investigation of χ_{ac} on the molecular nanomagnet Fe$_8$ is shown in Fig. 3.19.

An important advantage of ac susceptometry on measurements of the decay of the magnetization is that in the first case a stationary state is monitored. Experimental details, like the magnetic history of the sample, have no crucial effects on the results.

Also from the data such as those reported in Fig. 3.19 it is possible to gain some hints if a single τ describes the relaxation, or if a distribution is present. According to (3.25), at $T_{\max}, \chi'$ and χ'' should have the same value if

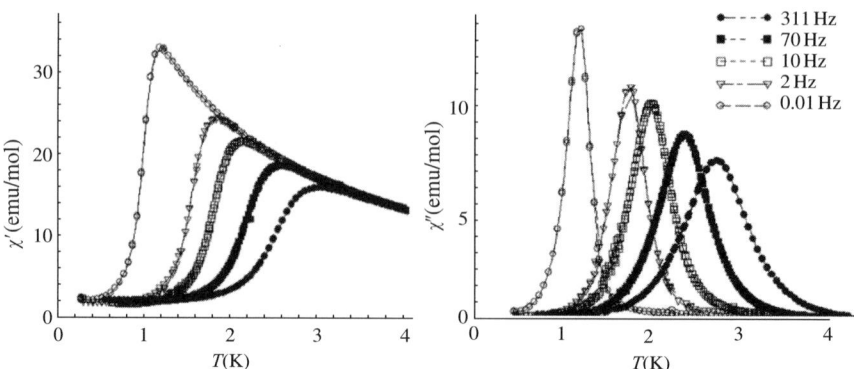

FIG. 3.19. Temperature dependence of the real and imaginary components of the magnetic susceptibility of the cluster Fe_8 measured in zero static field at five frequencies.

the adiabatic susceptibility is close to zero, as commonly observed in molecular nanomagnets.

We will not discuss here the different mechanisms of relaxation (direct, Raman, and Orbach) observed in paramagnets (Gorter 1947). It is important to stress that most paramagnetic systems show a divergence of the spin–lattice relaxation time at low temperature, and τ is for instance of the order of some ms for iron ammonium alum at 1.2 K (Casimir and Du Pré 1938). The observation of an imaginary component of χ is therefore a rather common result. However, simple paramagnets show a non-zero χ'' only if a static field larger than a few hundreds Oe is applied. The presence of a static field in fact induces a polarization, i.e. the spins precess around the direction of the static field. The application of the oscillating field involves a modification of the population of the two energy levels according to (3.23), and some spins reverse their precession to attain the equilibrium. This relaxation process involves an exchange of energy with the thermal bath through a coupling with the lattice. This energy transfer, however, is not necessary in zero static field. Each spin in fact precesses along a variable internal field generated by the surrounding spins. The small applied oscillating field adds to the internal one and the spins precess around a slightly different local field. In zero static field the relaxation process therefore involves the coupling with the other spins of the systems. In this case the equilibrium is attained in the time-scale of the spin–spin relaxation time, which does not diverge at low temperature, being temperature independent. In concentrated paramagnets the spin-spin relaxation time is of the order of 10^{-9} s while the frequency range commonly investigated with ac susceptibility goes from 1 Hz to 100 kHz. In zero static field therefore the isothermal limit with $\chi'' = 0$ is always observed.

A non-zero χ'' is observed in some cases, for instance at the magnetic phase transition when the magnetic order produces an internal field. As the dynamics follow a critical behaviour at T_c, no significant frequency dependence of $\chi(T)$

around T_c is observed. In some cases, as in spin glasses or disordered ferro-magnets, frequency dependence is more evident and it can become tricky to distinguish between the single-molecule magnet behaviour and the freezing of a spin-glass. Mydosh has suggested analysing the frequency-dependent shift of the temperature of the maxima of $\chi(\omega, T)$, i.e. the freezing temperature T_f , by evaluating:

$$ F = \frac{(\Delta T_f(\omega)/T_f(\omega))}{\Delta \ln(\omega)} \qquad (3.28) $$

where $\Delta T_f(\omega)$ is the difference between the highest and lowest blocking temper-atures corresponding to the extremes of the investigated frequency range that appears, on a logarithmic scale, in the denominator.

While for spin-glasses F ranges from 0.001 to 0.08, in the case of paramagnetic relaxation F has the value of 0.28 (Mydosh 1993). This can only be observed when no distribution of τ is observed, a difficult condition to be realized in real systems. For instance, Fe_8, probably the SMM that best approaches the ideal behaviour, has $F = 0.24$. Large deviations from the paramagnetic limit should, however, be looked at with a critical eye and considered as evidence of a large distribution of relaxation times that deserves further investigation.

Before closing this section it is interesting to pose a question: Why do SMMs, which are substantially paramagnets, have $\chi'' \neq 0$ in zero static field? We anti-cipate here a qualitative answer leaving the detailed treatment to Chapter 5. A SMM is a molecule with a spin ground state that is split in zero field by the magnetic anisotropy with the $m = \pm S$ states lying lowest. In zero field and low temperature the relaxation transfers population from one level to the other of the lowest doublet. These states are pure $m = \pm S$ in strictly axial symmetry while are admixed by transverse anisotropy or transverse field. The larger S is, the smaller is the admixing and the lower is the transition probability between these two states. If we neglect this very slow process, i.e. the tunnelling process, the relaxation requires the system to populate other levels at an energy separation which exceeds the spin spin interaction, and energy transfer with the lattice is necessarily involved. In other words, the magnetic anisotropy and the large spin value induce a sort of internal field that makes the fast spin–spin relaxation inefficient. $\chi'' \neq 0$ is thus observed also in zero static field, and its presence is considered as a fingerprint of SMM behaviour.

3.2   Specific heat measurements

3.2.1   *The specific heat and its magnetic part*

The specific heat is one of the thermodynamic properties most investigated in condensed matter physics, both from the experimental and theoretical points of view (Tari 2003). From its definition, the heat, per unit of mass, necessary to increase by a degree the temperature of the material, it is apparent that the specific heat provides information over all the possible excitations: electronic, vibronic, rotational, magnetic, nuclear, etc. If we focus, as usual, on molecular magnetic materials, we have to separate the magnetic contribution from the others. A common procedure is to measure the specific heat of an isostructural non-magnetic compound. The lattice contribution is thus estimated, and then subtracted from the specific heat of the investigated compound. Such a procedure is not applicable to molecular clusters because the diamagnetic analogue usually cannot be prepared. The lattice contribution to the specific heat of insulating compounds at intermediate temperature is found to follow (Ashcroft and Mermin 1976):

$$C = \frac{12\pi^4}{5} n k_B \left(\frac{T}{\theta_D} \right)^3 \tag{3.29}$$

where n is the number of atoms in the crystal, and θ_D is the Debye temperature. This parameter represents the temperature below which the vibrational modes start to be frozen and is therefore a measure of the 'stiffness' of the lattice. Its experimental determination is important to evaluate the phonon contribution to magnetic relaxation. Moreover, given the strong temperature dependence, well below θ_D the lattice contribution becomes negligible compared to the magnetic contribution, whose investigation becomes thus feasible at low temperature.

3.2.2 Magnetic specific heat at equilibrium

In magnetism it is convenient to use the specific heat at constant volume and at constant fields. It is given by the general formula

$$C = \frac{\partial}{\partial T} \langle \mathcal{H} \rangle \tag{3.30}$$

where the derivative is taken at constant volume and fields and

$$\langle \mathcal{H} \rangle = \frac{1}{Z} \mathrm{Tr} \mathcal{H} \exp(-\beta \mathcal{H}) \tag{3.31}$$

where $Z = \mathrm{Tr} \exp(-\beta \mathcal{H})$ is the partition function.

In (3.31) it is possible to split the Hamiltonian into a non-magnetic part, and a magnetic part, H_{mag}. Then one can define a magnetic specific heat:

$$C = \frac{\partial}{\partial T} \langle \mathcal{H}_{mag} \rangle \tag{3.30 bis}$$

where

$$\langle \mathcal{H}_{mag} \rangle = \frac{1}{Z} \mathrm{Tr} \mathcal{H}_{mag} \exp(-\beta \mathcal{H}). \tag{3.31 bis}$$

In this chapter the magnetic Hamiltonian will be assumed independent of non-magnetic degrees of freedom. In particular the spin–phonon interaction will be ignored in the calculation of the average values (although it is essential for the time evolution). Then (3.31 bis) reads:

$$\langle \mathcal{H}_{mag} \rangle = \frac{1}{Z} \mathrm{Tr} \mathcal{H}_{mag} \exp(-\beta \mathcal{H}_{mag}) \tag{3.31 ter}$$

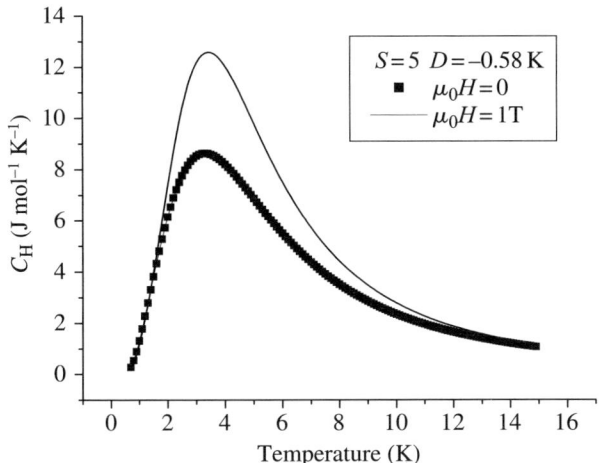

FIG. 3.20. Temperature dependence of the magnetic specific heat calculated for $S = 5$, $D/k_B = -0.58$ K in zero field and with $\mu_0 H = 1$ T, respectively directed along the easy axis of magnetization.

As an example we can evaluate the magnetic contribution to the specific heat of a system of N spins S in an axial magnetic anisotropy D and a magnetic field H. The average (magnetic) energy is given by:

$$\langle \mathcal{H}_{\text{mag}} \rangle = \frac{N \sum_{m=-S}^{S} (Dm^2 + hm) \exp[-(Dm^2 + hm)/(k_B T)]}{\sum_{m=-S}^{S} \exp[-(Dm^2 + hm)/(k_B T)]} \tag{3.32}$$

where $h = g\mu_B H$. The molar magnetic specific heat is easily obtained from (3.32) if N is the Avogadro number.

The calculated temperature dependence for the case $S = 5$ and $D/k_B = -0.58$ K is shown in Fig. 3.20. The curve has the typical behaviour with a broad maximum, also known as the Schottky anomaly, which always arises when discrete levels are separated by an energy difference Δ. At high temperature the specific heat goes to zero because all the states are thermally populated. At very low temperature fewer phonons are available to excite the modes and therefore the specific heat is zero. At intermediate temperature, a maximum is observed. Its height and position depend of course on the energy level spectrum and the degeneracy of each level but can easily be estimated from (3.30) and (3.32). For a two-level system $T_{\text{max}} \approx 0.42\Delta/k_B$.

In the case of a fluid, which is usually treated in textbooks of thermodynamics, one distinguishes a specific heat at constant volume and at constant pressure. Similarly, in magnetism, one might define a magnetic specific heat at constant magnetization as

$$C_M = \left(\frac{\partial W}{\partial T}\right)_M \tag{3.33}$$

where $W = \langle \mathcal{H} \rangle - H\langle M \rangle$. The correspondence with the fluid is obtained from the following substitutions

$$V \to -M$$
$$P \to H. \tag{3.34}$$

The quantity $\langle W \rangle$ may be called the internal energy as in a fluid, while $\langle \mathcal{H} \rangle$ is analogous to the enthalpy. The formula analogous to (3.32) from which C_M can be evaluated is:

$$\langle W \rangle = \frac{N \sum_{m=-S}^{S} Dm^2 \exp[-(Dm^2 + hm)/(k_B T)]}{\sum_{m=-S}^{S} \exp[-(Dm^2 + hm)/(k_B T)]}. \tag{3.35}$$

Specific heat measurements can therefore be used to estimate the energy spectrum, but the resolution is usually very low compared to other techniques. In the investigation of SMMs the technique has, however, been demonstrated to provide precious information. First of all, it has excluded the possibility that the slow magnetic relaxation observed in SMMs is due to the occurrence of magnetic order.

In fact a magnetic transition is revealed as an anomaly in the specific heat with typical λ shape. No anomalies of these type have been observed for $Mn_{12}ac$ or Fe_8 (Gomes *et al.* 1998, 2001).

The application of an external magnetic field turns out to be very useful in the characterization of SMMs. The Zeeman splitting removes the energy degeneracy of the $\pm m$ pairs of states and the effects on the specific heat are well shown in Fig. 3.20 with an increased specific heat compared to the zero-field case. The contribution due to the Zeeman splitting can be detected only if the system can reach equilibrium in the time-scale of the experiment. We will see later in detail that precious information on the magnetization dynamics can be extracted in this way.

3.2.3 *Measurement of the magnetic specific heat: the relaxation method*

We will briefly show here how the specific heat can be measured. The simplest way, at least in principle, to evaluate the specific heat is to decouple the sample from the surrounding, and send a known amount of heat to the sample. The temperature of the sample is monitored and its variation allows evaluating C. This type of measurement, schematized in Fig. 3.21, is however seriously affected by the non-perfect adiabatic condition. The wires of the thermometer and of the heater, as well as the sample holder, are responsible for the thermal leak.

Bachmann *et al.* (1972) have developed another method, also known as the relaxation method, where the thermal link of the sample with the thermal bath is taken into account. A heat pulse is sent to the sample and this results in a

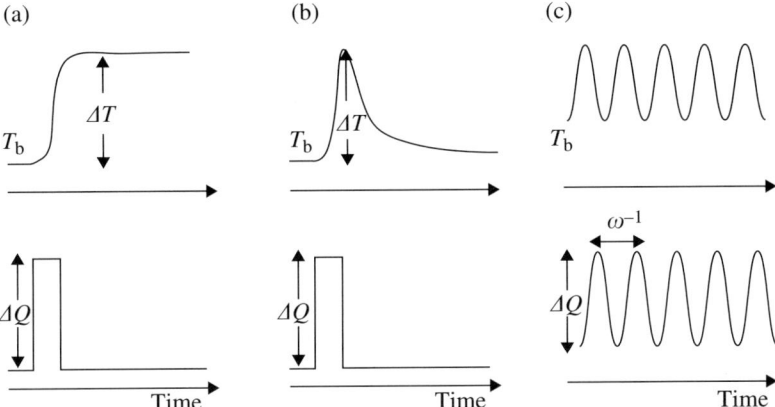

FIG. 3.21. Schematic view of the heat pulse (bottom) and temperature variation
(top) in: an adiabatic calorimeter (a); a semi-adiabatic calorimeter employing
the relaxation method (b); an ac calorimeter (c).

sudden temperature increase of the sample. Due to the thermal link with the
thermostat the temperature goes back to the equilibrium value, as shown in
Fig. 3.21. If the thermal conductance K_b between the sample and the bath is
constant over the investigated temperature range, and if the heat pulse gives a
temperature increase which is small compared to the sample temperature, the
time dependence of the relaxation follows the equation

$$\tilde{C}\frac{dT(t)}{dt} = K_b[T(t) - T_b] \tag{3.36}$$

where T_b is the bath temperature and $\tilde{C}$ is the thermal capacity, $\tilde{C} = MC$, of
the sample of mass M, with appropriate corrections due to the sample holder,
etc. Integration yields

$$T(t) = T_b + \Delta T(0)\exp\left(-\frac{t}{\tau_b}\right) \tag{3.37}$$

where

$$\tau_b = \tilde{C}/K_b \tag{3.38}$$

The thermal conductance K_b is known from a calibration of the system. Then
the measurement of the thermal relaxation τ_b yields the heat capacity $\tilde{C} = MC$
and thus the specific heat C.

The relaxation method presents several advantages: adiabaticity is not
required and the sample can also have poor thermal conductivity. Moreover,
heat pulses at the same thermostat temperature can be repeated, increasing the
sensitivity of the technique.

3.2.4 *Magnetic specific heat in an alternating current*

From (3.38) it is evident that measurements of the specific heat with the relaxation method provide information on the dynamics of the magnetization if the characteristic time of this last one is comparable to or longer than τ_b. Faster magnetic relaxation can be investigated using a method 'in alternating current'. The method, indeed quite old (Corbino 1911), is schematized in Fig. 3.21c. An alternate current is applied to the heater at a frequency $\omega/2$. Its effect is the generation of a heat pulses at frequency ω and thereby temperature oscillations at the same frequency. The amplitude of the temperature oscillation is related to the specific heat of the sample and the sample holder. This is only true if the frequency is sufficient large, $\omega\tau_b > 1$, where τ_b given by (3.38), is the relaxation time of the thermal link to the bath. This can be seen from the formula

$$\tilde{C}\frac{dT}{dt} = -K_b[T - T_b] + Q\cos\omega t + Q \tag{3.39}$$

which is an extension of (3.36) in the presence of a sinusoidal heat flux of frequency ω. The solution is easily seen to be

$$T = T_b + \frac{Q}{K_b} + B\cos(\omega t - \varphi) \tag{3.40}$$

where $\tan\varphi = \omega\tilde{C}/K_b$ and

$$|Q/B| = K_b\sqrt{1 + \omega^2\tau_b^2}. \tag{3.41}$$

The measurement quantities are Q and B, and therefore the ratio Q/B given by (3.41) can be evaluated. For $\omega = 0$ this ratio is independent of the heat capacity $\tilde{C}$. However if ω is of the order of τ_b or larger, the measurement of Q/B yields τ_b and therefore $\tilde{C}$ through (3.38). In practice, a large value $\omega\tau_b \gg 1$ is chosen, so that (3.41) reads

$$|Q/B| = \omega\tilde{C}. \tag{3.42}$$

This result turns out to be independent of K_b and yields directly $\tilde{C}$.

The condition $\omega\tau_b \gg 1$ implies a weak response B/Q, so that the system can be said to be 'quasi-adiabatic'. However, if ω is too large B/Q becomes too small and the accuracy decreases. Quasi-adiabaticity implies that the temperature T is almost in phase quadrature with the heat flux. The thermal relaxation time τ_b has to be large enough in order that the temperature is homogeneous throughout the sample.

The main advantage of the ac technique is that very small samples can be measured, of the order of μg in mass, provided that the contribution to the specific heat of the sample holder, thermometer and heater, the so-called 'addenda', are not much larger than that of the sample (Fominaya *et al.* 1997a). Another important advantage is the fact that ω can be varied and therefore the time dependence of the enthalpy can be investigated. This will be seen to be useful in the case of the investigation of the thermally activated magnetic relaxation of $Mn_{12}ac$ and Fe_8 discussed in Chapter 10. The use of a single crystal is of great interest if a magnetic field is applied, because its alignment with respect to the crystal axes can be chosen at will. More details on the specific heat in alternating currents can be found in the work of Sullivan and Seidel (1968a,b).

3.3 Magnetic resonances

3.3.1 *Electron paramagnetic resonance*

Electron paramagnetic resonance spectroscopy induces transitions within the m states belonging to a given S multiplet split by an external magnetic field (Abragam and Bleaney 1986; Pilbrow 1990; Bencini and Gatteschi 1990; Bencini and Gatteschi 1999). It provides information on the chemical environment of the paramagnetic centre(s) and also in principle on the spin dynamics. The former is associated to the spin–orbit coupling contribution and is given by the **g**, **D**, and **A** tensors described in Chapter 2. In the following we will consider only the EPR spectra of magnetically non-diluted systems. Under these conditions the hyperfine structure associated to the A tensor is wiped out by the dipolar and exchange interactions between the magnetic centres and is not experimentally available. Therefore we will neglect this important part of the EPR spectra and in the same way we will not mention other satellite techniques like ENDOR spectroscopy. On the other hand, the hyperfine interaction can be obtained from the NMR spectra described in Section 3.3.2.

An important piece of information which can be obtained from the EPR spectra is the anisotropy of the tensors. It can be obtained directly not only from experiments performed on single crystals but also from systems in which the tumbling ratio of the paramagnetic centres is slow compared to the EPR time-scale and all the orientations of the magnetic molecules are present. This can be achieved by using polycrystalline powders or frozen solutions, for instance. This opportunity has been largely exploited as outlined below.

Another important feature of the EPR experiment is that it can also provide information on the spin dynamics, particularly when pulsed techniques are used. This part of the use of the technique has so far been largely neglected in molecular magnetism and will not be treated further.

For a static magnetic field parallel to the z axis of the **g** tensor, in the case of $S = \frac{1}{2}$, the transition energy between $m = -\frac{1}{2}$ and $m = \frac{1}{2}$ is given by:

$$h\nu = g_z\mu_{\mathrm{B}}H \qquad (3.43)$$

where ν is the radiation frequency and H is the static field. The resonances are in the microwave range, and in this spectral region it is difficult to continuously sweep frequency over a large range. Therefore the resonance is measured by using a fixed frequency and sweeping the magnetic field. At the beginning klystrons were used for generators; now it is more common to use semiconductor diodes, like Gunn or IMPATT diodes. In the traditional set-up the static magnetic field and the oscillating magnetic field of the radiation used to induce

TABLE 3.1. Microwave frequencies and resonant fields for EPR spectroscopy.

Frequency (GHz)	Symbol	Resonant field $(g = g_e)$, T
9	X	0.3211
35	Q	1.2489
95	W	3.3899
200		7.1366
300		10.7048
500		17.8414

the transition are orthogonal to each other, so that selection rules of the type $\Delta m = \pm 1$ hold. The resonance fields for $g_z = g_e$ at various frequencies are given in Table 3.1. For historical reasons some frequency ranges are indicated by letters, as shown in Table 3.1. In order to use small fields, easily achieved with electromagnets, X-band spectrometers are widely spread out. Also it is possible to use commercial spectrometers operating at Q and, since a few years ago, W-band. All the other sources are implemented on home-made spectrometers. In fact in the last few years there has been a wide increase in the use of high-frequency spectrometers.

Equation (3.43) is sufficient for multiplets S if the zero field splitting is absent. If the field is applied parallel to x the resonance condition is the same, only g_z must be replaced by g_x. The spectra are recorded by modulating at low frequency the magnetic field in such a way that the output is the derivative of the absorption curve, usually a Gaussian or a Lorentzian. The spectra of polycrystalline powders of a compound with $S = \frac{1}{2}$ and anisotropic g are shown in Fig. 3.22.

For $g_x = g_y = g_z$ only one line is observed like in a single crystal. For $g_x = g_y = g_\perp$; $g_z = g_{//}$ two features are observed, that corresponding to $g_{//}$ being less intense because the crystallites with the z axis parallel to the applied field are less numerous than those with the x or y axes parallel to the field. For $g_x \neq g_y \neq g_z$ three lines are observed. The three g values are easily obtained from the spectra, but it is impossible to know their orientation in the molecular frame. For this, single-crystal experiments are needed.

The inclusion of a zero-field splitting complicates the spectra. The simplest case is that of an axial ZFS which is small compared to the Zeeman energy. The energy levels are quantized along the magnetic field, and resonance fields for the various $m \rightarrow m + 1$ transitions in a perturbation approach for axial symmetry are given by:

$$H(m \rightarrow m+1) = \frac{g_e}{g} \left[H_0 - D' \frac{2m+1}{2} (3\cos^2\theta - 1) \right] \qquad (3.44)$$

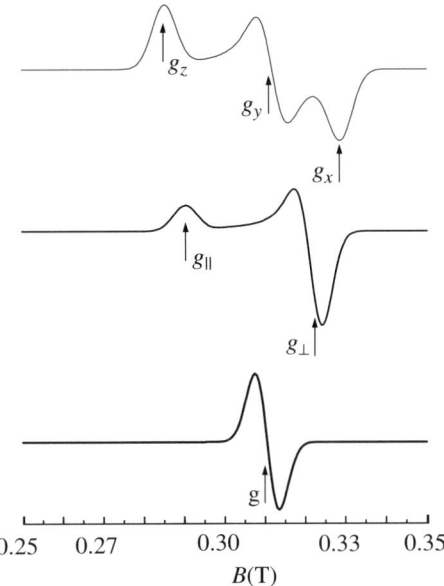

FIG. 3.22. Polycrystalline powder EPR spectra of a system with $S = \frac{1}{2}$ characterized by $g_x \neq g_y \neq g_z$ (top), $g_x = g_y \neq g_z$ (middle), and isotropic g value
(bottom).

where H_0 is the resonant field of the free electron, $D' = D/(g\mu_B)$, θ is the
angle between the unique axis and the external magnetic field. For a given orientation each Zeeman line is split into $2S$ components. The neighbouring lines
are separated by $D'(3\cos^2\theta - 1)$, therefore the ZFS parameter can be directly
extracted from the spectra. The associated structure of the spectra is called the
fine structure.

The advantages of increasing the operating frequency are (Eaton and Eaton
1999; Barra *et al.* 1998; Barra 2001):

(i) increased resolution;
(ii) increased sensitivity;
(iii) simplification of the spectra and of their assignment;
(iv) observation of the spectra in 'EPR' silent species;
(v) determination of the sign of the zero-field splitting anisotropy.

Point (iv) is particularly important in SMMs. In fact these often correspond to
systems with an integer S for which the separation between neighbouring m levels
in the absence of an applied magnetic field is large. For instance, in $Mn_{12}ac$ the
separation between the lowest $m = -10$ and the first excited $m = -9$ state in
zero field is about 14 K. Using a 9 GHz exciting frequency, which corresponds

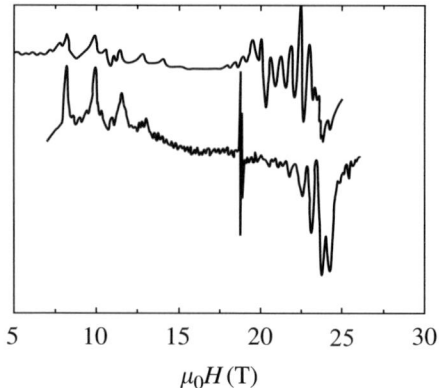

$\mu_0 H$ (T)

FIG. 3.23. EPR spectra of polycrystalline powder spectra of $Mn_{12}ac$ at 525 GHz and 30 K. Lower experimental spectra, upper calculated. The low-field features correspond to crystallites with the external field parallel to the tetragonal axis. The features correspond to the $-10 \rightarrow -9, -9 \rightarrow -8, -8 \rightarrow -7,$ and $-7 \rightarrow -6$ transitions. Reprinted with permission from (Barra *et al.* 1997b). Copyright (1997) by the American Physical Society.

to 0.5 K, the transition can never be observed (Blinc *et al.* 2001). In fact at zero field this exciting frequency is too small to induce the transition (14 K is about 300 GHz) and an increase in the field further increases the separation between the two levels. At X-band frequency only a few transitions are observed, corresponding to the high lying m levels, like $-2 \rightarrow -1, -1 \rightarrow 0$.

At the highest frequency so far used, 525 GHz (Barra *et al.* 1997b), the transitions corresponding to the lowest lying levels, namely $m = -10 \rightarrow -9; m = -9 \rightarrow -8; m = -8 \rightarrow -7; m = -7 \rightarrow -6$, are clearly resolved in the spectrum of Fig. 3.23. This spectrum corresponds to measurements on a polycrystalline powder pressed in a pellet. The low-field lines correspond to the resonances of crystallites with their tetragonal axis parallel to the external magnetic field, while the high-field lines correspond to crystallites with the tetragonal plane parallel to the field.

If all the m levels are equipopulated (high-temperature approximation) the intensities of the transitions can be calculated by using the Fermi golden rule:

$$I(m \rightarrow m + 1) \propto [S(S + 1) - m(m + 1)]. \tag{3.45}$$

A simple inspection of Fig. 3.23 shows that (3.45) is not followed in the high-frequency spectra. In fact (3.45) suggests for the allowed transitions the pattern of intensities given in Table 3.2.

The large deviation is originated by the breakdown of the high-temperature approximation, which requires that $g\mu_B H_0/k_B \ll T$. At 525 GHz, $g\mu_B H_0/k_B = 25$ K, therefore at low temperature only the lowest $m = -S$ level will be populated and one transition should be observed. At intermediate temperature the

TABLE 3.2. Calculated temperature dependence of the relative intensities of the $m \rightarrow m+1$ transitions for $S = 10$. The frequency is 525 GHz. The field is applied parallel to z. Only the second-order axial splitting $D/k_B = -0.72$ K is included.

	m = −10	m = −9	m = −8	m = −7
$I(T = \infty)$	1	1.9	2.7	3.4
$I(T = 300$ K$)$	1	1.75	2.27	2.6
$I(T = 100$ K$)$	1	1.47	1.60	1.52
$I(T = 50$ K$)$	1	1.14	0.95	0.68
$I(T = 35$ K$)$	1	0.92	0.61	0.34
$I(T = 20$ K$)$	1	0.53	0.20	0.06
$I(T = 5$ K$)$	1	0.01	0.00	0.00

intensities of (3.38) will be weighted according to the Boltzmann population of the m levels. It is interesting to note that even at 300 K the high-temperature limit is not reached.

Another important feature of HF-EPR spectra is that they also provide the sign of the D parameters. This should become apparent with the example worked out below. The calculated energy levels for $S = 3$ in a field parallel to z are shown in Fig. 3.24 for $D > 0$ and $D < 0$. It is evident that in the former case the transition involving the lowest energy m level occurs at high field, while in the latter at low field. Therefore the EPR spectra at high frequency provide the sign of the D parameter, or the sign of the magnetic anisotropy. It is an easy exercise using (3.44) to show that the reverse pattern is observed for the transitions perpendicular to the unique axis.

Useful as they are, and simple to measure, polycrystalline powder EPR spectra cannot provide the principal directions of the spin Hamiltonian tensors. In many cases this is an important piece of information, which can only be obtained by single-crystal spectra.

In order to show the potentialities of single-crystal EPR experiments we present the case of the first tetranuclear iron(III) cluster of the class of the 'iron stars' already discussed in Section 3.1. The present one has the formula [Fe$_4$ (OCH$_3$)$_6$(dpm)$_6$] and its structure (Barra *et al.* 1999) is shown in Fig. 3.25. The crystals are monoclinic, with a crystallographically imposed binary axis passing through Fe1 and Fe2. Each iron ion is coordinated to six oxygen atoms, defining a distorted octahedron. The oxygen atoms belong to dpm$^-$ ligands (dpm$^-$ is the anion of dipivaloylmethane) shown also in Fig. 3.25 and to methoxide groups.

As all the other Fe$_4$ clusters of this type it has an $S = 5$ ground state. The large zero-field splitting observed in the ground state grants SMM behaviour to

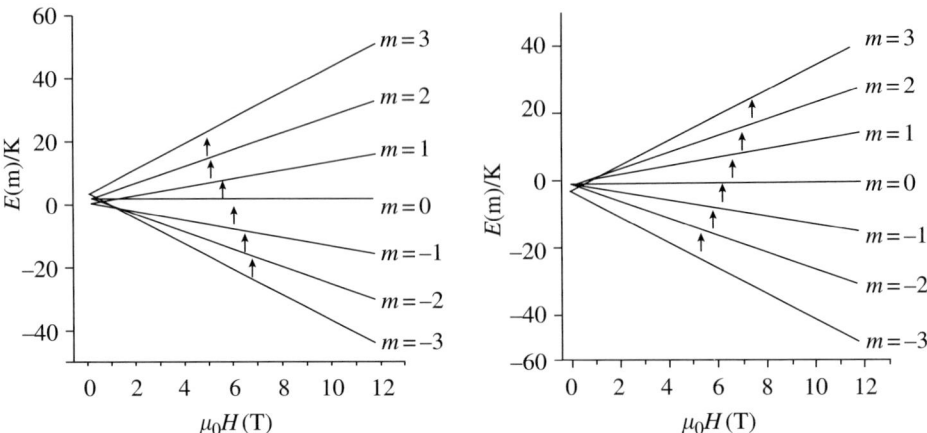

FIG. 3.24. Spin levels of an $S = 3$ state in a magnetic field parallel to the unique axis. Left, $D/k_B = 0.40$ K; right, $D/k_B = -0.40$ K. The transitions are calculated for a frequency of 150 GHz.

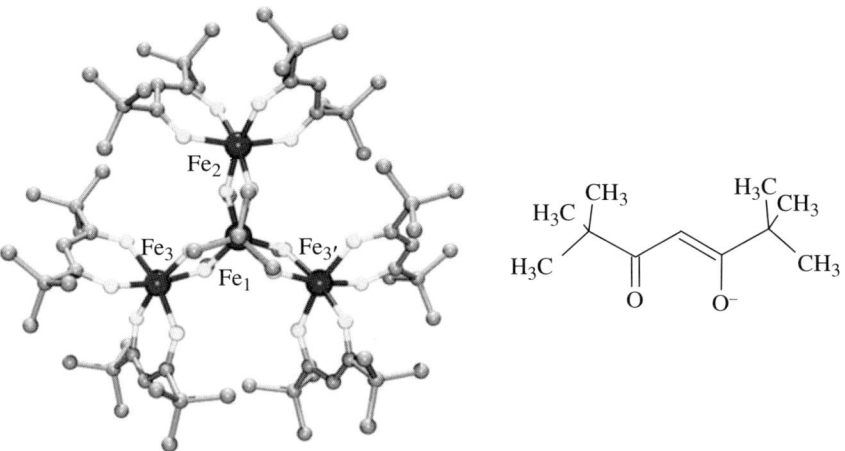

FIG. 3.25. View of the structure of the Fe_4 cluster $[Fe_4(OCH_3)_6(dpm)_6]$ and of the anion of dipivaloylmethane.

this molecule. Single-crystal W-band EPR spectra were measured (Bouwen *et al.* 2001), which showed the expected transitions for $S = 5$, as shown in Fig. 3.26.

However it is apparent that more sets of transitions are present, indicating the presence of different species in the lattice. In fact, in the crystal structure determination, disorder was observed, which is due to the wrapping of the organic moieties in the molecular environment, as shown in Fig. 3.27.

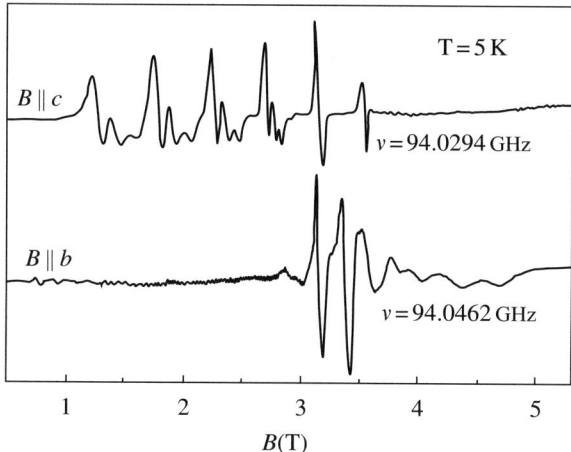

FIG. 3.26. Single-crystal W-band EPR spectra of $[Fe_4(OCH_3)_6(dpm)_6]$. Upper static field parallel to c; lower parallel to b. Reprinted with permission from Bouwen *et al.* (2001). Copyright (2001) American Chemical Society.

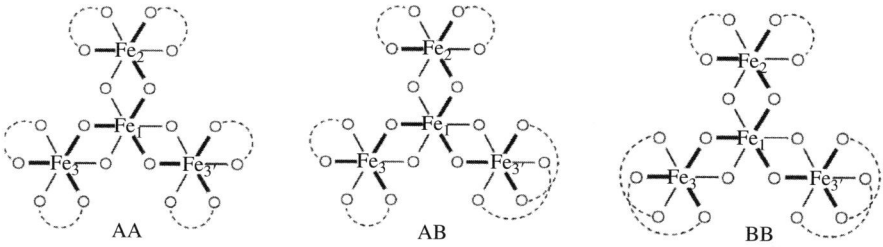

FIG. 3.27. Three different isomers present in the unit cell of Fe_4. Bold lines correspond to bonds emerging from the plane of the paper. They differ in the way the *bis*-chelate ligands wrap around the peripheral iron(III) ion.

The oxygen atoms of the dpm and methoxide ligands define two layers, like in a compact structure of an oxide. The iron ions related by the binary axis, Fe3 and Fe3′, are disordered in the way the organic moieties of dpm connect the oxygen atoms bound to the metal ions. In case A the two oxygen atoms of the dpm ligand are on different layers, while in case B they are on the same layer. There can be three different types of isomers, as sketched in Fig. 3.27, namely AA, AB, and BB. Experimentally the AA isomers correspond to 49% of the molecules, AB to 42% and the BB isomer to 9%. Although the bond distances between the iron(III) and oxygen atoms are very similar to each other, minor changes in the bond angles determine significant changes in the **D** tensors of the various

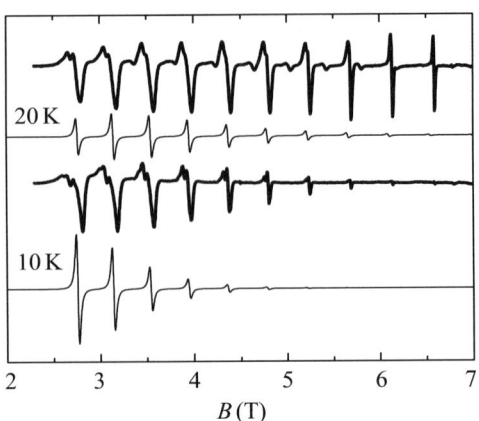

FIG. 3.28. EPR spectra of a single crystal of Fe$_8$ measured along the easy axis at 190 GHz. Thicker lines correspond to experimental spectra.

isomers. In fact the axial zero-field splitting parameter D/k_B is $-0.296(1)$ K for AA, $-0.273(2)$ for AB, and -0.252 K for BB.

In the case of Fe$_4$ the position of the principal axes of the **D** tensor is largely determined by the symmetry and shape of the molecule. Matters are more complex in a system like Fe$_8$ where no crystal symmetry requirement is present (Wieghardt *et al.* 1984). Single-crystal EPR spectra recorded at 190 GHz, shown in Fig. 3.28, correspond to the easy axis of the magnetization at low temperature (Barra *et al.* 2000). It is interesting to note that the linewidth of the $m \to m+1$ transitions decreases with $|m|$. This has been attributed to crystal strain effects leading to a small distribution of D parameter.

The use of high frequencies requires the use of high fields. The latter tend to orient the crystals along the easy direction. If averaged polycrystalline powders need to be used it is necessary to avoid orientation using the same techniques described in Section 3.1.1 for magnetic measurements. Alternatively the high field can be used to measure pseudo-single-crystal spectra.

In the recent years several developments have been made for the measurement of frequency swept experiments (Kozlov and Volkov 1998 ; Van Slageren *et al.* 2003). In fact, the possibility of employing backward-wave oscillators (BWO), originally developed in the 1960s, has been explored. The BWO is a vacuum tube similar to a klystron, where electrons are generated at a heated cathode and accelerated towards the anode. Their kinetic energy change generates microwaves. The power output of a BWO varies from several hundreds of mW (at about 100 GHz) to 1 mW at 1.5 THz. For a frequency swept experiment several different sources are used. The advantage of using frequency swept experiments is shown in Fig. 3.29, which shows the spectra of a pressed pellet (Mukhin *et al.* 1998) of Mn$_{12}$ac at 4 K.

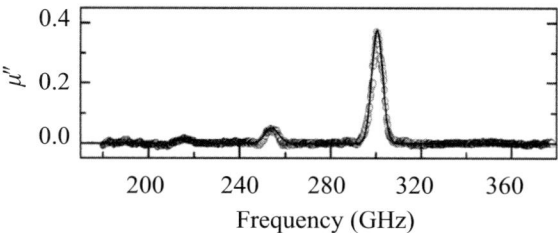

FIG. 3.29. Zero-field EPR spectra of a pressed pellet of $Mn_{12}ac$ at 5 K. Redrawn from Mukhin *et al.* (1998) with permission of EDP sciences.

The peak at about 300 GHz corresponds to the transition $\pm 10 \rightarrow \pm 9$; that at about 250 GHz to the $\pm 9 \rightarrow \pm 8$ transition and that at about 220 GHz to the $\pm 8 \rightarrow \pm 7$ transition. The relative intensities essentially depend on the different populations of the levels at low temperature. The zero-field splitting parameters obtained from the analysis of the spectrum of Fig. 3.29 compare well with those obtained from the analysis of the HF-EPR spectra. The spectrum of Fig. 3.29 clearly shows that the separation between neighbouring lines is not constant and suggests the need to include fourth-order terms of the type αS_z^4.

3.3.2 *Nuclear magnetic resonance*

Nuclear magnetic resonance can provide information on the properties of clusters in several ways. One is the type of structural information which is so familiar to chemists, through the so-called chemical shift. However, since in general the clusters are paramagnetic it is the less conventional paramagnetic NMR technique which is of relevance here (Bertini and Luchinat 1996; Köhler 2001). The chemical shift is contained in the nuclear Zeeman term, while additional information is contained in the hyperfine tensor **A**, defined in Section 2.1.2. The experiments are performed in solution and at room temperature. The molecules in solution tumble rapidly on the NMR time scale. Under these conditions the experimental output is the isotropic component of the hyperfine tensor defined in (2.19), the dipolar components being averaged to zero.

Experiments can be performed also on solid samples, both polycrystalline powders and single crystals. In both cases, information can be obtained on the anisotropy of the hyperfine tensor. Further it is possible to obtain information on the electron spin dynamics by performing pulsed experiments.

An example of the use of ^1H NMR in solution for obtaining structural information is provided (Eppley *et al.* 1995) by the spectra of $Mn_{12}ac$ shown in Fig. 3.30. The x axis corresponds to the shift (with a sign) from the absorption of a reference diamagnetic species like tetramethylsilane. The spectra of diamagnetic species are generally limited to shifts of a few parts per million (ppm). The large shifts observed here depend on the paramagnetic nature of the cluster.

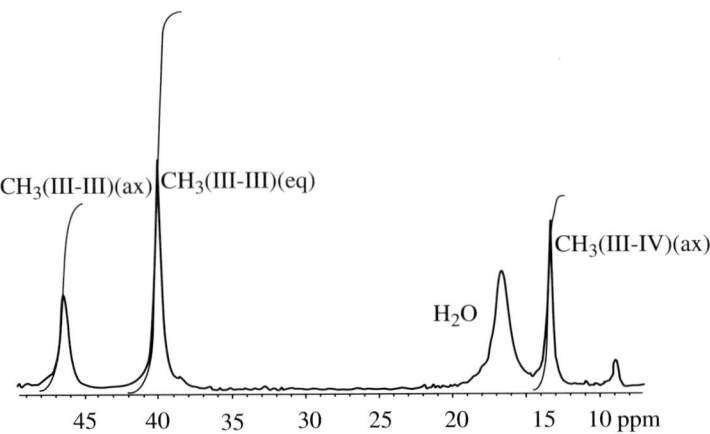

FIG. 3.30. ^{1}H NMR spectra of Mn$_{12}$ac in CD$_3$CN solution at room temperature and assignment of the peaks. The grey lines represent the integrated area of the peaks. The labels refer to the bridging made of the acetate ligand carrying the methyl group. Courteously provided by A. Cornia.

The first information which comes from these spectra is that the structure of the cluster is preserved in solution, because the signals of the protons correspond nicely to what is expected for a symmetric tetragonal structure, as will be shown below. The peak at ca. 18.7 ppm is broader than the others. It is assigned to the weighted average of the water molecules bound to the cluster and free. The fact that only one averaged signal is observed indicates that there is fast exchange between bound and free water. Fast means that the average residence time of the water in the coordination spheres of the manganese ions is short on the NMR time-scale. The other signals disappear if deuterated acetic acid is used in the synthesis of the cluster, therefore they correspond to the methyl groups of the acetates. The presence of three signals, with a pattern of intensities 1:2:1, agrees with the tetragonal structure of the cluster in the solid state. In fact this requires that there are four different CH$_3$ groups, all the others being reported by the symmetry elements of the cluster. Perusal of the structure shows that there are three types of acetate groups which bridge pairs of manganese(III) ions, and one bridge manganese (III)-manganese(IV) pairs. Two peaks are largely shifted from the corresponding diamagnetic position, while the other is closer to it. Therefore the largely shifted methyl signals are globally three times more intense than the other one. On this basis, the latter is assigned to the four acetates which bridge the manganese(III) and manganese(IV) ions, while the others are assigned to the manganese(III)-manganese(III) bridges. Among these the signal with double intensity is assigned to the eight equatorial acetates and the other to the axial ones. Equatorial means that they approximately lie in the tetragonal plane, while axial means that they point out of the plane as can be seen in Fig. 1.8.

These qualitative considerations should rely on quantitative consideration: The shifts observed correspond to the isotropic part of the hyperfine tensors of the various protons. In fact the anisotropic components are averaged to zero by the rapid tumbling of the molecules in solution. The isotropic part of the hyperfine tensor is given by the sum of two contributions: one is the contact and the other the pseudocontact one.

The contact contribution is given by the unpaired spin density on the nucleus N:

$$A(N) = \frac{\mu_0}{3S} g_e \mu_B \gamma_N \, |\psi(0)|^2 \tag{3.46}$$

where μ_0 is the vacuum permeability, S is the electron spin quantum number and $|\psi(0)|^2$ is the probability of finding the unpaired electron at the observed nucleus. Only s orbitals have a non-zero density on the nucleus. The observed shifts are related to the isotropic component by:

$$\delta_T^{con}(N) = A(N) \frac{g_{av} \mu_B S(S+1)}{3 \gamma_N kT}. \tag{3.47}$$

This expression is valid in the case of isotropic g and in the absence of zero-field splitting. A more accurate relation takes into account the relaxation of these conditions by introducing the magnetic susceptibility:

$$\delta_T^{con}(N) = A(N) \frac{1}{3 \gamma_N \mu_B} \left(\frac{\chi_{xx}}{g_{xx}} + \frac{\chi_{yy}}{g_{yy}} + \frac{\chi_{zz}}{g_{zz}} \right). \tag{3.48}$$

It is apparent that the NMR experiment provides the unpaired spin densities on the magnetic nuclei. In this sense it is similar to polarized neutron experiments to be discussed below. However for the latter the spin density is measured in all the lattice positions, while the NMR experiment can only monitor it at the discrete positions corresponding to the coordinates of the magnetic nuclei.

The contact term is not the only one giving rise to an isotropic shift. In the case of not completely quenched orbital moment one must also take into account the isotropic part of the orbital-dipolar contribution. It can be calculated in the ligand field approximation by extending the approaches outlined in Chapter 2. A convenient form of expressing it, in analogy to (3.48), is

$$\delta_T^{dip} = \frac{1}{24\pi} \frac{1}{r^3} \left\{ [2\chi_{zz} - (\chi_{xx} + \chi_{yy})](3\cos^2 \theta - 1) + 3(\chi_{xx} - \chi_{yy}) \sin^2 \theta \cos 2\Omega \right\} \tag{3.49}$$

where θ is the angle between the metal-nucleus vector $\mathbf{r}$ and the z molecular axis, and Ω is the angle between the projection of the $\mathbf{r}$ vector in the xy plane and the x axis.

Equations (3.48) and (3.49) are appropriate for defining the shifts in solution, because the terms arising from the dipolar interaction between the electron and nuclear spin are averaged to zero by the rapid tumbling of the molecules. This is, of course, no longer true when the experiment is performed in condensed phase, where the dipolar terms give a sizeable contribution. In fact this kind of experiment can provide first-hand information on the spin density of the magnetic clusters. For instance in $Mn_{12}ac$ it is possible to perform NMR experiments on ^1H $(I = \frac{1}{2})$, ^2H $(I = 1)$, ^{13}C $(I = \frac{1}{2})$, and ^{55}Mn $(I = \frac{5}{2})$. Perhaps the most exciting nucleus is ^{55}Mn because it is the closest to where the magnetic action occurs.

The ^{55}Mn NMR spectra of oriented polycrystalline powders of $Mn_{12}ac$ at 1.4 K have been reported (Furukawa et al. 2001a; Kubo et al. 2002). The experiments were performed both in zero applied magnetic field and at varying field. The spectra at zero field are shown in Fig. 3.31. The required field for the NMR experiment is provided by the magnetization of the clusters themselves which at this temperature are not fluctuating. The spectra clearly show three main signals of approximately the same intensity, suggesting that they correspond to the three non-equivalent manganese ions of the unit cell (two manganese(III) and one manganese(IV), respectively). The structure of the signals is due to the quadrupolar interaction of the ^{55}Mn nuclei. In fact they have $I = \frac{5}{2}$, splitting the signal into five lines of intensities 1:1:1:1:1. The assignment of the signals is made based on the extent of the quadrupolar splitting. The manganese(IV) ions have a

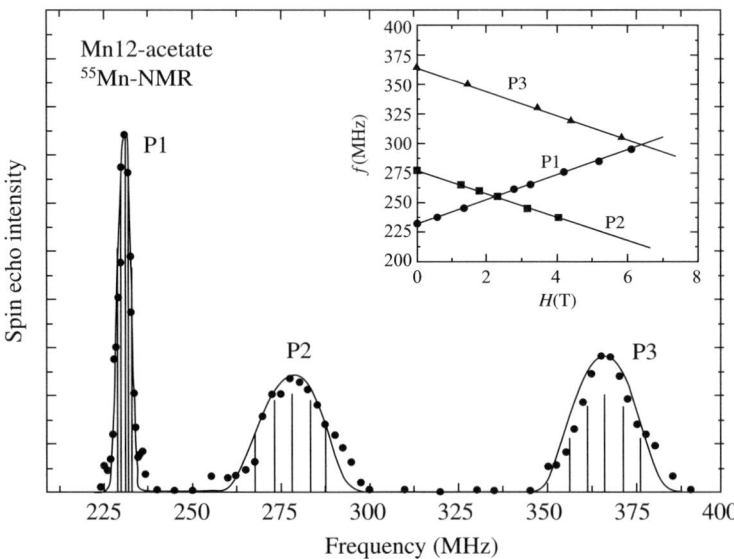

FIG. 3.31. ^{55}Mn NMR spectra of Mn12Ac at 1.4 K and zero applied magnetic field. The inset shows the field dependence of the signals. Reprinted with permission from Furukawa et al. (2001). Copyright (2001) American Physical Society.

coordination environment close to a regular octahedron, so they show a relatively narrow line, because in octahedral symmetry the quadrupolar splitting goes to zero. The Hamiltonian for the quadrupolar interaction is usually written as:

$$\mathcal{H} = \frac{e^2 Q q_{zz}}{4I(I-1)} \left[3I_z^2 - I(I+1) + \eta \left(I_x^2 - I_y^2 \right) \right] \tag{3.50}$$

where eQ is the quadrupole moment and eq_{zz} is the maximum principal value of the electric field gradient V_{zz} along the z axis. η is an asymmetry parameter defined by $(V_{xx} - V_{yy})/V_{zz}$. It is interesting to note the analogy of the Hamiltonian (3.50) with that of the zero-field splitting (2.7). The quadrupolar splitting for the signal attributed to manganese(IV) is $\Delta\nu_Q = 0.72(5)$ MHz. The broad signal at 279.4(1) MHz , P1, has a quadrupole splitting $\Delta\nu_Q = 4.3(1)$ MHz, while the signal at 364.4(1) MHz, P2, has $\Delta\nu_Q = 2.9(1)$ MHz. Since the deviations of the coordination octahedron of the two manganese(III) sites from regular octahedral symmetry are different it has been possible to assign the signal with larger quadrupolar moment to the more distorted Mn^{3+} sites.[1]

Another possible origin of the broad lines is the presence of several distorted isomers in the unit cell, a feature that will be discussed in Section 4.7.1. Better resolved spectra have indeed been observed on a different Mn_{12} derivative, namely with bromo-acetic acid, which does not show distorted isomers (Harter *et al.* 2005).

The addition of an external magnetic field parallel to the unique axis provides additional information on the relative orientation of the spins on the different manganese ions. In fact the signals corresponding to the manganese(III) ions move to lower frequency, suggesting that their magnetization is parallel to the external field, while the signal corresponding to manganese(IV) shifts to higher frequency, confirming that its magnetization is essentially antiparallel to the magnetic field. This is a confirmation of the ferrimagnetic nature of the ground state.

The hyperfine tensor of the manganese(IV) ion is isotropic and dominated by the contact contribution. In fact the dipolar contribution goes to zero in an octahedral environment. The unpaired electrons are to a good approximation in the d orbitals, but spin polarization effects induce some density also in the s orbitals. The calculated Fermi contact contribution for the manganese(IV) ion is -29.3 T, while the value observed in Mn_{12}ac is -21.84 T. The reduction in the value in bound metal ions is usually associated with covalency effects.

For the manganese(III) ions both the Fermi contact and the dipolar contribution are different from zero. The NMR frequency for the transition $m_I - 1 \leftrightarrow m_I$, ν_m, is given by:

$$\nu_m = \nu_F - \frac{1}{2}(3\cos^2\theta - 1)\left[\nu_d - \left(m_I - \frac{1}{2}\right)\nu_Q\right] \tag{3.51}$$

where ν_F is the Fermi, ν_d is the dipolar and ν_Q is quadrupolar frequency, respectively. By analysing the dependence of the observed hyperfine fields on

---

[1] This corresponds to Mn(2) in the labelling scheme of Fig. 4.20.

the orientation of the external magnetic field it was possible to obtain the Fermi contact and the dipolar component. It was found that the Fermi contact components are identical for Mn(2) and Mn(3), -41.24 T, the difference in resonance frequencies being determined by the dipolar component. The calculated value for the free manganese(III) ion is -48.5 T, while the experimental value measured by EPR for Mn^{3+} in TiO_2 is -41.48 T (Geritsen and Sabisky 1963).

Beyond providing information on the spin density on the clusters through the resonance frequencies, NMR can provide useful information through the relaxation times. In a modern NMR spectrometer pulsed techniques are used, associated with fast Fourier transforms. The spectra and the time dependence of the free induction decay, FID, are measured using different pulse techniques. In the case of SMM it is the so-called Hahn echo technique which is used, taking advantage of the extremely long relaxation times observed at low temperature (Slichter 1963).

The Hahn echo consists in sending a 90° pulse which sends the nuclear magnetization vector in the plane perpendicular to the magnetic field, Fig. 3.32. During the waiting time τ the magnetization starts to dephase in the plane until a 180° pulse refocuses it, yielding an echo signal. How the echo amplitude decreases on increasing τ provides information on the nuclear spin dynamics.

Echo experiments were performed on $Mn_{12}ac$ (Goto *et al.* 2003) and the magnetic relaxation was observed on the Mn^{4+} resonance. The sample was prepared by cooling the oriented powder in a field $H = 1.2$ T from 4.2 to 2.0 K. Under these conditions the resonance is observed at $\nu = 230.4+(\gamma_{Mn}/2\pi)H$. The relaxation was monitored by measuring the time variation of the intensity of the echo after switching the field to -1.2 T. The reversal of the external field can be achieved in a few seconds. Since the nuclear relaxation is much slower than this the measurements are not affected. The recovery follows a square-root time dependence (Fig. 3.33) as suggested by theoretical treatments to be discussed below. In the inset is shown the field dependence of the relaxation times at 2.0 K. A periodic behaviour is apparent, with minima at $H = nH_1$, where $n = 0, 1, 2, \ldots$ and $H_1 = 0.4$ T. The anomalies correspond to the fields where crossing of the m levels of the ground $S = 10$ multiplet occur, and are the signature of quantum tunnelling effects, as will be discussed at length below. Analogous results were obtained from measurements on the proton spectra.

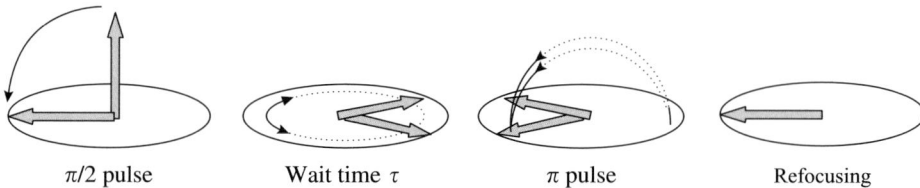

$\pi/2$ pulse   Wait time τ π pulse   Refocusing

FIG. 3.32. Scheme of the pulse sequence for the Hahn echo.

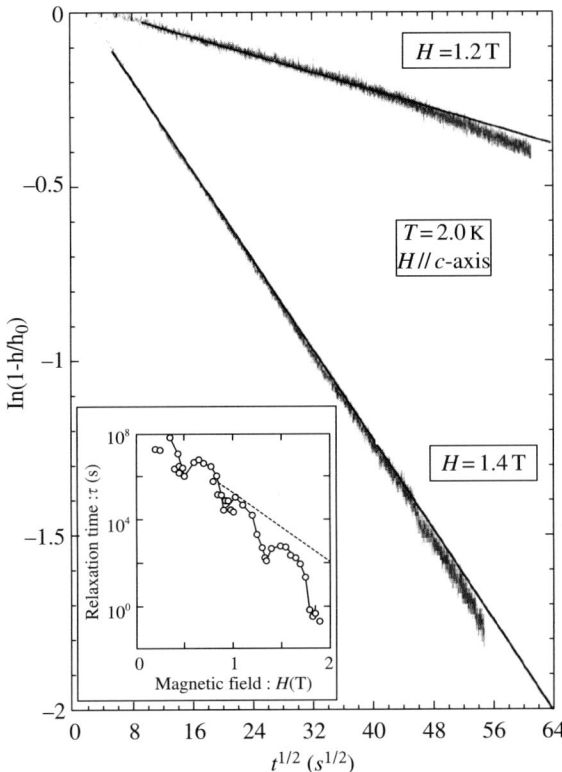

FIG. 3.33. ^{55}Mn spin echo amplitude of Mn^{4+} ions in Mn_{12}ac. The inset shows the relaxation time τ versus the applied magnetic field. Reprinted with permission from Kubo *et al.* (2002). Copyright (2002) American Physical Society.

3.3.3 *Muon spin resonance (μSR)*

Under the heading μSR several techniques are collected: μ stands for muon, S for spin, and R can be rotation, resonance, relaxation (Blundell 2001; Blundell and Pratt 2004). A muon, which has spin $S = \frac{1}{2}$, is a particle of mass and gyromagnetic ratio intermediate between that of the electron and the proton (Table 3.3).

It may be either positively or negatively charged, but the one which is mostly used for experiments on magnets is μ^+. The μ^+ are obtained by collision of a high-energy proton beam with a target which generates pions (Schenck and Gygax 1995). These decay rapidly to muons. In the simplest case (the only one to be considered in this book) the emerging muon beam is completely spin polarized, and this is indeed an important feature for the μSR technique. The muons enter a sample and localize at some particular site, which is generally not known. The implanted muon decays into a positron with emission of a neutrino

TABLE 3.3. Physical constants for proton, muon, electron, and neutron.

Particle	Mass (kg)	Charge	Spin	μ(J/T)	g	$\gamma/2\pi$(MHz/T)
Proton	$1.672 \cdot 10^{-27}$	$+1$	$\frac{1}{2}$	$1.410607 \cdot 10^{-26}$	5.585 695	42.5774813
Muon	$1.883 \cdot 10^{-28}$	± 1	$\frac{1}{2}$	$-4.490448 \cdot 10^{-26}$	$-2.002\ 331$	135.538817
Electron	$9.109 \cdot 10^{-31}$	-1	$\frac{1}{2}$	$-9.284764 \cdot 10^{-24}$	$-2.002\ 319$	28024.9532
Neutron	$1.675 \cdot 10^{-27}$	0	$\frac{1}{2}$	$-0.966236 \cdot 10^{-26}$	$-3.826\ 085$	29.1646958

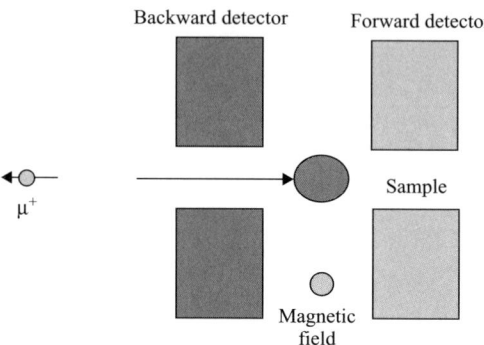

FIG. 3.34. Scheme of a µSR spectrometer showing the two positron detectors and the polarization of the spin of the muon that is antiparallel to the muon beam. The magnetic field is applied perpendicular to this direction.

and an antineutrino according to the equation:

$$\mu^+ \rightarrow e^+ + \nu_e + \bar{\nu}_\mu \tag{3.52}$$

with a mean lifetime of 2.2 µs. The emitted positrons are monitored with two detectors in order to provide information on the spin direction of decaying muons. Accelerator facilities where experiments can be performed are in Switzerland, England, Canada, and Japan. A very simple scheme of an apparatus is shown in Fig. 3.34.

The muons are spin polarized with their spin antiparallel to the beam direction. The positrons are emitted preferentially along the muon spin direction. When the muon is implanted in the sample the presence of a transverse magnetic field causes the precession of the muons and a change of the spin polarization. Short-lived muons have no time to change their spin polarization, therefore the emitted positrons will be captured by the backward detector, while the long-lived muons will invert their polarization giving rise to positrons which are captured by the forward detector. Therefore the information on the spin autocorrelation function $G(t) = A(t)/A_{\max}$, where $A(t)$ is the so-called asymmetry function, is given by

$$A(t) = \frac{N_B(t) - N_F(t)}{N_B(t) + N_F(t)} \tag{3.53}$$

and $N_B(t)$ and $N_F(t)$ are the numbers of positrons detected by the backward and forward detector, respectively.

For a muon site with a local field at an angle θ to the initial muon spin direction at the moment of implantation the spin autocorrelation function is given by

$$G(t) = \cos^2\theta + \sin^2\theta\cos(\gamma_\mu H t). \tag{3.54}$$

In the absence of an external field, if the direction of the local magnetic field H is entirely random while the modulus is the same everywhere then the averaging over all directions would yield

$$G(t) = \frac{1}{3} + \frac{2}{3}\cos(\gamma_\mu H t). \tag{3.55}$$

For a Gaussian distribution of the local field experienced by the muons of width Δ/γ_μ centred around zero, then

$$G(t) = \frac{1}{3} + \frac{2}{3}e^{-\Delta^2 t^2/2}(1 - \Delta^2 t^2). \tag{3.56}$$

This is the so-called Kubo–Toyabe equation (Kubo and Toyabe 1967).

μSR measurements were performed on Mn$_{12}$ac samples (Lascialfari *et al.* 1998), both in zero field and in the presence of applied magnetic fields ranging from 0.025 to 0.37 T. The time dependence of the asymmetry of the muon beam was fitted with a stretched exponential:

$$A(t) = A(0)\exp[-\lambda t]^\beta \tag{3.57}$$

where λ is the equivalent of a longitudinal relaxation time, defined more precisely in (3.59), and β is between 0 and 1. The need of a stretched exponential for fitting the experimental data is taken as an indication that there are several muon sites. The plot of the λ values, obtained from the analysis of the experimental data, as a function of the temperature at various fields is shown in Fig. 3.35.

For the sake of simplicity, the interaction between the muon spin $\mathbf{I}$ and a molecular spin $\mathbf{S}$ will be assumed to be isotropic and described by the parameter a:

$$\mathcal{H}_{int} = a\mathbf{S} \cdot \mathbf{I}. \tag{3.58}$$

Detailed formulae will be given in the case of an exponential relaxation ($\beta = 1$) resulting from the interaction of the muon with a single molecular spin S. Then, the relaxation rate of the muon spin can be expressed (White 1983; Lancaster *et al.* 2004) as

$$\lambda = \frac{a^2}{2\hbar^2}\int_{-\infty}^{\infty}\langle S^-(t)S^+(0)\rangle\exp(i\omega t)dt \tag{3.59}$$

where ω is the Larmor frequency of the muon. It vanishes if the magnetic field is zero. Formula (3.59) is well known in nuclear magnetic resonance, and gives the contribution of a paramagnetic impurity to the spin–lattice relaxation rate $1/T_1$

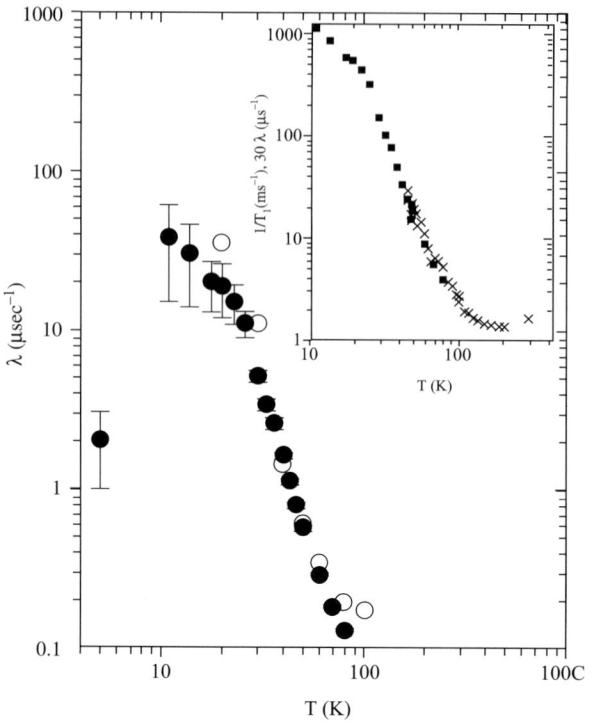

FIG. 3.35. Temperature dependence of the asymmetry of the muon beam in Mn$_{12}$ac. In the inset is shown the corresponding temperature dependence of the proton relaxation rate. Reprinted with permission from Lascialfari *et al.* (1998). Copyright (1998) American Physical Society.

(White 1983). It is valid when the electronic spin S relaxes much faster than the muonic or nuclear spin I.

If the external field and the muon spin are parallel to the easy anisotropy axis z,

$$\langle S^-(t)S^+(0)\rangle = \sum_{m,m'} \frac{1}{Z} \exp \frac{-E_m}{k_B T} \langle m|S^-(t)|m'\rangle\langle m'|S^+(0)|m\rangle \qquad (3.60)$$

where $Z = \sum_m \exp[-E_m/(k_B T)]$. In the absence of damping,

$$\langle m|S^-(t)|m'\rangle = \delta_{m,m'-1}\langle m|S^-|m'\rangle \exp \frac{i(E_m - E_{m'})t}{\hbar}$$

has an oscillating behaviour. In practice the oscillations are damped. A simple assumption is

$$\langle m|S^-(t)|m+1\rangle = \langle m|S^-|m+1\rangle \exp \frac{i(E_m - E_{m+1})t}{\hbar} \exp \frac{-t}{\tau_m^+}.$$

Then according to Salman (2002) and Lancaster *et al.* (2004)

$$\langle S^-(t)S^+(0)\rangle = \frac{1}{Z}\sum_m |\langle m+1|\, S^+\,|m\rangle|^2 \exp\frac{-E_m}{k_B T} \exp\frac{i(E_m - E_{m+1})t}{\hbar} \exp\frac{-t}{\tau_m^+}.$$

(3.61)

In the simplest case, $E_m = -|D|m^2 + g\mu_B H_z$. If the field is not parallel to the easy axis z, formula (3.61) should be replaced by a more complicated one derived by Lancaster *et al.* (2004). The expression for τ_m^+ as a function of the spin–phonon interaction will be given in Chapter 11.

3.4 Neutron techniques

Neutrons are very powerful probes for obtaining information on crystal structures, on the magnetization density and on the energies of the low-lying levels in condensed matter studies (Lovesey 1986; Williams 1988). For molecular nanomagnets the most appealing applications are the inelastic scattering, INS, which provides information on the low-lying spin levels that is largely complementary to that obtained by EPR as outlined above, and polarized neutron diffraction, PND, which provides information on the magnetization density in the clusters (Schweizer and Ressouche 2001; Gillon 2001; Basler *et al.* 2003). The latter therefore provides information which is analogous to that obtained through NMR, as shown in Section 3.3.2. However, magnetization density is available to NMR only on magnetic nuclei, while the use of neutrons allows the reconstruction of the complete magnetization map in all the points of space.

3.4.1 *Polarized neutron diffraction*

The neutrons interact in matter with nuclei, but also with magnetic moments, because neutrons have a spin. The key idea for using polarized neutron diffraction (PND) is to polarize all molecular magnetic moments in the same direction by a magnetic field, and to take advantage of the different cross-section for spin polarized neutron beams of different polarization. A quantity which is measured in a typical PND experiment is the ratio of the intensities I^+ and I^- of the scattered beams when the neutron spin is respectively parallel and opposite to the magnetization, i.e. to the magnetic field (Blume 1963; Tasset 2001). The quantity $R = I^+/I^-$ is called the *flipping ratio*. In the case of centric structure, it is given by:

$$R(\mathbf{K}) = \frac{1 + 2P\gamma\sin^2\alpha + \gamma^2\sin^2\alpha}{1 - 2P\gamma\sin^2\alpha + \gamma^2\sin^2\alpha}$$

(3.62)

where α is the angle between the field axis and the scattering vector $\mathbf{K}$, and

$$\gamma = \frac{F_{Mz}(\mathbf{K})}{F_N(\mathbf{K})}.$$

(3.63)

$\mathbf{K}$ is the vector difference between the propagation vectors of the incident and scattered beams. F_N is the nuclear structure factor

$$F_N(K) = \sum_j b_j e^{iK.r_j} e^{-W_j} \tag{3.64}$$

where the sum is over all the atoms of the unit cell, b_j is the nuclear scattering factor for the element j, $\mathbf{r}_j$ gives the coordinates in the unit cell, and w_j is the thermal Debye–Waller factor. F_M is the magnetic structure factor:

$$F_M(K) = \int_{\text{cell}} M(r) e^{iK.r} d^3 r. \tag{3.65}$$

In the case of a paramagnetic sample, where the value of γ is much smaller than 1, equation (3.62) can be approximated as $R \cong 1 + 4(P \sin^2 \alpha)\gamma$. The intensity scales linearly with γ in the polarized beam, while for the unpolarized case it scales as γ^2. Therefore magnetic moments as small as 10^{-3} μ_B can easily be detected.

The first step for a PND experiment is the measurement of the crystal structure in the absence of an applied magnetic field. Then a given number of additional reflections are measured in the presence of the field and with polarized neutrons. The interpretation of the data is far from being simple, and a short discussion is needed for critically understanding the reported 'experimental' results.

There are several different methods for obtaining the direct space magnetization starting from the reciprocal lattice data obtained from experiment. They can be reduced to two approaches, namely the direct (model-free) methods and the parameterized models. Among the latter the most common approaches are the so-called wavefunction approach and the multipole expansion. In the wavefunction approach the spin density is calculated starting from model wavefunctions, using Hartree–Fock type magnetic wavefunctions constructed as linear combinations of atomic orbitals. The parameters are the spin populations at each magnetic site, the coefficients of the linear combinations and the radial exponents of the Slater orbitals.

The multipole expansion provides more flexibility. It uses an expansion of the spin density using spherical harmonic functions centred at the atomic positions for the angular part and Slater-type functions for the radial part:

$$\mathbf{M}(\mathbf{r}) = \sum_{\text{atoms}} \sum_l R_l(\mathbf{r}) \sum_{m=-l}^{+l} P_{lm} Y_{lm}(\theta, \varphi) \tag{3.66}$$

where $R_l(\mathbf{r})$ is a radial function, P_{lm} is the statistical weight of the spherical harmonic, $Y_{lm}(\theta, \phi)$ is a spherical harmonic. The parameters are the radial exponents of the Slater functions and the populations of the spherical harmonics. It is apparent that these models depend dramatically on the starting model, given also the small excess of experimental data points compared with the parameters which are needed for the fit of the experimental data.

In order to avoid this inconvenience, in the last few years a model-free approach has been introduced, the so-called maximum entropy approach (Gull and Daniell 1978; Papoular and Gillon 1990). This is a general approach for the treatment of experimental data. Given the limited number of data points, and the error bars on them, there are infinite numbers of maps of spin density which approximately satisfy them. The maximum entropy approach defines a probability for each map, as the product of the likelihood, which is the agreement with the observed data, and the prior, which represents the intrinsic probability of the map and is related to its Boltzmann entropy. The best map is identified as the one which gives the maximum Boltzmann entropy and agrees with the experimental data.

PND data have been used for several molecular magnetic materials, and in the case of clusters they have provided useful information on the mechanism of exchange interaction which is difficult to obtain otherwise. Perhaps the most exciting result has been obtained from the analysis of Fe_8 (Pontillon *et al.* 1999). The magnetic data of this compound provide clear evidence of a $S = 10$ ground state. Since the cluster comprises eight iron(III) ions, each with $S = \frac{5}{2}$ (Wieghardt *et al.* 1984), it is a simple matter to suggest that the $S = 10$ ground state originated from having six individual spins parallel to each other ($S = 15$), and two antiparallel ($S = 5$). However it is by no means trivial to understand which are the spin-up and the spin-down ions. The analysis of the temperature dependence of the magnetic susceptibility suggested that the spin-down ions are

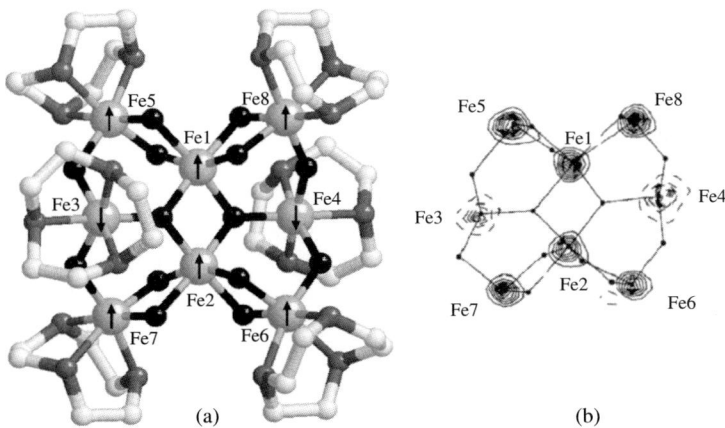

FIG. 3.36. (a) Structure of the Fe_8 cluster with the spin structure of the ground $S = 10$ state schematized by the arrows. (b) Spin density map obtained by polarized neutron diffraction experiments. Spin density contours are drawn at $0.7 \, \mu_B/\text{Å}^2$. Negative spin density is represented by dashed lines. Spin density map reprinted with permission from Pontillon *et al.* (1999). Copyright (1999) American Chemical Society.

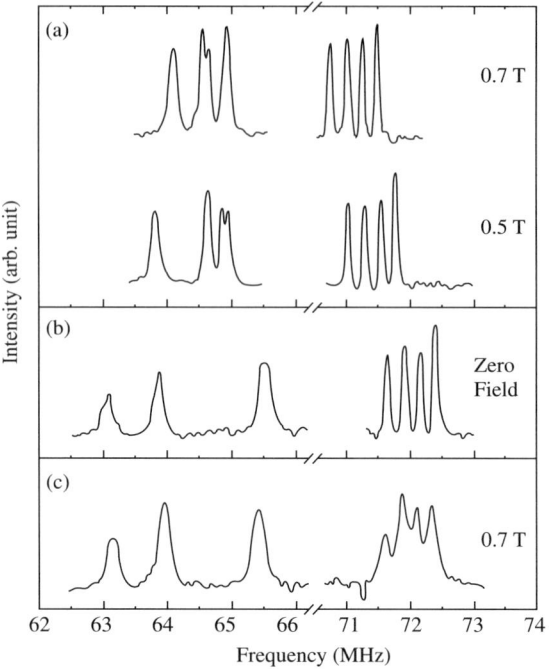

FIG. 3.37. Oriented-crystals ^{57}Fe NMR spectra of Fe$_8$ at 1.5K in zero field and
with the field applied along the easy axis (a) and perpendicular to it (c).
Kindly provided by F. Borsa. More information is available in Furukawa
et al. (2003).

the ones labelled as Fe3 and Fe4 in Fig. 3.36. The PND data were collected at
$T = 2$ K using a single crystal of $2.9 \times 1.8 \times 1.0$ mm size in magnetic field of 4.6
T. At the qualitative level the answer provided by the PND experiment is very
simple: the spins of the ions Fe1, Fe2, Fe5, Fe6, Fe7, Fe8 are up and the spins
of the Fe3 and Fe4 ions are down, thus confirming the results obtained by the
analysis of the magnetic data.

It is interesting to compare the PND data with those obtained by ^{57}Fe NMR
(Furukawa *et al.* 2003) shown in Fig. 3.37. The spectra are recorded on oriented
crystals of ^{57}Fe enriched Fe$_8$. Eight signals are observed in agreement with the
lack of symmetry elements in the Fe$_8$ molecule. The field dependence of the
signals clearly shows that two spins are antiparallel and six are parallel to
the applied field.

Another useful application of PND to magnetic clusters has been obtained
(Caneschi *et al.* 1997) with [Et$_3$NH]$_2$ [Mn(CH$_3$CN)$_4$(H$_2$O)$_2$][Mn$_{10}$O$_4$(biphen)$_4$
Br$_{12}$], Mn$_{10}$ (Goldberg *et al.* 1995). The compound comprises isolated
manganese(II) ions, and a cluster with ten metal ions, whose structure is sketched
in Fig. 3.38. The cluster has tetragonal symmetry, and can be described as a
central octahedron of manganese ions, with four faces capped by an additional
metal ion. The ten metal ions have a total positive charge corresponding to

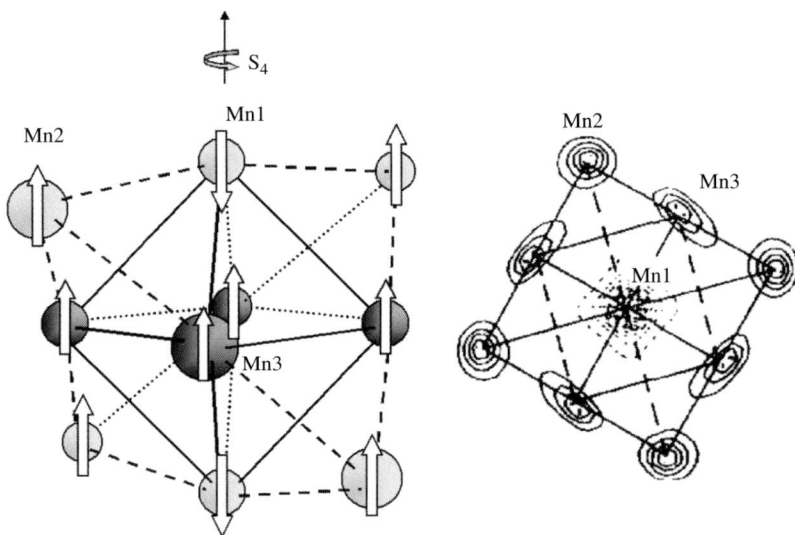

FIG. 3.38. Left: sketch of the structure of the core of Mn_{10}. The Mn^{3+} ions
are drawn as darker spheres and the arrows represent the spin structure.
Right: spin density map of Mn_{10} projected on the plane perpendicular to the
tetragonal axis. Broken lines are used to represent negative spin densities.
Redrawn from Caneschi *et al.* (1997). Copyright (1997) Elsevier.

24 electrons. Four manganese(III) ions, labelled as Mn3, define the equatorial
plane of an octahedron while and two manganese(II) ions, Mn1, define the axial
vertices of the octahedron. Four faces of the octahedron are capped by man-
ganese(II) ions, labelled as Mn2. The assignment of the oxidation states are
made on the basis of charge compensation and bond distances. The manganese
ions are connected by oxide and bromide bridges. The maximum spin value for
the ground state is $S = 23$, for parallel orientation of all the spins. In the case
of antiferromagnetic interactions many different ground states can be observed.

The temperature dependence of the magnetic susceptibility provides clear
evidence for the presence of antiferromagnetic interaction, but the nature of the
ground state could not be unambiguously determined, given also the presence
of an additional manganese(II) ion in the lattice. It was, however, suggested
that in the ground state the spin is quite high, $S \geq 12$. PND data provided
a pictorial view of the nature of the ground state, as shown in Fig. 3.38. In
fact the magnetization density, shown in the plane orthogonal to the tetragonal
axis, is positive in all the manganese ions, except the two manganese(II) on the
vertices of the octahedron. A simple account, considering only spin-up, spin-
down configurations suggest $S = 13$. Interestingly the spin density around the
Mn^{3+} sites appears much more distorted from spherical symmetry, in agreement
with the electronic configuration of a d^4 ion exhibiting Jahn–Teller distortion.

3.4.2 *Inelastic neutron scattering*

Inelastic neutron scattering, INS, is a very powerful tool for examining the low-lying energy levels of magnetic clusters. At the basis of this spectroscopy is the fact that neutrons have $S = \frac{1}{2}$ and they can induce transitions between spin states $|f\rangle$ and $|i\rangle$ according to a magnetic scattering function $I(Q, \hbar\omega)$ defined as:

$$I(Q, \hbar\omega) \propto F^2(Q) \sum_{i,f} p_i \left| \langle f | S_\perp | i \rangle \right|^2 P(\hbar\omega - \Delta_{fi}, \Gamma_{fi}) \qquad (3.67)$$

where $F(Q)$ is the magnetic scattering factor of the metal ion, p_n is the Boltzmann population of the n level, $p_n = \exp(-E_n/kT)/Z$ and $S_\perp$ is the spin component perpendicular to the neutron scattering vector $\mathbf{Q}$. $P(\hbar\omega - \Delta_{fi}, \Gamma_{fi})$ describes the lineshape for a peak of full width at half maximum Γ_{fi} and energy transfer centred at Δ_{fi}. The selection rules for the allowed transitions can be resumed as $\Delta S = 0, \pm 1, \Delta m = \pm 1$.

The first case of successful use of INS for the analysis of the low-lying levels of a cluster was provided by Fe$_8$ (Caciuffo *et al.* 1998), where they confirmed the splitting of the ground $S = 10$ state obtained through HF-EPR analysis. The spectra at various temperatures are shown in Fig. 3.39.

The assignment of the transitions is easily made on the basis of band position and temperature dependence of the intensity. Compared to HF-EPR at any rate INS has an additional advantage, that of providing not only the energies of the states between which transitions are observed, but also the nature of the wavefunctions. These provide extremely important information which can put on a firm basis the speculations on the mechanisms of magnetic relaxation.

A complete analysis of the INS data can be performed using a method developed by Borras-Almenar *et al.* (1999). This was first applied to a tetra-nuclear nickel(II) cluster $[Ni_4(H_2O)_2(PW_9O_{34})_2]^{10-}$, which has the structure sketched in Fig. 3.40 (Clemente-Juan *et al.* 1999).

The temperature dependence of the magnetic susceptibility and of the magnetization provided clear evidence that the ground state has $S = 4$, with a zero-field splitting leaving the $M = 0$ component lowest. The INS spectra were analysed using the spin Hamiltonian:

$$\mathcal{H} = -J(\mathbf{S}_1 \cdot \mathbf{S}_3 + \mathbf{S}_1 \cdot \mathbf{S}_4 + \mathbf{S}_2 \cdot \mathbf{S}_3 + \mathbf{S}_2 \cdot \mathbf{S}_4)$$
$$- J'(\mathbf{S}_1 \cdot \mathbf{S}_2) + D(\mathbf{S}_{z1}^2 + \mathbf{S}_{z2}^2) + D'(\mathbf{S}_{z3}^2 + \mathbf{S}_{z4}^2). \qquad (3.68)$$

The best fit of the INS data, which provide the scheme of energy levels sketched in Figure 3.41, requires $J/k_B = 19.3$ K; $J'/k_B = 9.1$ K; $D/k_B = 5.4$ K; $D'/k_B = 7.0$ K. The comparison between the observed and calculated data, and the nature of the wavefunctions, are shown in Table 3.4. The wavefunctions are labelled as $:|(S_{12})(S_{123})SM\rangle$, with M an eigenvalue of S_z.

The observed separations between the $|S \pm M\rangle$ levels of the ground $S = 4$ state follow a 1:2.9:6.64:11.6 pattern to be compared to the expected 1:3:5:7 for a

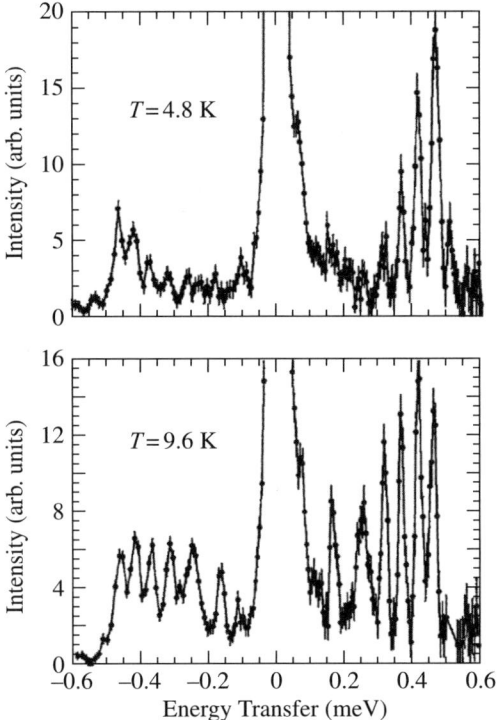

FIG. 3.39. Inelastic neutron scattering spectra of Fe_8 at 4.8 K (upper) and 9.6 K (lower). Adapted from Caciuffo *et al.* (1998). Copyright (1998) American Physical Society.

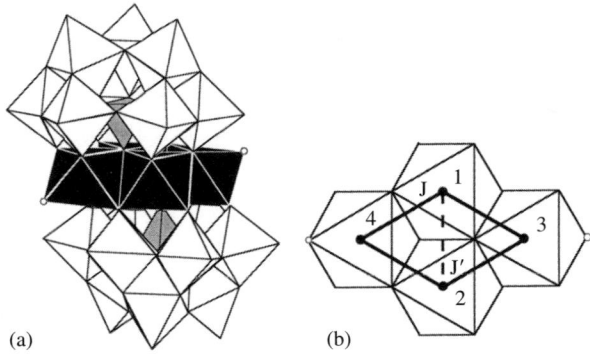

FIG. 3.40. Sketch of the structure of $[Ni_4(H_2O)_2(PW_9O_{34})_2]^{10-}$ (a) and exchange pathways (b). Reprinted with permission from Clemente-Juan *et al.* (1999). Copyright (1999) American Chemical Society.

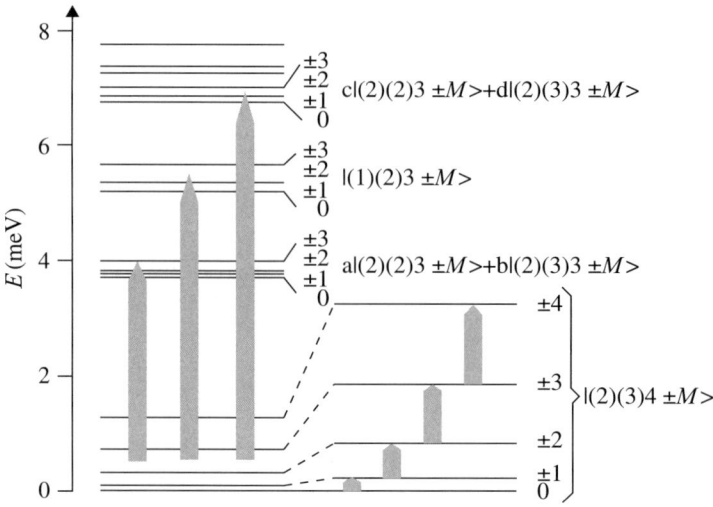

FIG. 3.41. Calculated energy level diagram derived from the fit of the INS data. Reprinted with permission from Clemente-Juan *et al.* (1999). Copyright (1999) American Chemical Society.

TABLE 3.4. Experimental and calculated energy levels for Ni_4 from the analysis of the INS data.

Experimental energy (meV)	Calculated energy (meV)	Main contributions to the functions $\|(S_{12})(S_{123})SM\rangle$
0	0	$0.997\|(2)(3)40\rangle$
0.11(1)	0.09	$0.997\|(2)(3)4 \pm 1\rangle$
0.32(2)	0.33	$0.999\|(2)(3)4 \pm 2\rangle$
0.73(2)	0.73	$1.000\|(2)(3)4 \pm 3\rangle$
1.28(2)	1.27	$1.000\|(2)(3)4 \pm 4\rangle$
Band centred at 3.6	3.75	$-0.576\|(2)(2)3\ 0\rangle + 0.815\|(2)(3)3\ 0\rangle$
	3.76	$\mp0.576\|(2)(2)3 \pm 1\rangle \pm 0.813\|(2)(3)3 \pm 1\rangle$
	3.82	$-0.575\|(2)(2)3 \pm 2\rangle + 0.813\|(2)(3)3 \pm 2\rangle$
	3.98	$\mp0.577\|(2)(2)3 \pm 3\rangle \pm 0.816\|(2)(3)3 \pm 3\rangle$
Band centred at 5.1	5.17	$\|(1)(2)3\ 0\rangle$
	5.20	$\pm0.990\|(1)(2)3 \pm 1\rangle$
	5.37	$0.996\|(1)(2)3 \pm 2\rangle$
	5.66	$\|(1)(2)3 \pm 3\rangle$
Band centred at 7.0	From 6.76 to 9.06	Linear combinations of $\|(2)(2)3 \pm M\rangle$, $\|(2)(3)3 \pm M\rangle, \|(2)(2)2 \pm M\rangle, \|(2)(3)2 \pm M\rangle$ and $\|(2)(1)2 + M\rangle$

pure $S = 4$ multiplet. Therefore a fitting of the $|S = 4 \pm M\rangle$ levels with only one parameter D_{cluster} cannot be made, and the Hamiltonian (3.67) appears to be more appropriate. The role of the admixtures of total spin levels with different S cannot be neglected. If we use the spin projection techniques of the individual spin on the total spin S we expect for the ground $S = 4$ state:

$$D_{\text{cluster}} = \frac{(D + D')}{14}. \qquad (3.69)$$

Using the best-fit D and D' values $D_{\text{cluster}} = 0.076$ meV is calculated which is significantly smaller than the energy difference between the $|(2)(3)4\ 0\rangle$ and $|(2)(3)4 \pm 1\rangle$ levels. The origin of the difference is due to the fact that the zero field splittings of the individual nickel(II) ions are not negligible compared to the isotropic coupling constants J and J'.

Another interesting example of the use of INS to get precious information on the magnetic exchange in molecular nanomagnets is a detailed study of an equatorial triangle of antiferromagnetic spins $S = \frac{1}{2}$ in the polyoxocluster known as V_{15} (Chaboussant *et al.* 2004). The magnetic properties of this cluster will be discussed in more detail in Section 14.3.2.

4

SINGLE-MOLECULE MAGNETS

In the physical literature on SMMs the molecules that are the object of the investigation are often over-schematized. Classes of molecules are nicknamed principally according to their metal nuclearity. More recently it has, however, emerged that what is around the magnetic centres plays a major role in the dynamics of the magnetization of these objects, especially in the quantum tunnelling regime. These aspects will be addressed in the following chapters. The aim of this chapter is rather to provide the reader with some basic tools to understand what is behind a long and unpronounceable formula.

The molecular aggregate object of this book represents a mesoscopic phase between the isolated paramagnetic metal ion, for instance the aquo-ion in acid aqueous solution, and the extended oxide or hydroxide lattices that precipitate at higher pH. This intermediate nuclearity in fact results from a complex balance of different interactions, and many parameters need to be controlled during the synthetic process. The many different groups involved in the synthesis of SMMs, or more generally of molecular clusters, have developed their own strategies. Some of these have led to aesthetically remarkable objects, some others have provided simple and predictable results that have, however, significantly improved our understanding of the dynamics of the magnetization in these materials.

When looking at a complex structure where tens of metal ions are wonderfully arranged to form rings, doughnuts, spheres, Archimedean solids, and in general a large variety of polyhedra (Alvarez 2005), a question comes to the mind: has that particular structure been predicted, or is it the outcome of a synthetic process that is mainly out of our control? We will try, as far as possible, to shed some light on this delicate point.

4.1 Serendipity versus rational design of SMMs

It is undeniable that chemists have the ambition to reach full control over the outcome of the synthetic process they are carrying on. This achievement can only occur through a long, and not always easy, process of rationalization of the factors that determine the molecular structure. Such a goal has certainly been achieved in some special, fortunate, cases, mainly where the polynuclear structure of the cluster results from the self-assembly of stable or, more precisely, inert building blocks. A well-known example, that we will discuss in detail in the next paragraph, is the use of polycyanometalate. For instance, the $Cr(CN)_6^{3-}$ building block has six dangling CN^- groups that can bind six metal ions to form a cluster which has the shape of an octahedron with the chromium in the centre. The outcome is in this case quite predictable and the number of metal atoms, the nuclearity, of these clusters can be controlled. Is this the winning strategy? Some scientists are not of this opinion (Winpenny 2002). The rational design approach has a major drawback: the outcome is limited by the scientist's imagination. The solutions nature can provide are instead much more numerous.

Instead on relying only on rationality, chemists have generally also some confidence in their good luck. This is called 'serendipity'. It has been claimed that the serendipity approach is a promising strategy for the discovery of compounds of pharmacological relevance. The casual discovery of the cytotoxicity of the complex $Pt(NH_3)_2Cl_2$, also known as *cis*-platin, during the investigation of the effect of an electric field on *Escherichia coli* bacteria, is a well-known example of serendipitous success (Rosemberg *et al.* 1969). After some decades of very active research in anticancer drugs, thousands of people have still to cope with the heavy side-effects of this powerful drug.

In a much less dramatic scenario, something similar has happened in the field of SMMs. $Mn_{12}ac$, obtained as an unexpected product of the reaction of Mn^{2+} and MnO_4^- in the presence of acetic acid (Lis 1980), still retains the record of the highest blocking temperature despite ten years of strenuous synthetic efforts towards SMMs with enhanced properties.

The serendipitous approach, even when successful, might appear as rather frustrating, minimizing the active contribution from scientists working in the field. It will be clearer in the following that, on the contrary, the ingenuity of chemists has played a major role, forcing nature to go towards unprecedented structures. Very successful examples are the synthesis and structure characterization of clusters that approach the size of small proteins (Müller *et al.* 2002; Tasiopoulos *et al.* 2004), or the formation of ring structures with an odd number of members thanks to the templating effect of cations of the appropriate size (Cador *et al.* 2004).

4.2 Synthetic strategies to SMMs

In order to design magnetic clusters it is necessary to have available both connecting blocks, which provide efficient bridges and determine the growth of the cluster, and terminal blocks, which stop the growth of the cluster at a finite size. The bridging blocks must not only provide the right connection between the metal ions but also provide efficient exchange pathways thus assuring strong magnetic coupling. The sign and intensity of the exchange interaction, parameters of paramount importance, depend dramatically on geometrical factors as already mentioned in Chapter 2.

In the design and synthesis of molecular clusters, indeed polynuclear coordination compounds, the choice of the appropriate ligand is probably the most important step. The ligand (from the latin word *ligare*, to bind) is any molecular moiety that has at least one donor atom, i.e. an atom with a non-bonding electron pair, like oxygen and nitrogen atoms in the molecules of Fig. 4.1.

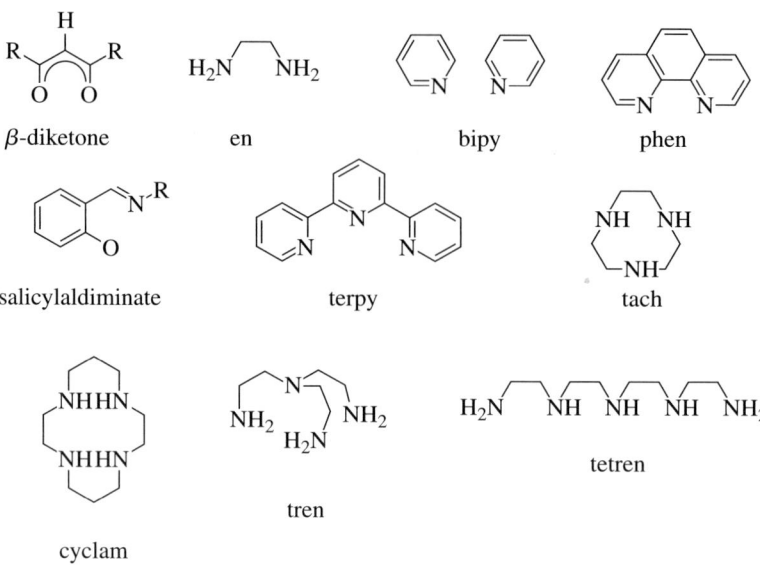

β-diketone   en   bipy   phen

salicylaldiminate   terpy   tach

cyclam

tren

tetren

FIG. 4.1. Schematic structure of some commonly employed chelating ligands and
    their abbreviated names.

A ligand is called mono-, bi-, tri-, etc., dentate, if it possess one, two, etc.,
donor atoms. The use of different types of ligands is often what differentiates the
research of the many groups involved in molecular magnetism. Depending on its
molecular and electronic structure the ligand can have different functions.

Monoatomic ligands, like O^{2-}, S^{2-}, F^-, Cl^-, etc., or ligands with one donor
atom, like OH^- or alkoxides, the anions of alcohols, can be coordinated to one
or more than one metal atom, and in the last case they act as bridges. They
are particularly efficient in transmitting the magnetic interactions, allowing a
significant overlap of the magnetic orbitals of the metal centres. Due to the small
steric hindrance they are suited to bridge more than two metal ions. The O^{2-}
ion can bridge up to six metal ions (Cornia et al. 1994), but two and three metal
ions are the most common cases, as is the case for OH^-- or OR^-, where R is a
generic organic group.

When these types of ligands are the only bridges the polynuclear structure tends to resemble that of a pure inorganic extended lattice. A remarkable
example is the cluster $[Fe_{19}(metheidi)_{10}(OH)_{14}(O)_6(H_2O)_{12}]^+$, shown in Fig. 4.2
(Goodwin et al. 2000). The OH^- and O^{2-} ions are the bridging building blocks,
while H_2O and the polydentate ligand metheidi, discussed later in more detail,
are the terminal ones. We focus on the central part of Fig. 4.2 neglecting for the
moment the metheidi ligand. The central iron(III) is connected to six surrounding iron(III) through six OH^- bridges, each OH^- bridging three iron(III) ions
in a layer structure that resembles that of $Mg(OH)_2$ in the Brucite mineral. The

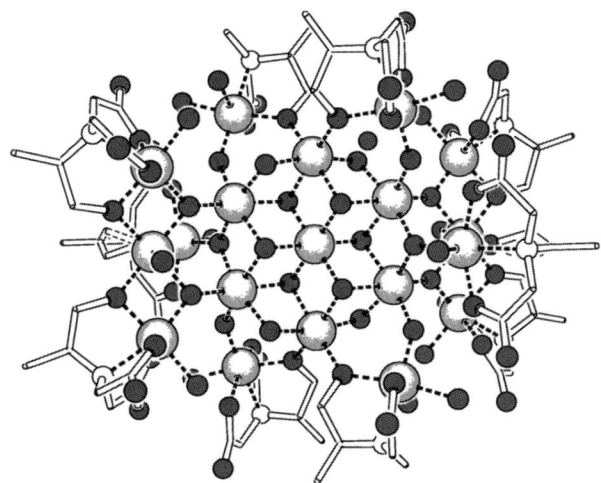

FIG. 4.2. View of the cluster $[Fe_{19}(\text{metheidi})_{10}(OH)_{14}(O)_6(H_2O)_{12}]^+$. The iron atoms are drawn as big spheres, nitrogen as white spheres, oxygen as dark grey spheres. The coordination bonds to the metal ions are represented as broken lines. Redrawn from Goodwin *et al.* (2000) by permission of the Royal Society of Chemistry.

Brucite structure is, however, confined to the central part of the cluster because, at the periphery, the organic ligand metheidi forms a sort of shell that encapsulates the inorganic core. The heptanuclear core with the Brucite-like structure is quite common in polynuclear clusters and is encountered, for instance, in an interesting mixed valence $[Fe_6^{II}Fe^{III}]$ characterized by $S = 29/2$ in the ground state (Oshio *et al.* 2003).

Commonly employed however are polydentate ligands that possess more than one donor atom. If the different donor atoms coordinate the same metal ion it is called a 'chelating' ligand, from the Greek word $\chi\eta\lambda\acute{\eta}$ for claw. Given the directionality of the coordinating bonds there are some structural conditions to fulfil in order to observe chelation. In octahedral complexes it is a common rule to consider that a polydentate ligand is almost exclusively acting as a chelating one if the metal ion with the two donor atoms and the other atoms that connect the donor ones form a five- or a six-member ring. If this is the case the complex formed by the chelating ligand is more stable than the analogue where the two donor atoms belong to different ligand molecules. The reason is quite intuitive. When the complex is formed its dissociation is less probable when the ligand is chelating. In fact this process requires that both claws are detached at the same time.

Looking at the thermodynamics of the process one can consider the two chemical reactions below. They both involve as starting material the hexa-aquo-ion, but the first one concerns the formation of a complex with six ammonia molecules as ligands while in the second two molecules of a generic L ligand containing three nitrogen atoms belonging to amine groups replace the coordinated water molecules:

$$[M(H_2O)_6]^{n+} + 6NH_3 \leftrightarrows [M(NH_3)_6]^{n+} + 6H_2O$$

$$[M(H_2O)_6]^{n+} + 2L \leftrightarrows [M(L)_2]^{n+} + 6H_2O. \tag{4.1}$$

The equilibrium is significantly more shifted towards the products, the right-hand side, in the second reaction. The reason is not a different enthalpic contribution to the free energy. In fact the number and type of chemical bonds that are formed and destroyed are the same in the two reactions. The entropic contribution to the free energy is, on the contrary, very different in the two cases. The formation of the chelate complex is in fact strongly favoured by the increase of the number of uncoordinated, and therefore more disordered, molecules when the six coordinated water molecules are released and replaced by only two molecules of a polydentate ligand. No significant gain in entropy is expected in the first reaction.

In Fig. 4.1 we have reported some commonly employed chelating ligands based on oxygen and nitrogen donor atoms. As chelating ligands have a tendency to coordinate one metal ion rather than to bridge more metal ions, they are used any time it is necessary to block two or more coordination sites around the metal, for instance to hamper the growth of an extended structure.

The library of ligands is practically unlimited, but looking at the few examples of Fig. 4.1 one can realize that each ligand is particularly well suited to a given function. For instance the ligand triazacyclononane, commonly abbreviated as tacn, has a rather rigid structure imposed by the closed ring that can only fit on a face of the coordination octahedron. The metal coordinated to tacn is expected to occupy a peripheral site of a cluster, because all the dangling bonds not occupied by tacn are pointing in the same hemisphere. On the contrary cyclam is better suited to occupy the four equatorial sites of an octahedron leaving the two *trans* coordination sites available for further connections that often develop in a linear extended structure (Mossin *et al.* 2004).

In some other cases the chelating mode is not the only possible one. The most common example is provided by the anion of carboxylic acids, $RCOO^-$. The negative charge is equally shared on the two oxygen atoms that can either be coordinated to the same metal ions or to more than one. In Fig. 4.3 some of the many possible coordination modes of carboxylates are represented.

In order to indicate the coordination mode two notations are commonly used. The first one uses the Greek letter μ to indicate the bridging mode with subscripts. These numbers indicate which atoms of the ligand are involved in the bridge. If the same oxygen atom is involved in the bridge, the bridging mode is indicated as $\mu_{1,1}$, while it is called $\mu_{1,3}$ if one oxygen is coordinated to one metal ion and the second oxygen to the other metal. The situation becomes more

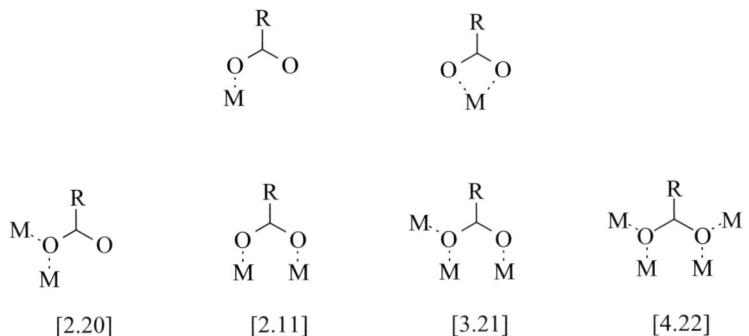

[2.20]   [2.11]   [3.21]   [4.22]

FIG. 4.3. Coordination and bridging modes of the carboxylate ligand in the Harris notation.

complex if the same polydentate ligand bridges more metal centres. A different notation, shown in Fig 4.3, has been developed by Harris (Coxall *et al.* 2000; Winpenny 2004). This notation uses a code [X.Y, Y′, Y″...] where X is the overall number of metal atoms bound by the whole ligand, and each value of Y refers to the number of metal ions attached to the different donor atoms. The ordering of Y is listed by the Cahn–Ingold–Prelog priority rules mainly based on decreasing atomic number, hence O before N (Stoll *et al.* 2004).

A well-known example of bridging carboxylate in the [2.11] mode is the dimeric form of copper(II) carboxylates. Bleaney and Bowers (1952) from EPR experiments deduced that the structure of copper(II) acetate consisted of dimeric units and for the temperature dependence of the magnetic susceptibility they derived the well-known homonym formula reported in (3.4). The dimeric structure was promptly confirmed by X-ray diffraction analysis (van Niekerk and Schoening 1953).

In most of the systems discussed in this book the carboxylates build the skeleton of the cluster acting as [2.11] bridges, but give rise to only a weak magnetic interaction, often overwhelmed by the one mediated by different bridging groups, like O^{2-}, OH^- or alkoxides, F^-, etc. Nevertheless carboxylates have been widely employed, also because the R group can be easily varied, enabling modification of important properties, like solubility, electron density on the donor atom, steric hindrance, etc.

Other widely used polydentate bridging ligands are shown in Fig. 4.4. It is important to stress that these types of ligands, even if they are not bridging in the $\mu_{1,1}$ mode, are quite efficient in transmitting the magnetic interaction, thanks to the conjugation and delocalization of the π orbitals. Cyanide and azide ions are well known for their capability to mediate a moderately strong magnetic interaction, that can be either ferro- or antiferromagnetic.

Oxalate and its derivative oxamidate (Girerd *et al.* 1980) are two bridging ligands used in molecular magnetism since the very beginning. Both of them

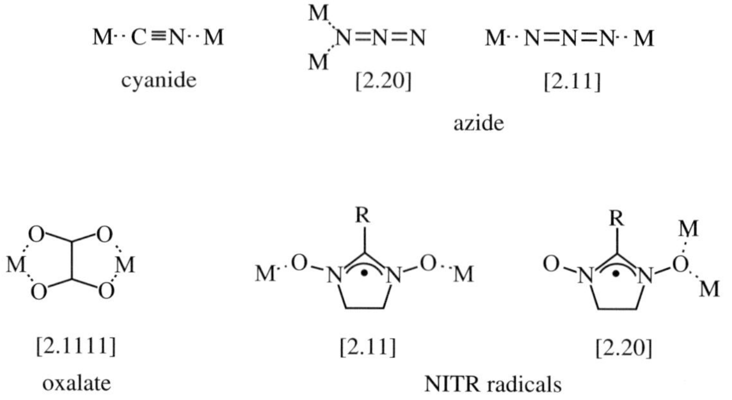

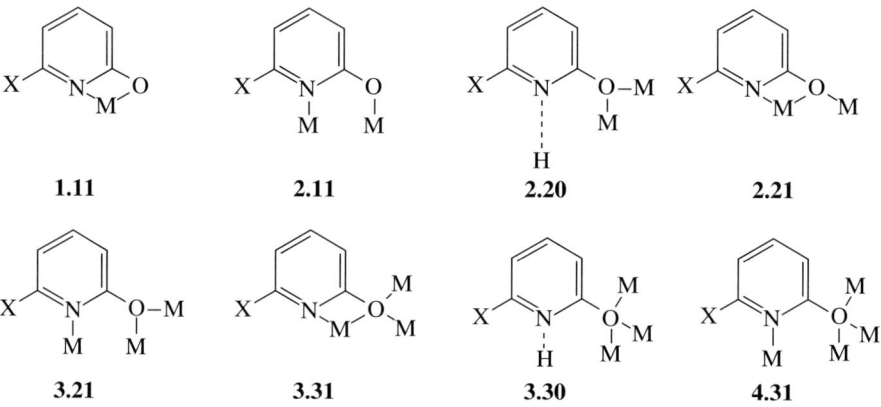

$M \cdots C \equiv N \cdots M$

cyanide

$\begin{matrix} M \\ \ \ \diagdown \\ \ \ \ \ N = N = N \\ \ \ \diagup \\ M \end{matrix}$ [2.20]

$M \cdots N = N = N \cdots M$ [2.11]

azide

[2.1111]

oxalate

[2.11]

[2.20]

NITR radicals

FIG. 4.4. Some examples of widely employed polydentate bridging ligands and their most common bridging modes listed according to the Harris notation.

1.11

2.11

2.20

2.21

3.21

3.31

3.30

4.31

FIG. 4.5. The possible chelating and bridging modes of the pyridonate ligand listed according to Harris notation. From Winpenny (2002) by permission of the Royal Society of Chemistry.

are, however, particularly well suited to connect only two metal ions, and are thus rarely used to obtain clusters, but mainly to obtain chains and extended structures. Very simple ligands, like the pyridonate, that carries one pyridine nitrogen, and one alkoxide oxygen as donor atoms, are able to provide a large variety of coordination and bridging modes, as shown in Fig. 4.5. For their versatility pyridonates have been widely exploited in the serendipitous approach to metal cages (Winpenny 2002).

An important class of polydentate ligands is constituted by polyalcohols, polymines, polycarboxylates or in general organic molecules where the different donor atoms are connected through aliphatic chains, as in the two ligands shown below:

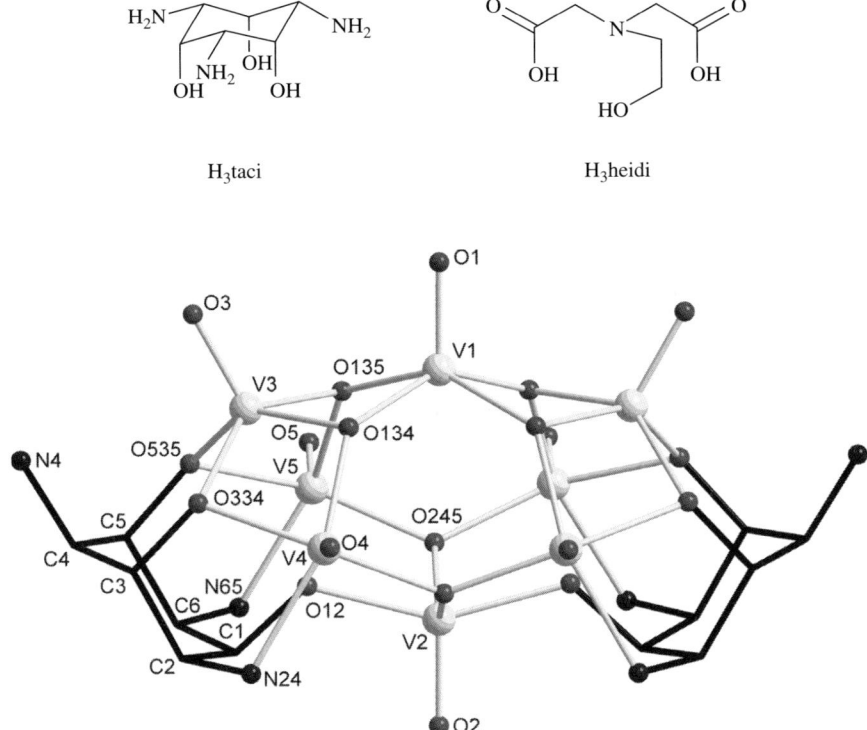

FIG. 4.6. View of the cluster $[V_8O_{14}(Htaci)_2]$, the first example of a V(IV)-oxo cluster with the spin of the ground state larger than 1/2. From Hegetschweiler *et al.* (2004) by permission of Wiley-VCH.

In this case a significant magnetic interaction is only present among the metal centres bridged by the same donor atom, that is in a $\mu_{1,1}$ bridging mode. However the geometrical constraints imposed by the ligand structure to the different bridging atoms play a major role in determining the structure of the cluster. This strategy, which employs structure-directing ligands, is the one that more closely mimics the biological processes of mineralization of clusters and aggregates (Mandel *et al.* 1999).

A nice example of ligand-directed structure is encountered with the ligand 1,3,5-triamino-1,3,5-trideoxy-*cis*-inositol (abbreviated as H$_3$taci and shown above) which contains six coordination sites, and three of them (i.e. the oxygen atoms) can act as bridging atoms. In the octanuclear vanadyl cluster $[V_8O_{14}(Htaci)_2]$, shown in Fig. 4.6, this ligand bridges four metals ($V^{IV} = 0$) and forces the vanadyl groups to assume an unprecedented orthogonal configuration that leads to a strong ferromagnetic interaction of the $S = 1/2$ of vanadium(IV) centres. It is worth noticing that the behaviour of any ligand strongly depends on the metal ions, in particular on its ionic radius. The same taci ligand, used with lanthanides ions, can bridge only three of them, giving rise to compounds with smaller nuclearity.

The heidi family of ligands, to which belongs also the metheidi ligand comprised in the cluster of Fig. 4.2 and schematized above, contains a central amine nitrogen atom, one alcohol, and two carboxylic groups. These last three groups, when deprotonated, are able to act as bridging units and the whole molecule is quite flexible. It is therefore not surprising that large and irregular aggregates can be formed as in the cluster shown in Fig. 4.2.

Rigid ligands can also be employed and in this case more regular and predictable structures are obtained. One of the most fascinating examples is the grid structures obtained with ligands based on a diazine backbone with inserted alkoxide fragments as in the ligand 2POAP shown in Fig. 4.7. In this type of ligand, contrary to H_3taci or heidi, contiguously arranged coordination pockets are present, where the metal can be stably chelated. The metal ions are bound in close proximity and can interact through a bridging atom of the ligand or through other fragments, i.e. the usual O^{2-} or OH^- bridges.

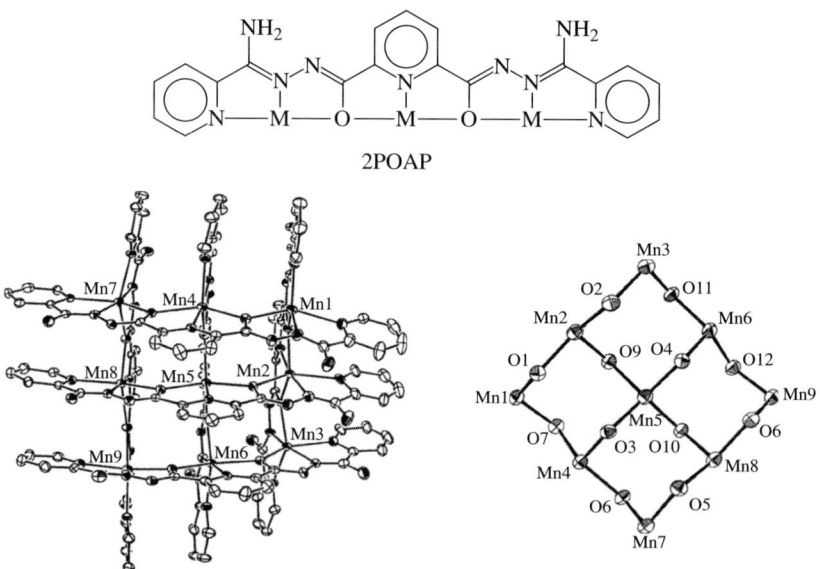

FIG. 4.7. Top: structure of the azine-based ligand 2POAP and its three chelating coordination modes. Bottom: structure of a nonanuclear manganese(II) cluster with a $[3 \times 3]$ grid arrangement of the metal ions. On the right, the metal–oxygen core is highlighted. Adapted from Thompson (2002). Copyright (2002) Elsevier.

The 2POAP ligand can therefore arrange three metal ions in a single array, as shown below, but each metal ion is only triply coordinated. The other three coordination sites can be occupied by another 2POAP ligand. Three linear trimers can thus be connected by orthogonal self-assembly of the ligands, as shown in Fig. 4.7, to form a $[3 \times 3]$ grid. This type of molecule has been recently investigated and a novel quantum effect in the magnetization has been observed and will be discussed in Chapter 14 (Waldmann *et al.* 2004).

Another interesting example of the use of structure-directing ligands has been provided by Pecoraro and co-workers (Dendrinou-Samara *et al.* 2003). They have used $MnCl_2$ in an alkaline solution of methanol with the ligand $pdol^{2-}$ = dipyridylketone-diolate, schematized below,

$$pdol^{2-} =$$

The outcome of this reaction is a very high nuclearity $Mn_4^{II}Mn_{22}^{III}$ cluster, which shows an out-of-phase signal of the ac susceptibility below 2.5 K. The most important feature of this cluster is the fact that the 12 pdol ligands and 10 manganese ions form four strands that are connected in a sort of giant tetrahedron. Each vertex of the tetrahedron corresponds to a Mn^{III} ion while the six edges are formed by pdol-Mn^{II}-pdol chains. These types of cages are also known as 'metallo-cryptands' because they can host and wrap up a secondary structure, thus forming a sort of crypt. Dendrinou-Samara *et al.* (2003) have shown that thanks to the dangling oxygen atom of the 12 $pdol^{2-}$ ligands an other 12 Mn^{III} ions can be hosted, which, in their turn, surround a cubane structure an other 4 Mn^{III} ions connected by oxide and methanolate bridges. The structure of the cluster can be schematized as that of an onion that allows a high density of metal ions, actually one of the major advantages of this synthetic strategy.

Some more complex ligands have been tailored on design to transmit a ferromagnetic exchange interaction when coordinating more than one metal ion. An interesting example is the ligand H_6talen = 2,4,6-tris(1-(2-salicylaldimino-2-methyl-propylimino)-ethyl)-1,2,5-trihydroxybenzene shown below (Glaser *et al.* 2003):

H_6talen =

The deprotonated form, talen^{6-}, can coordinate three metal ions thanks to the chelating pockets formed by two oxygen and two nitrogen atoms. The three pockets are inserted in meta-positions on the benzene ring. According to the spin polarization mechanism discussed in Chapter 2, the even number of bonds separating the pockets on the benzene ring induces spin polarization of the same sign and therefore ferromagnetic coupling.

In this short survey of coordination chemistry we have shown how chemists involved in the synthesis of SMMs have access to a sort of library from which they can select the building blocks that are better suited for the desired function. However, other parameters can be easily tuned to further increase the variability in the products that can be obtained. The first one is the different acidity of the various donor atoms of polydentate ligands. It is therefore a rather common strategy to play on the pH of the solution to selectively protonate or deprotonate some donor atoms, thus controlling their coordinating ability. Also the ligand to metal ratio in the reagents can modify the outcome of a reaction. A nice example has been provided by Saalfrank *et al.* (2001). The change of this ratio from 1.5 to 2 shifts the outcome of the reaction from a tetranuclear iron(III) cluster similar to that of Fig. 3.4 to a hexanuclear iron(III) ring. Moreover the transformation appears to be reversible.

We will survey in the next sections the most frequently employed rational synthetic strategies towards high nuclearity magnetic clusters.

4.3   The use of preformed building blocks

It is quite common, when describing the structure of a large and complex cluster, to split it into building blocks. In most cases these subunits are not present in the reaction environment. They are simply a tool to describe the structure and have no relation to the mechanism of cluster formation. A certain ambiguity about the meaning of the term building block is often encountered in the chemical literature.

4.3.1   *Cyanide-based clusters*

One of the best examples of the use of preformed building blocks that have a tendency to interact and organize in a regular structure, also known as *self-assembly* synthesis, is the preparation of polynuclear clusters based on a central polycyanometalate. Hexacyanochromate(III) and hexacyanocobaltate(III) are the most used among the first series transition metal ions, while octacyanomolybdate and octacyanotungstate have provided nice examples of very large spin molecules containing heavier elements.

The use of the cyanide ligand has several advantages: cyanide is a nonsymmetric bridging molecule, and can therefore bind selectively to two different metal ions. The polycyanometalates form stable building blocks, which in the case of chromium(III) are also inert (Sharpe 1976), in the sense that the complexes are not involved in a dynamic process of dissociation and reassociation.

The linear CN$^-$ bridge is very efficient in mediating the magnetic exchange interactions, and the sign of the interaction is often predictable simply on a symmetry basis (see also Chapter 2). The use of CN$^-$ is well known in solid state chemistry for its ability to form extended 3D structures analogous to the well-known Prussian blue (Verdaguer *et al.* 1999b). When paramagnetic metal ions are involved, long-range magnetic order is observed with critical temperatures that, in some cases, exceed room temperature (Ferlay *et al.* 1995). These insulating and transparent magnets have also been shown to be optically switchable, in the sense that their magnetic properties vary when the material is irradiated at a suitable frequency (Sato *et al.* 1996).

The growth of the extended cyanide-based lattice is blocked by the use of appropriate ligands on the peripheral ions. The use of neutral polyamines, like the tetradentate tren or the pentadentate tetren of Fig. 4.1, with pentaco-ordinate copper(II) or hexacoordinate manganese(II) and nickel(II), respectively, leaves only one free site to be coordinated by the cyanide (Marvaud *et al.* 2003a).

This strategy starts from an aqueous solution of $[M'L(H_2O)]^{2+}$, where L stands for the polyamine ligand. Adding the hexacyanometalate precursor yields directly the polynuclear complex of formula $[M(CN-M')_6]^{9+}$ complex ion, as shown in Fig. 4.8.

The high positive charge of the cluster also plays an important role, as has been stressed by Marvaud *et al.* (2003a). The compounds of this class are in fact highly soluble in polar solvents, thus allowing the slow precipitation of relatively large crystals. The presence of counter-ions in the lattice represents another parameter to play with: the size of the counter-ion controls the intercluster sep-aration, but it can also be exploited to selectively precipitate one of the several types of clusters that can be present in solution. It is in fact well known that the more the anions and cations have similar size the less the salt is soluble.

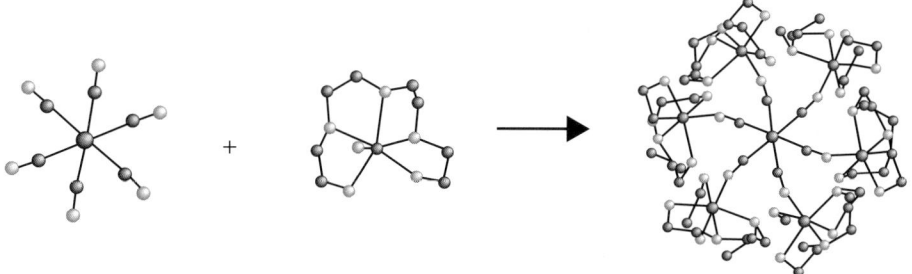

FIG. 4.8. Schematic view of the formation of the cluster $[Cr(CN-Mn(tetren)_6]^{9+}$. From Marvaud *et al.* (2003a). Reprinted with permission of Wiley-VCH.

As far as the magnetic properties are concerned when M and M′ in [M(CN-M′)$_6$]$^{9+}$ are paramagnetic a high-spin ground state is expected for either the ferro- or antiferro- M-M′ interaction. In this last case a ferrimagnetic spin structure of the cluster is warranted. The possibility of obtaining isostructural compounds based on the diamagnetic Co(CN)$_6^{3-}$ building block allows a better characterization of the magnetic properties of the peripheral metal ions.

In clusters of the [Cr(CN-M′)$_6$]$^{9+}$ type, a ground state spin as high as $S = 27/2$ has been observed for M′ = Mn^{II}, originating from the antiferromagnetic interaction of the six manganese(II) spins with the central chromium(III), $S = 6 \times 5/2 - 3/2$. Despite the many advantages of this approach, the octahedral symmetry of the cluster strongly quenches the magnetic anisotropy and so, even if their spin value is very large, the cluster does not behave as a SMM. It is possible anyhow to play with the ligand, to reduce the symmetry by reducing the nuclearity of the cluster. Marvaud *et al.* (2003b) have in fact used more bulky ligands, shown in Fig. 4.9, instead of tren and tetren to sterically hinder the M^{III} centre and to hamper the coordination of all six CN$^-$ groups (Marvaud *et al.* 2003b). An example is the hexanuclear complex with formula [Co(CN)}CN-Ni(dipropy$_2$)}$_5$]$^{7+}$ shown in Fig. 4.9 where one of the cyanide ions is terminal and not bridging.

A very interesting synthetic approach to obtain large cages based on cyanide has been developed by Long and co-workers. They employ the [(tacn)M(CN)$_3$] unit as a building block. The tacn ligand and its derivatives, as already mentioned, coordinate in the facial mode and therefore the three CN$^-$ groups are orthogonal to each other as shown in the scheme below.

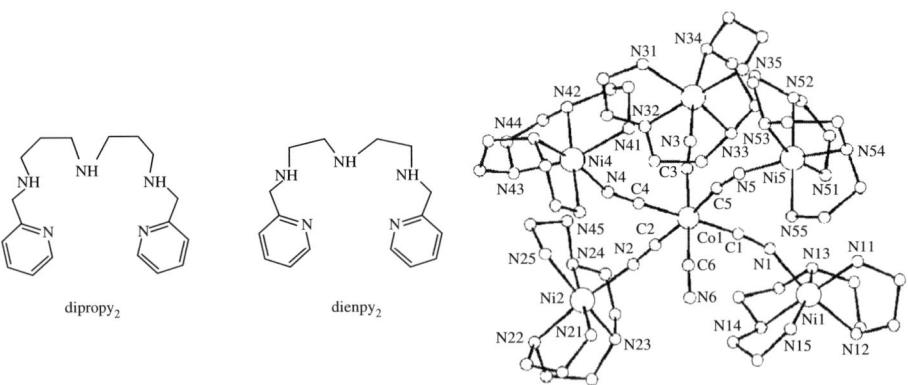

FIG. 4.9. Left: derivative of the tren and tetren ligands of Fig. 4.1 with more bulky pyridine groups replacing the terminal NH$_2$ groups. Right: structure of the hexanuclear [Co(CN-NiL)$_5$] cluster where one of the cyanide groups acts as a terminal ligand. Adapted from Marvaud *et al.* (2003b) with permission of Wiley-VCH.

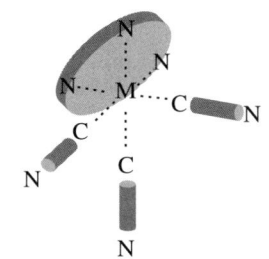

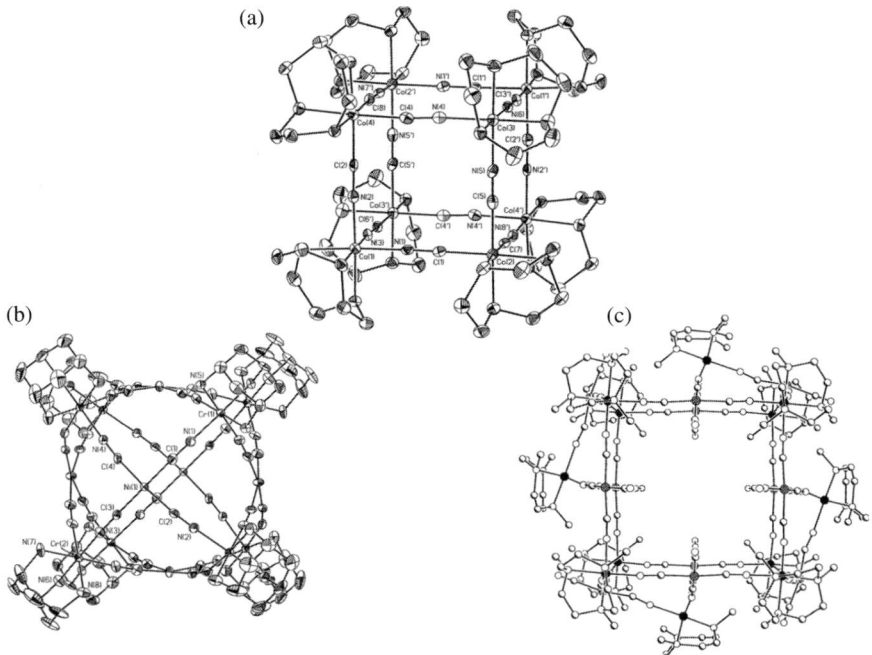

FIG. 4.10. Views of the structure of the clusters based on the combination of facial ligands tacn (and related ones) and cyanide bridging ligands: (a) $[(\text{tacn})_8\text{Co}_8(\mu\text{-CN})_{12}]^{12+}$; (b) $[(\text{tach})_8\text{Cr}_8\text{Ni}_6(\text{CN})_{24}]^{12+}$; (c) $[(\text{Me}_3\text{-tacn})_{12}\text{Cr}_{12}\,\text{Ni}_{12}\,(\text{CN})_{48}]^{12+}$. From Yang $et\ al.$ (2003) and Sokol $et\ al.$ (2002a). Copyright (2002 and 2003) American Chemical Society.

This building block is thus suited to form a vertex of a cube, where the edges are formed by the bridging cyanide, as shown in Fig. 4.10a for the cluster $[(\text{tacn})_8\text{Co}_8(\mu\text{-CN})_{12}]^{12+}$, which results from this simple reaction:

$$4[(\text{tacn})\text{Co}(\text{CN})_3] + 4[(\text{tacn})\text{Co}(\text{H}_2\text{O})_3]^{3+} \rightarrow [(\text{tacn})_8\text{Co}_8(\text{CN})_{12}]^{12+} + 12\text{H}_2\text{O}.$$
$$(4.2)$$

Clusters with higher nuclearity based on this cubic structure have been obtained. For instance in Fig. 4.10b is shown the cluster compound with formula $[(tach)_8Cr_8Ni_6(CN)_{24}]^{12+}$, with tach $= 1,3,5$-triaminocyclohexane, where every face of the cube is formed by a $M(CN)_4$ unit, raising the nuclearity to 14 (Yang *et al.* 2003). The nuclearity can be increased to 24 if each CN^- constituting an edge of the cube is replaced by a metal bridge $N\equiv C-M-C\equiv N$ as in the cluster $[(Me_3\text{-}tacn)_{12}Cr_{12}\,Ni_{12}\,(CN)_{48}]^{12+}$ shown in Fig. 4.10c (Sokol *et al.* 2002a).

Other types of interactions, often underestimated, can play a crucial role in determining the structure. For instance the interaction of terminal CN^- ligands with the K^+ cation in the crystal lattice seems to be at the origin of the unusual trigonal prismatic geometry around the $Mn(CN)_6^{4-}$ central ion in the compound with formula $K[(Me_3tacn)_6MnM_6(CN)_{18}](ClO_4)_3$, with $M=Cr^{III}$or Mo^{III}. AF interactions between the central $S=5/2$ spin and the $S=3/2$ peripheral spins stabilize an $S=13/2$ ground spin state. The clusters do not posses the cubic symmetry imposed by the octahedral environment and in the case of molybdenum they show a sizeable easy axis anisotropy with $D/k_B = -0.47$ K (Sokol *et al.* 2002b).[1] Non-zero χ'' in zero dc fields has been observed, suggesting that indeed the compound behaves as a SMM.

Tripodal phosphorus containing ligands have been employed by Dunbar and co-workers to obtain a cubane type cluster of abbreviated formula Mn_4Re_4, which also behaves like a SMM. The low-spin Re^{II} ions, $S=1/2$, and the high-spin Mn^{II} ions, $S=5/2$, are antiferromagnetically coupled to give a ground spin state with $S=8$ (Schelter *et al.* 2004). The use of heavier ($4d$ or $5d$) metals seems better suited for the synthesis of SMMs. Stronger exchange interactions are expected because the valence shell orbitals are very diffuse, thus allowing for a large overlap with the orbitals of the bridging ligands. Moreover the larger spin–orbit coupling, compared to $3d$ ions, induces a significantly larger magnetic anisotropy.

Other important building blocks employing $4d$ and $5d$ metal ions are octacyanomolybdate and octacyanotungstate. The most relevant example is the cluster with formula $[Mn^{II}\{Mn^{II}(MeOH)_3\}_8(\mu CN)_{30}\,\{Mo^V(CN)_3\}_6]\cdot 5MeOH\cdot 2H_2O$ prepared and characterized by Decurtins and co-workers (Larionova *et al.* 2000). The cluster, represented in Fig. 4.11 can be schematized as formed by a central $Mn(CN\text{-}Mo)_6$ octahedron, with each face capped with another Mn^{II} centre bridged by cyano groups to the three Mo^V defining the face. The crystals of this compound lose molecules of the crystallization solvent and therefore the assignment of the spin ground state is not straightforward. In the original work a ferromagnetic interaction between Mn^{II} and Mo^V spins was suggested thus leading to a ground spin state $S=9\times 5/2 + 6\times 1/2 = 51/2$, the highest spin state ever observed, only encountered later in a Mn_{25} cluster (Murugesu *et al.* 2004a).

---

[1] For the sake of simplicity, from now on the spin Hamiltonian parameters are given in temperature units without indicating that the parameter has been divided by k_B.

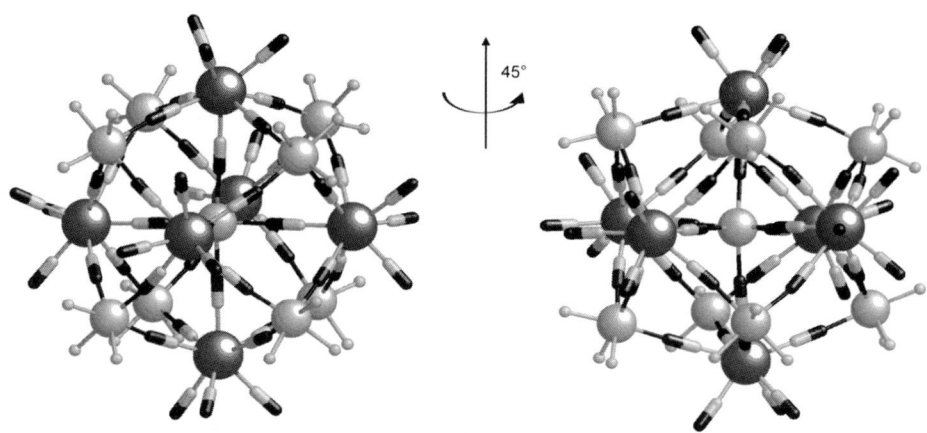

FIG. 4.11. Two schematic views (rotated by 45°) of the cluster $[Mn^{II}\{Mn^{II}(MeOH)_3\}_8(CN)_{30}\{Mo^V(CN)_3\}_6]$ (scheme: Mo, large dark; Mn, large light; N, black rod; C grey rod. The terminal methoxides are represented as small grey spheres.) Redrawn from Larionova *et al.* (2003) by permission of Wiley-VCH.

A successive investigation (Ruiz E. *et al.* 2005) of the spin density distribution in virgin crystals by means of polarized neutron diffraction and DFT calculations has revealed that Mn^{II} and Mo^V spins are antiparallel aligned to give $S = 9 \times 5/2 - 6 \times 1/2 = 39/2$, the same spin value encountered in the W^V derivative (Zhong *et al.* 2000).

Not all the cyanide ligands around the molybdenum atoms in the cyanide-based cluster of Fig. 4.11 are involved in a bridge. The loss of crystallization solvent molecules can lead to short intermolecular contacts of these terminal ligands probably responsible of the significant intercluster interactions and of the magnetically ordered state establised around 40 K.

The substitution of the rather isotropic Mn^{II} with Ni^{II} has again provided isostructural clusters characterized by $S = 12$ as a result of the ferromagnetic intracluster interaction (Bonadio *et al.* 2002). Despite the larger magnetic anisotropy of Ni^{II} that can exhibit D values as large as a few kelvin, HF-EPR spectra have shown that D of the cluster is only 0.022 K. The symmetry of the molecule has quenched, as in the case of hexacyanometalates, the magnetic anisotropy of the building blocks.

4.3.2 *The disruption of oxocentred carboxylate triangles*

Preformed building blocks are not always employed because they retain their structure in the final products. An example is the use, as starting species during

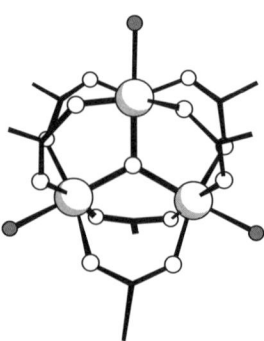

FIG. 4.12. Structure of the basic manganese carboxylate with general formula $[MnO(O_2R)_6(L)_3]^{n+}$. The metal atoms are represented by large shaded spheres, oxygen by empty spheres, and the donor atom of the L ligand by grey spheres.

the synthesis, of the trinuclear species with formula $[M_3O(RCOO)_6L_3]^{n+}$, where RCOO is a carboxylate and L a monodentate ligand. A schematic structure of the trinuclear complex is shown in Fig. 4.12.

These trinuclear species, also called 'basic metal carboxylates' are known for several metal ions of the transition first row, like vanadium, chromium, manganese and iron, which can be found in both the +3 and +2 oxidation states (Cannon and White 1988). Their magnetic properties have been extensively investigated because, in the presence of AF interactions, the triangular topology leads to spin frustration.

Christou and co-workers recognized that the addition of a chelating ligand, for instance the bipy ligand shown in Fig. 4.1, to a solution containing the trinuclear complex (Vincent *et al.* 1987) could not replace the mondentate ligand without strongly perturbing the structure of the trinuclear cluster (Christou 1989). An opening of the quite rigid trinuclear cycle is then expected, thus providing a route for the formation of higher nuclearity species. This is true if the chelating ligand is not added in strong excess, otherwise the monomeric product is obtained.

The first type of reaction that has been observed is the increase by one unit in the nuclearity of the cluster. Within the same Mn_4 nuclearity, for instance, several types of geometries have been observed, the most common of which are schematized in Fig. 4.13.

The cubane structure is encountered in a series of compounds of general formula $Mn_4O_3Cl_4(RCO_2)_3(L)_3$, an extensively investigated family of SMMs that we will discuss in more detail in the rest of this chapter.

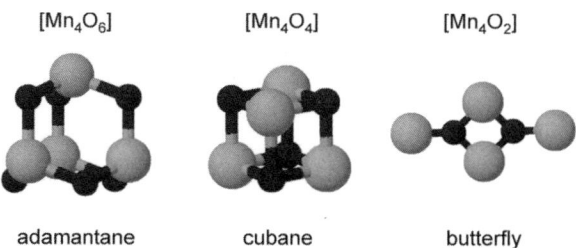

adamantane   cubane   butterfly

FIG. 4.13. Most common tetranuclear geometries obtained from the disruption of the $[M_3O(RCOO)_6L_3]^{n+}$ cluster.

The butterfly structure, common in Mn and Fe, is often encountered as a fragment of a larger structure, like in Fe_8, one of the most frequently investigated SMMs. The tetranuclear butterfly clusters on their own are characterized by a competition of AF interactions. The two metal atoms that are linked by two oxo groups are considered as the body of the butterfly while the two peripheral ones are the wings. If the body–wing AF interaction dominates the cluster has a singlet ground state, which is the case encountered in the case of iron(III) clusters. In the case of dominating body–body interaction intermediate spin states can be stabilized, as observed for the manganese(III) derivatives (McCusker *et al.* 1991).

The use of a trinuclear starting material has been, and remains, a successful strategy to obtain high-nuclearity clusters of mixed valent manganese ions (Aromi *et al.* 1998). To give an idea here below are schematized the different cluster nuclearities that have been obtained from the disruption of the $[Mn_3O(RCOO)_6L_3]^{n+}$, $(n=0,$ or $1)$ core. We can see that different nuclearity, as high as 22, has been obtained, by playing with parameters like the nature of the carboxylate, the starting ligand L or the added chelating ligand.

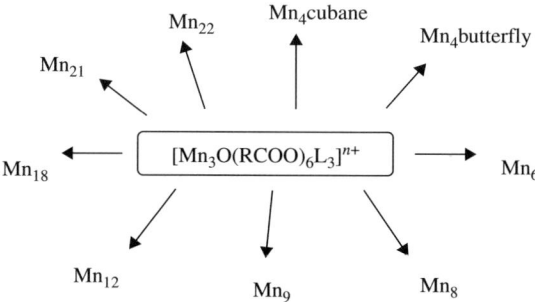

Such a large variety of products could appear as surprising but indeed it is not. In fact most reactions to form molecular clusters are carried at relatively low temperatures. At these temperatures not only the thermodynamically most stable compound is formed but also metastable compounds that are kinetically favoured. This is one of the major differences from conventional solid state chemistry.

4.4 Polyoxometalates

These clusters, usually abbreviated as POMs, can be visualized as formed by MO_n polyhedra that share vertexes or edges, seldom faces, to form a large polyanion. They are mainly based on high valence V, Mo, and W metal centres but can incorporate also S, P, and As oxoanions. They show a natural trend to form discrete species of high nuclearity and this field dates as far back as the eighteenth century with the studies of Scheele and Berzelius on the highly soluble blue compounds of molybdenum. The interest in POMs is rather general and goes from catalysis, to materials science. POMs have also found potential applications as antiviral agents. In the Literature, comprehensive reviews can be found (Müller *et al.* 1998; Müller and Roy 2003; Cronin 2004) and the aim of this chapter is only to present the very basic concepts of the chemistry of POMs.

The tendency of polyoxometalates to form discrete species can be partially attributed to the presence at the surface of the clusters of M=O double bonds that make the oxygen atom rather electron-deficient and therefore unable to act as a bridge by coordinating another metal atom. When the metal atoms are in their highest oxidation state the cluster is diamagnetic, but the metal ions can be easily reduced.

The formation of POMs is generally driven by a decrease of the pH. At very high pH the monomeric oxoanions are stable, but reducing the pH partial protonation and elimination of H_2O molecules induces the formation of the polynuclear species. Several parameters can be tuned to force the reaction towards a given product: the concentration of the reagents, the pH of the solution, the presence of reducing agents, the polarity of the solvent, and the nature of the cations. This approach has been extremely successful in providing very large nuclearity, up to 368 metal centres in a molybdenum-based cluster synthesized by Müller and co-workers (Müller *et al.* 2002). They have also shown that it is possible to organize such big objects in chains and monolayers, or to use them to encapsulate other clusters.

In many molecules of this class the magnetism is dominated by hetero-metal ions inserted in the POM′s structure, mainly Cu^{II} and Fe^{III} (Müller *et al.* 1998). In the case of POMs based on V^{IV}, a d^1 ion with a slightly anisotropic $S = 1/2$, interesting phenomena have been observed (Gatteschi *et al.* 1991; Chiorescu *et al.* 2000a). However low-spin ground states are usually observed, singlet or doublet states depending whether the cluster contains an even or odd number of V^{IV} centres, respectively. An exception is constituted by the octanuclear cluster of Fig. 4.6, which has a $S = 3$ ground state, but this results from the use of the organic ligand taci (Hegetschweiler *et al.* 2004). The most interesting magnetic properties observed in POMs will be briefly discussed in Chapter 14, while here we will focus on the formation mechanisms of these mesoscopic molecules.

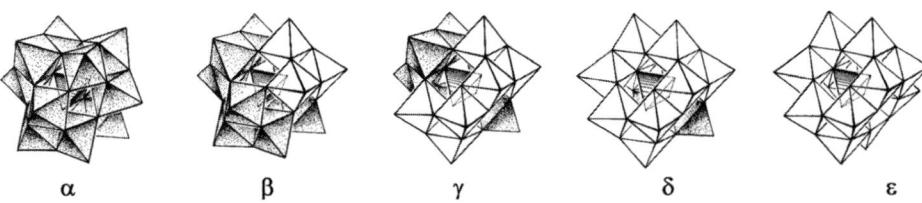

α   β   γ   δ   ε

FIG. 4.14. Polyhedral representation of the $[XM_{12}O_{40}]^{n-}$ Keggin ion. It can be schematized as a tetrahedron where each vertex is occupied by three vertex sharing octahedra forming a $\{M_3O_{13}\}$ fragment. The rotated $\{M_3O_{13}\}$ fragments are shown without shading to emphasize the differences between the five isomers. Reprinted with permission from Müller $et\ al.$ (1998). Copyright (1998) American Chemical Society.

Clusters of a wide class have a structure based on the so-called Keggin ions (Coronado and Gomez-García 1998). These ions have the general formula $[XM_{12}O_{40}]^{n-}$, e.g. X=P and M = Mo or W, and are formed by 12 vertexes and edges sharing octahedra, arranged in groups of three placed at the vertex of a tetrahedron as shown in Fig. 4.14. The Keggin ions can adopt up to five skeletal isomers $(\alpha, \beta, \ldots, \varepsilon)$ and these isomers are related to each other by the rotation by $\pi/3$ of one or more groups.

The process schematized in Fig. 4.15 shows how a big cluster can be assembled starting from the reduction of the Keggin ion, which becomes electron rich and therefore nucleophilic, thus attracting electrophilic MoO_3 groups. These, reduced in turn, act as templates towards electrophilic polyoxometalate fragments. Depending on the degree of protonation asymmetric structures, like the $\{Mo_{37}\}$ cluster of Fig. 4.15, can be obtained.

4.4.1 The role of pentagons

A fascinating example of designing or tailoring the final structure is represented by the family of Keplerate clusters (Müller $et\ al.$ 2001a). The name is given by the structural similarity to Kepler's early model of the universe, according to which the ratios between the successive orbits of the planets were equivalent to that of the spheres circumscribed around and inscribed within the five Platonic solids. Keplerates have a spherical shape and are based on 12 pentagons and 20 triangles or hexagons, as shown in Fig. 4.16.

The pentagon unit is based on the heptacoordinated MoO_7^{8-} unit, which is surrounded in the equatorial plane by five MoO_6 octahedra. It is well known that pentagons alone cannot cover a flat surface, and in fact none of the 230 crystallographic space groups is based on five-fold basic unit.

The use of pentagons as building blocks leads necessarily to a curved surface, like in the case of spheres or rings. In the case of Fig. 4.16b each of the 30 linkers corresponds to a Fe^{III} ion, thus forming a $\{Mo_{72}Fe_{30}\}$ cluster, an interesting antiferromagnetic molecule that will be discussed in detail in Chapter 14. In

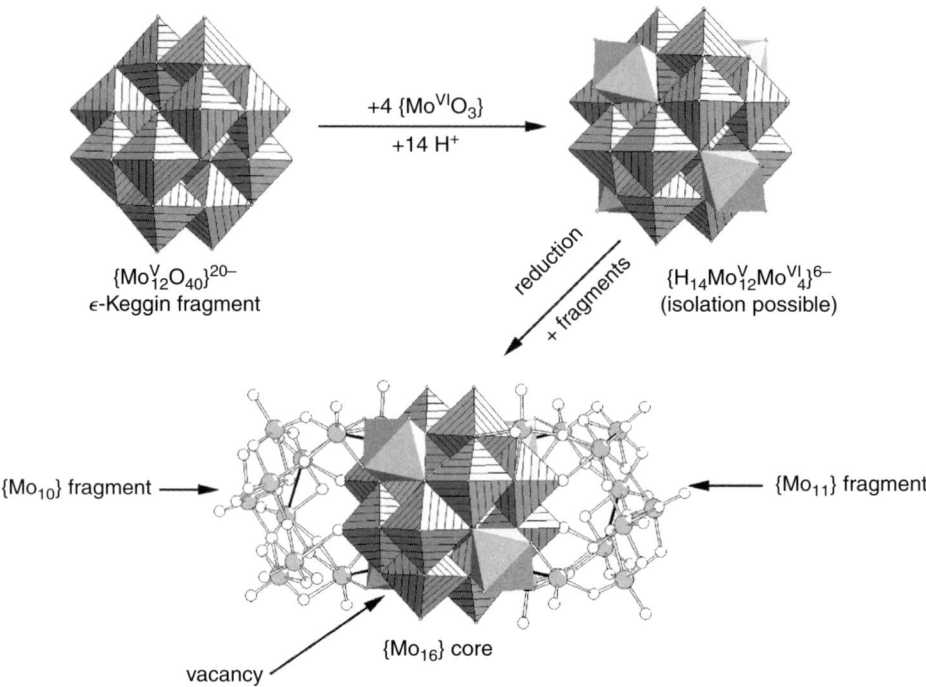

FIG. 4.15. Schematic view of the reaction that leads to the formation of a {Mo$_{37}$} cluster starting from the molybdenum-based Keggin ion. Adapted from Müller *et al.* (1998). Copyright (1998) American Chemical Society.

Fig. 4.16c the linkers are constituted by {Mo$_2$} units and the cluster has a {Mo$_{132}$} type nuclearity.

Starting from these simple building blocks it has been possible to encapsulate preformed clusters, like Keggin ions, inside the larger cavities of the Keplerates.

The interest in these giant molecules goes beyond their magnetic properties. For instance {Mo$_{72}$Fe$_{30}$} clusters have been used to catalyse the formation of nearly uniform single-walled carbon nanotubes (An *et al.* 2002; Huang *et al.* 2003).

It is important to stress that most POMs are obtained in a one-pot synthesis from a solution where many different building blocks are present and involved in rapid dynamic equilibria. Why such regularly shaped large molecules can be so easily isolated among hundreds of possible different fragments remains an open question. A possible explanation is that uniform, curved, and rigid structures do not present any site with highest reactivity; once they are formed they are harder to decompose, compared to less rigid fragments. Alternatively it could be

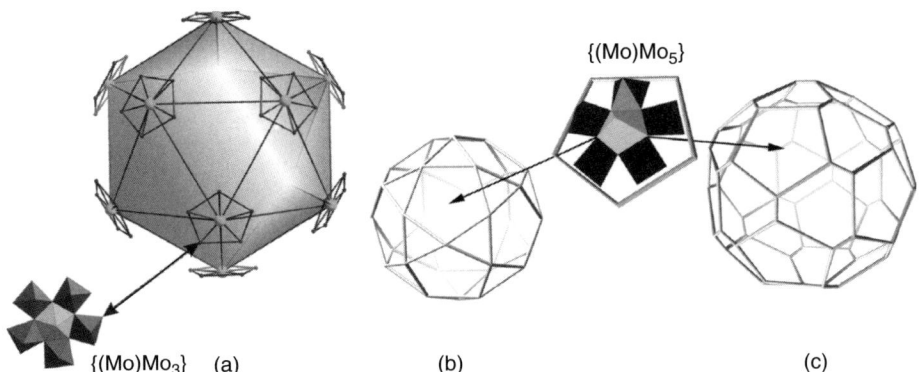

FIG. 4.16. Schematic view of the formation of spherical structures, also known as Keplerates, starting from the pentagonal fragment $\{(Mo)Mo_5\}$ (a). The 30 linkers can be constituted by a single metal centre, like in the case of $\{Mo_{72}Fe_{30}\}$ (b), or by 30 $\{Mo_2\}$ groups as for $\{Mo_{132}\}$ cluster (c). Adapted from Müller and Roy (2003). Copyright (2003) Elsevier.

argued that regular polyhedra and spheres pack efficiently and therefore have a greater tendency to crystallize than more irregular molecules.

To conclude this brief overview of POMs we want to quote the words of Leonardo da Vinci '*Where nature finishes producing its own species, man begins, using natural things and with help of this nature, to create infinity of species.*' that open a fascinating recent review of Müller and Roy (2003) to which the interested reader is referred.

4.5   The templating effect

We are accustomated to modelling our electronic documents using a template. The same can be done in synthetic chemistry if the interactions with an added chemical species, the templating agent, drive the formation of a complex molecule towards the desired structure. The use of such structure-directing agents is rather common in synthetic organic and inorganic chemistry, especially in the preparation of rings and closed structures. The forming polymeric species wrap around the template and this favours the proximity and reaction of the two edges and thus the formation of a closed structure.

In molecular magnetism the role of a templating cation has been shown in the formation of metal carboxylate-alkoxide ring structures, such as those described in detail in Chapter 14. Saalfrank and co-workers (1997) have shown that the size of iron(III) rings depends on the size of the alkali cation that is used in the template synthesis. They have reacted a tetradentate ligand

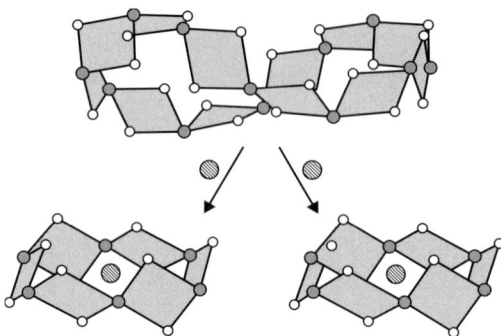

Fig. 4.17. Schematic view of the interconversion of the 12-member iron(III) cycle
in two six-member ones, due to the addition of alkali metal ions. Iron atoms
are black spheres, oxygen atoms are empty spheres, sodium or lithium ions
are represented by hatched spheres. Kindly provided by A. Cornia.

L, i.e. L = triethanolamine, with $FeCl_3$ and by using the strong base NaH to
deprotonate the ligand they have obtained a hexanuclear cluster of formula
$[Na \subset (Fe_8L_8)]^+$, where the symbol $\subset$ means that the Na^+ ions is in the middle
of the ring. On the contrary, an octanuclear cluster of formula $[Cs \subset (Fe_8L_8)]^+$
forms when Cs_2CO_3 is employed. Interestingly, if Fe^{3+} ions are in excess they
act as a templating agent and a tetranuclear cluster of formula $[Fe(FeL_2)_3]$ is
formed, which has the star structure already shown in Fig. 3.25. Like other iron
stars this tetranuclear cluster behaves as a SMM at low temperature.

The template effect can also affect the stability in solution of a cluster.
An interesting example is the conversion of the dodecanuclear iron(III) ring of
Fig. 4.17 in the hexanuclear derivative thanks to the addition of alkali ions, Li^+
or Na^+, to the solution containing the dodecanuclear species (Caneschi *et al.*
1999).

The driving force of this transformation is provided by the six M^+–O interactions that can be active if the hexanuclear ring is wrapped around the cation. In
the dodecanuclear cluster the hole in the middle is too large and the alkali ion
cannot interact at the same time with six oxygen atoms of the bridging ligands.

The templating agent can be also an anion, as has been found in a family of copper(II) large nuclearity clusters that are encapsulating a KCl_6^{5-} unit
(Murugesu *et al.* 2004b).

A sort of template synthesis has also been used by Winpenny and
co-workers to modify the octanuclear ring structure of the compound of formula $[Cr_8F_8(O_2CCMe_3)_{16}]$, where a fluoride and two pivalates bridge the Cr^{III}
ions (Van Slageren *et al.* 2002). Most rings with an even number of members
behave as antiferromagnets and have a diamagnetic ground state, as shown in
detail in Chapter 14. It would be therefore interesting to replace a metal ion

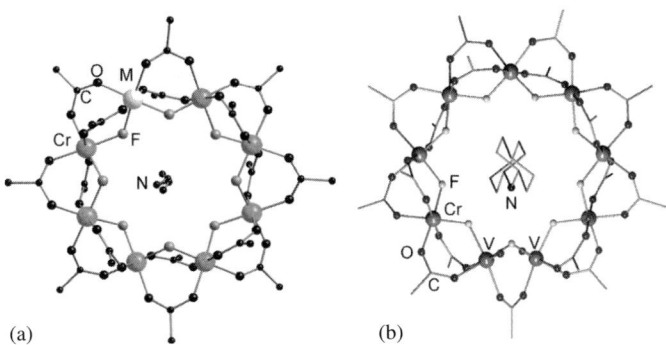

FIG. 4.18. Schematic view of the etherometallic rings: (a) $[(CH_3)_2NH_2]$ $[Cr_7MF_8(O_2CCMe_3)_{16}]$; (b) $[(C_6H_{11})_2NH_2][Cr_7(VO)_2(O_2CCMe_3)_{18}]$. Courteously provided by R. Winpenny.

with one having a different spin in order to have an uncompensated magnetic moment at low temperature. Doping with a tripositive diamagnetic ion is a possible route. These authors have, however, adopted a different approach: to replace a single Cr^{III} centre by a dication (M^{2+}) thus to form the monoanionic species $[Cr_7MF_8(O_2CCMe_3)_{16}]^-$ that can be precipitated as a salt of the dimethyl ammonium ion. For instance in the dimethyl-ammonium salt the cation is located inside the cycle and is hydrogen-bonded to three fluorine atoms, as shown in Fig. 4.18a (Larsen et al. 2003).

This finding has suggested that the use of larger substituents on the ammonium cation could increase the size of the ring as already observed in the template synthesis of iron rings. Reaction of hydrated chromium(III) fluoride with vanadyl acetate in pivalic acid in the presence of bulkier dicyclohexylamine, produced a nonanuclear ring of formula $[(C_6H_{11})_2NH_2][Cr_7(VO)_2F_9(O_2CCMe_3)_{17}]$, whose structure is shown in Fig. 4.18b (Cador et al. 2004). Odd-member rings containing more than three centres are extremely scarce compared to the richness of the even-member family. The first example is constituted by a ring of five copper(II) ions that surround an uranyl cation (Stemmler et al. 1996). A possible explanation for odd-member rings being so infrequent can be found in the fact that even if usually all the metal ions of the ring are on the same plane, this does not correspond to a mirror plane in respect of the bridging ligands. There is a sort of alternation of up- and down- arrangements of the bridges that results in an incommensurate structure in odd-member rings. Therefore only if additional interactions are present, like the interactions with the encapsulated cation, can this unfavourable structure be formed. Odd-member rings of antiferromagnetically interacting spins are extremely interesting to investigate spin-frustration effects in low-dimensional systems. Particularly interesting are the magnetic properties of the nonanuclear ring obtained by replacing vanadyl acetate with

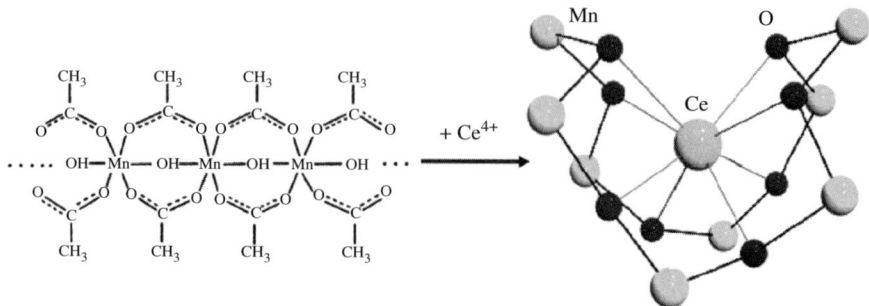

FIG. 4.19. Schematic view of the formation of the saddle-like core of the clusters $[Ce^{IV}Mn_8^{III}O_8(O_2CMe)_{12}(H_2O)_4]\cdot4H_2O$ starting from a Mn^{3+} chain structure and using Ce^{4+} as templating agent. Reprinted with permission from Tasiopoulos *et al.* (2003). Copyright (2003) American Chemical Society.

nickel(II) basic carbonate in the above reported synthesis. The compound with formula $[(C_6H_{11})_2NH_2][Cr_8NiF_9(O_2CCMe_3)_{18}]$ has been obtained, which has been described as the magnetic analogue of the Möbius ring, as discussed in more detail in Chapter 14.

Even if very interesting, the ring structures described in this section do not behave as SMMs. On the contrary a SMM with a large spin ground state, $S = 16$, has been obtained by Christou and coworkers (Tasiopoulos *et al.* 2003) through the conversion of a Mn^{III} carboxylate-hydroxide chain into an octanuclear cluster by adding oxophilic Ce^{4+} ions as a template agent. The chain wraps in a saddle-like closed structure, as shown in Fig. 4.19. Interestingly this cluster is one of the very few examples exhibiting ferromagnetic spin structure of the cluster. The $S = 16$ is associated with a moderate axial magnetic anisotropy, $D = -0.14$ K, and the behaviour of SMMs has been observed with magnetic hysteresis below 0.6 K The template approach seems to hold great potential for the synthesis of SMMs as it allows us to play with other parameters, like the size and chemical nature of the template agent.

4.6 Solvothermal synthesis

The conventional synthetic approaches described above are based on hydrolysis and condensation of fragments in the 'self-assembly' of complex products. They use soluble starting materials that allow low reaction temperatures favouring the kinetic trapping of a large variety of interesting metastable compounds. The reaction temperature is in fact limited by the boiling point of the solvent at atmospheric pressure.

Less exploited but quite interesting are higher temperature routes to form clusters using solvo-thermal techniques. This approach involves heating the solvent and reagents in a sealed vessel. This results in autogenous high pressure with increase of the boiling temperature of the solvent. When the solvent is water, temperatures in the range 140–260°C are usually employed, and the technique is called hydro-thermal. Lower temperatures (100–160°C) are instead reached with most common organic solvents, for which thick-walled quartz or glass tubes are usually employed. The technique is thus suitable for air and moisture sensitive reactions. A review of recent results obtained with this technique has been provided by Laye and McInnes (2004).

The advantages of solvothermal synthesis rely upon two main properties of super-heated solvents: the reduced viscosity and the different solubilizing properties. The lower viscosity increases the diffusion rates and consequently solid reagents are more reactive. It has also been suggested that it enhances the crystal growth from solution (Khan and Zubieta 1995).

The different solubility in super-heated solvents is determined also by major physical modification. For instance, the dielectric constant of water decreases rapidly with temperature and therefore under hydrothermal conditions a different chemical reactivity can be expected (Rabenau 1985). In general, the use of solvothermal conditions minimizes the different solubility of metal salts and organic ligands. A large variety of metal salts, unreactive at ambient condition, can be employed without the necessity to prepare intermediates. In some reactions the most thermodynamically stable products are not obtained because the kinetics of the reaction are too slow. The sensibly higher reaction temperatures used in solvothermal synthesis can lead to different kinetically trapped products, compared to those obtained at ambient pressure, without leading to the compounds formed at the higher temperatures employed in solid state inorganic chemistry.

In the field of solvothermal synthesis of large oxoclusters, incorporating also organic ligands, Zubieta and co-workers have done a remarkable systematic investigation of the reactivity of vanadium and molybdenum in the higher oxidation states (Khan and Zubieta 1995). Clusters with nuclearity as high as 18 and 42 for vanadium (Salta et $al.$ 1994) and molybdenum (Khan et $al.$ 1996) atoms, respectively, have been obtained. The advantages of this synthetic approach are relevant. For instance, the octanuclear cluster of Figure 4.6 has been obtained under solvothermal conditions. At ambient pressure the same reaction leads to a monomeric complex.

Examples of iron clusters have also been reported by the same group (Finn and Zubieta 2000). More recently McInnes and co-workers (Laye and McInnes 2004) have combined the use of preformed triangular clusters, the basic metal carboxylates already discussed in Section 4.3.2, in superheated alcohols that act at the same time as a solvent and as a source of bridging ligand. The quite inert $[Cr_3O(O_2CR)_6(H_2O)_3]^+$ species if heated in alcohol (R′OH) at 200°C yields the decanuclear cyclic cluster $[Cr_{10}(OR')_{10}(O_2CR)_{10}]$ (McInnes et $al.$ 2001), where the metal ions are now bridged by one carboxylate and two alkoxides. The use of solvothermal techniques has overcome the problems related to the kinetic inertness of chromium(III), problems not encountered in the synthesis of the iron(III) analogues that can be easily obtained at ambient conditions (Taft et $al.$ 1994).

As far as SMM behaviour is concerned solvothermal techniques have provided some examples of very high spin molecules. For instance by reacting $[Fe_3O(O_2CMe)_6(H_2O)_3]Cl$ in MeOH at 100°C in the presence of benzotriazole (btaH) the tetradecametallic Fe^{III} cluster of formula $[Fe_{14}O_6(bta)_6(OMe)_{18}Cl_6]$, where a ring of six iron ions is sandwiched between two $\{Fe_4(bta)_3\}$ moieties (Low *et al.* 2003). The competing antiferromagnetic interactions stabilize, in this compound, a large spin ground state that has been evaluated to range between 23 and 25.

4.7 A survey of the most investigated SMMs

Even taking into account only a limited number of synthetic strategies, an enormous variety of molecular clusters have been obtained. On the contrary, most physical studies have been focused on a small number of compounds, and till now the most frequently investigated compound is the archetypal $Mn_{12}ac$ cluster. Several reasons can be found for this choice. It has already been shown that the control of the nuclearity and the achievement of a very high spin for the ground state are not sufficient conditions to observe SMM behaviour. In Chapter 2 we have already shown that the magnetic anisotropy is the result of a tensorial combination of the contributions of the single ions and of the pair interactions. An additional control on the relative orientation of the single ion anisotropy axes is therefore necessary, but very hard to achieve, even employing the more rational synthetic strategies illustrated above.

Nevertheless a discrete number of molecular clusters have a magnetic anisotropy comparable to that of Fe_8, the other widely investigated SMM, but have only been partially characterized. To investigate the dynamics of the magnetization, and in particular the quantum effects on it, is necessary to have high-quality crystalline materials. Often solvent molecules of crystallization are weakly bound in the lattice and their release causes a disorder in the lattice that only in exceptional cases can be well modelled. Moreover in many crystal space groups the orientation of the molecule is not unique, even if symmetry related.

Another requirement is that intercluster interactions must be vanishingly small, otherwise bulk 3D magnetism is observed. Many polycyanometalates have terminal monocoordinated CN^- groups that can be involved in hydrogen bond networks, thus providing a pathway for intercluster interactions. This is not the only case. For instance, the cluster structure reported in Fig. 4.2 differs substantially from the structure of $Mn_{12}ac$ reported in Fig. 1.8. In fact, while in $Mn_{12}ac$ the inorganic core is completely surrounded by the hydrophobic shell of the acetate ligands, some oxygen atoms of the metheidi ligands of $\{Fe_{19}\}$ clusters are not coordinated to the metal ions but involved in hydrogen bonds that connect different clusters. An antiferromagnetic phase transition is thus observed around 1 K, hiding the single molecule magnetic behaviour expected for this slightly anisotropic $S = 31/2$ cluster (Affronte *et al.* 2002a).

4.7.1 The archetypal Mn$_{12}$ acetate cluster

As stated in the introduction $[Mn_{12}O_{12}(CH_3COO)_{16}(H_2O)_4]$ was the first SMM to be reported. Lis first reported in 1980 the structure of the molecule that is obtained by addition of permanganate to a solution of $Mn(CH_3COO)_2 \cdot 4H_2O$ (Lis 1980).

The actual formula of the compound is $[Mn_{12}O_{12}(CH_3COO)_{16}(H_2O)_4]$ $\cdot 2CH_3COOH \cdot 4H_2O$. The possibility of formation of dodecanuclear manganese acetate complexes was suggested as early as 1921 (Weinland and Fischer 1921) but it was only when X-ray structure determination became routine that it was possible to clearly prove the structure. The crystal has tetragonal symmetry (space group I$\bar{4}$) and the dodecanuclear cluster has S$_4$ symmetry.[2] The structure of the latter has already been shown in Fig. 1.8. The space group is acentric, but the molecule is not optically active.

Since the molecule has tetragonal symmetry there are three independent manganese ions, namely two manganese(III) and one manganese(IV), which are octahedrally coordinated as shown in Fig. 4.20. The manganese(III) ions can be easily recognized by the bond lengths and by the elongated structure typical of these distorted ions. The structural evidence is for a localized mixed-valence compound, where a well-defined charge can be associated to each metal ion. This situation is often referred to as Class I of the Robin and Day (Robin and Day 1967) classification (see Section 2.3). Class I behaviour is further corroborated by the bond valence sums (Brown and Wu 1976), which are close to 4 and to 3, respectively, for the two types of ions. Mn1, which corresponds to a manganese(IV), is coordinated to five oxide ions and to one oxygen of acetate. Mn2, which corresponds to a manganese(III), is bound to two oxide ions and to four oxygens of acetate molecules, while Mn3, which also corresponds to a manganese(III), is bound to two oxides, three oxygens of acetate molecules and a water molecule. All the oxides form μ_3 bridges. The manganese(III) ions show the typical elongated octahedral coordination seen in most complexes.

The distortion from octahedral symmetry is usually associated to the Jahn–Teller effect that removes the orbital degeneracy of the ^5E_g electronic state, as discussed in Section 2.2. The elongation axis of Mn2 makes an angle of 11° with the tetragonal axis, while that of Mn3 makes an angle of 37°. The value of these angles is important as they determine the height of the barrier to the reorientation of the magnetization as will be discussed below.

Another important feature of the structure is the presence of acetic acid and water molecules of crystallization. Thermogravimetric studies (Lis 1980) showed that loss of solvated molecules starts at 308 K and continues up to 463 K. Above this temperature decomposition of the compound sets in. The acetic acid molecules of solvation lie between adjacent clusters and close to a twofold axis and are therefore statistically distributed between the two symmetry

---

[2] Crystallographic cell parameters for Mn12ac at 83 K: $a = b = 17.1688(3)$Å, $c = 12.2545(3)$Å, $\alpha = \beta = \gamma = 90°$, $V = 3611.39(13)$Å^3.

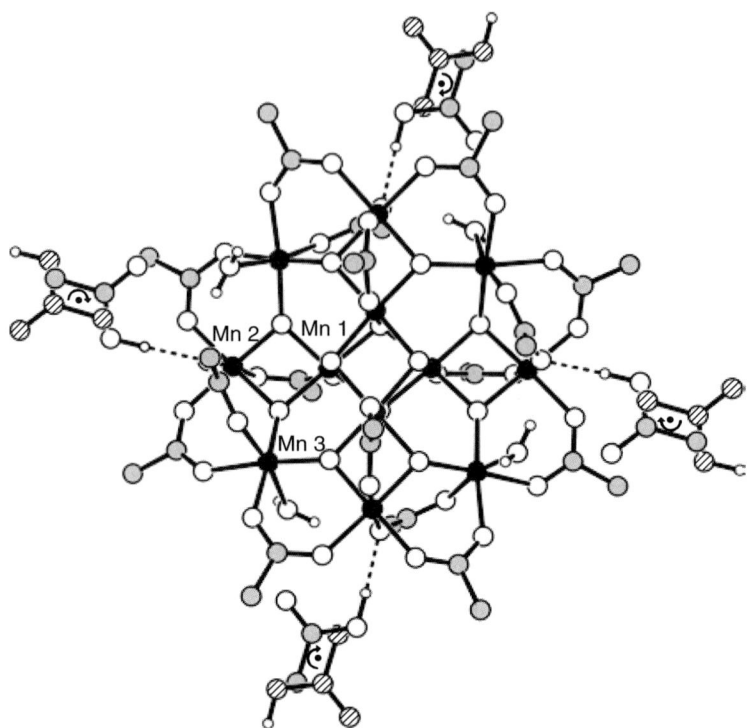

FIG. 4.20. Structure of the $[Mn_{12}O_{12}(CH_3COO)_{16}(H_2O)_4]$ ·$2CH_3COOH$ ·$4H_2O$ cluster at 83 K viewed slightly off the S_4 axis. The manganese atoms are represented as black spheres, oxygen atoms are white, carbon grey and hydrogen as smaller white spheres. The acetic acid of crystallization is disordered on two positions related by a binary axis (curved arrow). When the acetic acid occupies the position shown by the same colour scheme a hydrogen bond (dashed line) with the core of the cluster is formed. In the position shown by hatched spheres the same bond is formed with the neighbour cluster. The four symmetry-related acetic acid molecules independently occupy one or the other position. The water molecules of crystallization are not shown.

equivalent positions. They are at hydrogen-bond distances from the coordinated water molecule and the acetate ligand, as shown by low-temperature neutron diffraction studies (Langan *et al.* 2001). The original crystal structure solution of Lis (1980) showed some disorder of this acetate ligand. This disorder has later been well characterized thanks to a low-temperature (83 K) X-ray data collection (Cornia *et al.* 2002a).

In fact, through the low-temperature X-ray crystal structure, it has been possible to model the disorder induced by the presence of acetic acid. Two different positions, A and B, of the acetate ligand bridging Mn2 and Mn3 have been

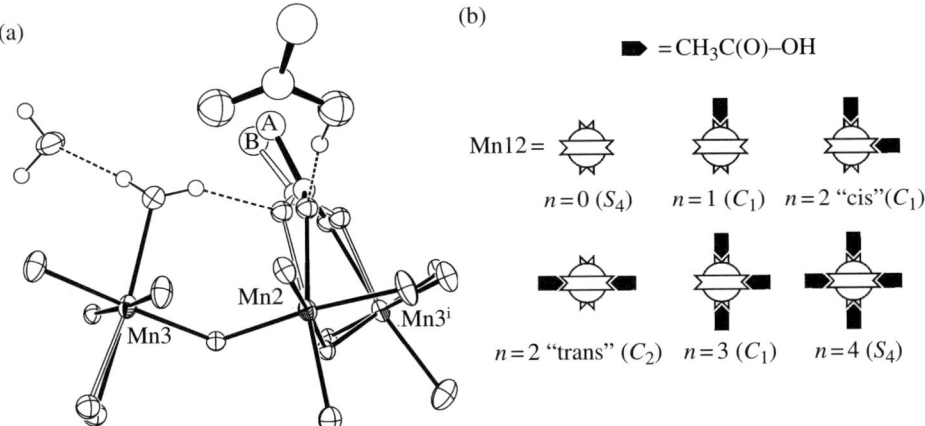

FIG. 4.21. (a) View of the two sites of the acetate ligand involved in the hydrogen bond with the disordered acetic acid of crystallization. When the acid is present, site A is occupied. (b) Schematic view of the six possible isomers of $Mn_{12}ac$. The black arrow indicates the presence of the acetic acid and therefore the A coordination mode.

assumed as shown in Figure 4.21a. The site occupation factors are very close to 0.5 like that of the acetic acid of crystallization, which, when present, forms a hydrogen bond with the acetate ligand slightly bending it (B position). Six possible isomers can be formed, as shown in Fig. 4.21b depending on the number and position of H-bonds that are formed. Only in two of them the fourfold symmetry is retained. The main effect of the disorder is thus to reduce the real symmetry of most molecules present in the crystals, even if the crystals maintain the tetragonal symmetry due to random orientation of the distorted species. We will discuss later in more detail these findings for their relevance to the tunnelling behaviour.

We briefly describe here the magnetic properties of $Mn_{12}ac$. To compare the experimental magnetic susceptibility χ with the Curie law, $\chi \propto 1/T$, it is convenient to plot the product of χ by the temperature. This is shown in Fig. 4.22 for a polycrystalline sample of $Mn_{12}ac$. At room temperature it has a value that is smaller than expected for eight $S = 2$ and four $S = 3/2$ uncoupled spins (31.5 emu mol^{-1} K). This indicates the presence of antiferromagnetic interactions. On decreasing temperature χT goes through a round minimum typical of ferrimagnetic behaviour originated by the antiferromagnetic interaction with non-compensation of the individual magnetic moments. On further lowering the temperature χT goes through a plateau at about 55 emu mol^{-1} K. This value is very far from that expected for ferromagnetic coupling ($S = 22$, 253 emu mol^{-1} K), but corresponds to a spin $S = 10$, confirming that the cluster has a ferrimagnetic spin arrangement.

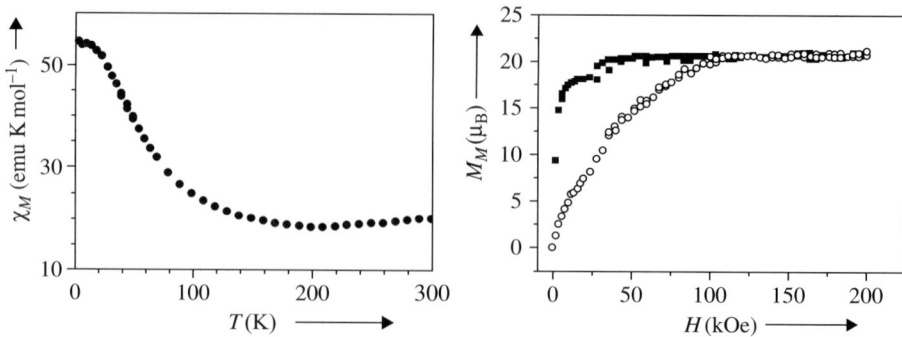

FIG. 4.22. Left: temperature dependence of the product of the magnetic susceptibility with temperature for a powder sample of Mn_{12}ac. Right: magnetization
curves recorded at 2 K on an oriented single crystals of Mn_{12}ac. Solid squares
correspond to the data with the magnetic field applied along the tetragonal
c axis.

The $S = 10$ ground state has been confirmed by high-field magnetization studies, which show saturation of the magnetization around the value of 20 μ_B
(Caneschi *et al.* 1991). Single-crystal magnetization data, as shown in Fig. 4.22,
provide evidence of a very large easy axis type magnetic anisotropy (Novak and
Sessoli 1995). Both the parallel (to the tetragonal axis) and perpendicular magnetization reach a saturation value of about 20 μ_B but for the last one a field as
high as 100 kOe is required, while the parallel magnetization goes to saturation
for much smaller fields.

The $S = 10$ ground state can be modelled at the simplest level by assuming that all the manganese(III) spins are up and the manganese(IV) spins are
down. Polarized neutron diffraction experiments performed on a single crystal
confirmed this view (Robinson *et al.* 2000). Even if the calculated total magnetic
moment well agrees with the 20 μ_B expected for $S = 10$, the spin density on each
metal centre is significantly reduced compared to the spin-only value. Such a
trend was indeed predicted by density functional theory calculations (Pederson
et al. 2000; Pederson and Khanna 1999). No significant spin density was found
on non-metal atoms.

Several attempts have been made to calculate the complete spectrum of the
spin levels of Mn_{12}ac (Sessoli *et al.* 1993b; Tupitsyn and Barbara 2002; Raghu
et al. 2001). The task is far from being simple, because the total number of spin
states is 100,000,000. Initially several attempts have been made by using some
ad hoc assumptions. For instance, it was assumed (Sessoli *et al.* 1993b) that
the J_1 coupling constant, defined in the scheme of Fig. 4.23 and corresponding
to the interaction between manganese(III) and manganese(IV) mediated by two
bridging O^{2-} ions (double oxo bridge), is strongly antiferromagnetic. In the same
scheme J_2 is the coupling constant between manganese(III) and manganese(IV)

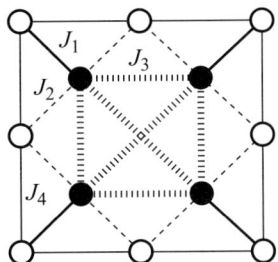

FIG. 4.23. Scheme and labelling of the intracluster exchange interactions in $Mn_{12}ac$: Mn^{IV} sites are in black. J_1 corresponds to the interaction mediated by the double μ-oxo-bridges between Mn1 and Mn2.

ions mediated by a single oxo-bridge, J_3 is that between manganese(IV) ions, while J_4 corresponds to the exchange between manganese(III) ions.

In this approximate model it may be assumed that the four Mn^{III}–Mn^{IV} pairs are in the ground $S = 1/2$ state, and that the contributions from the excited $S = 3/2$, $5/2$, and $7/2$ states may be neglected. Using this assumption sample calculations were performed, by varying the J_2, J_3, and J_4 constants in a relatively small range. It was observed that the calculated ground states could vary their S from 8 to 10, depending on the actual values of the parameters.

Recently more powerful calculation techniques have been implemented (Raghu et al. 2001). These methods first of all use an efficient system for representing a state in a computer by using a single number. Further the spatial symmetry (S_4 in the case of $Mn_{12}ac$) is exploited by using the Valence Bond method, which employs the Rumer–Pauling rule (Soos and Ramasesha 1990). Using the complete set it was shown that the values of the constants obtained by the previous approximate calculations fail to give the correct ground state. Sample calculations showed that, fixing $J_1 = -214$ K, $J_2 = -85$ K, the spin of the ground state is very sensitive to J_4 for a fixed value of J_3. It is a pity that the data have not been used to calculate the temperature dependence of the susceptibility, which might have provided a check of the goodness of the parameters. The chosen values of the parameters were based on the assumption of a $S = 10$ ground state with an $S = 9$ state lying 35 K above. In the reported calculations, with $J_3 = -85$ K, $J_4 = 85$ K. Calculations have also been performed (Regnault et al. 2002) in the Lanczos formalism (Cullum and Willoughby 1985) providing the best fit parameters: $J_1 = -89$ K, $J_2 = -88$ K, $J_3 = 0$ K, $J_4 = -17$ K. A similar set of exchange coupling constants ($J_1 \approx J_2 = -65$ K and $J_3 \approx J_4 = -7$ K) has been obtained through the simulation of inelastic neutron scattering experiments (Chaboussant et al. 2004). In particular these authors have observed a signal corresponding to $\Delta E/k_B \approx 61$ K that has been attributed to the $|10, \pm10\rangle \rightarrow |9, \pm9\rangle$ transition. The presence of a second $S = 9$ spin state at slightly higher energy has also been shown by these experiments.

The strong Ising-type magnetic anisotropy observed in the magnetization curves discussed above has been confirmed by HF-EPR spectra. The spectrum reported in Fig. 3.23 provides a first estimation of the axial anisotropy parameter D by simply measuring the separation between the low-field lines as $D \approx 1/2 \, g\mu_B \Delta H$. Actually the separation between pairs of neighbouring lines is not a constant, indicating that the higher order terms of the anisotropy play some relevant role here. Single-crystal HF-EPR spectra (Hill et al. 1998), inelastic neutron scattering (Hennion et al. 1997; Mirebeau et al. 1999; Zhong et al. 1999), and torque magnetometry (Cornia et al. 2000) have confirmed the necessity of using fourth-order terms in (2.15). In tetragonal symmetry only B_4^0 and B_4^4 are different from zero.

The spin Hamiltonian that describes the magnetic anisotropy becomes

$$\mathcal{H}_A = D\mathbf{S}_z^2 + B_4^0\mathbf{O}_4^0 + B_4^4\mathbf{O}_4^4 \tag{4.3}$$

where $\mathbf{O}_4^0$ and $\mathbf{O}_4^4$ are the Steven's operators listed in Appendix A.5.

A comparison of the second and fourth order parameters evaluated with the above-mentioned techniques is listed in Table 4.1.

The largest contribution to the magnetic anisotropy comes from the single ion anisotropy of the Jahn–Teller distorted manganese(III) ions. In fact dipolar interactions are too weak, and so are the anisotropic exchange contributions, due to the small spin–orbit coupling of the manganese ions. However, to evaluate how the single-ion anisotropy reflects on the anisotropy of the ground state the projection techniques introduced in Sections 2.1.3 and 2.5.2 must be used. Assuming that the ground state is well described by a coupling scheme in which the eight manganese(III) ions are ferromagnetically coupled to give a total spin $S_A = 16$, and the four manganese(IV) to give a total spin $S_B = 6$, the iterative use of the procedure described in Section 2.5.2 gives for the $S = 10$ ground state:

$$D_{S=10} = a_2 D_2 + a_3 D_3 \tag{4.4}$$

where $a_2 = a_3 = 0.02845$ and the index refers to the manganese(III) site, having neglected the contribution of manganese(IV) ions. If, for the sake of simplicity,

TABLE 4.1. Magnetic anisotropy parameters of $\mathrm{Mn_{12}ac}$ evaluated from six different experiments.

D(K)	B_4^0 (K)	B_4^4 (K)	Ref
$-0.66(2)$	$-3.2(2) \times 10^{-5}$	$\pm 6(1) \times 10^{-5}$	(Barra et al. 1997b)
$-0.657(3)$	$-3.35(6) \times 10^{-5}$	$\pm 4.3(7) \times 10^{-5}$	(Mirebeau et al. 1999)
-0.76	-2×10^{-5}	-1.2×10^{-4}	(Hill et al. 1998)
$-0.66(1)$	-3.4×10^{-5}	—	(Cornia et al. 2000)
-0.66	-3.2×10^{-5}	—	(Mukhin et al. 1998)
-0.65	-3.0×10^{-5}	$\pm 4.6 \times 10^{-5}$	(Del Barco et al. 2005)

the single-ion anisotropy of the two metal sites is assumed identical, i.e. $D_2 = D_3$, its value can be easily extracted from the experimental D value using (4.4). The estimated value using the available data of Table 4.1 is $D_{2,3} \approx -3.9$ K, in good agreement with the experimental D parameters observed in isolated manganese(III) complexes exhibiting a Jahn–Teller elongation of the octahedron (Barra *et al.* 1997a; Krzystek *et al.* 2001).

Analogous relations have been worked out also for the projection of the single-ion fourth-order terms (Hartmann-Boutron *et al.* 1996), although in this case no safe values are available for the individual ions.

In (4.4) the magnetic anisotropy is treated as a scalar quantity, while expression (2.26) involves tensors. In a more sophisticated treatment it is thus necessary to know also how the anisotropy tensors of the local spins are oriented in the crystal reference frame. This information can be retrieved experimentally only in very few cases, for instance when each metal site lies on a symmetry element of the crystal space group, or when the single-ion contribution can be experimentally investigated by doping the diamagnetic analogue cluster. Unfortunately a diamagnetic analogue of $Mn_{12}ac$ is not available and thus the orientation of the single-ion anisotropy tensor can only be conjectured. In this respect a very powerful technique is the angular overlap model (Bencini *et al.* 1998) as outlined in Section 2.2 It has been shown (Barra *et al.* 1997a) that the magnetic anisotropy of Mn^{III} ions can be predicted with great accuracy by taking into account the crystal field generated by the ligands and introducing the real geometry of the coordination sphere. The application to $Mn_{12}ac$ has been particularly useful to evaluate the effects of the disorder in the structure described in Fig. 4.21 on the magnetic anisotropy (Cornia *et al.* 2002b). The four different coordination environments of the Mn^{III} ions provided by X-ray analysis have been used to calculate the magnetic anisotropy using the AOM approach. The values of the e_σ and e_π parameters for each donor atom, taken from literature data, have been corrected for the actual metal–ligand distance assuming an exponential dependence.

In Table 4.2 the results obtained from the AOM calculation are reported. As can be seen, the interaction with the acetic acid molecule induces only minor

TABLE 4.2. Calculated second-order magnetic anisotropy parameters for the Mn^{III} ions in $Mn_{12}ac$.

Site	D (K)	E (K)	$\delta(°)$[a]
Mn2A	−4.92	0.40	11.6
Mn2B	−5.27	0.27	10.7
Mn3A	−4.57	0.10	37.2
Mn3B	−4.40	0.07	37.1

[a] δ is the angle between the easy-axis direction of each manganese site and the crystallographic c axis.

changes in the single-ion D and E parameters as well as in the direction of the local easy axis. However the effects on the deviation from tetragonal symmetry of the cluster can be sizeable. In fact four of the six different isomeric forms of $Mn_{12}ac$ in Fig. 4.21 have symmetry lower than tetragonal. The ZFS parameter of the six different isomers can be easily calculated using (4.4) where the $D_{2,3}$ parameters are now replaced by tensors:

$$\mathbf{D}_{\text{tot}} = d_2 \sum_{i=1}^{4} \mathbf{R}_i^T \mathbf{D}_2^{\alpha(i)} \mathbf{R}_i + d_3 \sum_{i=1}^{4} \mathbf{R}_i^T \mathbf{D}_3^{\alpha(i)} \mathbf{R}_i \qquad (4.5)$$

where $\alpha(i) = A$ or B differentiate the two environments observed in the presence or absence of hydrogen-bonded acetic acid, respectively. In (4.5) $\mathbf{D}_2^{\alpha(i)}$ and $\mathbf{D}_3^{\alpha(i)}$ are the single-ion ZFS tensors for the Mn2 and Mn3 sites generated by the i-th symmetry operation of the S_4 point group, described by the rotation matrix $\mathbf{R}_i$. The resulting $\mathbf{D}_{\text{tot}}$ tensor turns out to be axial and diagonal in the crystal axes reference frame only for $n=0$ and $n=4$. In the other four cases non-zero off-diagonal terms are present and diagonalization of the matrices provided the D and E parameters reported in Table 4.3. It is also possible to evaluate the angle θ between the easy axis and the crystallographic c axis and the angle φ formed by the hard axis with the crystallographic a axis. It is interesting to notice here that the principal axes of the second-order transverse anisotropy, due to its extrinsic nature, do not necessarily coincide with those of the fourth order ones. Indeed the origin of the four-fold anisotropy remains to be clarified. What is certain is that the principal transverse anisotropy axes do not necessarily coincide with the crystallographic axes.

The easy axis of the different species does not deviate significantly from the c crystallographic axis. The six species have very similar D values (within $\pm 2\%$),

TABLE 4.3. Calculated second-order magnetic anisotropy of the six disorder-induced isomers of $Mn_{12}ac$.

Isomer	Concentration %	D (K)	E (K)	$\theta(°)$[a]	$\varphi(°)$[b]
n = 0	6.25	0.759	0	0	—
n = 4	6.25	0.797	0	0	—
n = 1	25	0.769	2.34×10^{-3}	0.3	50
n = 2 cis	25	0.778	1.87×10^{-4}	0.4	60
n = 2 trans	12.5	0.778	4.70×10^{-3}	0	50
n = 3	25	0.788	2.35×10^{-3}	0.3	50

[a] θ is the angle between the easy-axis direction of each species and the crystallographic c axis.
[b] φ is the angle between the hard-axis direction and the crystallographic a axis. Given the four-fold symmetry of the crystal the hard axis is encountered at $n\pi/2 + \varphi$, with $n = 0, 1, 2, 3$.

in acceptable agreement with the experimental values reported in Table 4.1, considering the approximations involved (definition of the ground-state wave-functions and the related projection coefficients, neglect of higher-order terms, etc.). The E parameters for the low symmetry species are very small but their effect will be far from negligible at low temperature, as discussed in Chapter 12.

A high-resolution INS study has demonstrated that an average E term, ranging from 6 to 8 mK, is necessary to satisfactorily reproduce the spectra (Bircher *et al.* 2004). The order of magnitude well compares to the estimated values of Table 4.3.

The magnetic anisotropy of Mn_{12}ac has been calculated by using a first-principles treatment based on the DFT approach (Pederson and Khanna 1999). Interestingly the calculated ground state of the complete cluster is $S = 10$, as experimentally observed. The unpaired spin density is essentially localized on the manganese ions, being negative for the inner tetrahedron and positive for the outside ring, but significant spin density is also calculated on the bridging oxo-groups. Introducing spin–orbit coupling in the calculations the authors estimated the second-order contribution to the barrier in very good agreement with the experimental data. A refined calculation has allowed workers also to evaluate the fourth-order terms of the magnetic anisotropy (Pederson *et al.* 2002). The same authors have also taken into account the presence of different isomers induced by the acetic acid of crystallization. Their calculated values are also in good agreement with those reported in Table 4.3. In particular the same trend in the D, E, and θ parameters for the different isomers is observed (Park 2004).

Other contributions, including that from spin–spin interactions, to the magnetic anisotropy can be taken into account. Among these the intracluster dipolar interactions seem to play only a minor role, as stated above, while those determined by antisymmetric exchange may provide a significant contribution. Detailed calculations have been performed assuming that isotropic J_1 is dominant (Katsnelson *et al.* 1999). In order to justify the experimental zero-field splitting an antisymmetric exchange interaction ranging from -15 to $+1.5$ K was included. Unfortunately this result cannot be compared with any independent estimation of the antisymmetric contribution to the exchange.

As already mentioned, Mn_{12}ac is the first compound to have shown slow relaxation of the paramagnetic magnetization. The ac susceptibility measured in zero field is characterized by the appearance of an out-of-phase signal, typical of the slowing down of the relaxation rate that becomes comparable to the ac frequency, as discussed in Section 3.1.5.

The typical outcome of an ac susceptibility experiment on Mn_{12}ac is reported in Fig. 4.24. The position of the peaks shifts to higher temperature on increasing the frequency as the relaxation rate increases with temperature. The data, if reported on the Argand plot, suggest a small dispersion in the relaxation times, the parameter α being defined in (3.27) smaller than 0.1. However, below 5–10 K, the temperature region of the main frequency-dependent peak, a second and much smaller peak appears in Fig. 4.24. This peak has been associated with

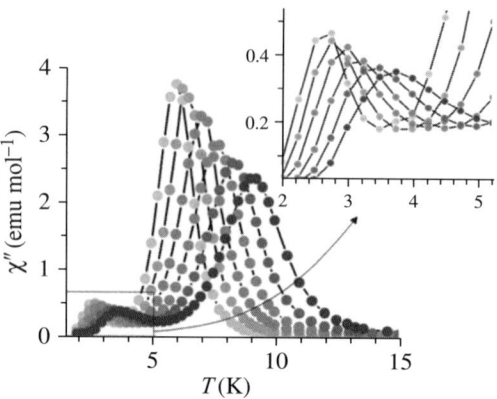

FIG. 4.24. Imaginary component of the ac susceptibility of $Mn_{12}ac$ measured here with frequency ranging from 200 Hz (light grey) to 20 kHz (black). In the inset an enlargement of the low-temperature peaks is shown.

the presence in the crystals of a small fraction of molecules (2–5%) exhibiting a significantly faster relaxation (Evangelisti and Bartolomé 2000). This fast relaxing species cannot be associated to one of the six isomers of Table 4.3 because they present a significantly reduced barrier. They probably correspond to the presence in the crystals of some molecules with a significantly smaller magnetic anisotropy originating in the flipping of the elongation axis of a manganese(III) octahedron. Given the small concentration, this type of disorder cannot be seen in the crystal structure analysis. It has, however, been observed in other Mn_{12} derivatives and it will be discussed in more detail in the next paragraph.

On lowering the temperature the relaxation time increases and can be directly evaluated from the time decay of the magnetization experiments described in Section 3.1.2. The relaxation time, τ, extracted from a single-exponential decay, together with that derived from ac susceptibility, are shown in Fig. 4.25. They are well reproduced by an Arrhenius law (1.2). The physical meaning of this observation will be the main topic of the next chapter.

It is interesting here to stress that at sufficiently low temperature, ca. 2.5 K, the magnetization relaxes so slowly that a magnetic hysteresis appears (Sessoli *et al.* 1993a). The width of the hysteresis loops increases on lowering the temperature and depends on the rate at which the field is swept. The origin of the opening of a hysteresis cycle is not due to a collective behaviour, like the irreversible displacement of magnetic domains. Experiments on frozen diluted solutions of $Mn_{12}ac$ in organic solvents like CH_3CN or CH_2Cl_2 have also been performed (Sessoli 1995; Cheesman *et al.* 1997; McInnes *et al.* 2002; Domingo *et al.* 2004). They have shown that the slow relaxation is retained even when the clusters are very far apart one from the other. The hysteresis thus appears to be a property of the isolated molecule.

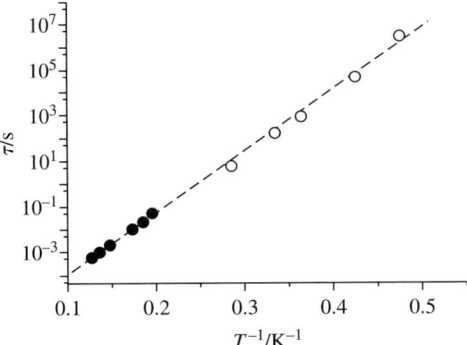

FIG. 4.25. Temperature dependence of the relaxation time on a log scale of Mn$_{12}$ac extracted from ac susceptibility data (solid symbols) in the frequency range 1-270 Hz and from time decay of the magnetization (empty symbols). The line corresponds to the Arrhenius law $\tau = \tau_0 \exp(\Delta E/T)$ with $\tau_0 = 2.1 \times 10^{-7}$ s and $\Delta E = 62$ K.

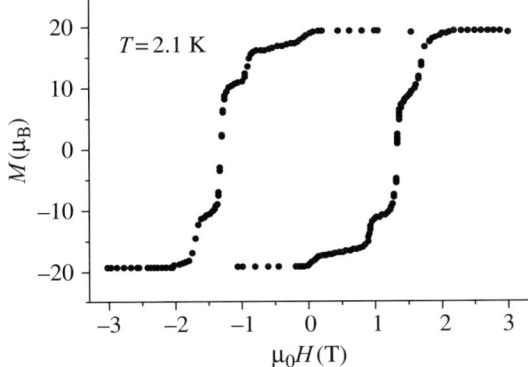

FIG. 4.26. Typical stepped hysteresis loop recorded on a single crystal of Mn$_{12}$ac with the applied field parallel to the tetragonal c axis.

Since the very early stages of research on molecular nanomagnets, anomalies in the field dependence of the relaxation time have been noted in ac susceptibility experiments (Novak and Sessoli 1995). It was only later, from accurate hysteresis loop measurements on a bunch of field aligned crystals (Friedman et $al.$ 1996) and then on a single crystal (Thomas et $al.$ 1996) that well-defined regularly spaced steps have been observed by applying the field parallel to the c axis. This unusual shape of the hysteresis, shown in Fig. 4.26, is now commonly considered the fingerprint of quantum tunnelling of the magnetization and will be extensively discussed in the next few chapters.

4.7.2 The Mn_{12} family

From the chemical point of view one of the most exciting properties of Mn_{12} clusters is their stability in solution, as shown by proton NMR measurements (Eppley *et al.* 1995) as discussed in Section 3.3.2.

Other Mn_{12} derivatives with different carboxylates are therefore generally obtained by treating a slurry of the acetate derivative in a suitable solvent with the desired carboxylic acid. Christou and co-workers demonstrated that it is possible to exploit the different acidity of the carboxylic acid to substitute the acetate, as the equilibrium is shifted towards the production of the weakest acid. Carboxylate groups in axial position are in general more prone to substitution by less basic incoming ligands as compared with their equatorial counterparts (Soler *et al.* 2001a; Chakov *et al.* 2003). However, the synthesis of mixed-ligand derivatives often results in partial substitution of both axial and equatorial positions (Boskovic *et al.* 2001).

A typical reaction is that described for the preparation of the benzoate derivative (Sessoli *et al.* 1993b). Treatment of Mn_{12}ac with a 100% excess of benzoic acid in CH_2Cl_2 leads to a majority, but not all, of the CH_3COO^- groups being exchanged. The black solid obtained, which contains both acetate and benzoate, is further treated with a ten-times excess of benzoic acid to give the completely substituted derivative. A different strategy employs, as starting material, the derivative $[Mn_{12}O_{12}(^tBu\text{-}COO)_{16}(H_2O)_4]$ because the *tert*-butyl group, $-C(CH_3)_3$, strongly increases the solubility of the cluster in organic solvents (Gerbier *et al.* 2003). It is interesting to note that not all the numerous Mn_{12} derivatives seems to be characterized by an $S = 10$ ground state. This spin state in fact results from competition of antiferromagnetic interactions, and minor changes of their relative strength can stabilize smaller spin states, commonly $S = 9$ (Sessoli *et al.* 1993b).

In Table 4.4 are reported some of the Mn_{12} derivatives whose structures have been fully characterized. They all have the general formula $[Mn_{12}O_{12}(RCOO)_{16}(H_2O)_x]\cdot Y$, where $Y =$ solvent molecules (Aromi *et al.* 1998). The main structural difference is the presence of derivatives with a different number of coordinated water molecules. An example of the latter category is provided by $[Mn_{12}O_{12}(Et\text{-}COO)_{16}(H_2O)_3]$, which crystallizes either in a monoclinic space group, without solvent molecules, or in a triclinic space group, with four solvated water molecules (Aubin *et al.* 2001).

Also the clusters with $x = 4$ show significant structural differences. In all cases the manganese(III) ions are either doubly bridged to one Mn^{IV}, type I, or singly bridged to two Mn^{IV}, type II. The water molecules are bound to type II ions. The coordination in Mn_{12}ac is such that one water molecule per type II ion is present (Lis 1980). The manganese–water direction roughly indicates the Jahn–Teller distortion axis. In other derivatives some of the type II manganese ions have no water molecule, and some have two. In Table 4.4, as in the chemical literature, a notation has been adopted that indicates the water coordination

TABLE 4.4. Some Mn_{12} derivatives with reported structures. General formula $[Mn_{12}O_{12}(RCOO)_{16}(H_2O)_x] \cdot Y$.

	R group/L[a]	x	Space group	Solvent. Molecules of crystallization	Water coord.	Ref
1	CH_3	4	$I\bar{4}$	$2CH_3COOH \cdot 4H_2O$	1:1:1:1	(Lis 1980)
2	CH_2CH_3	3	$P\bar{1}$	$4H_2O$	1:1:1	(Eppley et al. 1995)
3	CH_2CH_3	3	$P2_1/c$		1:1:1	(Aubin et al. 2001)
4	CH_3; CH_2CH_3	4	$I\bar{4}$	$2H_2O \cdot 4EtCOOH$	1:1:1:1	(Wei et al. 1997)
5	$CH_2C(CH_3)_3$	4	$P\bar{1}$	$CH_2Cl_2 \cdot CH_3NO_2$	1:2:1	(Sun et al. 1998)
6	$CH_2C(CH_3)_3$	4	$P\bar{1}$	$CH_2Cl_2 \cdot CH_3CN$	1:2:1	(Soler et al. 2003)
7	C_6H_5	4	$P\bar{1}$		2:0:2	(Sessoli et al. 1993b; Boyd et al. 1988)
8	C_6H_5	4	$Fdd2$	$2C_6H_5COOH$	2:0:2	(Takeda et al. 1998)
9	C_6H_4-p-CH_3	4	$C2/c$	HO_2C-C_6H_4-p-CH_3	1:2:1	(Aubin et al. 2001)
10	C_6H_4-p-CH_3	4	$I2/a$	$3H_2O$	1:1:2	(Aubin et al. 2001)
11	C_6H_4-p-Cl	4	$C2/c$	$8CH_2Cl_2$	2:0:2	(Aubin et al. 2001)
12	C_6H_4-m-Cl	4	$P\bar{1}$	HO_2C- C_6H_4-m-Cl	1:1:2	(An et al. 2000)
13	C_6H_4-o-Cl	4	$Pnn2$	$CH_2Cl_2 \cdot 5H_2O$	1:1:2	(Ruiz-Molina et al. 1998)
14	CH_2-C_6H_5	4	$P\bar{1}$		1:2:1	(Sun et al. 1998)
15	$CHCHCH_3$	4	$Ibca$	H_2O	2:0:2	(Ruiz-Molina et al. 2002)
16	CF_3	4	$P\bar{1}$	$2.5H_2O$	1:1:2	(Gomez-Segura et al. 2005)
17	CF_3	4	$I\bar{4}$	$2CF_3COOH \cdot 4H_2O$	1:1:1:1	(Zhao et al. 2004)
18	CF_3	4	$P2_1/n$	$CF_3COOH \cdot 7H_2O$	2:0:2	(Zhao et al. 2004)
19	CH_2Cl; $CH_2C(CH_3)_3$	3	$P\bar{1}$	$CH_2Cl_2 \cdot H_2O$	2:0:1	(Soler et al. 2001a)
20	CH_2Cl; CH_2CH_3	3	$P\bar{1}$	CH_2Cl_2	2:0:1	(Soler et al. 2001a)
21	CH_2Br	4	$I\bar{4}2d$	$4CH_2Cl_2$	1:1:1:1	(Tsai et al. 2001)
22	$CH_2C(CH_3)_3/NO_3$	4	$C2/c$	CH_3NO_2	2:0:2	(Artus et al. 2001)
23	CH_3/Ph_2PO_2	4	$P4_2/n$	$12CH_2Cl_2$	1:1:1:1	(Boskovic et al. 2001)
24	$CH_3/PhSO_3$	4	$P\bar{1}$	$4CH_2Cl_2$	2:0:2	(Chakov et al. 2003)
25	CH_3/CH_3SO_3	3	$Pbca$	$3CH_3CN_2 \cdot 4H_2O$	1:1:1	(Kuroda-Sowa et al. 2004)
26	CH_3/Ph_2PO_2	–	$P2_1/n$	$6.1\,CH_2Cl_2 \cdot 0.4H_2O$	—	(Bian et al. 2004)
27	C_6H_4-p-SCH_3	4	$I\bar{4}$	$8CHCl_3$	1:1:1:1	(Zobbi et al. 2005)
28	$C_6H_5/$ adc[b]	4	$I4_1/amd$	$8CH_2Cl_2$	1:1:1:1	(Pacchioni et al. 2004)

[a] When appropriate L indicates the non carboxylate ligand.
[b] adc = 10-(4-acetylsulfanylmethyl-phenyl)-anthracene-1,8-dicarboxylic acid.

number of neighbouring sites. The 1:1:1:1 isomer type is encountered in $Mn_{12}ac$ (Lis 1980). The other three different isomers have been experimentally observed, corresponding to the schemes 1:2:1 (Sun *et al.* 1998; Aubin *et al.* 2001), 1:1:2, (Aubin *et al.* 2001; Ruiz *et al.* 1998), and 2:0:2 (Sessoli *et al.* 1993b; Boyd *et al.* 1988; Takeda *et al.* 1998; Aubin *et al.* 2001; Zhao *et al.* 2004; Artus *et al.* 2001; Chakov *et al.* 2003).

There are some exceptions to the rule that the elongation axis of the Mn^{III} ions does not involve the bridging oxide ligands. For instance, the Jahn–Teller distortion axis in compound 9 of Table 4.4, $[Mn_{12}O_{12}(OOC\text{-}C_6H_4\text{-}p\text{-}CH_3)_{16}(H_2O)_4]\cdot HO_2C\text{-} C_6H_4\text{-}p\text{-}CH_3$, involves one of the bridging oxide ions in such a way that for one Mn^{III} ion the elongation axis is almost orthogonal to the elongation axes of the other Mn^{III} ions (Aubin *et al.* 2001). It is interesting to note that with the same carboxylate a different formulation, $[Mn_{12}O_{12}(OOC\text{-}C_6H_4\text{-}p\text{-}CH_3)_{16}(H_2O)_4]\cdot 3H_2O$, can be obtained (compound 10 in Table 4.4), where the elongation axes are now essentially parallel to each other (Aubin *et al.* 2001). A view of the cores of compounds 9 and 10 of Table 4.4 given in Fig. 4.27. The two clusters therefore can be considered as distortion isomers as frequently observed for instance in copper(II). Distortion isomerism is also encountered for the pairs 5–6 and 17–18 of Table 4.4. A distortion isomer of this type may be responsible of the presence of the fast-relaxing species observed even in single crystals of $Mn_{12}ac$.

The magnetic properties are strongly affected by the tilting of the elongation axis of one manganese ion. The magnetic anisotropy of the whole cluster results from a tensorial sum, as in (4.5), and the tilted Mn^{III} ion provides a contribution which is of easy-plane rather than easy-axis type, even if the single

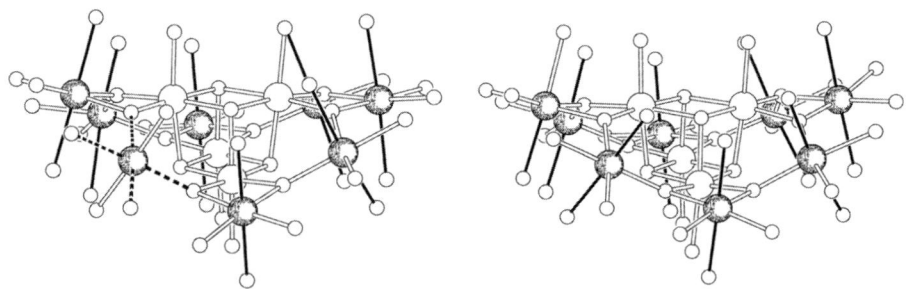

FIG. 4.27. Views of the structure of two distortion isomers of $[Mn_{12}O_{12}(OOC\text{-}C_6H_4\text{-}p\text{-}CH_3)_{16}(H_2O)_4]$ with Mn^{III} ions evidenced by the shadow. Cluster compounds 9 and 10 of Table 4.4 are shown on the left and on the right, respectively. The elongation axes of the Mn^{III} coordination polyhedra are indicated by full bonds. The dashed lines identify an unusual orientation for the elongation axis in 9. Redrawn with permission from Aubin *et al.* (2001). Copyright (2001) American Chemical Society.

spin anisotropy remains of Ising type. A strongly reduced anisotropy is thus encountered for compound 9, and therefore a reduced barrier for the reorientation of the magnetization.

The compounds with $x = 3$ have one manganese(III) which is five-coordinated (An *et al.* 2000). The overall temperature dependence of χT is not much different from that observed in $Mn_{12}ac$, but an $S = 9$ ground spin state was suggested from the analysis of the magnetization. The fit of the reduced magnetization versus H/T is not really satisfactory, so the ground state is not unequivocally established. A somewhat better agreement was observed in the case of $R = -C_6H_5$, $Y = -C_6H_5COOH \cdot CH_2Cl_2$. It must also be mentioned that several derivatives show some chemical instability, like solvent loss, which provide ambiguity in the interpretation of experimental data.

Interesting variations on the theme of the carboxylate derivatives has been also reported. By treating $[Mn_{12}O_{12}(RCOO)_{16}(H_2O)_4]$ derivatives with nitric acid in acetonitrile, new compounds of formula $[Mn_{12}O_{12}(NO_3)_4(RCOO)_{12}(H_2O)_4]$ were obtained, with $R = CH_2tBu$, Ph. The four nitrate groups are not disordered and are bound in bridging modes at positions occupied by bridging carboxylate groups in the parent compound. The ground state is still $S = 10$, and SMM behaviour is observed at low temperature. The zero-field splitting and the barrier are very similar to those observed in $Mn_{12}ac$ (Artus *et al.* 2001). Eight acetates have been replaced by diphenylphosphinates ($Ph_2PO_2^-$) (Boskovic *et al.* 2001) but these last ones occupy both equatorial and axial positions. Christou and co-workers have succeeded in selectively substituting the eight axial acetates with sulfonates as $PhSO_3^-$. Again an $S = 10$ state has been assigned to the ground state and the D parameter has been evaluated as -0.49 K (Chakov *et al.* 2003).

The easy substitution of the carboxylate ligands in Mn_{12} clusters can also be exploited to insert additional functional groups, for instance sulphur-containing groups able to interact with a gold surface (Cornia *et al.* 2003; Pacchioni *et al.* 2004; Zobbi *et al.* 2005).

4.7.3   *The reduced species of Mn_{12} clusters*

An interesting feature of Mn_{12} clusters is their rich redox chemistry. Differential pulsed voltametry investigations have been reported for $Mn_{12}ac$ in acetonitrile, showing two reversible processes, an oxidation and a reduction, and four redox processes altogether (Sessoli *et al.* 1993b). Very similar results were observed for the benzoate derivative in CH_2Cl_2. The oxidation process occurs at 0.79 V and the reduction process at 0.11 V. The values are referred to the Cp_2Fe/Cp_2Fe^+ pair where Cp stays for cyclopentadiene. The corresponding processes are summarized according to the scheme below:

$$[Mn_{12}]^+ \rightarrow [Mn_{12}] \rightarrow [Mn_{12}]^- \rightarrow [Mn_{12}]^{2-} \rightarrow [Mn_{12}]^{3-}. \qquad (4.6)$$

$[Mn_{12}]$ is a shorthand notation for $[Mn_{12}O_{12}(RCOO)_{16}]$ so the uncharged species correspond to $Mn_{12}ac$. A series of benzoate derivatives, which have similar solubility in a given solvent, allowed a comparative analysis of the role of the carboxylate in the redox potentials (Aubin *et al.* 1999). It was found that the potentials satisfactorily correlate with the electron withdrawing properties of the *para* substituent in the benzene ring. An electron withdrawing substituent causes the carboxylate to become less basic, reducing the electron density on the metal ions thereby making the cluster easier to reduce and at the same time harder to oxidize.

The first reduction potential is so easily accessible that mild reductants like iodide can be used. The PPh_4^+ derivatives of the R = Et and R = Ph were directly obtained by adding the iodide to the appropriate unreduced Mn_{12} clusters. The cation is magnetic with a moment corresponding to one unpaired electron. $(PPh_4)[Mn_{12}O_{12}(Et\text{-}COO)_{16}(H_2O)_4]$ crystallizes in the monoclinic $P2_1/c$ space group (Eppley *et al.* 1995). The overall structure of the anion is very similar to that of the unreduced species. The reduction yielded a valence-localized species, one of the external Mn^{III} ions being reduced to manganese(II). The identification of the reduced ion has been made on the basis of structural features and confirmed by bond valence sums (Brown and Wu 1976).

It has been suggested that the reduction of Mn^{III} rather than of Mn^{IV} is due to the fact that the reduction of the latter would introduce a distorted Mn^{III} site creating a strain in the apparently rigid $[Mn_4O_4]$ structure of the core. The temperature dependence of the χT product is qualitatively similar to that of Mn_{12}ac, suggesting a ferrimagnetic ground state. The ground state is suggested to correspond to $S = 19/2$, with a large negative zero-field splitting. The magnetic data require, however, a physically unreasonable $g = 1.74$. It is apparent that the ground spin state cannot be simply obtained by changing one of the $S = 2$ spins to a $S = 5/2$ spin, because one would expect either $S = 21/2$, for parallel or $S = 11/2$ for antiparallel alignment of this spin with the manganese(III) ones. A more frustrated structure, analogous to that which gives rise to $S = 9$ in some Mn_{12} derivatives is dominant for $[Mn_{12}]^-$. Similar results were obtained for the R = Ph derivative. HF-EPR spectra have been used to obtain the zero-field splitting parameter D (Aubin *et al.* 1999). The spectra were recorded on loose polycrystalline samples. It was assumed that a complete orientation of the powders was achieved and the spectra were assigned on the assumption of the strong field limit. With this simplified treatment the zero-field splitting D was found to be -0.63 K.

Beyond derivatives with diamagnetic cations recently the structure of a compound comprising an organic cation radical, $m\text{-}N$-methylpyridinium nitronyl nitroxide, $[m\text{-}MPYNN]^+$, and the $[Mn_{12}O_{12}(O_2CC_6H_5)_{16}(H_2O)_4]$ anion (Takeda and Awaga 1997), or the paramagnetic $[Fe(C_5Me_5)_2]^+$ cation and the $[Mn_{12}O_{12}(O_2CC_6F_5)_{16}(H_2O)_4]$ anion (Kuroda-Sowa *et al.* 2001), have also been reported. The ground state for the latter was found to be $S = 21/2$, with axial zero-field splitting parameters $D = -0.52$ K and $B_4^0 = -5.2 \times 10^{-7}$ K, as determined through HF-EPR spectroscopy.

Using carboxylates with more electron withdrawing derivatives it was also possible to obtain the $[Mn_{12}]^{2-}$ derivatives (Soler *et al.* 2000; Soler *et al.* 2001b) $[Cat]_2[Mn_{12}O_{12}(O_2CR)_{16}(H_2O)_4]$, where $Cat^+ = PPh_4^+$, $NnPr_4^+$, $R = CHCl_2$, C_6F_5, $2,4\text{-}C_6H_3(NO_2)_2$. The second reduced manganese(II) is also localized on the external ring. The ground state is reported to be $S = 10$, with an axial zero-field splitting $D \sim -0.45$ K. A different approach to obtain the reduced form of Mn_{12} clusters is based on an exchange reaction between the acetate precursor and a carboxylate carrying a positive charge as the quaternary ammonium cation in the betaine cation schematized below, which is associated with the hexafluorophosphate anion (Coronado *et al.* 2004). The reaction leads to $[Mn_{12}O_{12}(bet)_{16}(EtOH)_4](PF_6)_{14}$ clusters, with only 14 PF_6^- anions, which means a two-electron reduced species is formed. The spin ground state appears to be $S = 11$ but the presence of Mn^{II} reduces the anisotropy to $D \sim -0.32$ K.

= betaine

4.7.4 *Fe$_8$ clusters*

The second molecule that has been intensively investigated for its behaviour as an SMM is commonly indicated as Fe$_8$ and its structure has already been presented in Fig. 3.36. The complete formula is $[Fe_8O_2(OH)_{12}(tacn)_6]Br_8(H_2O)_9$, (tacn = 1,4,7-triazcyclononane), and the compound is prepared by controlled hydrolysis of $Fe(tacn)Cl_3$ in a mixture of water and pyridine with the addition of sodium bromide. Only a cluster molecule is contained in the elementary cell, as the compound crystallizes in the acentric P1 space group of the triclinic system (Wieghardt *et al.* 1984).[3]

The internal iron(III) ions are octahedrally coordinated to the two oxides and to four hydroxo bridges. Fe3 and Fe4 coordinate three nitrogen atoms of the tacn molecules two hydroxides and one oxide ion, while the external iron(III) coordinate three nitrogen atoms and three hydroxides. The presence of three different Fe sites has also been confirmed by Mössbauer spectroscopy (Barra *et al.* 1996; Cianchi *et al.* 2002).

The oxides form μ_3 bridges, while the hydroxide ligands are involved in μ_2 bridges. Fe1, Fe2, Fe3, Fe4 form a structure often encountered in polynuclear metal complexes, which has been indicated as a butterfly structure (see

---

[3] Crystallographic cell parameters for Fe$_8$ at 113 K: $a = 10.521(1)$Å, $b = 14.088(1)$Å, $c = 15.089(1)$Å, $\alpha = 89.83(1)°$, $\beta = 109.80(1)°$, $\gamma = 109.43(1)°$, $V = 1968.8(3)$Å^3.

Fig. 4.13). From the magnetic point of view this structure gives rise to spin frustration effects, which make the prediction of the most stable spin arrangement difficult.

The temperature dependence of χT, shown in Fig. 4.28, clearly indicates a ferrimagnetic structure as already observed for $Mn_{12}ac$. The high-temperature value of χT is in fact much lower than the value expected for eight uncoupled $S = 5/2$ spins, $\chi T = 35$ emu K mol^{-1} for $g = 2$. Strong antiferromagnetic interactions must therefore be active, but at low temperature χT increases and goes through a round maximum at $\chi T \approx 54$ emu K mol^{-1}. Such behaviour can be rationalized with an $S = 10$ ground state confirmed by high-field magnetization measurements. In the simplest possible approach this ground state can be justified by putting six $S = 5/2$ spins up and two down. In idealized D_2 symmetry only four different exchange interactions are present, as schematized in the inset of Fig. 4.28.

The Heisenberg exchange Hamiltonian becomes:

$$H_{ex} = -J_1\mathbf{S}_1 \cdot \mathbf{S}_2 - J_2(\mathbf{S}_1 \cdot \mathbf{S}_3 + \mathbf{S}_1 \cdot \mathbf{S}_4 + \mathbf{S}_2 \cdot \mathbf{S}_3 + \mathbf{S}_2 \cdot \mathbf{S}_4)$$
$$- J_3(\mathbf{S}_1 \cdot \mathbf{S}_5 + \mathbf{S}_1 \cdot \mathbf{S}_8 + \mathbf{S}_2 \cdot \mathbf{S}_6 + \mathbf{S}_2 \cdot \mathbf{S}_7)$$
$$- J_4(\mathbf{S}_3 \cdot \mathbf{S}_5 + \mathbf{S}_3 \cdot \mathbf{S}_7 + \mathbf{S}_4 \cdot \mathbf{S}_6 + \mathbf{S}_4 \cdot \mathbf{S}_8). \tag{4.7}$$

Using the irreducible tensor operators approach described in Section 2.5 it has been possible to evaluate the magnetic susceptibility even if a true fitting procedure up to now has never been attempted. Starting from magneto-structural correlation (Gorun and Lippard 1991) it can be expected that between the AF interactions mediated by the oxo-ions J_2 dominates because it corresponds to the largest Fe–O–Fe angle (128.8° versus 96.8°). This angular dependence of the exchange interaction in Fe(III) polynuclear compounds is now well established (Weihe and Güdel 1997b; Le Gall et al. 1997). Similar reasoning suggests that

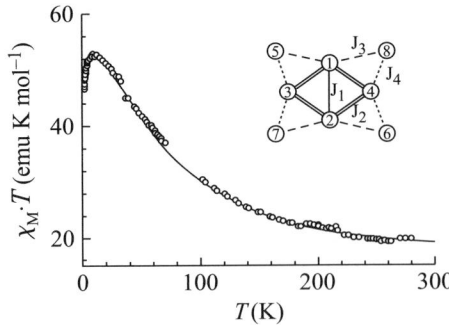

FIG. 4.28. Temperature dependence of the χT product for Fe_8. The solid line corresponds to the calculated values using the exchange Hamiltonian (4.7) and the labelling scheme is shown in the inset of the figure.

the interaction described by J_4 is stronger than J_3. A reasonable agreement with the experimental data of Fig. 4.28 has been obtained with the parameters $J_1 = -200$ K, $J_2 = -36$ K, $J_3 = -26$ K, $J_4 = -59$ K. The low temperature decrease of χT is reproduced by introducing magnetic anisotropy effects (*vide infra*). With this set of parameters the spin structure of the ground state is that described in Fig. 3.36, with the spins of Fe3 and Fe4 oriented antiparallel to the other spins.

As already mentioned in Chapter 3, the fitting of the magnetic susceptibility with many independent parameters, four in the present case, often results in overparametrization with a large uncertainty on the parameters. Polarized neutron diffraction provides unambiguous information on the spin structure through the reconstruction of a spin density map such as that shown in Fig. 3.36 for Fe_8 (Pontillon *et al.* 1999), which confirm the proposed spin structure.

The ground $S = 10$ state is largely split in zero field as shown by the HF-EPR spectra reported in Fig. 3.28 (Barra *et al.* 1996; Barra *et al.* 2000), inelastic neutron scattering data reported in Fig. 3.39 (Caciuffo *et al.* 1998; Amoretti *et al.* 2000), and far-infrared spectroscopy (Mukhin *et al.* 2001). The temperature dependence of the far-infrared spectra is shown in Fig. 4.29. Four clear transitions are observed corresponding to energy values, divided by k_B, of 3.2, 3.5, 4.2, and 5.3 K. The four peaks show different temperature dependence with the one at higher energy strongly gaining intensity at low temperature. It has then been associated to the transition involving the ground doublet and thus to the $m = \pm 10 \rightarrow m' = \pm 9$ transitions.

Since the cluster has no symmetry the spin Hamiltonian describing the magnetic anisotropy is more complex than that used for Mn_{12}ac. The experimental data were fit with the following zero-field Hamiltonian defined in the basis set of the $S = 10$ functions of the ground state:

$$H_A = D\mathbf{S}_z^2 + E(\mathbf{S}_x^2 - \mathbf{S}_y^2) + B_4^0\mathbf{O}_4^0 + B_4^2\mathbf{O}_4^2 + B_4^4\mathbf{O}_4^4 \qquad (4.8)$$

which assumes D_2 symmetry. The values of the relevant parameters obtained by fitting of data provided by different experimental techniques are shown in Table 4.5.

The differences between HF-EPR and INS data have been attributed to the presence of low-lying excited spin states that can be admixed with the ground state, an effect that has been named S-mixing (Liviotti *et al.* 2002). Single-crystal HF-EPR spectra have also provided information on how the directions of the principal axes of the **D** tensor are oriented in respect of the molecular reference frame as shown in Fig. 4.30. The easy axis of magnetization is almost perpendicular (ca. 10°) to the normal to the plane of the 'butterfly' core of the cluster defined by Fe1, Fe2, Fe3, and Fe4, while the intermediate axis passes through the Fe1 and Fe2 ions, this direction being an idealized binary axis of the molecule. The hard axis is obviously orthogonal to the other two.

The above results show that the magnetic anisotropy has a large rhombic component, mainly determined by the large value of the ratio $|E/D|$. Consequently

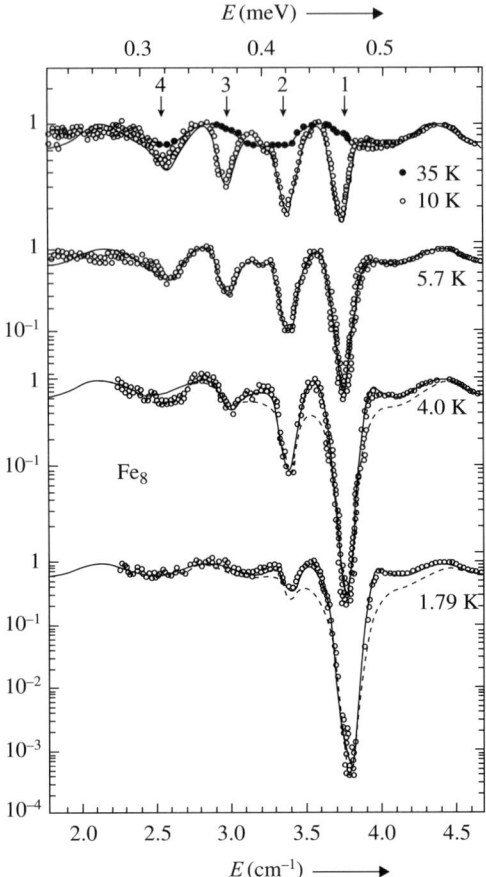

FIG. 4.29. Far-infrared spectra in zero field of Fe_8 at different temperatures. The peaks labelled as 1, 2, 3 and 4 correspond to the $m \to m'$ transitions, $\pm 10 \to \pm 9$, $\pm 9 \to \pm 8$, $\pm 8 \to \pm 7$, $\pm 7 \to \pm 6$, respectively. From Mukhin *et al.* (2001). Copyright (2001) American Physical Society.

TABLE 4.5. Magnetic anisotropy parameters for Fe_8 evaluated from HF-EPR, neutron scattering, and far-infrared spectroscopy. The spin Hamiltonian is defined in (4.8) where the Zeeman term has been added to reproduce the HF-EPR experiments.

D (K)	$|E/D|$	B_4^0 (K)	B_4^2 (K)	B_4^4 (K)	Ref
-0.295	0.19	2.3×10^{-6}	-7.2×10^{-6}	-1.2×10^{-5}	(Barra *et al.* 2000)
-0.292	0.16	1.0×10^{-6}	1.2×10^{-7}	8.6×10^{-6}	(Caciuffo *et al.* 1998)
-0.295	0.15	2.0×10^{-6}	1.2×10^{-7}	8.6×10^{-6}	(Mukhin *et al.* 2001)

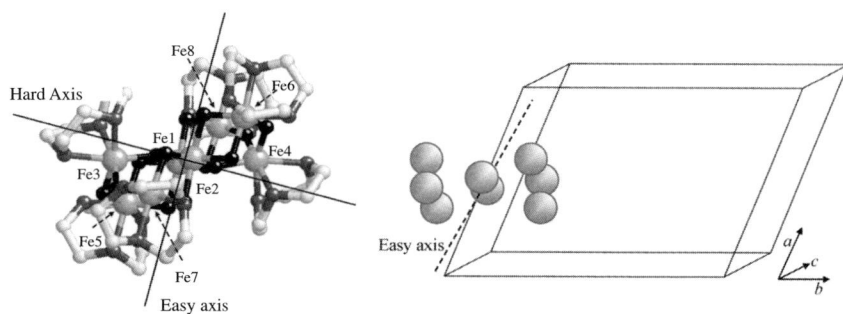

FIG. 4.30. Left: view of the molecular structure of the Fe_8 cluster superimposed with the principal axes of the magnetic anisotropy. The intermediate axis, not shown, passes through Fe1 and Fe2. Right: orientation of the easy axis in respect of the unit cell. The shape of the crystals reflects the triclinic unit cell and is that of thin slabs with the faces $(0, 0, \pm 1)$ being the largest ones.

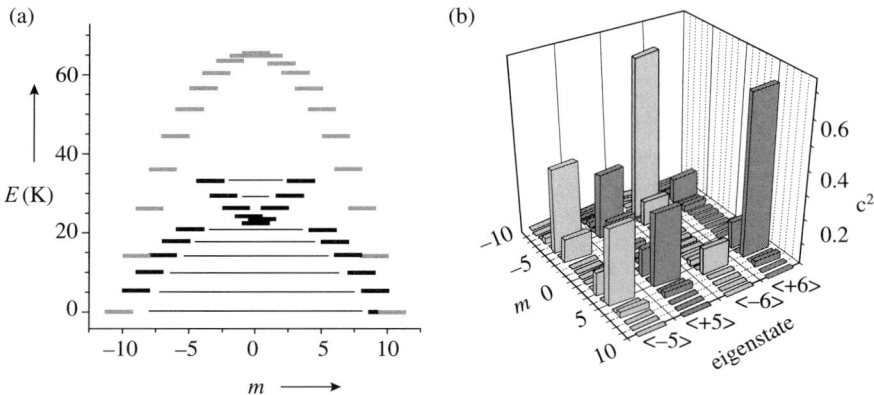

FIG. 4.31. (a) Calculated energy levels of the split components of the $S = 10$ ground multiplet of $Mn_{12}ac$ (grey) and Fe_8 (black). The levels of the latter in the upper part are strongly admixed and their m assignment in the graph is arbitrarily chosen for the sake of clarity. (b) Wavefunction composition of the formally labelled $\langle +6 \rangle$, $\langle -6 \rangle$, $\langle +5 \rangle$, and $\langle -5 \rangle$ eigenstates of Fe_8 in the m basis. A weak longitudinal field has been considered to localize the eigenstates on one side of the graph.

the energies of the m levels are very different from those calculated for $Mn_{12}ac$, as shown in Fig. 4.31a. The lowest lying levels are almost degenerate in pairs $\pm m$, but approaching the top of the barrier the levels are strongly admixed, as shown in Fig. 4.31b, and the m labelling loses any significance. In fact above the levels which can be loosely indicated as $+5$ and -5 there are three non-degenerate levels, admixtures of states with small m. Above these states three pairs of quasi-degenerate levels are again observed but they do not correspond, not even approximately, to $\pm m$ pairs.

The origin of the magnetic anisotropy responsible of the observed zero-field splitting is presumably a mixture of dipolar and single-ion contributions. The intracluster dipolar contributions can be easily and rather accurately calculated assuming that the magnetic moments are localized on the iron ions and using the point dipolar approximate formula (2.50) with the generalization outlined in section 2.5.2. In this frame the calculated zero-field splitting parameters D and E are much smaller than the experimental values, suggesting that the single-ion anisotropy of the Fe^{III} centres is relevant. This is not an unexpected result, because similar trends were observed in other polynuclear iron(III) compounds (Abbati et al. 2001).

Contrary to Mn_{12}, Fe_8 does not seem to be stable in solution. This has prevented the synthesis of different derivatives as encountered in the Mn_{12} family. Only minor variations are possible for Fe_8, involving either pseudopolymorphs,[4] i.e. crystallization with different unit cell and with different solvation molecules, or replacement of the Br^- counterions with ClO_4^- (Barra et al. 2001).

4.7.5 Mn$_4$ clusters

Several different topologies are encountered in tetranuclear manganese clusters and some of them are shown in Fig. 4.13. The most frequently investigated class, showing SMM behaviour at relatively high temperature (Aubin et al. 1998), comprises a mixed valence cubane core of formula $[Mn_4(\mu_3\text{-O})_3(\mu_3\text{-X})]^{6+}$ schematized in Fig. 4.32 (Wemple et al. 1993). Structural evidence suggests a $[Mn^{IV}Mn_3^{III}]$ localized charge distribution. Three carboxylates bridge the Mn^{IV} centre to the Mn^{III} ones. The coordination octahedron of Mn^{III} centres is completed either by a chelating ligand, like β-diketonates, or by two monodentate ligands, usually a Cl^- and an N-based neutral ligand such as pyridine. The general formula is therefore $[Mn_4(\mu_3\text{-O})_3(\mu_3\text{-X})(\mu_2\text{-O}_2CR)_3L_3]$ or $[Mn_4(\mu_3\text{-O})_3(\mu_3\text{-X})(\mu_2\text{-O}_2CR)_3L_3\,L_3']$, respectively, where X is a mono-negative ion like F^-, Cl^-, Br^-, N_3^- etc.

The clusters possess either strict trigonal symmetry, when they crystallize in trigonal space groups, or idealized trigonal symmetry. In some clusters, like $[Mn_4O_3(O_2C\text{-}C_6H_4-p\text{-}R')_4(dbm)_3]$ with $R' = H, CH_3, OCH_3$, the unique bridge X^- is replaced by a carboxylate. This last one acts as a [3.2,1] bridge destroying the three-fold symmetry as shown in Fig. 4.32.

The first Mn_4 cluster of this type was obtained in 1987 in a synthetic effort aimed at the modelling of the photosynthetic oxidation centre in photosystem II (Bashkin et al. 1987). From the very beginning it was noticed that these

---

[4] Crystals of Fe_8 with a more complex morphology can be obtained and have the unit cell parameters: $a = 13.27$ Å, $b = 13.57$ Å, $c = 15.08$ Å, $\alpha = 114.8°$, $\beta = 108.1°$, and $\gamma = 101.4°$ (unpublished results corteously provided by C. Sangregorio).

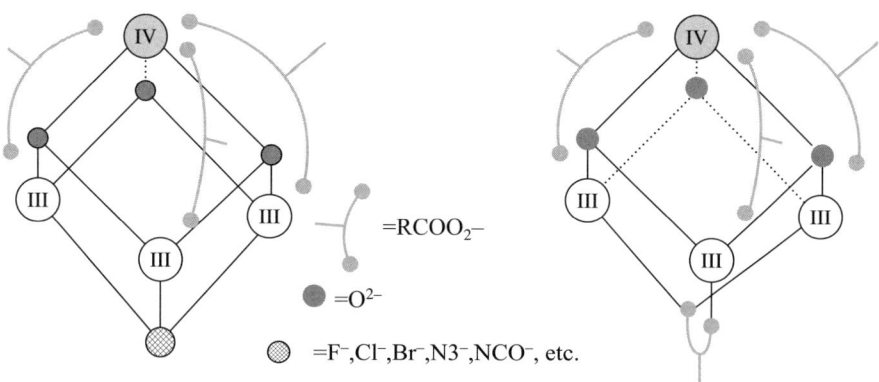

FIG. 4.32. Schematic view of the structure of $[\text{Mn}_4(\mu_3\text{-O})_3(\mu_3\text{-X})(\mu_2\text{-O}_2\text{CR})_3]^{3+}$ clusters (right) and of $[\text{Mn}_4(\mu_3\text{-O})_3(\mu_3\text{-O}_2\text{CR})(\mu_2\text{-O}_2\text{CR})_3]^{3+}$ clusters (left). The Mn^{IV} centre is shown in grey.

clusters are all characterized by a ground state with spin $S = 9/2$. This ferrimagnetic structure results from the ferromagnetic interaction between Mn^{III} ions and antiferromagnetic coupling with Mn^{IV}. Typical exchange interactions range between 20 K and 45 K for $J_{\text{III}-\text{III}}$ and -55 to -90 K for $J_{\text{III}-\text{IV}}$. The ground spin state is well isolated, being the first excited state, characterized by $S = 7/2$, at least 250 K above in energy.

The ground spin state is split by the magnetic anisotropy, and the major contribution is provided by the single-ion anisotropy of Mn^{III} centres, which show a pronounced Jahn–Teller elongation along the M-X bond. The ground state ZFS tensor $\mathbf{D}$ is related to the single-ion contribution according to

$$\mathbf{D}_{\mathbf{S}=\mathbf{9/2}} = d_i \sum_{i=1}^{3} \mathbf{D}_i. \tag{4.9}$$

The d_i coefficients can be calculated using the projection techniques presented in Chapter 2, provided that the spin ground state wavefunction is known. In the present case the $S = 9/2$ state is well represented by a coupling scheme where the Mn^{III} spins are first coupled ferromagnetically to each other to give an intermediate spin $S_{\text{Mn}}^{\text{III}} = 6$, and then antiparallel to that of Mn^{IV}. This results in $d_i = 35/242$ for every i. As already emphasized above (4.9) involves tensors and therefore their relative orientation is of paramount importance. In fact, if the clusters had strict cubic symmetry the M-X bonds would be perpendicular to each other and the magnetic anisotropy would be quenched. In trigonal symmetry (4.9) can be rewritten in a simpler form:

$$D_{\mathbf{S}=\mathbf{9/2}} = \frac{3}{2}\frac{35}{242} D_{\text{Mn}^{\text{III}}} (3\cos^2\alpha - 1) \tag{4.10}$$

where D is now a scalar quantity representing the axial anisotropy parameter and α is the angle formed by the elongation axis of Mn^{III} and the trigonal axis passing through Mn^{IV} and X.

The larger the deviation from the magic angle, $\alpha = 54.7356°$ (for which arcos $\alpha = 1/\sqrt{3}$ and the term in parenthesis of (4.10) goes to zero) the larger is the projection of the single-ion anisotropy on the ground state.

The large variety of similar clusters has stimulated many physical characterizations in order to establish correlations between structure and magnetic properties. In particular a INS study has been performed on four derivatives of the $[Mn_4O_3X(O_2CMe)_3(dbm)_3]$ series, with $X^- = F^-$, Cl^-, Br^-, and $MeCO_2^-$ while dbm^- is the anion of dibenzoylmethane (Andres *et al.* 2000). None of these compounds has rigorous trigonal symmetry and the spin Hamiltonian that has been used to fit the data includes a rhombic E term, even if $|E/D|$ does not exceed 0.04. Good agreement between the trends of α and D in the four compounds has been observed. D ranges between -0.545 K for $X = F$ and -0.761 K for $X = Cl$, while α is $54.2°$ and $45.1°$, respectively. Of course the crystal field generated by the different X^- ligands plays a role not only in determining the α parameter but also the amplitude of the single-ion magnetic anisotropy. The

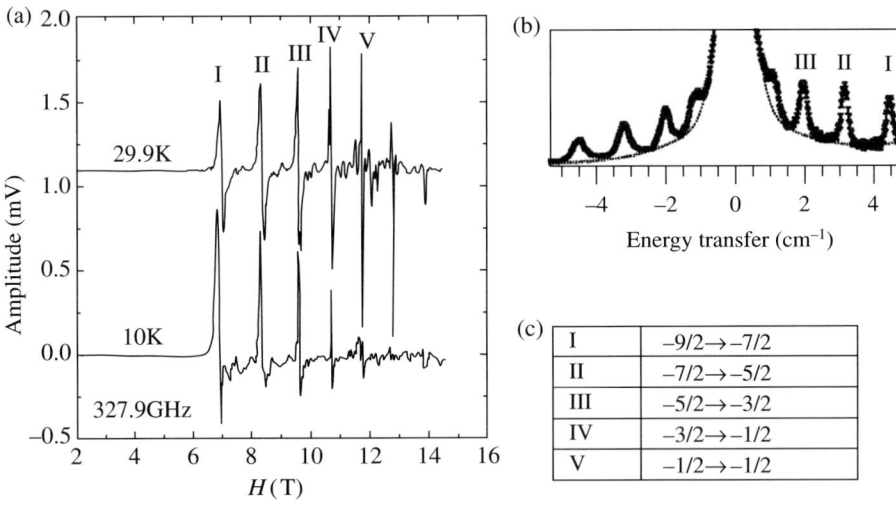

FIG. 4.33. (a)   HF-EPR   spectra   of   an   oriented   sample   of $[Mn_4O_3Cl(O_2CMe)_3(dbm)_3]$ recorded at 327.9 GHz with the applied field parallel to the easy axis (after Aubin *et al.* 1998). (b) Inelastic neutron scattering data for the same compound at 18 K (after Andres *et al.* 2000). (c) Assignment of the peaks with $m \to m'$ transitions. In the case of INS data the assignment only refers to the absolute value of m. Copyright (1998 and 2000) American Chemical Society.

blocking temperature, as determined from ac susceptibility, scales well with the D values.

In Fig. 4.33 are reported, for comparison, the HF-EPR spectra on an oriented sample of the $X = Cl$ derivative (Aubin *et al.* 1998) and the INS spectra (Andres *et al.* 2000), together with the assignment of the observed peaks. What is evident in both experiments is the non-regular separation of the lines, while a regular spacing should indeed be observed if only axial second-order magnetic aniso-tropy is present. As line separation decreases on going towards the centre of the spectra, i.e. decreasing $|m|$ of the involved transitions, a sizeable fourth-order axial term, $\mathbf{O_4^0}$, is expected to be active and with a negative B_4^0 parameter. Very good agreement between parameters estimated from the two techniques has been observed: $D = -0.76$ K, $B_4^0 = -1.1 \times 10^{-4}$ K and $D = -0.76$ K, $B_4^0 = -9.4 \times 10^{-5}$ K, for HF-EPR and INS, respectively. The better agreement between the two techniques, compared to that observed for Fe$_8$, could reflect the presence of a much weaker S-mixing, given the large separation observed in Mn$_4$ clusters between the ground and the first excited spin state. In the case of neutron data an E term, $E = 0.032$ K, has been included in the analysis, but this term could not be evaluated from HF-EPR spectra because these last ones have been recorded with the applied field along the easy axis of magnetization.

5

THERMALLY ACTIVATED MAGNETIC RELAXATION

5.1 Relaxation and relaxation time

In the previous chapters, especially in Section 3.1, it was shown how the dynamic properties of a magnetic material can be experimentally investigated. An essential feature is the relaxation time τ, which in the simplest case is defined by (3.9). In the present chapter a quantitative interpretation of the experimental results will be presented. The simplest case is when the relaxation is so slow that the decay of the magnetization can be directly measured. This is generally so at low enough temperature. Typically, one takes a magnetic material and applies a strong magnetic field H_1 in the easy magnetization direction. One waits some time until the magnetization has reached its equilibrium value. Then, at time $t = 0$, one switches the field off, or one applies a field H in the opposite direction, and one measures the magnetization $\mathbf{M}(t)$ at time t. The field H_1 is such that the initial magnetization is negative and close to its saturation value. The field H is weak or even equal to 0. In this and the next chapters, the material will be assumed to be a single, perfectly periodic crystal.

This chapter is mainly devoted to situations where the relaxation is exponential,

$$M(t) = M_{\text{eq}}(H) + \delta M_0 \exp(-t/\tau). \tag{5.1}$$

This relation generalizes (3.9) and defines the *relaxation time* τ. The equilibrium magnetization $M_{\text{eq}}(H)$ has been introduced. The materials which will be considered are generally paramagnetic, so that if the field H vanishes, then $M_{\text{eq}}(0) = 0$.

If τ is measured at different temperatures T, a decrease is observed which is often well represented by the Arrhenius formula

$$\tau = \tau_0 \exp \frac{T_0}{T}. \tag{5.2}$$

Since one likes to have a straight line, it may be of interest to plot $1/\ln(\tau/\tau_0)$ as a function of T. If the Arrhenius formula (5.2) holds, then

$$\frac{1}{\ln(\tau/\tau_0)} = \frac{T}{T_0}. \tag{5.3}$$

Figure 5.1 gives a schematic picture of measurements on $Mn_{12}ac$ (Paulsen *et al.* 1995; Hernández *et al.* 1996) and Fe_8 (Caneschi *et al.* 1998; Barra *et al.* 2001;

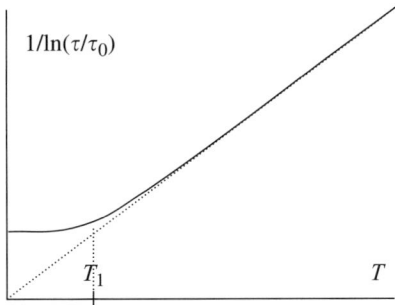

FIG. 5.1. Typical behaviour of the relaxation time τ of the magnetization as a function of temperature. The Arrhenius law is verified at high enough temperature while τ is constant at low temperature. The cross-over value T_1 is of the order of 2 K in $Mn_{12}ac$, 0.5 K in Fe_8.

Gatteschi and Sessoli 2003). It has been shown in Section 4.7.1 that this behaviour is indeed observed for $Mn_{12}ac$, with $\tau_0 \simeq 10^{-7}$ s and $T_0 \simeq 60$ K. It follows from (5.3) that the relaxation of the magnetization becomes slow at low temperature.

Formula (5.3) is only satisfied for temperatures larger than a threshold T_1. The reason for that will be seen in the next few chapters. The present chapter is devoted to temperatures larger than T_1 but appreciably lower than T_0, say between $T_0/20$ and $T_0/2$. The corresponding relaxation times are moderate. The experimentalist can measure the time with an ordinary watch, but does not need to spend several weeks for each measurements.

5.2 Potential barrier

The Arrhenius law (5.2) suggests the presence of a 'potential barrier'. It will now be argued that such a barrier does exist for the magnetization of each molecular group $Mn_{12}ac$.

Inside such a molecular group, the dominant magnetic interaction is the exchange. We are in the strong exchange limit defined in Chapter 2. At low enough temperature, only the states with lowest exchange energy can be reached. Since the isotropic exchange Hamiltonian commutes with the total spin $\mathbf{S}$, these states have a well-defined value of $\mathbf{S}^2$, which in the following chapters will be called $s(s+1)$ (to avoid confusion between the operator $\mathbf{S}$ and the number s). Thus, the magnetic properties of the molecule are those of a spin of modulus s. As seen in the preceding chapter, $S = 10$ in the case of $Mn_{12}ac$, and also in the case of Fe_8 which also corresponds to the strong exchange limit. This spin is subject to an anisotropy, or crystal field which, in the case of a tetragonal crystal like $Mn_{12}ac$, is well described by the Hamiltonian (2.10), $\mathcal{H}_{an} = DS_z^2$. In $Mn_{12}ac$, the constant D turns out to be negative, as discussed in Section 4.7.1.

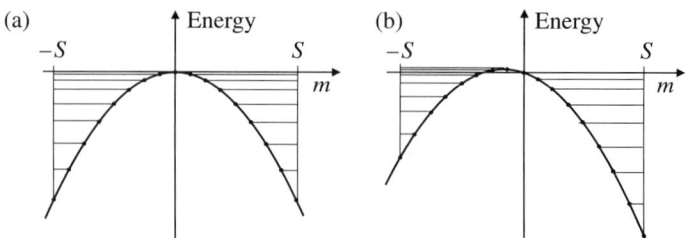

FIG. 5.2. Right-hand side of (2.10) as a function of S_z, in zero magnetic field (a) and in a negative magnetic field parallel to the easy axis (b). Horizontal lines indicate the energies $-|D|m^2 + g\mu_B H_z m$ of the quantum states. The graph (a) can be considered as the quantum analogue of Fig. 1.4.

Therefore, the curve representing DS_z^2 as a function of S_z is a parabola (Fig. 5.2) whose maximum at $S_z = 0$ is indeed a potential barrier. At low temperature, the spin should be in one of the two sides of a double potential well. This description is similar to that of a particle in a double potential well.

In this chapter and in the following ones until chapter 12, the properties of a large spin s subject to a double well anisotropy will be studied. Moreover, D will be replaced by $-|D|$, not to forget its sign.

The eigenvectors of the Hamiltonian (2.10) are the eigenvectors $|m\rangle$ of S_z The eigenvalues of the Hamiltonian (2.10) are therefore quantized and equal to

$$E_m = -|D|m^2 \tag{5.4}$$

where $m = -s, -(s-1), -(s-2), \ldots, (s-1), s$ designates the eigenvalues of S_z, therefore $S_z|m\rangle = m|m\rangle$. There are $(2s+1)$ eigenvectors and $(2s+1)$ eigenvalues.

Formula (2.10) is only an approximation. In a crystal of symmetry lower than tetragonal, the anisotropy Hamiltonian is given by (2.5) at lowest order and it is not easy to obtain its eigenvalues and eigenvectors.

The simplest idea is to treat the E-term in (2.5) as a perturbation, although in practice $|E/D|$ is not always very small, as seen from Table 4.5. In particular, for the states $|m\rangle$ with small values of m, the level spacing is of the order of D while the off-diagonal perturbation is of the order of Es^2 which is much larger! As seen in Chapter 4, numerical calculations confirm that formula (5.4) is quite a bad approximation for small values of $|m|$, although it is still acceptable for the lowest states.

It is easily understood that the relaxation of the magnetization becomes slow at low temperature. Indeed the spin can hardly get out of its half-potential well to go into the other one. On the other hand, it can more easily do that at higher temperature, because it receives a lot of energy from phonons, which help it to climb to the top of the barrier and then go down on the other side. This process is called *thermal activation*.

At equilibrium, the probability p_m^0 that the spin is in state $|m\rangle$ is given by (3.21), $p_m^0 = (1/Z)\exp[-\beta(E_m)]$. Intuitively, it may be expected that the relaxation rate $1/\tau$ is proportional to the probability to be at the top of the barrier. Therefore, at low temperature and in vanishing field, introducing a proportionality constant $1/\tau_0$,

$$1/\tau = (1/\tau_0)p_0^0 = (1/Z)(1/\tau_0)\exp(-\beta E_0)$$
$$\approx (1/\tau_0)\exp[-\beta(E_0 - E_s)] \approx (1/\tau_0)\exp(-\beta|D|s^2).$$

This is exactly the Arrhenius formula (5.2) with

$$k_B T_0 = |D|s^2. \tag{5.5}$$

This quantity $k_B T_0$ is called the activation energy. The above formulae hold for an integer spin. If s is half-integer, then in vanishing field $k_B T_0 = |D|s^2 - |D|/4$.

In the presence of a field H_z parallel to the easy axis, the energy as a function of $S^z = m$ is

$$E_m^{(0)} = -|D|m^2 + g\mu_B H_z m \tag{5.6}$$

which has its maximum for $2|D|m = g\mu_B H_z$ or

$$m_{\max} = (g\mu_B H_z)/(2|D|). \tag{5.7}$$

This energy maximum is

$$E_{\max} = (g\mu_B H_z)^2/(4|D|). \tag{5.8}$$

The spin has been initially placed in the state $m = -s$ where its energy is

$$E_{-s} = -|D|s^2 - g\mu_B H_z s. \tag{5.9}$$

Formula (5.5) must therefore be replaced by

$$k_B T_0 \approx |D|s^2 + g\mu_B H_z s + (g\mu_B H_z)^2/(4|D|) = |D|[s + g\mu_B H_z/(2|D|)]^2. \tag{5.10}$$

Therefore, the field lowers the potential barrier for a spin situated in the 'wrong' well ($m < 0$ if $H_z < 0$).

Formulae (5.5)–(5.10) will be modified in Chapter 10. They clearly give no information on the prefactor τ_0 of formula (5.2). To obtain this information, it is necessary to take time explicitly into account. The simplest way to do that is to postulate the existence of transition probabilities.

5.3 Transition probabilities and the master equation

In this section we wish to formulate a simple picture of the evolution of a particular molecular spin **S**. As a first approximation, its energy or Hamiltonian is a function of its components S_x, S_y, S_z only. This Hamiltonian is a result of anisotropy. It may have the simple form (2.10) and then its eigenvectors are the eigenvectors $|m\rangle$ of S_z. Or it may have a more complicated form as (2.5), but the eigenvectors

$$|m^*\rangle = \sum_{m'} \varphi_{m'}^{(m)} |m'\rangle \qquad (5.11)$$

will be assumed not to be very different from $|m\rangle$. This is only correct for low energy states ($|m|$ large) as seen in Fig. 4.31. It should be stressed that the asterisk modifies the state vector, not the number m. For instance, the notation $|10^*\rangle$ has a meaning, but 10^* would mean nothing else than the complex conjugate of 10, which is of course also 10.

In this chapter, the following picture will be used. At a given time t, the spin has a certain probability $p_m(t)$ of being in state $|m^*\rangle$. But in a short time dt it has a certain probability $\gamma_m^{m'} dt$ to make a transition to some other state $|m'^*\rangle$. If the transitions are independent of each other, the probabilities $p_m(t)$ evolve according to the equation

$$\frac{d}{dt} p_m(t) = \sum_q \left[\gamma_q^m p_q(t) - \gamma_m^q p_m(t) \right] \qquad (5.12)$$

which is called the *master equation*. Such an evolution is called a *Markov process*. In the present case the master equation is a system of $(2s + 1)$ equations, or an equation for the matrix of the coefficients $p_m(t)$.

Though it is intuitive, equation (5.12) cannot be easily justified. 'We face a fundamental problem: the (classical or quantum) equations which determine the motion of the system at the microscopic scale are invariant by time reversal; however the macroscopic evolution of the system, which follows the master equation (5.12), has not this property' (Diu *et al.* 1989). This problem has been much debated since the time of Ludwig Boltzmann and Henri Poincaré, and the debate involves highly complicated mathematics. A flavour of the solution in a particular case is given in Appendix C.

In fact, equation (5.12) is only acceptable for times larger than a microscopic time τ_{col}, necessary for the establishment of irreversibility in a macroscopic system. Irreversibility is typical of a large, macroscopic system. For short times, each microscopic element (spin, molecule, atom) is reacting with its neighbours and does not know that it is in a large system. For instance, in a gas, irreversibility is a result of the collisions. For short times, the evolution is a result of the collisions which are taking place, and are microscopic processes described by time-reversible equations. In that case τ_{col} is the duration of a collision. Only for $t > \tau_{col}$ can the Boltzmann equation be applied.

The reality of time irreversibility is unfortunately obvious in daily life. *Fugit irreparabile tempus*, as Latins said.

Coming back to spin, its transitions in (5.12) are effects of its interaction with its environment, and especially collisions with phonons. The transition probabilities γ_q^m arising from the interaction with phonons will be calculated in Section 5.6.

Independently of the interaction mechanism, equation (5.12) has general properties. A simple one is that the equilibrium value $p_m^0 = (1/Z)\exp(-\beta E_m)$ of p_m (see formula 3.21), is a trivial solution of (5.12),

$$\sum_q \left[\gamma_q^m p_q^0 - \gamma_m^q p_m^0\right] = 0. \tag{5.13}$$

Another general property is the principle of detailed balance. It states that not only the sum (5.13), but each individual term vanishes. Thus, $\gamma_m^{m'} p_m^0 = \gamma_{m'}^m p_{m'}^0$, or

$$\gamma_m^{m'}/\gamma_{m'}^m = p_{m'}^0/p_m^0 = \exp[\beta(E_m - E_{m'})]. \tag{5.14}$$

The above argument looks *a priori* rather poor. Clearly, although the sum (5.13) must vanish at equilibrium, this does not oblige each term of that sum to vanish! However, in second-order perturbation theory, the detailed balance condition can be proven, and will be shown in Sections 5.6 and 5.7.

5.4   Solution of the master equation

The reader, who hopefully has full confidence in the authors of this book, will probably expect to find a solution of (5.12) which has the exponential form (5.1). As a matter of fact, anybody who has some knowledge of linear differential equations will easily find $(2s + 1)$ exponential solutions of (5.12), with $(2s + 1)$ characteristic times τ_k labelled by an index $k = 0, 1, ..., 2s$. Indeed, substituting the exponential form

$$p_m(t) = \varphi_m^{(k)} \exp(-t/\tau_k) \tag{5.15}$$

into (5.12), one obtains

$$-\frac{1}{\tau_k}\varphi_m^{(k)} = \sum_q \left[\gamma_q^m \varphi_q^{(k)} - \gamma_m^q \varphi_m^{(k)}\right] = \sum_q \left[\gamma_q^m - \delta_q^m \sum_{q'} \gamma_m^{q'}\right] \varphi_q^{(k)} \tag{5.16}$$

where the Kronecker symbol δ_q^m (=1 if $q = m$, while $\delta_q^m = 0$ if $q \neq m$) has been introduced. Equation (5.16) expresses the fact that $-1/\tau_k$ is an eigenvalue of the 'master' matrix defined by its $(2s + 1) \times (2s + 1)$ elements

$$\Gamma_q^m = \gamma_q^m - \delta_q^m \sum_{m'} \gamma_m^{m'}. \tag{5.17}$$

Thus there are $(2s+1)$ characteristic times τ_k which are the roots of an algebraic equation of order $(2s + 1)$, namely $\det(\Gamma - 1/\tau)=0$. One of the characteristic times, $\tau_k = \tau_0$, is infinite according to (5.13). It corresponds to equilibrium. If the system is at equilibrium at some time, it will still be in the same state at $t = \infty$.

The eigenvector in (5.16) is

$$
\begin{bmatrix}
\varphi_{-s}^{(k)} \\
\varphi_{-s+1}^{(k)} \\
\cdots\cdots\cdots \\
\varphi_{s}^{(k)}
\end{bmatrix}.
\tag{5.18}
$$

The master matrix (5.17) is a square, $(2s + 1) \times (2s + 1)$ matrix which is *not* Hermitian. This is unfortunate because Hermitian matrices have simple properties. However, the eigenvalue equation (5.16) can be written in terms of the Hermitian, real matrix

$$
\tilde{\Gamma}_m^{m'} = \Gamma_m^{m'} \exp[\beta(E_{m'} - E_m)/2] = \tilde{\Gamma}_{m'}^m.
\tag{5.19}
$$

The Hermiticity of that matrix results from (5.14).

The eigenvalue equation (5.16) can be written as

$$
\sum_q \tilde{\Gamma}_q^m \tilde{\varphi}_q^{(k)} = -\frac{1}{\tau_k} \tilde{\varphi}_m^{(k)}
\tag{5.20}
$$

where

$$
\tilde{\varphi}_m^{(k)} = \varphi_m^{(k)} \exp(\beta E_m/2).
\tag{5.21}
$$

Since $\tilde{\Gamma}$ is Hermitian, it has $(2s+1)$ real eigenvalues $1/\tau_k$ and $(2s+1)$ eigenvectors. The matrix (5.16) has obviously the same property.

It follows from (5.13) that one eigenvalue is equal to 0. The corresponding eigenvector of (5.17) is p_m^0. It is shown in Appendix D that all other rates $1/\tau_k$ are positive.

The general solution of (5.12) is a linear combination of all exponential solutions,

$$
p_m(t) = \sum_k \lambda_k \varphi_m^{(k)} \exp(-t/\tau_k)
\tag{5.22}
$$

where the parameters λ_k are determined by the initial values $p_m(0)$.

Now, the magnetization $M(t)$ has the form $M(t) = \sum p_m(t) M_m$ which was already encountered in Chapter 3. Substituting in this formula $p_m(t)$ as expressed in (5.22), $M(t)$ is seen to be a sum of exponentials $\exp(-t/\tau_k)$ and an additive constant (since $1/\tau_0$ is 0). This seems to contradict the experimental result (5.1), according to which there is a *single* exponential. However, at low temperature, most of the exponential functions vanish rapidly, except one of them which vanishes slowly, in agreement with (5.1). In other words, one of the negative eigenvalues $-1/\tau_k$ is very small at low temperature.

This property is a consequence of the shape of the potential, more precisely of the existence of a potential barrier. Since the spin requires a very long time to jump over that barrier, one eigenvalue must be very small while the $(2s - 1)$ other ones correspond to spin motion inside the left or right hand well, and have no reason to depend much on temperature. They give rise to transients which are only important at short times.

Thus, the experimental result (5.1) is understood. The constant corresponds to the vanishing eigenvalue of (5.17), and the exponential corresponds to the very small eigenvalue.

The derivation of the Arrhenius law (5.2) from the master equation is done in Appendix E. The conclusion is in agreement with (5.5).

The Arrhenius law is not correct at too high temperature because the solution of the master equation is a sum of several exponential functions of time. It is not correct at too low temperature either, because of tunnelling, as will be seen in the next chapter.

The theory which has been presented in this section is purely phenomenological. If one wants to go further and to calculate the prefactor τ_0 of the Arrhenius law (5.2), it is necessary to know the source of the energy which allows the spin to reach the top of the barrier. This energy comes from phonons.

5.5 Spin–phonon interaction

5.5.1 *Basic features*

The mechanism of the interaction between molecular spins and lattice vibrations ('phonons') can be described as follows. Under the effect of phonons, each molecular group undergoes a time-dependent transformation which can be analysed as the superposition of a local rotation and a local strain. These two effects produce a time-dependent modification (or 'modulation') of the anisotropy energy. The new terms of the Hamiltonian induce transitions between the eigenstates $|m^*\rangle$ of the static Hamiltonian $\mathcal{H}_{\mathrm{an}}$ defined by (2.10) or (2.5) or any more accurate approximation. In this chapter, it will be assumed that $\mathcal{H}_{\mathrm{an}}$ is well approximated by (2.10), so that its eigenstates are those of S_z, namely $|m\rangle$.

As a matter of fact, phonons modulate all terms of the spin Hamiltonian, including the intramolecular isotropic exchange which is larger than the anisotropy. However, the spin–phonon interaction resulting from the modulation of exchange is of minor importance. Indeed, as will be seen in Section 5.9, it does not induce transitions inside the $(s = 10)$ space of lowest exchange energy.

The above statements will now be translated into equations. This just requires a slight reformulation of the first chapter of any elasticity textbook. Elastic waves ('phonons') transform any vector $\mathbf{r}$ into $\mathbf{r} + \mathbf{u}(\mathbf{r})$, where the displacement is $\mathbf{u}(\mathbf{r})$. To first order in $\mathbf{u}$, the change of an infinitesimal length in this transformation depends only on the symmetric combinations

$$\epsilon_{\alpha\gamma} = (1/2)(\partial_\alpha u_\gamma + \partial_\gamma u_\alpha) \tag{5.23}$$

of the derivatives $\partial_\alpha u_\gamma$, which will be assumed to be small. The tensor of components $\epsilon_{\alpha\gamma}$ is called the strain. The antisymmetric combinations $(\partial_\alpha u_\gamma - \partial_\gamma u_\alpha)$ do not modify lengths, and therefore correspond to rotations. To determine the relation between these combinations and the rotation vector Ω, one can write, for a homogeneous rotation in the absence of strain, $u_x = z\Omega_y - x\Omega_z$ and similar relations for u_y and u_z. Hence, $\Omega_y = \partial_z u_x = -\partial_x u_z$. It follows that $\Omega_y = (\partial_z u_x - \partial_x u_z)/2$, which is still correct in the presence of a strain. The corresponding vector equation is

$$\boldsymbol{\Omega} = -(1/2)\mathbf{rot\ u}. \qquad (5.24)$$

Formulae (5.23) and (5.24) hold to first order for small values of the derivatives $\partial_\alpha u_\gamma$. Thus $\boldsymbol{\Omega}$ is a small quantity. In this book, terms of second or higher order with respect to elastic displacements will generally be ignored (as they are in standard elasticity). Such high-order terms will be briefly discussed in Section 5.8, and it will be seen that they can be important in certain cases.

5.5.2   Local rotation

For the sake of simplicity, tetragonal symmetry will be assumed. If the crystal is rotated by a rotation vector $\boldsymbol{\Omega}$, the new easy axis Z is parallel to a unit vector $\mathbf{e}_Z$ related to the fixed basis vectors $\mathbf{e}_\alpha$ by $\mathbf{e}_Z = \mathbf{e}_z + \boldsymbol{\Omega} \times \mathbf{e}_z = \mathbf{e}_z + \Omega_y \mathbf{e}_x - \Omega_x \mathbf{e}_y$. The anisotropy energy which replaces (2.10) is thus, to first order,

$$\mathcal{H}_{\mathrm{rot}} = -|D|S_Z^2 = -|D|(\mathbf{e}_Z.\mathbf{S})^2 = -|D|(S_z + S_x\Omega_y - S_y\Omega_x)^2. \qquad (5.25)$$

Inserting (5.24) and retaining only linear terms, (5.25) reduces to the original anisotropy term (2.10) plus a correction

$$\delta\mathcal{H}_{\mathrm{rot}} = |D/2|\{S_z, S_x\}(\partial_x u_z - \partial_z u_x) + |D/2|\{S_z, S_y\}(\partial_y u_z - \partial_z u_y). \qquad (5.26)$$

In this equation, valid for a tetragonal crystal only, $\mathbf{S}$ is the spin $\mathbf{S}(\mathbf{R})$ of the molecule situated in the unit cell $\mathbf{R}$, $\mathbf{u}$ is the displacement $\mathbf{u}(\mathbf{R})$ in the unit cell $\mathbf{R}$, and $\delta\mathcal{H}_{\mathrm{rot}}$ is a Hamiltonian $\delta\mathcal{H}_{\mathrm{rot}}(\mathbf{R})$ acting on the spin $\mathbf{S}(\mathbf{R})$. The true effect of local rotations is the sum $\sum \delta\mathcal{H}_{\mathrm{rot}}(\mathbf{R})$ on all $\mathbf{R}$. For the sake of simplicity, this complication has been and will be ignored.

In the derivation of (5.25), the rotation has been assumed uniform. The formula holds as well for a local rotation resulting from phonons (the case of interest here) but in that case there is another contribution to the energy, which results from the strain and will be calculated in the next section.

5.5.3   Local strain

For a tetragonal crystal, the anisotropy energy $-|D|S_z^2$ (formula 2.10) depends on a single parameter D if restricted to quadratic terms. When writing the anisotropy energy $-|D|S_z^2 + E(S_x^2 - S_y^2)$ (formula 2.5) for a triclinic crystal, two energetic parameters E and D were sufficient, although there are also geometric parameters which define the orientation of the easy and hard axes. In the presence of phonons, there is a local strain tensor $\epsilon_{\alpha,\gamma}(\mathbf{r})$, and it is reasonable to expand the magnetic anisotropy energy as a power series of this strain. Only linear terms in ϵ will be kept. On the other hand, they will be assumed to be quadratic in the spin operators. Indeed, for a *static* strain, time-reversal symmetry imposes that linear terms vanish in vanishing magnetic field, in agreement with Chapter 2. It will be assumed that the magnetic field is never strong enough to introduce an appreciable spin–phonon interaction linear in the spin operators. As to the applicability of the rule valid for static strains, this will be examined in Section 5.5.4. Under these hypotheses, and keeping only terms linear in the displacements, the spin–phonon interaction, involving a spin at point $\mathbf{r}$, is

$$\mathcal{H}_{\text{s-ph}} = \sum_{\alpha,\gamma,\xi,\zeta} \tilde{\Lambda}_{\alpha,\gamma,\xi,\zeta} \, \partial_\alpha u_\gamma(\mathbf{r}) \, S_\xi S_\zeta. \tag{5.27}$$

This formula takes both strain and rotation into account.

Each index $\alpha, \gamma, \xi, \zeta$ can take three values x, y, z, so that the number of coefficients $\Lambda_{\alpha,\gamma,\xi,\zeta}$ cannot be larger than $3^4 = 81$, which is already quite large. The reduction of these coefficients by various symmetries has been discussed by Lüthi (1980). See also Hartmann-Boutron *et al.* (1996). As to the order of magnitude of the Λ coefficients, it is presumably that of D; anyway it cannot be smaller, as follows from Section 5.5.2.

Hamiltonian (5.27) is a perturbation to be added to the steady anisotropy term (2.10) and to the Hamiltonian of free phonons.

Hamiltonian (5.27) contains a term in S_z^2 which just modifies quantitatively the static Hamiltonian, e.g. (2.10), but is unable to produce transitions. More interesting terms are those in $S^z S^x$ and $S^z S^y$, which change m by ± 1, and terms in S_x^2, S_y^2 and $S^x S^y$ which change m by ± 2. Thus, the spin can climb the s steps, either step after step, or skipping a step from time to time. Thus

$$|m - m'| = 1 \text{ or } 2. \tag{5.28}$$

This selection rule is exact if the Hamiltonian is the sum of the spin Hamiltonian (2.10) and a spin–phonon Hamiltonian which one treats perturbatively at lowest order as explained in the following section. However, if the spin Hamiltonian contains terms which do not commute with S^z as in (2.5), its eigenvectors are not $|m\rangle$, but $|m^*\rangle$, and the selection rule (5.28) is only approximate. For instance, in Fe_8, a single phonon has some probability to produce a transition from $|m^*\rangle$ to $|(m+3)^*\rangle$. This probability is weak if $E/|D|$ is small. The method of working with the true eigenvectors $|m^*\rangle$ in a numerical treatment is explained in Appendix F.

The terms (5.27) have been derived for a constant strain, following the scheme of the Born–Oppenheimer approximation. In the next section, it will be seen that another type of spin–phonon interaction exists, which cannot be derived from the Born–Oppenheimer approximation.

In addition to (5.27), there may be terms containing spin operators $S_R^\alpha \, S_{R'}^\gamma$ corresponding to two different spins. These terms are presumably less important. They will be addressed in Section 5.9.

5.5.4 *Terms linear in the spin operators*

This section addresses a question of fundamental interest which, however, is probably not essential for the interpretation of experimental data. It can be skipped by the uninterested reader.

In this section, the magnetic field will be assumed to vanish. Thus, the Hamiltonian should be even with respect to angular momenta. However, angular momenta do not reduce to their electronic part, responsible for magnetism. In the presence of phonons, each magnetic molecule has an additional angular momentum, which is mainly that of nuclei, and is a sum of phonon creation and annihilation operators. It is proportional to the rotation vector $\boldsymbol{\Omega}$ (defined by the fact that the velocity of an atom at point $\mathbf{r}$ is $-\boldsymbol{\Omega} \times \mathbf{r}$).

While, in vanishing field, invariance under time reversal prevents the Hamiltonian from containing terms linear with respect to angular momenta, it does not forbid a spin–phonon interaction of the form

$$\mathcal{F}_{\mathrm{gyr}} = -\sum_{\alpha\gamma} \lambda_{\alpha\gamma}\Omega_{\alpha}S_{\gamma} \qquad (5.29)$$

which has been proposed by Chudnovsky (1994). The subscript 'gyr' refers to 'gyromagnetic' effects, i.e. the interaction between rotation and magnetism (Barnett 1935; Landau and Lifshitz 1969).

Actually, a term of the form (5.29) appears if one wishes to write the equations of motion of a mechanical system in a rotating frame of coordinates. As shown in textbooks (Landau and Lifshitz 1960), if a particle has a kinetic moment $\mathbf{M}$, its Hamiltonian in the rotating frame is

$$\mathcal{H}_{\mathrm{rot}} = \mathcal{H}_0 - \mathbf{M} \cdot \boldsymbol{\Omega} \qquad (5.30)$$

where $\mathcal{H}_0$ is the Hamiltonian in the fixed frame and $\boldsymbol{\Omega}$ is the rotation vector, so that a point $\mathbf{r}$ of the rotating frame has velocity $\boldsymbol{\Omega} \times \mathbf{r}$.

The second term of the right-hand side of (5.30) is the same as would result form a magnetic field $\boldsymbol{\Omega}$. This is easily understood in the particular case $\mathcal{H}_0 = 0$. Then the magnetic moment is immobile in the fixed frame, but precesses in the rotating frame. This is the same motion as the Larmor precession in a magnetic field.

Conversely, if a magnetic moment is in a magnetic field $\mathbf{H}$, one can simplify its equation of motion by writing it in a rotating frame, so that the term $-\mathbf{M} \cdot \boldsymbol{\Omega}$ of (5.30) cancels the Zeeman term $-\mathbf{M} \cdot \mathbf{H}$. This trick is commonly used in the theory of magnetic resonance.

However, the term $-\mathbf{M} \cdot \boldsymbol{\Omega}$ is only present in the rotating frame. Although it has been proposed to write the spin–phonon interaction in such a time-dependent and space-dependent frame (Chudnovsky 1994, Chudnovsky and Martinez-Hidalgo 2002) this would lead to high algebraic complications. To avoid them, the spin–phonon interaction should be written in the fixed frame of reference, where the Hamiltonian is $\mathcal{H}_0$. This does not exclude terms of the form (5.29) in the spin–phonon Hamiltonian, but these terms are much more difficult to derive than (5.30).

This section will be closed by a digression about gyromagnetic effects (Barnett 1935; Landau and Lifshitz 1969; Cagnac and Pébay-Peyroula 1983). The simplest gyromagnetic effects are the Barnett effect and the Einstein–de Haas effect, which is the inverse effect. The Barnett effect is the following phenomenon: a magnetic material (e.g. Fe) which undergoes a rapid rotation acquires a magnetization. The experimental observation is in agreement with (5.30). As Barnett wrote, 'rotating a body at 100 revolutions a second is equivalent to putting it in a magnetic field which is one fifteen-thousandth as intense as the earth's field'.

5.6   Transition probabilities and the golden rule

The existence of a spin–phonon interaction, which does not commute with the single-spin Hamiltonian (2.10) or (2.5) or else, implies the possibility of transitions from a state $|m^*\rangle$ to a state $|m'^*\rangle$ through phonon absorption or emission. An intuitive picture is that the spin has a certain probability per unit time of capturing a phonon of wavevector $\mathbf{q}$, and another probability of emitting a phonon. More precisely, the system (spin + phonons) is characterized by the state $|m^*\rangle$ of the spin, and the number $n_{q,\rho} = b_{q\rho}^+ b_{q\rho}$ of phonons of each 'mode'. A phonon mode is characterized by its wavevector $\mathbf{q}$ and an additional index ρ. The definition of a phonon mode results from the phonon Hamiltonian which can be written as

$$\mathcal{H}_{\mathrm{ph}} = \sum_{q\rho} \hbar\omega_{q\rho} b_{q\rho}^+ b_{q\rho} \qquad (5.31)$$

where $b_{q\rho}^+$ and $b_{q\rho}$ are phonon creation and annihilation operators and satisfy Bose commutation rules.

The number ν of phonon modes is equal to three times the number of atoms in the unit cell. However, we are interested in relatively low temperatures (more precisely, much lower than the Debye temperature) where only low-frequency phonons are available, i.e. acoustic phonons of small wavevector. Since a three-dimensional crystal has three acoustic phonon branches, $\rho=1$, 2 or 3.

Although there are many spins, their interaction is sufficiently weak to be neglected in the phonon emission or absorption process, so that attention can be focused on a single spin which is initially in state $|m^*\rangle$. Also, there are many phonon modes, but the interaction will be treated in second-order perturbation theory, so that attention can be first focused on the mode $\mathbf{q}, \rho$ of the absorbed or emitted phonon, and then one should sum over all modes. Thus, a state $|m^*, n_{q\rho}\rangle$ of the total system (spin + phonons) is characterized by the state $|m^*\rangle$ of the spin and the number $n_{q\rho}$ of phonons. It has a certain probability per unit time, $p(m, n_{q\rho} \to m', n_{q\rho} \pm 1)$, of making a transition to state $|m'^*, n_q \pm 1\rangle$ by absorbing or emitting a phonon. This stochastic picture is not very easy to justify but, if it is accepted, the probability $p(m, n_{q\rho} \to m', n_{q\rho} \pm 1)$ can be obtained from perturbation theory. The formula is Fermi's *golden rule* (Cohen-Tannoudji *et al.* 1986).

$$p(m, n_{q\rho} \to m', n_{q\rho} \pm 1) = \frac{2\pi}{\hbar} \left| \langle m, n_{q\rho} \,|\, \mathcal{H}_{\mathrm{s\text{-}ph}} | m', n_{q\rho} \pm 1 \rangle \right|^2$$

$$\times \, \delta(E_{m'} - E_m \pm \hbar\omega_{q\rho}) \qquad (5.32)$$

where $\hbar\omega_{q\rho}$ is the energy of a phonon. The asterisks have been omitted to simplify both the typography and the underlying algebra.

Apart from the factor $2\pi/\hbar$, formula (5.32) is easy to remember. The delta function expresses energy conservation and the first factor vanishes if there is no interaction.

As already stated, $|E_{m'} - E_m|$ will be assumed to be so small that only acoustic phonons can safisfy the relation

$$E'_m - E_m = \pm\hbar\omega_q \tag{5.33}$$

implied by the delta function in (5.32).

Formula (5.27) may be more conveniently written in terms of the creation and annihilation operators $b^+_{q\rho}$ and $b_{q\rho}$ as

$$\mathcal{H}_{\text{s-ph}} = N^{-1/2}\sum_{q\rho} U_{q\rho}(\mathbf{S})b_{q\rho} + N^{-1/2}\sum_{q\rho} U^*_{q\rho}(\mathbf{S})b^+_{q\rho} \tag{5.34}$$

where N is the number of unit cells, $\mathbf{q}$ designates the vectors of the reciprocal space, ρ designates the various phonon modes and

$$U_{q\rho}(\mathbf{S}) = i(\omega_{q\rho}M/\hbar)^{-1/2}\sum_{\alpha\xi\zeta} q_\alpha\Lambda_{\alpha\rho\xi\zeta}S_\xi S_\zeta \tag{5.35}$$

where M is the mass of the unit cell. The factor $(\omega_{q\rho}M/\hbar)^{-1/2}$ has been introduced in order that the coefficients $\Lambda_{\alpha\rho\xi\zeta}$ have the same order of magnitude as $\tilde{\Lambda}_{\alpha,\gamma,\xi,\zeta}$ in (5.27), as seen in Appendix F.

For the reader who is not familiar with such formulae, it may be of interest to notice that for a macroscopic crystal, the factor $N^{-1/2}$ is very small but the number of terms in the sum over q is very large. In the final expressions of physical quantities, both infinities cancel and the summation over q can be replaced by an integration.

Insertion of (5.34) into (5.32) yields

$$p(m, n_{q\rho} \to m', n_{q\rho} - 1) = \frac{2\pi}{\hbar N}\left|\langle m \mid U^*_{q\rho}(\mathbf{S})|m'\rangle\right|^2 \langle n_{q\rho} \mid b^+_{q\rho}b_{q\rho}|n_{q\rho}\rangle$$

$$\times \delta(E_{m'} - E_m - \hbar\omega_{q\rho}). \tag{5.36}$$

The delta function implies $E_{m'} > E_m$, i.e. the final state has a higher energy $E_{m'}$ than the initial state. Therefore we are calculating an absorption probability. To obtain the transition probability $\gamma_m^{m'}$ which appears in (5.12), one must

(i) multiply (5.36) by the probability $P(n_q)$ that the number of phonons of wavevector $\mathbf{q}$ is n_q, then sum over all values of n_q; in other words one should average n_q;

(ii) sum over $\mathbf{q}$ and over modes ρ.

The effect of (i) is to replace the second matrix element by the average value $\langle n_{q\rho} \rangle$. Thus

$$\gamma_m^{m'} = \frac{2\pi}{\hbar N} \sum_{q\rho} \langle n_{q\rho} \rangle \left| \langle m \,|\, U_{q\rho}^*(\mathbf{S})|m' \rangle \right|^2 \delta(E_{m'} - E_m - \hbar\omega_{q\rho}).$$

The sum over $\mathbf{q}$ can be transformed into an integral through the formula $(1/V)\sum_q = (2\pi)^{-3} \int d^3q$, where V is the total volume. Thus

$$\gamma_m^{m'} = \frac{v}{4\pi^2\hbar} \sum_\rho \int d^3q \langle n_{q\rho} \rangle \left| \langle m \,|\, U_{q\rho}^*(\mathbf{S})|m' \rangle \right|^2 \delta(E_{m'} - E_m - \hbar\omega_{q\rho}) \qquad (5.37)$$

where v is the volume of the unit cell. Replacing $\langle n_{q\rho} \rangle$ by its average value according to Bose-Einstein statistics

$$\langle n_{q,\gamma} \rangle = \frac{1}{\exp(\beta\hbar\omega_{q,\gamma}) - 1} \qquad (5.38)$$

one obtains for $m' > m$

$$\gamma_m^{m'} = \frac{v}{4\pi^2\hbar} \sum_\rho \int \frac{d^3q}{\exp(\beta\hbar\omega_{q,\rho}) - 1} \left| \langle m' \,|\, U_{q\rho}(\mathbf{S})|m \rangle \right|^2 \delta(E_{m'} - E_m - \hbar\omega_{q\rho}).$$
$$(5.39)$$

The same calculation for $m' < m$ yields

$$\gamma_m^{m'} = \frac{v}{4\pi^2\hbar} \sum_\rho \int \frac{d^3q}{1 - \exp(-\beta\hbar\omega_{q,\rho})} \left| \langle m' \,|\, U_{q\rho}(\mathbf{S})|m \rangle \right|^2 \delta(E_{m'} - E_m + \hbar\omega_{q\rho}).$$
$$(5.40)$$

Formulae (5.39) and (5.40) satisfy the detailed balance relation (5.14).

When writing (5.38), it is assumed that the dynamics of the phonon bath are much faster than that of the magnetization. The occupation number of each phonon state thus depends only on the temperature of the bath. This approximation neglects the possibility of avalanches, in which the heat which is locally produced by the reversal of a spin does not diffuse rapidly enough, and produces reversal of neighbouring spins.

5.7 Qualitative formulae

These general formulae are rather complicated. In all calculations where phonons appear (e.g. the specific heat of a solid) it is usual to make the (qualitatively acceptable) assumption $\omega_{q,\rho} = c_s q$, where c_s is the sound velocity, assumed to be the same for the three acoustic modes. This is Debye's model, which yields correct orders of magnitude. In (5.39) and (5.40) all contributing vectors $\mathbf{q}$ lie on the sphere of radius $q = (E_{m'} - E_m)/\hbar$. The squared matrix element of $U_{q\rho}(\mathbf{S})$ can just be replaced by its average value on that sphere. As seen from (5.35), it is proportional to $\hbar\sqrt{q}/(c_s M)$ times a quadratic form of the spins which involves anisotropy constants $\tilde{D}_{\alpha\gamma}$, presumably of the order of D. Replacing d^3q by $4\pi q^2 dq$, one finds

$$\gamma_m^{m'} = \frac{3v}{\pi c_s M} \int \frac{q^3 dq}{\exp(\beta\hbar c_s q) - 1} \left| \langle m' | \sum_{\alpha\gamma} \tilde{D}_{\alpha\gamma} S_\alpha S_\gamma | m \rangle \right|^2 \delta(E_{m'} - E_m - \hbar c_s q)$$

or after integration with referred to q

$$\gamma_m^{m'} = \frac{3v}{\pi\hbar^4 M c_s^5} \frac{(E_{m'} - E_m)^3}{\exp[\beta(E_{m'} - E_m)] - 1} \left| \langle m' | \sum_{\alpha\gamma} \tilde{D}_{\alpha\gamma} S_\alpha S_\gamma | m \rangle \right|^2. \tag{5.41}$$

This formula is valid for both signs of $(E_{m'} - E_m)$.

Analogous formulae (using for instance particular expressions of the $\tilde{D}_{\alpha\gamma}$ coefficients) have been written by Abragam and Bleaney (1986), Villain *et al.* (1994), Garanin and Chudnovsky (1997), Leuenberger and Loss (1999), and Chudnovsky and Garanin (2002).

Two observations can be made.

(i) When m and m' are close to 0 (i.e. near the top of the barrier), the matrix elements $\langle m | S_\xi S_\zeta | m \pm 1 \rangle$ (which correspond to $\xi, \zeta = z, x$ or $\xi, \zeta = z, y$) are small in comparison with the matrix elements $\langle m | S_\xi S_\zeta | m \pm 2 \rangle$ (which correspond to $\xi, \zeta = x, y$). The situation is just opposite when $|m|$ and $|m'|$ are close to s.

(ii) When $E_m - E_{m'}$ is small (which also corresponds to the top of the barrier) the transition probability is small because of the factor $(E_m - E_{m'})^3$, which mainly reflects the fact that there are few phonon states of very low energy.

These two qualitative properties demonstrate the interest of formula (5.41). However, its derivation contains many shortcomings. Moreover the coefficients $\tilde{D}_{\alpha\gamma}$ are not known. Therefore, (5.41) is only appropriate for a qualitative evaluation. In the following attempt, the unknown coefficients $\tilde{D}_{\alpha\gamma}$ will be assumed to be of the order of $|D|$ and for the sake of simplicity the external field will be assumed to vanish. The population of the various levels of the left-hand well will be assumed to satisfy $p_m(t) = p_{-s}(t) \exp[-\beta(E_m - E_{-s})]$ as would be the case at equilibrium. The escape rate from the left-hand well is equal to the escape rate from the upper levels of the left-hand well, which is roughly given by $p_{-1}\gamma_{-1}^0$. A contribution proportional to $p_{-2}\gamma_{-2}^0$ is ignored in this very rough evaluation. Thus the relaxation time τ is roughly given by $1/\tau \approx \gamma_{-1}^0 \exp[-\beta(E_{-1} - E_{-s})]$. Using formula (5.41) with $m = -1$ and $m' = 0$, one obtains

$$1/\tau \approx \frac{k_B}{c_s^5} \frac{v}{M} \left(\frac{k_B}{\hbar} \right)^4 \left(\frac{\delta E}{k_B} \right)^3 \left(\frac{|D|s^2}{k_B} \right)^2 \exp[-\beta(E_0 - E_{-s})] \tag{5.42}$$

where $\delta E = E_0 - E_{-1} = |D| \simeq 0.6$ K for Mn$_{12}$ac. The Boltzmann constant has been introduced because it is customary to evaluate energies in kelvin.

Unfortunately, in (5.42), the magnitude of the sound velocity c_s in $Mn_{12}ac$ is, to our knowledge, not known. Since it appears with the power 5, this is the main cause of uncertainty. Tentatively, the value $c_s \simeq 1000$ m/s, which is the right order of magnitude for silicon, will be used. The ratio v/M will be taken equal to 0.001 m³/kg, as for water. Then (5.42) yields, in s^{-1},

$$\tau \approx 10^{-7} \exp \frac{T_0}{T}. \tag{5.43}$$

For $T_0 = 60$ K and $T = 2$ K, the Arrhenius factor is 10^{13} and the relaxation time is 10^6 seconds $\cong 10$ days, to compare with the experimental value (2 months). Since the above argument contains a number of guesses one can only conclude that the relaxation is extremely slow, in agreement with experiment.

The detailed formula (5.39) might allow for a precise experimental check. This would require a precise knowledge of the elastic constants and of the spin–phonon interaction.

Formulae (5.39) and (5.41) provide a qualitative idea of the effect of a magnetic field and of temperature (Fort *et al.* 1998; Leuenberger and Loss 1999). This point will be addressed again in Section 10.3. Information on the spin–phonon interaction can be obtained by nuclear magnetic resonance (Furukawa *et al.* 2001a).

5.8 Multiphonon processes

In (5.41) and in the following formulae, there is a factor $(E_{m'} - E_m)^3$ which becomes very small if the energy difference is small (as it is at the top of the barrier). This factor results in particular from the fact that multiphonon processes have been ignored, so that the phonon density of states near zero energy (which is weak) appears. This factor is not present in the two-phonon contribution.

Actually, the theory developed in the previous section is that of the so-called 'direct' relaxation processes (Abragam and Bleaney 1986). Other processes are familiar to the experts of electron paramagnetic resonance. If the spin–phonon interaction (5.27) is just expanded to second order with respect to the strain, and if the golden rule is applied, one obtains the so-called first-order 'Raman processes'. Another possibility addressed in textbooks corresponds to the excitation to an intermediate state followed to the deexcitation to the final state. This is called an 'Orbach process'. As a matter of fact, the relaxation process described in this chapter is a high-order Orbach process since it involves successive excitations to higher states followed by a cascade of deexcitations.

The decision to ignore Raman processes is reasonable at low temperature. It is also motivated by the wish to simplify the presentation. Two-phonon relaxation processes can be handled in the same way as direct processes. The simple theory presented in this chapter is useful, but qualitative. The number of parameters is indeed very large. Including multiphonon processes into elementary transition probabilities would just make the situation worse.

5.9   Spin–phonon interactions resulting from exchange

As a matter of fact, spin–phonon interactions can arise from the modulation by phonons of any term of the Hamiltonian which act on spins: the anisotropy energy, the exchange and dipole interactions. So far, we have focused our attention on the anisotropy although isotropic exchange inside a molecule is stronger. In this section it will be shown that isotropic exchange modulation is indeed of minor importance. Anisotropic exchange and dipole interactions are very small.

Such a modulation is represented by a Heisenberg Hamiltonian (2.53), i.e. a sum of terms $J_{ij}\mathbf{s}_i \cdot \mathbf{s}_j$ with exchange integrals J_{ij} which depend on space and time. The indices i, j denote the atoms of the same molecule and s_i is the spin of the atom i. The problem is whether a modulation $J_{ij}(t)\mathbf{s}_i \cdot \mathbf{s}_j$ can induce transitions between two eigenstates of the static spin Hamiltonian. These states will be denoted $|s, m, p\rangle$, where $s(s+1)$ and m are the eigenvalues of $\mathbf{S}^2$ and S^z, $\mathbf{S} = \sum_i \mathbf{s}_i$ and p designates the other parameters which are necessary to specify the spin states of the molecule (6^8 in the case of Fe$_8$). In other words, p labels the various values of the exchange energy inside a molecule; one might say, the various 'exchange multiplets'.

The transition probability per unit time from $|s, m, p\rangle$ to $|s', m', p'\rangle$ induced by the modulation of J_{ij} by phonons is given by a golden rule analogous to (5.32), and vanishes if the matrix element $\langle s, m, p \,|\, \mathbf{s}_i \cdot \mathbf{s}_j | s', m', p'\rangle$ is zero. Moreover, it also vanishes if the energy difference between $|s, m, p\rangle$ and $|s', m', p'\rangle$ is larger than the energy of thermal phonons. If the temperature is not too high (typically, less than 50 K), this implies that $p = p'$. In practice both states should correspond to the lowest exchange energy, and therefore to $s = 10$ in Mn$_{12}$ac and Fe$_8$.

Now, for any pair (i, j), the product $\mathbf{s}_i \cdot \mathbf{s}_j = s_i^z s_j^z + (1/2)s_i^+ s_j^- + (1/2)s_i^- s_j^+$ commutes with $S_z = \sum s_i^z$. Of course, it also commutes with S_x, S_y, $S^+\,S^-$ and $\mathbf{S}^2 = S_x^2 + S_y^2 + S_y^2$. Therefore,

$$\langle s, m, p \,|\, [\mathbf{s}_i \cdot \mathbf{s}_j, S^z]|s', m', p'\rangle = (m' - m)\,\langle s, m, p \,|\, \mathbf{s}_i \cdot \mathbf{s}_j|s', m', p'\rangle = 0,$$

and

$$\langle s, m, p \,|\, \mathbf{s}_i \cdot \mathbf{s}_j|s', m', p'\rangle = 0 \text{ if } m' \neq m.$$

Similarly, $\langle s, m, p \,|\, \mathbf{s}_i \cdot \mathbf{s}_j|s', m', p'\rangle = 0$ if $s' \neq s$.

The only non-vanishing elements are therefore $\langle s, m, p \,|\, \mathbf{s}_i \cdot \mathbf{s}_j|s, m, p'\rangle$. Moreover, $S^+|s, m, p\rangle$ is equal to $\sqrt{(s + m + 1)(s - m)}|s, m+1, p\rangle$ and therefore $\langle s, m + 1, p \,|\, \mathbf{s}_i \cdot \mathbf{s}_j|s, m+1, p'\rangle$ is equal to $[(s + m + 1)(s - m)]^{-1}\,\langle s, m, p \,|\, S^- \mathbf{s}_i \cdot \mathbf{s}_j S^+|s, m, p'\rangle$. The scalar product commutes with S^-, so that this expression can be written as $[(s + m + 1)(s - m)]^{-1}\,\langle s, m, p \,|\, \mathbf{s}_i \cdot \mathbf{s}_j S^- S^+|s, m, p'\rangle$, which is equal to $\langle s, m, p \,|\, \mathbf{s}_i \cdot \mathbf{s}_j|s, m, p'\rangle$. By recursion, this matrix element is independent of m and

$$\langle s, m, p \,|\, \mathbf{s}_i \cdot \mathbf{s}_j|s', m', p'\rangle = A_{ij}^{pp'}\delta_{mm'}\delta_{ss'}$$

In particular, inside the lowest exchange multiplet (with $s = 10$ in $Mn_{12}ac$ and Fe_8), $\mathbf{s}_i \cdot \mathbf{s}_j$ reduces to a constant A_{ij} times the unit matrix. The modulation of exchange integrals by phonons can therefore not induce transitions between the $(2s+1)$ states of the lowest exchange multiplet. The transitions allowed by the matrix element are to excited exchange multiplets with the *same* s, which are presumably at such high energies that no phonons are available except at quite high temperatures.

5.10   Effect of photons

5.10.1   *Phonons and photons*

In the previous paragraphs, the lifetime of excited states (and, more generally, the transition probabilities) has been assumed to result from acoustic phonon emission (and adsorption, at higher temperature). Of course, photon emission is also possible. If the photons are at thermal equilibrium, the transition probability may be calculated in the same way as that which results from phonons, and it will be shown in the next subsection that it is much weaker.

On the other hand, while phonons are generally at thermal equilibrium, it is possible to irradiate a sample with a high photon flux. This is done in EPR, already addressed in Section 3.3.

5.10.2   *Photons at thermal equilibrium*

The interaction of a spin with the electromagnetic field is given by (2.1), where the $\mathbf{g}$ tensor will be approximated by a scalar $g \simeq 2$, i.e.

$$\mathcal{H}_{\text{int}} = g\mu_B \delta \mathbf{H} \cdot \mathbf{S} = g\mu_B \mathbf{S} \cdot \text{rot}\mathbf{A}. \tag{5.44}$$

Here, $\delta\mathbf{H}$ is the variable part of the field and $\mathbf{A}$ is the corresponding vector potential. The uniform part is irrelevant for photons. The field $\delta\mathbf{H}$ may be emitted by the spin or it may be the microwave field in an EPR experiment.

According to textbooks, e.g. Cohen-Tannoudji *et al.* (2001) in Section I.C.6, the magnetic field can be expressed in terms of photon creation and annihilation operators $\alpha_{q,\rho}$ and $\alpha_{q,\rho}^+$ as

$$\mathbf{H}(\mathbf{r}) = iN^{-1/2} \sum_{q,\rho} \left[\frac{\hbar\mu_0 cq}{2v}\right]^{1/2} \left[\alpha_{q,\rho}\frac{\mathbf{q}}{q} \times \mathbf{e}_{q,\rho}e^{i\mathbf{q}\cdot\mathbf{r}} - \alpha_{q,\rho}^+\frac{\mathbf{q}}{q} \times \mathbf{e}_{q,\rho}e^{-i\mathbf{q}\cdot\mathbf{r}}\right] \tag{5.45}$$

where the polarization index ρ is 1 or 2, v is the volume of the unit cell, c is the speed of light, while $\mathbf{e}_{q,1}$ and $\mathbf{e}_{q,2}$ are two orthogonal unit vectors which are perpendicular to $\mathbf{q}$.

Formulae (5.44) and (5.45) yield

$$\mathcal{H}_{\text{int}} = iN^{-1/2}g\mu_B \sum_{q,\rho} \left[\frac{\hbar\mu_0 cq}{2v}\right]^{1/2} \alpha_{q,\rho}\left[\left(\frac{\mathbf{q}}{q} \times \mathbf{e}_{q,\rho}\right) \cdot \mathbf{S}\right]e^{i\mathbf{q}\cdot\mathbf{r}}$$

$$- iN^{-1/2}g\mu_B \sum_{q,\rho} \left[\frac{\hbar\mu_0 cq}{2v}\right]^{1/2} \alpha_{q,\rho}^+\left[\left(\frac{\mathbf{q}}{q} \times \mathbf{e}_{q,\rho}\right) \cdot \mathbf{S}\right]e^{-i\mathbf{q}\cdot\mathbf{r}}. \tag{5.46}$$

This has just the form (5.34). Of course the phonon operators $b_{q\rho}$ must be replaced by photon operators $\alpha_{q\rho}$. The operators $U_{q\rho}(\mathbf{S})$ are now defined by

$$U_{q\rho}(\mathbf{S}) = g\mu_B \left[\frac{\hbar\mu_0 cq}{2v}\right]^{1/2} \left[\left(\frac{\mathbf{q}}{q} \times \mathbf{e}_{q,\rho}\right) \cdot \mathbf{S}\right] \qquad (5.47)$$

where the spin has been assumed to be at the origin, $\mathbf{r} = 0$.

The photon absorption probability can be obtained by insertion of (5.47) into (5.36), which can also be written as

$$p(m, n_{q\rho} \rightarrow m', n_{q\rho} - 1) = \frac{2\pi}{\hbar N} \left|\langle m \,|\, U_{q\rho}^*(\mathbf{S})|m'\rangle\right|^2 \langle n_{q\rho}\rangle \delta(E_{m'} - E_m - \hbar\omega_{q\rho}). \qquad (5.48)$$

If photons are at thermal equilibrium, averaging over all photons yields (5.40). The calculation of Section 5.7 can then be reproduced. Qualitatively, the result can be written as

$$\tilde{\gamma}_m^{m'} \approx \frac{w_0 v s^2}{\pi \hbar^4 c^3} \frac{(E_m - E_{m'})^3}{1 - \exp[-\beta(E_m - E_{m'})]} \qquad (5.49)$$

where v is the volume per spin (i.e. the volume of the unit cell, assumed to contain a single magnetic molecule) and w_0 is of the order of magnitude of the dipole energy between two neighbouring spins.

The ratio $\tilde{\gamma}_m^{m'}/\gamma_m^{m'}$ of the photon-induced and phonon-induced transition probabilities are given by the ratio of expressions (5.49) and (5.41), namely

$$\frac{\tilde{\gamma}_m^{m'}}{\gamma_m^{m'}} = \frac{w_0}{|\tilde{D}|s^2} \frac{\rho v c_s^2}{|\tilde{D}|} \frac{c_s^3}{c^3} \qquad (5.50)$$

where $\tilde{D}$ is the typical value of the coefficients $\tilde{D}_{\alpha\gamma}$ which appear in (5.41). This expression is presumably dominated by the last factor $(c_s/c)^3$, which is very small, so that spontaneous photon emission is negligible.

5.10.3 The beauty of light

The interest of photons with respect to phonons is that they are more versatile. It is easier to have them out of thermal equilibrium, with a high number of photons of a particular frequency ω. If their energy $\hbar\omega$ is equal to the energy difference $(E_{m-1} - E_m)$ between two states $|m\rangle$ and $|m - 1\rangle$ of the spin, and if the lowest level (one can suppose it is $|m\rangle$) is populated, then a transition to state $|m - 1\rangle$ is induced, and this is detected as an absorption of the electromagnetic wave. This is electron paramagnetic resonance (EPR) and has been addressed in Section 3.3.1. The standard method is to apply a magnetic field and to change the strength of this field while irradiating at a constant frequency. The electronic levels are strongly modified and the interpretation of the experimental data requires a model. A more direct method is to apply a constant field but to change the radio frequency. This requires more elaborate instruments

(Schwartz *et al.* 1995) but allows for the determination of the various resonances for a given field. It is of interest to observe that the above theory is applicable, and formula (5.48) holds.

Another advantage with respect to phonons is that the photons emitted from a solid can be analysed. In particular, they have a well-defined polarization. Even though photon emission is usually weak, it has been suggested that it might be enhanced by means of resonant cavities (Amigó 2003a,b). It should be necessary in that case to take this polarization into account. The electromagnetic field may be decomposed into plane waves of all wavevectors $\mathbf{q}$, and the spin–photon interaction is the sum of all interactions $\mathcal{H}_\mathbf{q}$ with each wave. As an example, if $\mathbf{q}$ is parallel to the easy axis z, it follows from (5.46) that

$$
\mathcal{H}_\mathbf{q} = (1/2)(i/c)g\mu_B N^{-1/2} \left[\frac{\hbar c q}{2\epsilon_0 v}\right]^{1/2}
$$

$$
\times (S_x \mathbf{e_x} + S_y \mathbf{e_y}) \cdot (\alpha_{\mathbf{q},x}\mathbf{e_y} - \alpha_{\mathbf{q},y}\mathbf{e_x} - \alpha_{\mathbf{q},x}^+\mathbf{e_y} + \alpha_{\mathbf{q},y}^+\mathbf{e_x})
$$

$$
= (1/2)(i/c)g\mu_B N^{-1/2} \left[\frac{\hbar c q}{2\epsilon_0 v}\right]^{1/2}
$$

$$
\times [S^+(\alpha_{\mathbf{q},x} - \alpha_{\mathbf{q},x}^+ - i\alpha_{\mathbf{q},y} + i\alpha_{\mathbf{q},y}^+) + S^-(-\alpha_{\mathbf{q},x} + \alpha_{\mathbf{q},x}^+ - i\alpha_{q,y} + i\alpha_{q,y}^+)]
$$

(5.51)

where $\mathbf{e}_x$ and $\mathbf{e}_y$ are the unit vectors along the x and y axes, v is the volume of the unit cell, N is the number of unit cells, and $\alpha_{\mathbf{q},x}$ and $\alpha_{\mathbf{q},y}$ are photon annihilation operators.

In order to simplify the argument, let the action of this Hamiltonian on the state $|m-1, 0\rangle$ be considered, where $|m-1, 0\rangle$ designates the state with 0 photon and the spin in state $|m - 1\rangle$. One has to introduce the state $|\mathbf{q}, x\rangle$ with one photon of wavevector $\mathbf{q}$ and magnetic field along x. The photon energy $\hbar\omega_q$ will be assumed to be close to $(E_m - E_{m-1})$. Assuming the spin–photon interaction (5.44) to be small, one finds

$$
\mathcal{H}_\mathbf{q}|m - 1, 0\rangle = -(1/2)(1/c)g\mu_B N^{-1/2} \left[\frac{\hbar c q}{2\epsilon_0 v}\right]^{1/2} [|m, \mathbf{q}, y\rangle - i|m, \mathbf{q}, x\rangle] \quad (5.52)
$$

where states of the form $|m - 2, \mathbf{q}, y\rangle$ have been neglected because their energy is very different from that of the initial state $|m - 1, 0\rangle$.

Thus, $\mathcal{H}_\mathbf{q}$ transforms $|m - 1, 0\rangle$ into $|m, \mathbf{q}, y\rangle - i|m, \mathbf{q}, x\rangle$. The vector $|m, \mathbf{q}, y\rangle$ corresponds to a magnetic field proportional to $\mathbf{e_y}\exp(iqz - i\omega_q t)$, where $\mathbf{e_y}$ is the unit vector along the y axis. Similarly $|m, \mathbf{q}, x\rangle$ corresponds to a magnetic field proportional to $\mathbf{e_x}\exp(iqz - i\omega_q t)$. Multiplying the 'ket' by i is equivalent to multiply the exponential by i. Therefore, $|m, \mathbf{q}, y\rangle - i|m, \mathbf{q}, x\rangle$ corresponds to a magnetic field proportional to

$$
[\mathbf{e_y} - i\mathbf{e_x}]\exp(iqz - i\omega_q t)
$$

and therefore to circularly polarized light.

5.10.4 *Classical electromagnetism and quantum electrodynamics*

Some of the above results may be derived from an at least partly classical treatment: a quantum spin interacts with a classical electromagnetic field, which appears as a scalar in the equations. The book of Abragam and Bleaney (1986) uses this approach. Also, the power emitted by a magnetic moment can be calculated from classical electromagnetism. A molecular spin in an excited state $|m\rangle$ undergoes a Larmor precession with a frequency ω_m, where $\hbar\omega_m = E_m - E_{m'}$, as in the case of phonons. Here, $|m'\rangle$ designates the state just below $|m\rangle$ ($m' = m+1$ in the right-hand well). The rotating spin emits electromagnetic radiation of frequency ω_m, as an antenna would do. This emission has a certain power $\mathcal{P}_0$. Coming back to quantum mechanics, one can say that the probability per unit time of emitting a photon is $\mathcal{P}_0/(\hbar\omega_m)$. Indeed the emitted photons have frequency ω_m. The power is given in textbooks (for instance Toraldo di Francia and Bruscaglioni 1988). One can thus retrieve (5.50), at least at low temperature (when there is no stimulated emission). To summarize, the quantization of the electromagnetic field is not absolutely necessary to obtain the results necessary for this book, but it provides a unified, coherent, and elegant formalism which is finally simpler than the description of quantum phenomena by semiclassical methods. Of course, certain phenomena which are outside the scope of this book, such as the anomalous magnetic moment of the electron, require quantum electrodynamics.

5.10.5 *Coherence and superradiance*

A privilege of photons is the possibility of coherence. It has been suggested by Chudnovsky and Garanin (2002) that photon emission can become important when a large number of spins are in phase, and thus emit coherent light. Indeed the electromagnetic field emitted by N spins is proportional to N, and therefore the power is proportional to N^2. This is *superradiance* (Dicke 1954). However, it looks incredible that the spins can remain in phase during a time comparable to the period $1/\omega$ of the emitted light. The most intrinsic source of dephasing is the dipole interaction between spins. In the more usual case of *electric* dipole emission too, interactions between atoms generally (but not always) destroy superradiance (Gross and Haroche 1982). In the absence of external radiation (present in magnetic resonance) it is unlikely that the factor $(c_s/c)^3$ of formula (5.50) can be compensated by amplification schemes, e.g. resonant cavities, as proposed by Tejada *et al.* (2003).

5.11 Limitations of the model

The present chapter suffers from various approximations.

- The molecule has been considered as a spin ($s = 10$ in $Mn_{12}ac$ and Fe_8). This is strictly correct only at low enough temperature. The effect of higher exchange energies will be addressed in Section 10.7.

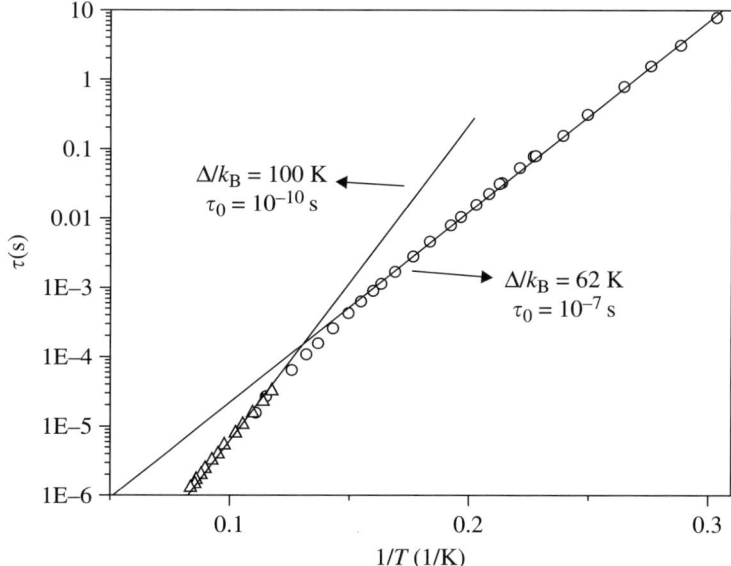

FIG. 5.3. Fit of the experimental relaxation time of $Mn_{12}ac$ by two different Arrhenius laws. From Novak *et al.* (2005).

- The evolution of the spin has been described by transition probabilities between its eigenstates. This disregards the quantum-mechanical possibility of being in a coherent superposition of two or more eigenstates. A full description should make use of the density matrix as will be seen in Chapter 11. However, this is only necessary at short times and low temperatures.
- Spin relaxation does not always require climbing to the top of the barrier by phonon absorption. At low temperature, when few phonons are available, it can be easier to cross the barrier by quantum tunnelling. This effect will be studied in the next chapters.

On the other hand, a more precise treatment reported in Appendix E, as well as a numerical treatment discussed in Appendix F, suggest deviations from the Arrhenius law (5.2) at high temperatures. Experimental data for $Mn_{12}ac$ obtained over a wider temperature range by using high-frequency ac suscepto-metry can be fitted by two different Arrhenius laws with different parameters τ_0 and T_0 as shown in Fig. 5.3 (Novak *et al.* 2005).

6

MAGNETIC TUNNELLING OF AN ISOLATED SPIN

6.1 Spin tunnelling

6.1.1 *Particle tunnelling: a reminder*

In this section, as in the preceding one, we consider a spin subject to a double potential well (Fig. 5.2) which results from an anisotropy Hamiltonian, e.g. $\mathcal{H} = -|D|S_z^2$. However attention will be focused on low temperatures, when few phonons are available, so that the spin can hardly jump over the potential barrier.

This spin is comparable to a particle in the double potential well, treated in textbooks of quantum mechanics. This model has natural realizations in the hydrogen bond of ice and certain ferroelectrics.

Textbooks of quantum mechanics tell us that a particle in a *symmetric* double potential well $V(x)$ (analogous to that of Fig. 5.2) undergoes a 'tunnel effect' or 'tunnelling'. This means that (1) if the particle is put into the left-hand side, it goes to the right-hand side, then back to the left-hand side, etc., with a frequency ω_T which depends on the shape of the potential. (2) The eigenfunctions of the Hamiltonian are *delocalized*, i.e. they have a component in the left-hand side, and a component in the right-hand side, with equal probabilities of being in either side. This delocalization is a consequence of the symmetry $V(x) = V(-x)$. The solutions of the Schrödinger equation $[V(x) - (1/2)(\hbar^2/m)d^2/dx^2 - E]\psi(x) = 0$ are symmetric or antisymmetric, i.e. $\psi(x) = \psi(-x)$ or $\psi(x) = -\psi(-x)$. In both cases they are delocalized.

As will be seen in the following sections, a spin subject to a spin Hamiltonian can possess the same properties. The spin has, however, an additional feature, namely its sensitivity to a magnetic field.

In the case of a spin as in that of a particle, it is possible to define a wavefunction $\varphi(m)$. This definition is just contained in (5.11), $|m^*\rangle = \sum_{m''} \varphi(m'')\,|m''\rangle$. This function satisfies the equation

$$\sum_{m'} \langle m\,|\,\mathcal{H}\,|\,m'\rangle\,\varphi(m') = E\varphi(m) \qquad (6.1)$$

which is nothing but the Schrödinger equation with a discrete variable m.

Spin tunnelling is especially important at low temperature, when the spin is in its ground state with a high probability.

The following pages do not contain many experimental results. This is because
an isolated spin does not really exist. However, its study is an unavoidable
preliminary to that of realistic situations.

6.1.2 An example

An example is a spin subject to the Hamiltonian (2.10) if a transverse magnetic
field (i.e. perpendicular to the easy axis) is added. Adding also a longitudinal
field (i.e. parallel to the easy axis), the Hamiltonian is now

$$\mathcal{H} = -|D|S_z^2 + g\mu_B H_z S_z + g\mu_B H_x S_x \tag{6.2}$$

The field will generally be assumed small, $g\mu_B|H_\alpha| \ll |D|s$.

If $H_x = 0$, the eigenvectors are the eigenvectors $|m\rangle$ of S_z. If $H_x = H_z = 0$,
the eigenvectors $|m\rangle$ and $|-m\rangle$ are degenerate. More generally, if $H_x = 0$,
degeneracy of two eigenvectors $|m\rangle$ and $|m'\rangle$ is possible for certain values of H_z.
Indeed the energy is then given by (5.6), $E_m^{(0)} = -|D|m^2 + g\mu_B H_z m$. This value
will be called unperturbed. For $H_x = 0$, degeneracy occurs (Fig. 6.1a) if

$$E_m^{(0)} - E_{m'}^{(0)} = 0. \tag{6.3}$$

This condition implies that H_z is equal to

$$g\mu_B H^{(mm')} = |D|(m + m'). \tag{6.4}$$

In the important experimental cases, $m + m'$ is small, $0, 1, 2\ldots$, Fig. 6.2 shows
the levels in the double well when (6.3) is satisfied for $m = -s$ and $m' = s - 1$.

An essential point, as will now be seen, is that these degeneracies can be
removed by a perturbation which does not commute with S_z. They are replaced

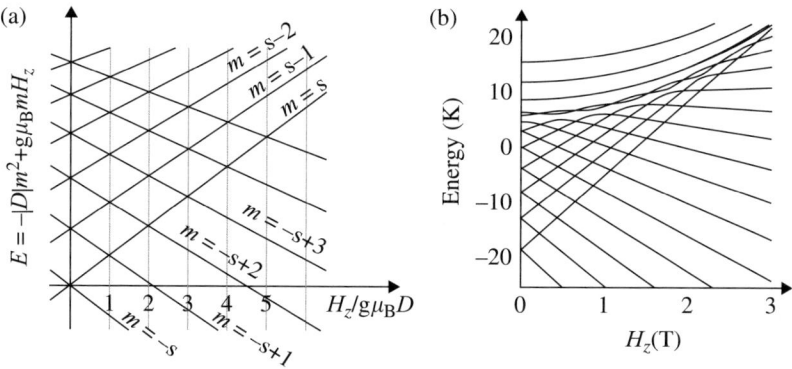

FIG. 6.1. Level crossing: (a) for $\mathcal{H} = -|D|S_z^2 + g\mu_B H S_z$; (b) for $\mathcal{H} = -|D|S_z^2 + E(S_x^2 - S_y^2) + g\mu_B H S_z$, with the values of D and E taken from Table 4.5, which correspond to Fe$_8$. Part of the level crossings are suppressed and replaced by maxima and minima. For low energies the distance between maxima and minima is too small to be visible.

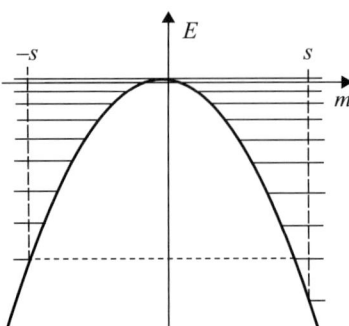

FIG. 6.2. Energy levels of a spin subject to a (negative) longitudinal field which
satisfies condition (6.4) with $m = -s$ and $m' = s-1$. Tunnelling is possible (in
the presence of an appropriate transverse field H_x or H_y) between the lowest
state of the left-hand well and the first excited state of the right-hand well.

by tunnelling. Figure 6.1b shows the effect of a low-symmetry perturbation of
the anisotropy. In the remainder of this section, The Hamiltonian (6.2) will be
investigated. Then, the perturbation which does not commute with S_z is the
transverse field $H_x \neq 0$.

6.1.3   *The case $s = 1/2$*

The absence of degeneracy in a non-vanishing field is easily seen in the simple,
but instructive case of a spin $1/2$.

The first term $-|D|S_z^2$ of (6.2) reduces to the constant $-|D|/4$. The direction
of the total magnetic field $(H_x, 0, H_z)$ may be called Z, and (6.2) reads

$$\mathcal{H} = g\mu_B H S_Z - |D|/4 \qquad (6.5)$$

where $H = \sqrt{H_x^2 + H_z^2}$. The Hamiltonian (6.2) has two eigenvalues

$$E^\pm = \pm(g\mu_B/2)\sqrt{H_x^2 + H_z^2} = \pm(g\mu_B/2)H. \qquad (6.6)$$

These eigenvalues are never degenerate except if all field components
H_x, H_y, H_z vanish. If one plots the energies as a function of H_z for a fixed value
of H_x, one obtains Fig. 6.3. The curves $E^+(H_z)$ and $E^-(H_z)$ do not cross at
$H_z = 0$. Instead, their difference has a minimum. One of the eigenvalues has a
maximum and the other one has a minimum.

6.1.4   *Case of an arbitrary spin*

The above property has no reason to be special to spin $1/2$. The generalization
is the following.

For a given values of $|D|$ and s, the eigenvalues of (6.2) are non-degenerate,
except for a finite set of values of H_x and H_z.

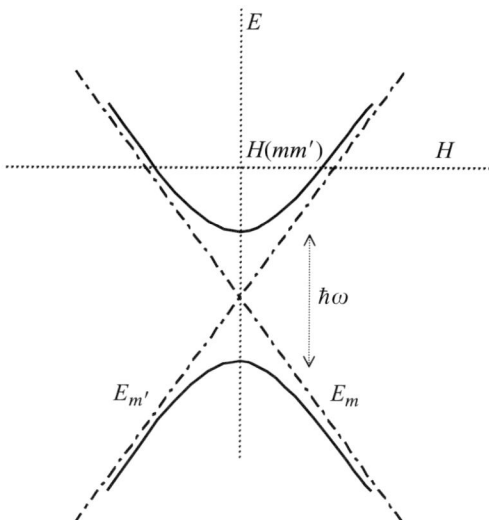

FIG. 6.3. Behaviour of two eigenvalues E_m and E'_m of a spin Hamiltonian as functions of H_z near the value $H^{(mm')}$ where these values would be equal for $H_x = H_y = 0$.

More generally, for a given anisotropy Hamiltonian, the eigenvalues of the Hamiltonian in an external field are non-degenerate, except for a finite set of values of the field H_x, H_y, H_z.

As will be seen in Section 6.8, this property is a special case of a theorem of Wigner and von Neumann (1929). In the case of a spin $1/2$, it may be viewed as a consequence of the fact that the Hamiltonian matrix is a 2×2 Hermitian matrix. Its eigenvalues are equal if and only if $\langle 1/2 \,|\, \mathcal{H} \,|\, 1/2 \rangle = \langle -1/2 \,|\, \mathcal{H} \,|\, -1/2 \rangle$ and if the real and imaginary parts of the off-diagonal elements vanish, $\Re \langle 1/2 \,|\, \mathcal{H} \,|\, -1/2 \rangle = \Im \langle 1/2 \,|\, \mathcal{H} \,|\, -1/2 \rangle = 0$. These three conditions determine the three variables H_x, H_y, H_z.

As seen in Section 6.1.2, the Hamiltonian (6.2) is degenerate for $H_x = 0$ if H_z satisfies (6.3). The theorem of Wigner and von Neumann implies that this degeneracy is removed by any small field $H_x \neq 0$. This will be checked by perturbation theory in Section 6.4.

The non-degeneracy of the Hamiltonian in the general case will be now be assumed, and its justification postponed to Sections 6.4 and 6.8. Let the consequences of this property be addressed.

6.1.5 Delocalization

Let two eigenvalues E_m and $E_{m'}$ of (6.2) be plotted as a function of H_z (Fig. 6.3) for a small, fixed value of H_x. The case of interest in this section is when H_z is close to the level crossing value $H^{(mm')}$ given by (6.4). One can assume $m < 0$ and $m' > 0$.

It follows from the previous section that the two curves do not cross if $H_x \neq 0$. Instead, one of the eigenvalues has a maximum and the other one has a minimum. This is called the *anticrossing*. The difference between the maximum and the minimum will be called the *tunnel splitting* and denoted $2\hbar\omega_{\mathrm{T}}^{(mm')}$ or $\Delta_{(mm')}$. It vanishes for $H_x = 0$. The quantity $\omega_{\mathrm{T}}^{(mm')}$ will be called the tunnel frequency, for reasons which will become clear in Section 6.5, when kinetic properties will be investigated.

For the sake of simplicity, the indices m, m' will often be omitted in $\omega_{\mathrm{T}}^{(mm')}$. The notation ω_{T} will generally denote $\omega_{\mathrm{T}}^{(-s,m')}$, where m' is related to the magnetic field by the level-crossing condition (6.3), with $m = -s$.

Rather than calling the eigenvalues E_m and $E_{m'}$, it is preferable to call the lower eigenvalue E_ℓ and the larger eigenvalue E_L. Indeed, for $H \approx H^{(mm')}$, the two eigenvalues cannot be assigned to m or m'.

The absence of actual crossing has important consequences on the corresponding eigenvectors $\mid \Phi_\ell \rangle$ and $\mid \Phi_L \rangle$ and on the wavefunctions $\varphi_\ell(m'')$ and $\varphi_L(m'')$, related to the eigenvectors by

$$\mid \Phi_\ell \rangle = \sum_{m''} \varphi_\ell(m'') \mid m'' \rangle \qquad (6.7)$$

and a similar formula for $\mid \Phi_L \rangle$.

For $H \ll H^{(mm')}$, $\varphi_\ell(m'')$ is localized near $m'' = m$, i.e. in the left-hand well. For $H \gg H^{(mm')}$, $\varphi_\ell(m'')$ is localized near $m'' = m'$, i.e. in the right-hand well. What can happen for $H \approx H^{(mm')}$? Continuity implies (Fig. 6.4 and 6.5) that $\varphi_\ell(m'')$ is delocalized on both sides, with two maxima of $|\varphi_\ell(m'')|$ at $m'' = m$ and $m'' = m'$. An analogous argument shows that $\varphi_L(m'')$ is also delocalized for $H \approx H^{(mm')}$.

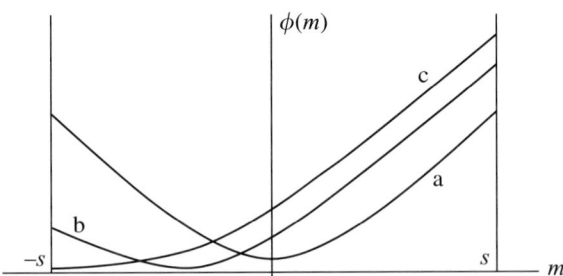

FIG. 6.4. Schematic representation of the ground state wavefunction $\varphi(m)$ of Hamiltonian (6.2). (a) $H_z = 0$. (b) H_z is slightly negative and favours positive values of $S_z = m$. (c) H_z is more strongly negative. Since the spin modulus s is large, m is approximated by a continuous variable.

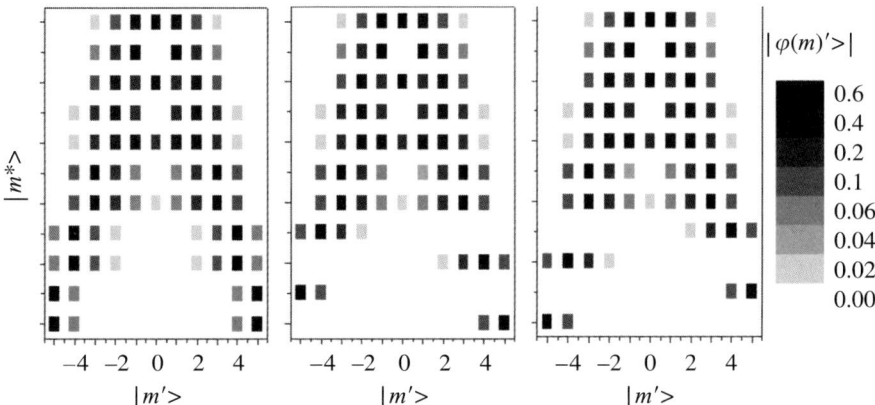

FIG. 6.5. Numerically calculated eigenfunctions $\varphi(m)$ of Hamiltonian (6.2) for a spin $s = 5$, with $D/k_B = 1$ K and $\mu_0 H_x = 0.4$ T. Left: $H_z = 0$. Centre: $\mu_0 H_z = -0.1$ mT is slightly negative and favours positive values of $S_z = m$. Right: $\mu_0 H_z = +0.1$ mT. The two lowest lines show the amplitude of the two lowest eigenfunctions, which are $|5^*\rangle$ and $|-5^*\rangle$ for $\mu_0 H_z = \pm 0.1$ mT and linear combinations of $|5^*\rangle$ and $|-5^*\rangle$ for $H_z = 0$. The next two lines show the next two eigenfunctions, etc. Very weak amplitudes do not appear although they are very important for the tunnelling process.

In practice, it will generally be assumed that

$$m + m' \ll s, \quad s + m \ll s, \quad s - m' \ll s \tag{6.8}$$

which is the situation of interest at low temperature. In that case, the wavefunction is very small near the top $m_{\max}$ of the barrier given by (5.7).

This behaviour of the spin wavefunction $\varphi(m'')$ (equal localization in both wells, with two maxima of $|\varphi(m'')|$ separated by a minimum) is called the spin tunnelling or magnetic tunnel effect. This expression is used only for large s; it would not make sense for $s = 1/2$. Magnetic tunnelling is present for $H_z = 0$, disappears if a longitudinal magnetic field H_z is applied, then reappears for a particular value, then disappears again, etc. The field values which allow tunnelling are the 'level crossing' values (6.4). The level scheme is illustrated in Fig. 6.2.

6.1.6 *Large spin = classical spin?*

As said above in the case of the Hamiltonian (6.2), the tunnel frequency ω_T vanishes when the Hamiltonian has no term which does not commute with S_z, i.e. when $H_x = 0$ in the case of the Hamiltonian (6.2). What is perhaps more unexpected is that it also vanishes for a given value of $H_\alpha/|D|$, if s goes to ∞. As will be seen in Section 6.4, the ground state splitting even vanishes exponentially as $[H_x/(4|D|s)]^{2s}$. Thus, ω_T is small if s is large and the typical transverse Zeeman energy $H_x s$ is smaller, but not necessarily *much* lower than the anisotropy energy $|D|s^2$. This is related to the general fact that a very large spin behaves as a classical spin. Indeed, tunnelling is a quantum effect, related to the representation of non-commuting physical quantities by non-commuting operators. The quasi-classical nature of large spins results from the fact that the commutator $[S_\alpha, S_\gamma]$ of two components is of the order of s while the anticommutator is of the order of s^2.

6.1.7 *Approximate localized eigenstates*

As seen in Section 6.1.5, tunnelling is characterized by exact eigenfunctions $\varphi_\ell(m'')$ and $\varphi_L(m'')$ which are delocalized on both sides of the double well. The modulus $|\varphi_\ell(m'')|$ or $|\varphi_L(m'')|$ has two maxima at $m < 0$ and $m' > 0$. Figure 6.6 gives a schematic representation of $\varphi_\ell(m'')$ and $\varphi_L(m'')$ in vanishing magnetic field.

This picture is not always convenient, and it is often preferable to express the exact eigenfunctions in terms of approximate eigenfunctions which are localized in the left-hand well ('left-localized') or in the right-hand well ('right-localized'). This approximation becomes excellent when the tunnel splitting becomes very small. More details on the quality of the approximation will be given in a special case in Section 6.7.2.

Since inequalities (6.8) are satisfied, $\varphi_\ell(m'')$ and $\varphi_L(m'')$ are very small at the top $m_{\max}$ of the barrier given by (5.7). Therefore, it is reasonable to write them as sums of a left-localized part $\varphi^{(g)}(m'')$ (which vanishes for $m'' > m_{\max}$)

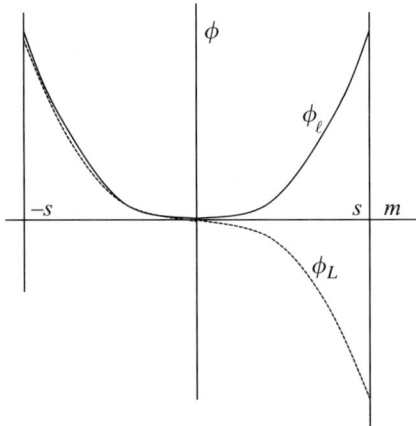

FIG. 6.6. Schematic representation of the two wavefunctions of lowest energy in vanishing magnetic field. They are approximated by continuous curves as in Fig. 6.4. The wavefunction is assumed to be real, which is only possible for particular spin hamiltonians such as (6.2).

and a right-localized part $\varphi^{(d)}(m'')$.

$$\varphi_\ell(m'') = \varphi_\ell^{(g)}(m'') + \varphi_\ell^{(d)}(m'') \tag{6.9}$$

$$\varphi_L(m'') = \varphi_L^{(g)}(m'') + \varphi_L^{(d)}(m''). \tag{6.10}$$

The four localized functions $\varphi_\ell^{(g)}(m'')$, $\varphi_\ell^{(d)}(m'')$, $\varphi_L^{(g)}(m'')$ and $\varphi_L^{(d)}(m'')$ are approximate solutions of the Schrödinger equation (6.1), i.e.

$$\sum_{m'''} \langle m'' \mid \mathcal{H} \mid m''' \rangle \, \varphi(m'') = E\varphi(m'').$$

Actually, for all values of m'', except near the the top of the barrier, $m'' \simeq m_{\max}$, this equation is exactly satisfied, however with slightly different energies E_ℓ and E_L. For $m'' \simeq m_{\max}$, the above equation is not exactly satisfied, but only approximately since $\varphi(m''')$ and $\varphi(m'')$ are very small.

Moreover, $\varphi_\ell^{(g)}(m'')$ and $\varphi_L^{(g)}(m'')$ must be approximately proportional to each other, because they are two approximate solutions of the same Schrödinger equation with almost the same eigenvalue and the same localization. This situation is indeed analogous to a particle in a square or harmonic well. There cannot be two independent eigenfunctions with two almost equal energies. Analogously, $\varphi_\ell^{(d)}(m'')$ and $\varphi_L^{(d)}(m'')$ are proportional. They have a single maximum for $m'' = m'$ while $\varphi_\ell^{(g)}(m'')$ and $\varphi_L^{(g)}(m'')$ have a single maximum for $m'' = m$.

From (6.9), (6.10) and the approximate proportionality of $\varphi_\ell^{(g)}(m'')$ and $\varphi_L^{(g)}(m'')$, and of $\varphi_\ell^{(d)}(m'')$ and $\varphi_L^{(d)}(m'')$, it follows that $\varphi_\ell^{(g)}(m'')$, $\varphi_L^{(g)}(m'')$, $\varphi_\ell^{(d)}(m'')$ and $\varphi_L^{(d)}(m'')$ are approximate linear combinations of the exact eigenfunctions $\varphi_\ell(m'')$ and $\varphi_L(m'')$. It is, however, preferable to introduce exact linear combinations which are approximately localized. The corresponding vectors will be called $\mid m^* \rangle$ and $\mid m'^* \rangle$. The linear combinations will be written for normalized vectors, $\langle \Phi_\ell \mid \Phi_\ell \rangle = \langle \Phi_L \mid \Phi_L \rangle = \langle m^* \mid m^* \rangle = \langle m'^* \mid m'^* \rangle = 1$. They will assume the orthogonality of $\mid m^* \rangle$ and $\mid m'^* \rangle$, and take into account the orthogonality of $\mid \Phi_\ell \rangle$ and $\mid \Phi_L \rangle$. Moreover, each of the vectors $\mid \Phi_\ell \rangle$, $\mid \Phi_L \rangle$, $\mid m^* \rangle$, $\mid m'^* \rangle$ can be multiplied by any constant $\exp(i\theta)$ of modulus 1. Using this possibility, the appropriate linear combinations can be chosen as

$$\mid m^* \rangle = \mid \Phi_\ell \rangle \cos\phi - \mid \Phi_L \rangle \sin\phi \tag{6.11}$$

and

$$\mid m'^* \rangle = \mid \Phi_\ell \rangle \sin\phi + \mid \Phi_L \rangle \cos\phi \tag{6.12}$$

where the parameter ϕ should be chosen such that it optimizes the localization of $\mid m^* \rangle$ and $\mid m'^* \rangle$. The reciprocal relations of (6.11) and (6.12) are

$$\mid \Phi_\ell \rangle = \mid m^* \rangle \cos\phi + \mid m'^* \rangle \sin\phi \tag{6.13}$$

and

$$\mid \Phi_L \rangle = - \mid m^* \rangle \sin\phi + \mid m'^* \rangle \cos\phi \tag{6.14}$$

The localization of $|m^*\rangle$ in one well and of $|m'^*\rangle$ in the other well is approximate. Both vectors have a small component in the wrong well. Far from level crossing, $\phi = 0$ on one side, $\phi = \pi/2$ on the other side. At level crossing, $\phi = \pi/4$ and delocalization has its maximum.

Formulae (6.13) and (6.14) demonstrate the possibility of defining localized approximate eigenvectors $|m^*\rangle$ and $|m'^*\rangle$ of $\mathcal{H}$ near a level crossing. However, if $|\Phi_\ell\rangle$ and $|\Phi_L\rangle$ are not known, they do not provide an explicit determination of $|m^*\rangle$ and $|m'^*\rangle$. This task will be done in Sections 6.4 and 6.7.

In the case of a real spin, which interacts with its environment, the nature of tunnelling is deeply modified. The main features of tunnelling are kinetic. Anticipating Section 6.5, tunnelling of an isolated spin would be characterized by oscillations of S_z between positive and negative values. For a real spin, these oscillations are generally not observed. However, tunnelling is observed as a faster relaxation. Indeed, as will be seen in Chapter 10, the spin does not need to jump to the top of the barrier, but to a state where tunnelling is possible. This state is lower if the field is close to a level crossing. Thus, the characteristic property (the 'signature', as is sometimes said) of magnetic tunnelling is a succession of sharp maxima (Fig. 6.7) of the relaxation rate (i.e. minima of the relaxation time) as H_z is varied (Friedman *et al.* 1996; Thomas *et al.* 1996; Fominaya *et al.* 1997b). These maxima, which will be called *resonances*, are well described by formula (6.4).

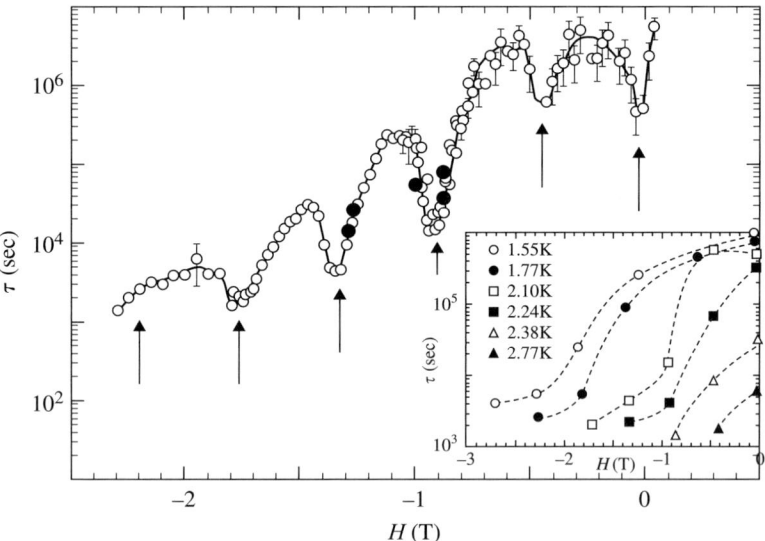

FIG. 6.7. Relaxation time τ as a function of the longitudinal field in Mn$_{12}$ac at 2.1 K. The sharp minima demonstrate tunnelling. The insert shows τ as a function of temperature (from Thomas *et al.* 1996). Reprinted with permission of Nature (http://www.nature.com).

6.2   Symmetry and selection rules for tunnelling

Tunnelling can be produced by any terms in the Hamiltonian which do not commute with S_z. These terms can result from a transverse magnetic field as seen before. However, in Fe_8, which is the material where magnetic tunnelling has been most clearly observed at low temperature, the dominant source of tunnelling is the biaxial anisotropy. The relevant Hamiltonian is generally assumed to have the form (2.9) or, in a 'longitudinal' magnetic field,

$$\mathcal{H} = -|D|S_z^2 + (B/2)(S_x^2 - S_y^2) + g\mu_B H_z S_z. \tag{6.15}$$

The second term does not commute with S_z, and therefore can produce tunnelling. However the Hamiltonian (6.15) has particular symmetry properties. The consequences are that tunnelling does not occur for any field value which satisfies (6.4).

The peculiarity of (6.15) is that its matrix elements $\langle m \,|\, \mathcal{H} \,|\, m' \rangle$ vanish if $|m - m'|$ is odd. It follows that the general expression (6.7) of the eigenvectors has either the form

$$| \Psi \rangle = \sum_p \varphi(s - 2p) \,|\, s - 2p \rangle \tag{6.16}$$

or the form

$$| \Psi \rangle = \sum_p \varphi(s - 2p - 1) \,|\, s - 2p - 1 \rangle . \tag{6.17}$$

The sum is over the values of the argument $(s - 2p$ or $s - 2p - 1)$ between $-s$ and s, and p is an integer.

It will now be assumed that, for $B = 0$, two eigenvectors $|\, m \rangle$ and $|\, m' \rangle$ have equal, or approximately equal unperturbed energies, $E_m^{(0)} - E_{m'}^{(0)} \approx 0$. Is tunnelling possible for $B \neq 0$? The wavefunction should then contain both a component $\phi(m) \,|\, m \rangle$ and a component $\phi(m') \,|\, m' \rangle$. This is only compatible with (6.16) and (6.17) if

$$\text{either } m = s - 2p \text{ and } m' = s - 2p';$$

$$\text{or } m = s - 2p - 1 \text{ and } m' = s - 2p' - 1.$$

Therefore, tunnelling between $|\, m \rangle$ and $|\, m' \rangle$ occurs only if the *selection rule*

$$m - m' = 2k \tag{6.18}$$

is satisfied, where k is an integer.

Only half of the level crossings of Fig. 6.1a correspond to tunnelling. In half of the cases, there is an exact degeneracy.

In particular, in zero field ($H_z = 0$), tunnelling is impossible if s is half-integer. All states are twice degenerate. This is Kramers' degeneracy.

Let the case of tetragonal symmetry now be considered. As in Chapter 2, fourth-order terms will be introduced into the spin Hamiltonian, but for simplicity, only those which do not commute with the main term will be retained. The

Hamiltonian reads

$$\mathcal{H}_{\mathrm{an}} = -|D|S_z^2 + C(S_+^4 + S_-^4) + g\mu_{\mathrm{B}}H_zS_z. \qquad (6.19)$$

This formula includes the familiar, quadratic anisotropy, the effect of the field and the quartic anisotropy (2.18) in which the S_z^4 term has been omitted for simplicity, because it is a mere correction to the first term. On the contrary, the other quartic terms have the essential property that they do not commute with S_z.

The eigenstates of the first and last terms in (6.19) are $|\,m\rangle$, and the matrix elements of the second term between $|\,m\rangle$ and $|\,m'\rangle$ vanish unless $m - m'$ is a multiple of 4. Therefore, tunnelling is possible only between $|\,m\rangle$ and $|\,m'\rangle$ if the selection rule

$$m - m' = 4k \qquad (6.20)$$

is satisfied, where k is an integer.

The selection rules obtained in this section are the effect of symmetry. Experimentally, they are in many cases not satisfied. The only explanation seems to be that symmetry is destroyed by crystal defects, e.g. dislocations, impurities, etc. This problem will be considered in Chapter 12.

6.3   Tunnelling width for an isolated spin

In formula (6.13), even though $|\,m^*\rangle$ and $|\,m'^*\rangle$ are not precisely known, it may be assumed that they do not vary much through the resonance, while ϕ varies from 0 to $\pi/2$.

If the Schrödinger equation $\mathcal{H}\,|\,\Phi_\ell\rangle = E\,|\,\Phi_\ell\rangle$ is multiplied by $\langle m^*\,|$, then by $\langle m'^*\,|$, one obtains the system of two equations

$$\begin{cases} [\langle m^*\,|\,\mathcal{H}\,|\,m^*\rangle - E]\cos\phi + \langle m^*\,|\,\mathcal{H}\,|\,m'^*\rangle \sin\phi = 0 \\[2mm] -\langle m'^*\,|\,\mathcal{H}\,|\,m^*\rangle \cos\phi + [\langle m'^*\,|\,\mathcal{H}\,|\,m'^*\rangle - E]\sin\phi = 0. \end{cases} \qquad (6.21)$$

It is appropriate to introduce the quantity δE defined by

$$E = \frac{\langle m^*\,|\,\mathcal{H}\,|\,m^*\rangle + \langle m'^*\,|\,\mathcal{H}\,|\,m'^*\rangle}{2} + \delta E \qquad (6.22)$$

and the quantities

$$\Delta = \frac{\langle m^*\,|\,\mathcal{H}\,|\,m^*\rangle - \langle m'^*\,|\,\mathcal{H}\,|\,m'^*\rangle}{2} \qquad (6.23)$$

and

$$\hbar\omega_{\mathrm{T}}^{(mm')} = |\,\langle m^*\,|\,\mathcal{H}\,|\,m'^*\rangle\,| = |\,\langle m'^*\,|\,\mathcal{H}\,|\,m^*\rangle\,|. \qquad (6.24)$$

It will be seen that (6.24) is consistent with the definition of $\omega_{\mathrm{T}}^{(mm')}$ given in Section 6.1.

The energy E is obtained by noting that the determinant of the coefficients of (6.21) vanishes. This yields

$$\delta E = \pm\sqrt{\Delta^2 + \hbar^2\omega_{\mathrm{T}}^2}\tag{6.25}$$

where the indices m and m' have been omitted. Formula (6.25) generalizes (6.6). The two eigenvalues are E_ℓ and E_L, so that their difference is $2|\delta E|$.

Formula (6.25) is useful in a field interval where the matrix element (6.24) does not vary much. Then, ω_{T} may be regarded as a constant. The minimum separation of the two levels corresponds to $\Delta = 0$ and is equal to $2\hbar\omega_{\mathrm{T}}$, in agreement with the statement of Section 6.1.

If the states $|\Phi_1\rangle$ and $|\Phi_2\rangle$ of interest correspond to the crossing of the unperturbed states $|m\rangle$ and $|m'\rangle$, (6.23) can be replaced by the approximate expression

$$\Delta \approx \frac{\langle m\,|\,\mathcal{H}\,|\,m\rangle - \langle m'\,|\,\mathcal{H}\,|\,m'\rangle}{2} = g\mu_{\mathrm{B}}(m - m')\delta H_z/2\tag{6.26}$$

where δH_z is the difference between the field and its value at level crossing. This approximation of order 0 is not very ambitious, but sufficient for current experimental needs. The quantity Δ will be called the 'Zeeman splitting' It vanishes at level crossing.

It results from (6.21) that

$$|\tan\phi| = \frac{\hbar\omega_{\mathrm{T}}}{|\delta E - \Delta|}.\tag{6.27}$$

This quantity characterizes the localization of the wavefunction (6.13).

If $|\Delta| \gg \hbar\omega_{\mathrm{T}}$ and E corresponds to the $-$ sign in (6.25), then $|\tan\phi| \simeq 0$, and the spin is localized on the left.

If $|\Delta| \gg \hbar\omega_{\mathrm{T}}$ and E corresponds to the $+$ sign in (6.25), then $|\tan\phi| \gg 1$, and the spin is localized on the right.

If $|\tan\phi| = 1$, the spin is completely delocalized and has the same probability to be in the left or right hand side of the well. This case corresponds to $\Delta = 0$.

The wavefunctions are localized for all values of H_z, except near an intersection of unperturbed levels in a field interval of width $\delta H_0^{(mm')}$ given by the following equation where the indices m and m' have been restored.

$$g\mu_{\mathrm{B}}\delta H_0^{(mm')}s \simeq \hbar\omega_{\mathrm{T}}^{(mm')}\tag{6.28}$$

since condition (6.8) implies $m' - m \simeq 2s$.

It will be seen in Chapter 9 that the 'bare' width $\delta H_0^{(mm')}$ given by (6.28) is actually smaller than other contributions, which result, in particular, from interactions with nuclear spins and other molecular spins.

When the longitudinal field H_z is close to the value $H_z^{(mm')}$ which corresponds to level crossing, and when $2|H_z - H_z^{(mm')}|$ is less than the width, it will be said that there is a *resonance*.

Formula (6.25) shows that the true levels generally do not cross. They only cross if both Δ and ω_T vanish. This requires many conditions, which are generally not satisfied. This point will be addressed in Section 6.8.

The essential results of this section are the level spacing (6.25), and formula (6.27) which says how localized or delocalized the wavefunction is. The variation of the matrix elements (6.24) is proportional to the variation of the magnetic field, and therefore it is weak in the resonance region if its width is small, i.e. if the tunnel splitting is small.

In the following sections, it will be seen how the tunnel splitting can be calculated.

6.4   Tunnel splitting according to perturbation theory

The spin Hamiltonians considered in the previous sections have the form

$$\mathcal{H} = \mathcal{H}_0 + \delta\mathcal{H} \tag{6.29}$$

where the 'unperturbed' Hamiltonian $\mathcal{H}_0$ commutes with S^z, e.g.

$$\mathcal{H}_0 = -|D|S_z^2 + g\mu_B H_z \tag{6.30}$$

while $\delta\mathcal{H}$ may assume various forms. Important examples are

(i) the effect of a transverse magnetic field

$$\delta\mathcal{H} = g\mu_B H_x S_x; \tag{6.31}$$

(ii) an anisotropy of symmetry lower than tetragonal, see formula (2.8)

$$\delta\mathcal{H} = (B/4)(S_+^2 + S_-^2); \tag{6.32}$$

(iii) the same tetragonal anisotropy as in (6.19)

$$\delta\mathcal{H} = C(S_+^4 + S_-^4). \tag{6.33}$$

All these examples have the property that diagonal elements vanish in the basis which diagonalizes S^z,

$$\langle m \,|\, \delta\mathcal{H} \,|\, m \rangle = 0. \tag{6.34}$$

This simplifies the calculation.

What happens if one tries to calculate by standard perturbation theory the coefficients $\varphi^{(m)}(m')$ defined by (5.11)?

To order 0, $\varphi^{(m)}(m') = \delta_{mm'}$. To first order, textbooks tell us that

$$\varphi^{(m)}(m') = \delta_{mm'} + (1 - \delta_{mm'})\frac{\langle m' \,|\, \delta\mathcal{H} \,|\, m \rangle}{E_m^{(0)} - E_{m'}^{(0)}} \tag{6.35}$$

Generally, a denominator $E_m^{(0)} - E_{m'}^{(0)}$ appears to all orders. Therefore, the method fails near a level crossing, defined by (6.4). However, at low orders, if $|m - m'|$

is large, and if only terms of low order with respect to spin operators are taken into account in $\delta\mathcal{H}$, then the numerator also vanishes. In the case of Hamiltonian (6.32), a non-vanishing tunnel frequency $\omega_{\mathrm{T}}^{s;-s}$ between $|-s\rangle$ and $|s\rangle$ (ground doublet) can only be obtained by treating the anisotropy in perturbation theory to order s. More generally, the calculation of $\omega_{\mathrm{T}}^{(mm')}$ requires us to go to order $|m'-m|$. How to do this has been explained by Korenblitt and Shender (1978) and Garanin (1991). Hartmann-Boutron (1996) has given the formulae relevant for the three cases (6.31), (6.32) and (6.33),

The Hamiltonian matrix $\mathcal{H}$ can be conveniently written in the basis of the eigenvectors $|m'\rangle$ of S_z, however, writing the lines and columns m and p as the first lines.

$$\mathcal{H} = \begin{bmatrix} \mathcal{M} & \mathcal{B}^* \\ \mathcal{B} & \mathcal{A} \end{bmatrix} \tag{6.36}$$

where $\mathcal{M}$ is a 2×2 diagonal matrix of elements $\langle m | \mathcal{H} | m \rangle$ and $\langle p | \mathcal{H} | p \rangle$, $\mathcal{A}$ is a $(2s-1) \times (2s-1)$ square matrix of elements $\langle m' | \mathcal{H} | p' \rangle$ $(m', p' \neq m, p)$ and $\mathcal{B}$ is a $(2s-1) \times 2$ matrix of elements $\langle m | \mathcal{H} | m' \rangle$ and $\langle p | \mathcal{H} | m' \rangle$. The eigenvalues E of (6.36) and the eigenvectors $(|\varphi\rangle, |\Phi\rangle)$ are given by

$$\mathcal{H} = \begin{cases} (\mathcal{M} - E)|\varphi\rangle + \mathcal{B}^*|\Phi\rangle = 0 \\ \\ \mathcal{B}|\varphi\rangle + (\mathcal{A} - E)|\Phi\rangle = 0. \end{cases} \tag{6.37}$$

Eliminating $|\Phi\rangle$, it follows that

$$[\mathcal{M} - E + \mathcal{B}^*(E - \mathcal{A})^{-1}\mathcal{B}]|\varphi\rangle = 0. \tag{6.38}$$

Thus, the energy E is the eigenvalue of a 2×2 matrix which is the sum of a diagonal part $\mathcal{M}$ and a small perturbation $\mathcal{B}^*(E - \mathcal{A})^{-1}\mathcal{B}$. In analogy with (6.24), one can write

$$\hbar\omega_{\mathrm{T}}^{(mm')} = \langle m | \mathcal{B}^*(E - \mathcal{A})^{-1}\mathcal{B} | m' \rangle. \tag{6.39}$$

The problem is that E appears in this expression. This would make equation (6.39) difficult to solve exactly. However, the problem simplifies in perturbation theory, if we stop at the lowest order that yields a non-vanishing tunnel splitting, hereafter called the lowest significant order. Then, E can just be replaced by its zero-order approximation $E_m^{(0)}$, given by (5.6) in the case of the Hamiltonian (6.30).

Let $\mathcal{A} = \mathcal{A}_0 + \delta\mathcal{A}$ be split into its diagonal part $\mathcal{A}_0$ and a small off-diagonal part $\delta\mathcal{A}$. The matrix elements of $\mathcal{M}$ and $\mathcal{A}_0$ can be identified with those of $\mathcal{H}_0$, while the non-vanishing elements of $\mathcal{B}$ and $\delta\mathcal{A}$ are those of $\delta\mathcal{H}$. Then

$$(E - \mathcal{A})^{-1} = (E - \mathcal{A}_0)^{-1} + (E - \mathcal{A}_0)^{-1}\delta\mathcal{A}(E - \mathcal{A}_0)^{-1}$$
$$+ (E - \mathcal{A}_0)^{-1}\delta\mathcal{A}(E - \mathcal{A}_0)^{-1}\delta\mathcal{A}(E - \mathcal{A}_0)^{-1} + \dots \tag{6.40}$$

This infinite sum looks horrible, but only the lowest non-vanishing term will be kept. Which one is it? This depends on m, m' and $\delta\mathcal{H}$. For definiteness, the remainder of the calculation will be restricted to the zero-field ground state splitting, which means that we assume $m = -s$ and $m' = s$. The general case is treated in Appendix G. Moreover, $\delta\mathcal{H}$ will be assumed to have the form (6.31). Replacing E in (6.39) by $E_0 = E_s^{(0)} = E_{-s}^{(0)} = -|D|s^2$, one obtains a first term

$$\langle -s \,|\, \mathcal{B}^* \,|\, -s+1\rangle \, \langle -s+1 \,|\, (E_0 - \mathcal{A}_0)^{-1} \,|\, s-1\rangle \, \langle s-1 \,|\, \mathcal{B} \,|\, s\rangle$$

which is 0 except if $s = 1$, since $\mathcal{A}_0$ is diagonal. The second term is

$$\langle -s \,|\, \mathcal{B}^* \,|\, -s+1\rangle \, \langle -s+1 \,|\, (E_0 - \mathcal{A}_0)^{-1} \,|\, -s+1\rangle$$
$$\langle -s+1 \,|\, \delta\mathcal{A} \,|\, s-1\rangle \, \langle s-1 \,|\, (E_0 - \mathcal{A}_0)^{-1} \,|\, s-1\rangle \, \langle s-1 \,|\, \mathcal{B} \,|\, s\rangle \,.$$

If $\delta\mathcal{A}$ is given by (6.31), its matrix element vanishes unless $s = 3/2$. For a general value of s, it is easily seen that the first non-vanishing term is that of order $(2s-2)$ in $\delta\mathcal{A}$. This term is of degree $2s$ in H_x because $\mathcal{B}$ is also linear in H_x. Ignoring the higher order terms, (6.39) reduces to

$$\hbar\omega_{\mathrm{T}}^{(-s,s)} = \hbar\omega_{\mathrm{T}} = \langle -s \,|\, \mathcal{B}^*[(E_0 - \mathcal{A}_0)^{-1}\delta\mathcal{A}]^{2s-2}\mathcal{B} \,|\, s\rangle \,. \qquad (6.41)$$

The calculation is straightforward and performed in Appendix G. It yields (Hartmann-Boutron 1996) the zero-field ground state splitting as $2\hbar\omega_{\mathrm{T}}$ where

$$\hbar\omega_{\mathrm{T}} = 4|D|s^2 \left(\frac{g\mu_{\mathrm{B}}H_x}{2|D|}\right)^{2s} \frac{1}{(2s)!} \,. \qquad (6.42)$$

When $\delta\mathcal{H}$ is given by (6.32), the formula for integer s is, as seen in Appendix G,

$$\hbar\omega_{\mathrm{T}} = 4|D|s^2 \left(\frac{B}{16|D|}\right)^{s} \frac{(2s)!}{(s!)^2} \,. \qquad (6.43)$$

When $\delta\mathcal{H}$ is given by (6.33), the formula for even s is (Hartmann-Boutron 1996)

$$\hbar\omega_{\mathrm{T}} = 4|D|s^2 \left(\frac{C}{16|D|}\right)^{s/2} \frac{(2s)!}{[(s/2)!]^2} \,. \qquad (6.44)$$

Of course, in agreement with the selection rules (6.18) and (6.20), $\omega_{\mathrm{T}}^{(-s,s)}$ vanishes in case (6.32) when s is half-integer and in case (6.33) when s is half-integer or odd.

For large s, the above formulae can be simplified by using Stirling's formula $s! \simeq s^{s+1/2}e^{-s}\sqrt{2\pi}$. For instance (6.43) reduces to

$$\hbar\omega_{\mathrm{T}}^{(-s,s)} = 4s^2|D| \left(\frac{B}{4|D|}\right)^{s} \frac{(2s)^{2s+1/2}e^{-2s}\sqrt{2\pi}}{\left[2^s s^{s+1/2}e^{-s}\sqrt{2\pi}\right]^2} = \frac{4|D|}{\sqrt{\pi}} s^{3/2} \left(\frac{B}{4|D|}\right)^{s} \,. \qquad (6.45)$$

Perturbation theory also allows us to treat the case where the off-diagonal terms of the Hamiltonian are linear combinations of (6.31), (6.32), (6.33),...

An example is given in Appendix G. It is also possible to obtain a rough determination of the maximum value of C below which C can be neglected if the perturbation is a linear combination of (6.32) and (6.33):

$$\delta\mathcal{H} = (B/4)(S_+^2 + S_-^2) + C(S_+^4 + S_-^4).$$

For $C = 0$, ω_T has a value $\omega_T^{(B)}$ given by (6.43). For $B = 0$, ω_T has a value $\omega_T^{(C)}$ given by (6.44). The ratio of the two values is

$$\omega_T^{(B)}/\omega_T^{(C)} = \frac{B^s}{(16|D|C)^{s/2}}\left(\frac{(s/2)!}{s!}\right)^2.$$

In the case $s = 10$, this is equal to

$$\omega_T^{(B)}/\omega_T^{(C)} = (6 \times 7 \times 8 \times 9 \times 10)^{-2}\left(\frac{B^2}{16|D|C}\right)^{s/2}$$

or

$$\omega_T^{(B)}/\omega_T^{(C)} = \frac{1}{914\,457\,600}\left(\frac{B^2}{16|D|C}\right)^5. \tag{6.46}$$

The correction C can be neglected if expression (6.46) is much larger than 1. This implies that a very small value of C has sizeable effects. This is not too surprising since C acts at order 5 and B only at order 10.

It follows from (6.42), (6.43) and (6.44) that, in all cases, ω_T is small if s is large, even if the ratio $B/|D|$ is not particularly small, provided it is not too large (<4). An analogous result may be established if the tunnel effect is produced by a transverse field as assumed in Section 6.1.

As a conclusion to this section, perturbation theory is an extremely powerful and effective method. It can be tested numerically or by the analytical methods presented in the following sections. These tests are positive. This was not obvious *a priori*.

Indeed, as pointed out in Section 5.2, for the states $|m\rangle$ with small values of m, the off-diagonal perturbation is of the order of Es^2 which is much larger than the level spacing. Now, the perturbative calculation of tunnelling between states $|m\rangle$ and $|m'\rangle$ with *large* values of m, m' involves all states $|m''\rangle$ with m'' between m and m', and these states are strongly perturbed. The success of high-order perturbation theory is therefore rather strange, and it is of great interest to compare this method with other ones as explained below in Section 6.7, as well as in the next chapter.

6.5   Time-dependent wavefunction: magnetic tunnelling

To observe the evolution of a spin, the experimentalist puts it initially in one of the wells, e.g. the left-hand one. At low temperatures, at which tunnelling takes place, the spin is then with a high probability in the lowest state of this well, i.e. $|-s^*\rangle$. In this section, more generally, the spin will be assumed to be initially in a quantum state $|m^*\rangle$ of the left-hand well ($m < 0$). The case of interest in this section is when the unperturbed energy $E_m^{(0)}$ is close to another unperturbed energy $E_{m'}^{(0)}$ ($m' > 0$). Then, as seen in Section 6.3, $|m^*\rangle$ is the combination of two eigenvectors $|\Phi\rangle$ and $|\Phi'\rangle$ of energies $E_0 \pm \delta E$, where δE, given by (6.25), is small. The wavefunction at time t is therefore

$$|\Psi(t)\rangle = \lambda_1 |\Phi_\ell\rangle \exp[i(E_0 - \delta E)t/\hbar]$$
$$+ \lambda_2 |\Phi_L\rangle \exp[i(E_0 + \delta E)t/\hbar]$$

where $|\Phi_\ell\rangle$ and $|\Phi_L\rangle$ are given by (6.13) and (6.14) and the constants λ_1 and λ_2 are given by the initial condition $|\Psi(0)\rangle = |m^*\rangle$. One obtains

$$|\Psi(t)\rangle = [|m^*\rangle \cos\phi + |m'^*\rangle \sin\phi] \cos\phi \exp[i(E_0 - \delta E)t/\hbar]$$
$$+ [|m^*\rangle \sin\phi - |m'^*\rangle \cos\phi] \sin\phi \exp[i(E_0 + \delta E)t/\hbar]$$

or

$$\exp[-iE_0t/\hbar] |\Psi(t)\rangle = |m^*\rangle [\cos(t\delta E/\hbar) - i\cos(2\phi)\sin(t\delta E/\hbar)]$$
$$- i|m'^*\rangle \sin(2\phi)\sin(t\delta E/\hbar). \qquad (6.47)$$

This formula shows that:

(i) At resonances, $\cos(2\phi) = 0$ and $\sin(2\phi) = \pm 1$, the spin oscillates between the left and right hand sides of the double potential well of Fig. 5.2. This is the magnetic tunnel effect. The oscillation frequency is then, according to (6.25), equal to ω_T. This justifies the expression 'tunnel frequency'.

(ii) Far from any resonance, $\sin(2\phi) = 0$, the spin remains in its initial position, assumed to be the left-hand side.

(iii) Near a resonance, the spin is the sum of a component which remains in the initial well, and an oscillating component.

Formula (6.47) can also be deduced from Schrödinger's equation

$$i\hbar \frac{d}{dt} |\Psi(t)\rangle = \mathcal{H} |\Psi(t)\rangle. \qquad (6.48)$$

If

$$|\Psi(t)\rangle = x(t) |m^*\rangle + y(t) |m'^*\rangle \qquad (6.49)$$

(6.48) can be written as

$$\begin{cases} \dot{x}(t) = \frac{1}{i\hbar}(E_0 + \Delta)x(t) - i\omega_T y(t) \\ \dot{y}(t) = -i\omega_T x(t) + \frac{1}{i\hbar}(E_0 - \Delta)y(t). \end{cases} \qquad (6.50)$$

The solution of this system just reproduces (6.47), i.e. $x(t) = \exp[iE_0t/\hbar]$ $[\cos(t\delta E/\hbar) - i\cos(2\phi)\sin(t\delta E/\hbar)]$ and $y(t) = -\exp[iE_0t/\hbar]\sin(2\phi)\sin(t\delta E/\hbar)$. But the differential equations (6.50) will turn out to be useful.

6.6 Effect of a field along the hard axis

This case has already been treated in Section 6.2 in the case of quadratic, tetragonal anisotropy. The simplest Hamiltonian is

$$\mathcal{H} = -|D|S_z^2 + g\mu_B H_x S_x \tag{6.51}$$

which is a special case of (6.2) when $H_z = 0$. Tunnelling is absent when $H_x = 0$, and present when $H_x \neq 0$. Numerical calculation shows that ω_T increases uniformly with $|H_x|$. The resulting tunnel frequency ω_T can be made so high that oscillations at this frequency are directly observed. The experiment was done by Bellessa $et~al.$ (1999) in Mn$_{12}$ac; the resulting tunnel frequency ω_T can be made so high that oscillations at this frequency are directly observed.

The uniform increase of ω_T with $|H_x|$ can look intuitive. However, a surprise appears in the case of a lower symmetry, when it is necessary to add a term BS_x^2 to (6.51). The result is that ω_T oscillates as $|H_x|$ increases! Only for sufficiently high values of $|H_x|$, is the increase of ω_T uniform. This effect, observed in Fe$_8$ (Wernsdorfer and Sessoli 1999) will be studied in Section 6.8. It is a typical quantum effect, somewhat related to Bohm–Aharonov oscillations of the current in a loop, or to the current oscillations of a Josephson junction with magnetic field. This analogy appears in a particularly precise way through the path integral formalism which will be presented in Chapter 7.

6.7 Evaluation of the tunnel splitting for large spins

6.7.1 General methods

The tunnel splitting $2\hbar\omega_T$ becomes very small for large s. It can be calculated numerically for all experimentally accessible values of s, and even for s up to, say, 50. However, since the numerical evaluation has no transparency, it is of interest to have an analytic formula. Such a formula does exist for particular forms of the Hamiltonian. An example is Hamiltonian (2.9) in a field along the intermediate axis y. The anisotropy can be written (Schilling 1995) as

$$\mathcal{H}_{an} = -D'S_z^2 + BS_x^2 + g\mu_B H_y S_y. \tag{6.52}$$

In that case, Schilling (1995) demonstrated that, neglecting corrections which are small for large s,

$$2\hbar\omega_T = \frac{16D'}{\sqrt{\pi}} s^{3/2} \frac{a^{5/4}b^{3/4}}{(b-a)^{1/2}} \left(\frac{\sqrt{b-a}}{\sqrt{a}+\sqrt{b}}\right)^{2s+1} \cosh\left(\pi s \frac{H_y}{H_c}\sqrt{\frac{D'}{B}}\right)$$

$$\times \exp\left[-2s\frac{H_y}{H_c}\sqrt{\frac{D'}{B}}\arctan\left(\frac{H_y}{H_c}\sqrt{\frac{D'b}{Ba}}\right)\right] \tag{6.53}$$

where

$$g\mu_B H_c = 2D'S, \quad a = 1 - \left(\frac{H_y}{H_c}\right)^2, \quad b = 1 + \frac{B}{D'}.$$

In vanishing field, $H_y = 0, a = 1$ and

$$\hbar\omega_\mathrm{T} = K s^{3/2} \left[1 + \frac{2D'}{B} + 2\sqrt{\left(1 + \frac{D'}{B}\right)\frac{D'}{B}}\right]^{-s} \tag{6.54}$$

where

$$K = \frac{8D'}{\sqrt{\pi}} \frac{(1 + B/D')^{3/4}}{1 + \sqrt{1 + B/D'}} \tag{6.55}$$

is independent of s.

These formulae are in agreement with the result (G.12) of perturbation theory. They have been reproduced by Garg (1999) by a 'discrete WKB method'.

In the case of Fe$_8$, $\hbar\omega_\mathrm{T}/k_\mathrm{B} \simeq 10^{-10}$K or $\omega_\mathrm{T} \simeq 10$ rad s^{-1}. As will be seen in Section 8.3.2, the experimental result is 1000 times as large. This presumably indicates that the Hamiltonian (2.5) is inadequate for the description of tunnelling in Fe$_8$.

Schilling's derivation of (6.54) is based on a path integral method whose principle will be explained in the next chapter.

Another method is to calculate the wavefunction directly. This calculation can only be approximate. Since the spin s is large, and therefore quasiclassical as seen in Section 6.1, Van Hemmen and Sütö (1986,1995) proposed to apply the quasiclassical (WKB) approximation of Wenzl, Kramers and Brillouin. However, the energy levels are discrete and this typically quantum property requires an extension of the original WKB method. This point was clarified by Braun (1993) and Garg (1999).

However, the dominant factor of (6.54), an exponential function of s, can be obtained rather easily as will now be seen. The wavefunction $\varphi(m)$ should satisfy Schrödinger's equation (6.1),

$$\sum_{m'} \langle m \,|\, \mathcal{H} \,|\, m'\rangle\, \varphi(m') = E\varphi(m).$$

The spin s is assumed to be large. Then, most of the equations of this system satisfy the condition

$$s - |m| \gg 1. \tag{6.56}$$

If this condition is satisfied, the coefficients of (6.1) have a weak relative variation when m changes by one unit, $m \to m \pm 1$. This suggests looking for solutions of (6.1) which also have a weak relative variation of the ratio

$$\varphi(m+1)/\varphi(m) \simeq \varphi(m)/\varphi(m-1) = \xi(m). \tag{6.57}$$

Then, (6.1) approximately reads

$$\sum_{m'} \langle m \,|\, \mathcal{H} \,|\, m'\rangle\, \xi^{m'-m}(m) = E. \tag{6.58}$$

This is an algebraic equation whose solution can be easy or not according to the form of the Hamiltonian.

6.7.2 Example: the Hamiltonian (2.5)

In the remainder of this chapter, particular attention will be given to the aniso-tropy Hamiltonian (2.5), in which E will be replaced by $B/2$ in order to avoid confusion with the energy. Thus the anisotropy is $\mathcal{H}_{\mathrm{an}} = -|D|S_z^2 + (B/4)(S_+^2 + S_-^2)$. Now the left-hand side of (6.58) has only three terms:

$$\langle m \,|\, \mathcal{H} \,|\, m - 2 \rangle \, \xi^{-2}(m) + \langle m \,|\, (\mathcal{H} - E) \,|\, m \rangle + \langle m \,|\, \mathcal{H} \,|\, m + 2 \rangle \, \xi^2(m) = 0. \quad (6.59)$$

The problem of interest is the tunnelling process, which mainly depends on the behaviour of the wavefunction inside the barrier, when $s - |m| \gg 1$. Then the first and the last coefficient in (6.59) are nearly equal,

$$\langle m \,|\, \mathcal{H} \,|\, m - 2 \rangle \simeq \langle m \,|\, \mathcal{H} \,|\, m + 2 \rangle \simeq (B/4)(s^2 - m^2). \quad (6.60)$$

The other coefficient in (6.59) is $\langle m \,|\, (\mathcal{H} - E) \,|\, m \rangle$. Attention will be focused on the ground state, and E will be approximated by the unperturbed eigenvalue (5.6), i.e. $E = -|D|s^2$, Then, for $s - |m| \gg 1$

$$\langle m \,|\, \mathcal{H} - E \,|\, m \rangle \simeq |D|(s^2 - m^2). \quad (6.61)$$

For $s - |m| \gg 1$, according to (6.60) and (6.61), (6.59) reads

$$\xi^{-2} + (4|D|/B) + \xi^2 = 0. \quad (6.62)$$

This second-degree equation in ξ^2 does not even depend on m! It has two solutions

$$\xi^2 = -\exp[\pm 2\kappa_0] \quad (6.63)$$

where

$$\exp[\pm 2\kappa_0] = (2|D|/B) \pm 2\sqrt{(4|D|^2/B^2) - 1}. \quad (6.64)$$

If the upper sign is chosen, two increasing wavefunctions are obtained. In other words, they are (almost completely) localized in the right-hand well ($m > 0$). These functions read

$$\varphi_d^{\pm}(m) = (\pm i)^{s-m} \varphi_d(s) \exp[-\kappa_0(s - m)]. \quad (6.65)$$

If the lower sign is chosen in (6.64), two wavefunctions of decreasing modulus are obtained, which are localized in the left-hand well ($m < 0$).

$$\varphi_g^{\pm}(m) = (\pm i)^{s-m} \varphi_g(s) \exp[-\kappa_0(s + m)]. \quad (6.66)$$

The vectors $|\,\varphi_g^{\pm}\rangle$ and $|\,\varphi_d^{\pm}\rangle$ do not have the form (6.16) or (6.17) which is required for the Hamiltonian (2.5). However, it is easy to form linear combina-tions of the two solutions (6.65), which do satisfy, respectively, (6.16) and (6.17), namely

$$\varphi_4(m) = \varphi_4(s) \exp[-\kappa_0(s - m)] \cos[\pi(s - m)/2] \quad (6.67)$$

and

$$\varphi_3(m) = \varphi_3(s-1)\exp[-\kappa_0(s-m)]\sin[\pi(s-m)/2] \qquad (6.68)$$

and linear combinations of the two solutions (6.66), which satisfy (6.16) and (6.17), namely

$$\varphi_1(m) = \varphi_1(-s)\exp[-\kappa_0(s+m)]\cos[\pi(s+m)/2] \qquad (6.69)$$

and

$$\varphi_2(m) = \varphi_2(-s+1)\exp[-\kappa_0(s+m)]\sin[\pi(s+m)/2]. \qquad (6.70)$$

These formulae are approximate, but exact values can be obtained using Schrödinger's equation (6.1) for $m'' \neq \pm s, \pm(s-1)$. Applying (6.1) for $m'' = -s$ to φ_1 yields a good approximation of the corresponding eigenvalue E_0. This approximation does not take tunnelling into account. Tunnelling can be obtained if an eigenfunction of (6.1) is assumed to have the form

$$\psi(m) = \varphi_1(m) + \epsilon\varphi_2(m) \qquad (6.71)$$

where, in the situation of interest, $\epsilon = \pm 1$. Insertion into (6.1) yields, replacing m by s and m' by m,

$$\sum_m \langle s \,|\, \mathcal{H} \,|\, m \rangle \, \varphi_1(m) - E\varphi_1(s) = \epsilon[E\varphi_2(s) - \sum_m \langle s \,|\, \mathcal{H} \,|\, m \rangle \, \varphi_2(m)]. \qquad (6.72)$$

The left-hand side is not very sensitive to small variations of E. It can be written as $kE_0\varphi_1(s)$, where k is a constant, presumably of order unity. The right-hand side vanishes for $E = E_0$ and may be assumed to be proportional to $\delta E = E - E_0$. It will be written as $k'\epsilon\delta E\varphi_2(s)$, where k' is a constant, presumably of order unity. Thus (6.71) can be written as $kE_0\varphi_1(s) = k'\epsilon\delta E\varphi_2(s)$, or, since $\varphi_2(s) = \varphi_1(-s)$,

$$\epsilon\delta E = E_0\frac{k}{k'}\frac{\varphi_1(s)}{\varphi_1(-s)}. \qquad (6.73)$$

If now (6.1) is applied to the case $m = -s$, one obtains in a similar way

$$\delta E = E_0\epsilon\frac{k}{k'}\frac{\varphi_1(s)}{\varphi_1(-s)} \qquad (6.74)$$

Equations (6.73) and (6.74) yield $\epsilon = \pm 1$ as already known, and

$$\delta E = \pm E_0\frac{k}{k'}\frac{\varphi_1(s)}{\varphi_1(-s)} \qquad (6.75)$$

or, replacing $\varphi_1(s)$ by its approximation (6.69),

$$\hbar\omega_T \approx |D|\exp(-2\kappa_0 s) \qquad (6.76)$$

where the symbol '$\approx X$' means 'equal to X multiplied by a factor whose divergence with s is weaker than an exponential of s'. Formula (6.76) is in agreement with Schilling's more precise formula (6.54).

The factor $s^{3/2}$ present in (6.54), and lacking in (6.76), can be obtained from (6.75) if a better approximation than (6.69) is used for $\varphi_1(s)$. The proof is available in Anna Fort's (2001) thesis.

One would like to identify the vectors φ_1 and φ_4 with the localized vectors $|-m^*\rangle$ and $|m^*\rangle$. For the sake of definiteness, s will be assumed to be integer and even, and E will be assumed to be the ground state energy, so that the relevant localized vectors are $|-s^*\rangle$ and $|s^*\rangle$. The identification of $|-s^*\rangle$ and $|s^*\rangle$ requires some care because these vectors must be orthogonal and φ_1 and φ_4 are not. Therefore the required formulae are

$$|-s^*\rangle = C\left(|\varphi_1\rangle - \epsilon|\varphi_4\rangle\right)$$
$$|s^*\rangle = C\left(|\varphi_4\rangle - \epsilon|\varphi_1\rangle\right) \tag{6.77}$$

where ϵ is given by

$$(1 + \epsilon^2)\langle\varphi_1\mid\varphi_4\rangle = \epsilon\left(\langle\varphi_1\mid\varphi_1\rangle + \langle\varphi_4\mid\varphi_4\rangle\right) \tag{6.78}$$

and C is given by

$$C\left((1+\epsilon^2)\langle\varphi_1\mid\varphi_1\rangle - 2\epsilon\langle\varphi_1\mid\varphi_4\rangle\right) = 1. \tag{6.79}$$

The tunnel splitting is now given by equation (6.24),

$$\hbar\omega_{\mathrm{T}}^{(-s,s)} = |\langle -s^*\mid\mathcal{H}\mid s^*\rangle|$$
$$= |C|^2\left(\langle\varphi_1\mid\mathcal{H}\mid\varphi_4\rangle + \langle\varphi_1\mid\mathcal{H}\mid\varphi_4\rangle - \epsilon\langle\varphi_1\mid\mathcal{H}\mid\varphi_1\rangle - \epsilon\langle\varphi_4\mid\mathcal{H}\mid\varphi_4\rangle\right). \tag{6.80}$$

If $|\varphi_4\rangle$ and $|\varphi_1\rangle$ are substituted from (6.67) and (6.69) one obtains, by another method, formula (6.76).

6.8 Diabolic points

6.8.1 *Degeneracy with and without symmetry*

In the previous sections, it has been seen that the levels of a Hamiltonian which acts on a spin are generally not degenerate. In the present section, attention will be paid to exceptions to this general rule.

Some exceptions have already been encountered in Sections 6.1 and 6.2. They are related to some particular symmetry. For instance

- If s is half-integer and the magnetic field is 0, degeneracy is possible. This is Kramers' degeneracy.
- If the field is longitudinal, the states $|m^*\rangle$ and $|m'^*\rangle$ are degenerate for appropriate field values of the longitudinal field if $m + m'$ is odd.
- For tetragonal symmetry, the states $|m^*\rangle$ and $|m'^*\rangle$ are degenerate for appropriate values of the longitudinal field unless $m + m'$ is a multiple of 4.

A symmetry-breaking perturbation often has the effect of raising an existing degeneracy. This occurs for instance in the Jahn–Teller effect, when an elastic distortion raises an electronic degeneracy. Another example is the raising of the Kramers degeneracy by a magnetic field (which breaks the time-reversal symmetry). For that reason, one might believed that a degeneracy is always a consequence of some symmetry. This is wrong, as will now be seen.

6.8.2   The von Neumann–Wigner theorem

Indeed, it may happen that the tunnel splitting $2\hbar\omega_{\mathrm{T}}^{(mm')}$ between two localized states $|\,m^*\rangle$ and $|\,m'^*\rangle$ vanishes although there is no symmetry reason. This problem has been clarified by a theorem of von Neumann and Wigner (1929) according to which two adjustable parameters are necessary (but not always sufficient) to produce the degeneracy of a real (Hamiltonian) matrix. For a complex matrix, three parameters are necessary.

In the case of a spin $1/2$ treated in Section 6.1.3, this property is a consequence of the fact that the Hamiltonian matrix has two eigenvalues which are equal if and only if $\langle 1/2\,|\,\mathcal{H}\,|\,1/2\rangle = \langle -1/2\,|\,\mathcal{H}\,|-1/2\rangle$, $\Re\,\langle 1/2\,|\,\mathcal{H}\,|-1/2\rangle = 0$ and $\Im\,\langle 1/2\,|\,\mathcal{H}\,|-1/2\rangle = 0$. These three conditions determine the three variables H_x, H_y, H_z.

This argument may be extended to a general spin s. The question is the possible degeneracy of two eigenvectors contained in a two-dimensional space (defined for instance by two localized vectors $|\,m^*\rangle$ and $|\,m'^*\rangle$). This degeneracy implies the conditions $\langle m'^*\,|\,\mathcal{H}\,|\,m'^*\rangle = \langle m^*\,|\,\mathcal{H}\,|\,m^*\rangle$, $\Re\,\langle m'^*\,|\,\mathcal{H}\,|\,m^*\rangle = 0$ and $\Im\,\langle m'^*\,|\,\mathcal{H}\,|\,m^*\rangle = 0$. These three equalities between quantities which are real (because the Hamiltonian is Hermitian) determine three parameters, which in the case of a spin can generally be identified with the components of the magnetic field.

In practice, H_z is defined, at least approximately, by the level-crossing condition $E_m^{(0)} = E_{m'}^{(0)}$. The other two components are determined by the condition $\omega_{\mathrm{T}} = 0$.

A point of the parameter space where degeneracy occurs without symmetry reasons is called a *diabolic* point, probably because it is an unexpected phenomenon which can only be an effect of the Devil.

A simple case is the Hamiltonian (2.8) in the presence of a field which has components on the easy and hard axes:

$$\mathcal{H} = -|D|S_z^2 + (B/2)(S_x^2 - S_y^2) + g\mu_{\mathrm{B}}H_zS_z + g\mu_{\mathrm{B}}H_xS_x. \qquad (6.81)$$

With the usual representation of spin operators $\langle s, m+1, p\,|\,S^+\,|\,s, m, p\rangle = \sqrt{(s+m+1)(s-m)}$, this Hamiltonian is real. The von Neuman–Wigner theorem therefore states that it can be degenerate for certain values of H_x and H_z. Actually, it is, as proven by (Garg 1993). Later, Wernsdorfer and Sessoli (1999) experimentally observed that the tunnel frequency in Fe$_8$ was an oscillating function of H_x (Fig. 6.8). Their next task, to explain this strange result, was to extract Garg's article from a literature which was becoming thick!

How can the diabolic points of (6.81) in the (H_x, H_z) plane be calculated? Of course, these points must correspond to a level crossing, and therefore be on

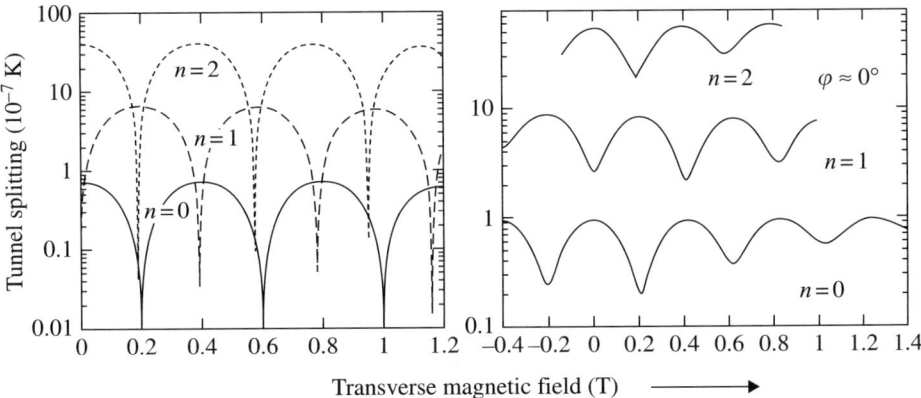

FIG. 6.8. Tunnel frequency ω_T as a function of H_x according to Wernsdorfer and
Sessoli (1999): (a) numerical calculation; (b) experiment on Fe_8. The field
component along the easy axis is either 0 (curve '$n = 0$') or given by (6.87)
with $p = 0$ (curve '$n = 1$') or given by (6.86) with $p = 1$ (curve '$n = 2$').

lines which are approximately given by (6.4). This is a condition on H_z. Then,
one has to determine the field H_x along the hard axis. Garg (1993) and then
Chiolero and Loss (1998) did that by using path integrals. This method, which
uses the concept of Berry phase often associated to the oscillations of Fig. 6.8,
will be described in the next chapter. However, partial results can be obtained
from the method of Section 6.7 as will now be seen.

6.8.3 *The quest of the Devil*

In Section 6.7, Schrödinger's equation (6.1) was applied to the Hamiltonian (6.81)
with $\mathbf{H} = 0$. It was found that, if restricted to the region $s - |m| \gg 1$, equations
(6.1) have four independent solutions $\varphi_r(m)$ given by (6.67)–(6.70). These solutions exist for any value of E. The eigenvalue E can only be determined if the
whole system (6.1) is solved for $-s \leq m \leq s$. However, in the region $s - |m| \gg 1$,
the solution of this system is a linear combination of the four functions $\varphi_r(m)$.
Their definition by (6.67)–(6.70) is only approximate, but more precise formulae
(which depend on E) can be deduced from the exact formulae (6.59). Four functions $\varphi_r(m)$ are thus obtained, which satisfy relations (6.1) exactly, except for
$m = \pm s$ and $m = \pm(s - 1)$. In zero field, assuming s to be even, these functions
are real and such that:

- $\varphi_1(m)$ is localized to the left ($m < 0$) and is 'even-valued', i.e. $\varphi_1(m) \neq 0$
  implies that m is even;
- $\varphi_2(m)$ is localized to the left ($m < 0$) and is 'odd-valued';
- $\varphi_3(m)$ is localized to the right and 'odd-valued';
- $\varphi_4(m)$ is localized to the right and 'even-valued'.

These properties hold for $\mathbf{H} = 0$. When H_z and H_x do not vanish, the method of Section 6.7 can be applied too (Villain and Fort 2000). As a first approximation, (6.58) can still be used. It is an equation of fourth degree in $\xi(m)$. If $\mathbf{H}$ is small, the following results are easily obtained:

- The system (6.1) (still truncated from its first two and last two equations) has, for $E \approx -|D|s^2$, four independent, real solutions $\varphi_r(m)$.
- $\varphi_1(m)$ and $\varphi_2(m)$ are left-localized, $\varphi_3(m)$ and $\varphi_4(m)$ are right-localized, as in case $\mathbf{H} = 0$.
- The field does not modify each function $\varphi_r(m)$ much in the region where it is large, i.e. in the bottom of the left-hand well for $\varphi_1(m)$ and $\varphi_2(m)$, in the bottom of the left-hand well for $\varphi_3(m)$ and $\varphi_4(m)$.
- However, for $H_x \neq 0$, the functions $\varphi_r(m)$ have a phase which depends on m (Fig. 6.9). For instance, (6.69) is replaced by

$$\varphi_1(m) = \varphi_1(-s)\exp[-\kappa_0(s+m)]\cos[\pi(s+m)/2 + \Phi_1(m) - \Phi_1(-s)]. \quad (6.82)$$

The phase $\Phi_1(m)$ varies slowly with m if H_x is small. Its presence has spectacular consequences if

$$\Phi_1(s) - \Phi_1(-s) = \pi/2 + k\pi \quad (6.83)$$

where k is an integer. Indeed, in that case, $\varphi_1(m)$, which on the left-hand side has even values, has odd values on the right-hand side. In particular $\varphi_1(s) \simeq \varphi_1(s-2) \simeq 0$ since s is even.

Let the case $H_z = 0$ be considered first. It corresponds to a level crossing between φ_1 and φ_4. Generally, Schrödinger's equation (6.1) can only be satisfied for $m = s$, if the wavefunction is a combination of φ_1 and φ_4 (the sum or the difference). This follows from (6.73) and (6.74). However, if (6.83) is satisfied, then $\varphi_1(s)/\varphi_1(-s) = 0$, and (6.73) and (6.74) are satisfied for $\delta E = \epsilon = 0$. This implies that both $\varphi_1(m)$ and $\varphi_4(m)$ are eigenfunctions of the Hamiltonian for the same energy E_0. In reality, a more refined derivation is necessary because (6.73) and (6.74) were derived for $H_x = 0$, when the Hamiltonian has no matrix elements between φ_1 (or φ_4) and φ_3 (or φ_2). If $H_x \neq 0$, the wavefunction is therefore not exactly φ_1 or φ_4, but contains small corrections proportional to φ_3 and φ_2. The calculation is slightly more complicated but the result is the same: there is an exact degeneracy for particular values of H_x.

These values can be deduced from (6.83) using the calculation of the function $\Phi_1(s)$ (Villain and Fort 2000) and one obtains

$$g\mu_{\mathrm{B}}H_x = (2n+1)\sqrt{B(|D| + B/2)}. \quad (6.84)$$

This relation was obtained for the first time by Garg (1993) by means of path integrals.

The above argument can be generalized to all fields which correspond to the crossing of two even values m, m' or of two odd values, so that $m + m'$ is even

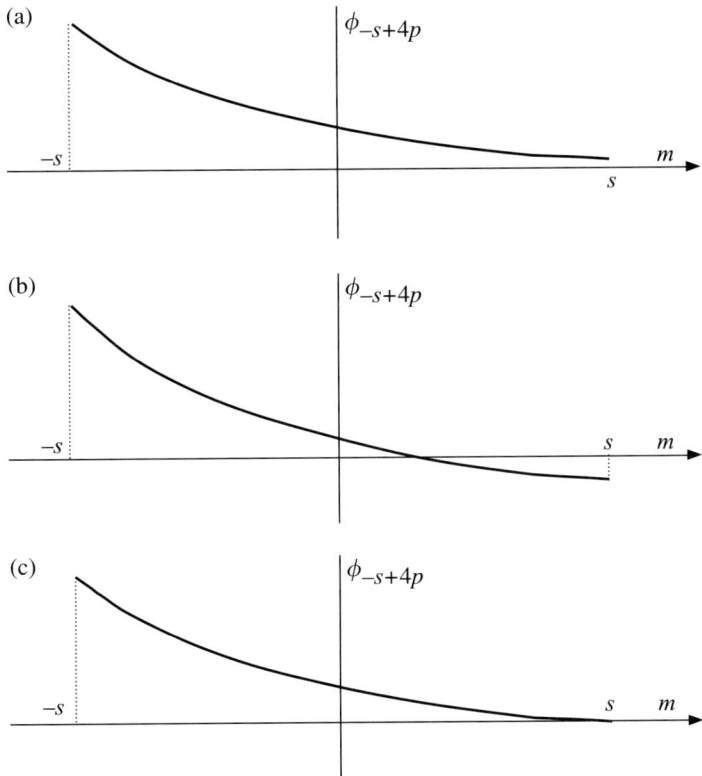

FIG. 6.9. Schematic representation of the lowest, left-localized wavefunction $\varphi_1(m)$. Only the values for $m = -s + 4p$ are represented. The field H_z is close to 0. (a) $H_x = 0$; (b) H_x larger than the first 'diabolic' value. (c) H_x equal to the first diabolic value. At resonance ($H_z = 0$) the eigenfunction is obtained by adding to $\varphi_1(m)$ its symmetric $\varphi_4(m)$. This figure is schematic and the proportions are not respected.

in (6.4). That equation and (6.84) define (in an approximate way) a family of diabolic points. It is easily seen that another family of diabolic points (for s an even integer) is defined for *odd* values of $m + m'$ by (6.4) and

$$\Phi_1(s) - \Phi_1(-s) = k\pi. \tag{6.85}$$

To summarize, diabolic points of the Hamiltonian (6.81) are approximately given by

$$\begin{cases} g\mu_{\mathrm{B}} H_z = 2p(|D| - B/2) & (a) \\ g\mu_{\mathrm{B}} H_x = (2n+1)\sqrt{B(|D| + B/2)} & (b) \end{cases} \tag{6.86}$$

and

$$\begin{cases} g\mu_{\text{B}}H_z = (2p+1)(|D| - B/2) & (a) \\ g\mu_{\text{B}}H_x = 2n\sqrt{B(|D| + B/2)} & (b) \end{cases} \qquad (6.87)$$

In these formulae, n cannot be larger than s.

The derivation of (6.86b) and (6.87b) is not exact, but numerical calculations confirm that they are very good approximations, for the anisotropy (2.5). Is the anisotropy really given by (2.5) in real materials? As seen from Fig. 6.8, the experimental results of Wernsdorfer and Sessoli (1999) agree with formulae (6.86b) and (6.87b) except for a constant factor of the order of 1.5. Moreover, the highest value compatible with Wernsdorfer's experimental facilities has not allowed to verify that all the 10 zeros predicted by the theory are indeed observable.

Is theory able to explain that? Yes indeed, according to Keçecioğlu and Garg (2002 and 2003). They introduce into the Hamiltonian (6.81) an additional term $C(S_+^4 + CS_-^4)$, with $C/k_{\text{B}} = 29 \times 10^{-6}$K. They obtain four zeros of the tunnel splitting for $H_x > 0$, with a distance of the order of the experimentally observed one. Furthermore, the value of the tunnel splitting is multiplied by a factor of the order of 1000, and becomes quite comparable with the experimental value. This last result can be compared with perturbation theory. Using (6.46), and the values $|D|/k_{\text{B}} = 0.292$ K, $B/k_{\text{B}} = 0.092$ K, $C/k_{\text{B}} = 29 \times 10^{-6}$ K, the tunnel splitting which results from (6.44) (if B is neglected) is actually much larger than the one calculated from (6.43) if C is ignored, as can be seen from (6.46).

Thus, Keçecioğlu and Garg drastically reduce the number of diabolic points by introducing a fourth-order anisotropy, and in that way obtain good agreement with experiment. A nasty question would be: what happens with sixth-order anisotropy? This would attract the nasty reply: try to do the calculation! That of Keçecioğlu and Garg introduces qualitatively new features, such as discontinuities in the quasiclassical trajectory.

For $H_z = 0$, the Hamiltonian (6.81) reduces to $\mathcal{H} = -D'S_z^2 + BS_x^2 + g\mu_{\text{B}}H_xS_x$. This Hamiltonian is easy to study in the limit $0 < D' \ll B$ (Weigert 1994). This case has perhaps no important experimental applications, but it is interesting because the number of diabolic points for given values of B and D can be determined by a simple, analytical argument.

The interested reader will discover many fascinating features of tunnelling in condensed matter in the books of Kagan and Leggett (1992) and Razavy (2003).

7

INTRODUCTION TO PATH INTEGRALS

7.1 General ideas

This chapter describes an alternative approach to the calculation of the tunnelling frequency. It makes use of mathematical methods which are very much used by theoreticians. They are technically very efficient, but make use of various mathematical tricks, e.g. an imaginary time, so that it is hard to retain contact with physical reality. On the other hand, they help to make a link with other quantum phenomena.

Path integrals were introduced by Feynman (1948) as providing an alternative presentation of quantum mechanics. Feynman and Hibbs (1965) published a complete book on this approach, where no discrete variables are *a priori* introduced. It is particularly appropriate to the study of nearly classical systems such as large spins.

Following the article of Schilling (1995) the method will first be explained in the case of a one-dimensional anharmonic oscillator. From there to a spin, there are well-known transformations, the best known of them (but not the most appropriate for us) being that of Holstein and Primakoff. The forthcoming presentation is neither rigorous not complete. It only aims at an explanation of the main concepts without giving mathematical details.

7.2 The anharmonic oscillator

Let a particle of mass[1] m in one-dimensional space be subject to a potential $V(x)$. We wish to know the probability amplitude

$$G(x',t'|x,t) = \langle x' \,|\, \exp[i(t'-t)\mathcal{H}/\hbar] \,|\, x\rangle \tag{7.1}$$

that the particle is at point x' at time t' if it was at point x at time t. The corresponding probability is of course the square of the absolute value of (7.1).

Feynman and Hibbs (1965) have shown that $G(x',t'|x,t)$ (also called the 'propagator') can be expressed as the 'path integral'

$$G(x',t'|x,t) = \int \mathcal{D}[x''(t'')] \exp\{iS([x''(t'')])/\hbar\}. \tag{7.2}$$

In this formula, $[x''(t'')]$ represents a path from x to x', i.e. a function of t'' subject to conditions

$$x''(t) = x \qquad x''(t') = x', \tag{7.3}$$

---

[1] Pay attention to the notation m which, in this chapter, is a mass, not the S_z component of a spin!

while

$$S([x''(t'')]) = \int_t^{t'} du \mathcal{L}(\dot{x}(u), x(u)) \tag{7.4}$$

is the Maupertuis action which involves the classical Lagrangian

$$\mathcal{L}(\dot{x}(u), x(u)) = \frac{m}{2} \dot{x}^2 - V(x). \tag{7.5}$$

Finally, $\int \mathcal{D}[x''(t'')]$ is an integral on all paths from x to x' which satisfy (7.3). The definition of this integral is similar to that of the usual Riemann integral: the interval $[t, t']$ is discretized into N equal intervals t, $t_1, \ldots, t_{N-1}$, t' and $\int \mathcal{D}[x''(t'')]$ is defined as the limit for infinite N of the N-fold integral.

$$\prod_{n=1}^{N-1} \int_{-\infty}^{\infty} dx_n$$

where $x_n = x(t_n)$. More precisely

$$\int \mathcal{D}[x''(t'')] = \mathrm{Lim}_{N \to \infty} \int_{-\infty}^{\infty} \frac{dx_n}{A} \tag{7.6}$$

where

$$A = \sqrt{\frac{2\pi\hbar(t'-t)}{mN}}. \tag{7.7}$$

According to Maupertuis' principle, the classical motion of the particle is determined by the minimization of the action (7.4). This formulation implies

$$\frac{m d^2 x}{dt^2} = \frac{-dV(x)}{dx} \tag{7.8}$$

but it is slightly more precise since (7.4) might also correspond to a maximum of the action.

Let Maupertuis' principle be applied to formula (7.2) in the classical limit $\hbar \to 0$. When the action in (7.2) is much greater than $\hbar$, the phase is rapidly variable and produces destructive interference except for the stationary classical trajectory.

If the action were not multiplied by i, the integral would be dominated by the path which minimizes it, which is the classical trajectory. One might then improve the classical approximation by replacing the action by a Taylor expansion near its minimum. This is the saddle-point method. As a matter of fact, there is a simple possibility to get rid of the i factor, namely one can take an imaginary time, $t = i\tau$.

7.3   Tunnel effect and instantons

We shall now address the case where the potential $V(x)$ has two equal minima, and thus allows tunnelling. An example is

$$V(x) = -Bx^2 + Ax^4 \tag{7.9}$$

where A and B are real and positive. In classical mechanics, there are two equilibrium positions at $x = \pm x_0$, where $x_0 = \sqrt{B/(2A)}$.

To calculate the tunnelling frequency ω_T and the ground state energy E_0, a possible way is the calculation of the propagator (7.1). Indeed, using (6.47), it is easy to see that

$$G(-x_0, t|x_0, 0) = \exp(iE_0 t/\hbar) \sin(\omega_T t). \tag{7.10}$$

or for an imaginary time $t = i\tau$

$$G(-x_0, i\tau|x_0, 0) = i \exp[-E_0\tau/\hbar] \sinh(\omega_T \tau). \tag{7.11}$$

To calculate this propagator near the classical limit, it is appropriate to solve the classical equation of motion (7.8). A first integral, familiar to those who remember the pendulum equation, can be obtained by multiplying both sides by dx/dt. Integration then yields

$$m(dx/dt)^2 = -2V(x) + C \tag{7.12}$$

where C is an integration constant. In classical mechanics, there is equilibrium when $dx/dt = 0$, which implies $C = 2V(x_0)$. If this value is chosen, (7.12) and (7.9) yield

$$\begin{aligned} (m/2)(dx/dt)^2 \\ = Bx^2 - Ax^4 - Bx_0^2 + Ax_0^4 \\ = (x^2 - x_0^2)\left[B - A(x^2 + x_0^2)\right] \\ = -A\left(x^2 - x_0^2\right)^2. \end{aligned} \tag{7.13}$$

The goal is to describe tunnelling, i.e. a path from the left-hand minimum, $x = -x_0$, to the right-hand minimum, $x = x_0$. With a real time, this is impossible since (7.13) has no solution. This is no surprise. The surprise is that (7.13) has a solution if $t = i\tau$ is imaginary. Introducing an integration constant τ_1, one obtains

$$\tau = \sqrt{m/(2A)} \int_0^x \frac{du}{x_0^2 - u^2} + \tau_1 = (1/x_0)\sqrt{m/(2A)} \ln \frac{x_0 + x}{x_0 - x} + \tau_1 \tag{7.14}$$

which corresponds to $x = -x_0$ for $\tau = -\infty$ and to $x = x_0$ for $\tau = \infty$.

The solution (7.14) of the equation of motion (7.8) is called an *instanton* because the particle needs a short instant $(1/b)\sqrt{m/(2A)}$ to go from the neighbourhood of point $-b$ to the neighbourhood of point b. The instanton solution is mathematically identical to that which is called a *soliton* in other physical problems.

The next task is the evaluation of the propagator (7.2) for $x = -x_0$, $x' = x_0$, and $t' - t$ large with respect to the duration $\sqrt{m/(2A)}$ of the instanton. One may be tempted to choose $t' - t = \infty$, but this is not possible because the integral (7.2) would diverge. The reason for this divergence is just that there are an infinity of possible choices of the instanton position τ_1. For a finite value of $t' - t$, the number of choices is proportional to $t' - t$, and therefore the propagator should contain a factor $t' - t$. As a matter of fact, this is only true if $t' - t$ is not too large, otherwise more than one soliton may be present.

The propagator also contains a factor $\exp(-S_E/\hbar)$, where the 'Euclidean' action S_E of the instanton[2] can be deduced from (7.4) by the substitutions $t = -i\infty$, $t' = i\infty$, $t'' = i\tau$. Thus

$$S_E = \int_{-\infty}^{\infty} d\tau \mathcal{L}_E \left(\dot{x}(\tau), x(\tau) \right) \tag{7.15}$$

where the 'Euclidean' Lagrangian $\mathcal{L}_E$ is obtained by multiplying the first term of the usual Lagrangian (7.5) by $i^2 = -1$, i.e.

$$\mathcal{L}_E(\dot{x}, x) = -\frac{m}{2}(dx/d\tau)^2 - V(x) \tag{7.16}$$

and $dx/d\tau$ is given by (7.13). The result is

$$S_E = \int_{-\infty}^{\infty} d\tau[(m/2)(dx/d\tau)^2 + V(x) - V(b)] \tag{7.17}$$

where an additive constant has been inserted between the brackets. The result is a change of the energy origin and a modification of the phase of the wavefunction, and has no physical consequences. Using (7.12) with $C = 2V(b)$, (7.17) becomes

$$S_E = m \int_{-\infty}^{\infty} d\tau(dx/d\tau)^2 = m \int_{-b}^{b} dx(dx/d\tau) = m \int_{-x_0}^{x_0} dx \sqrt{2m[V(x) - V(b)]}. \tag{7.18}$$

The imaginary-time ('Euclidean') propagator $G_E(-x_0, \tau'|x_0, \tau)$ is thus dominated, for small $(\tau' - \tau)$, by a factor $(\tau' - \tau) \exp(-S_E/\hbar)$. This factor is expected to be the first term of the short time expansion of the hyperbolic sine in (7.11). Actually, the full formula has the form (Schilling 1995)

$$G_E(-b, \tau'|b, \tau) = C \exp[-(1/2)(\tau' - \tau)\sqrt{V''(x_0)/m}] \sinh[(\tau' - \tau)\Delta \exp(-S_E/\hbar) \tag{7.19}$$

where C and Δ have complicated expressions which will not be reproduced here. Comparison with (7.11) shows that

$$\omega_T = \Delta \exp(-S_E/\hbar) \approx \exp(-S_E/\hbar) \tag{7.20}$$

where the sign $\approx$ means that factors are ignored, which, in the classical limit $m \to \infty$, go less rapidly to 0 than $\exp(-S_E/\hbar)$, or go less rapidly to ∞ than $\exp(S_E/\hbar)$.

---

[2] The word 'Euclidean' is used when the time $t = i\tau$ is imaginary, because the relativistic $ds^2 = dx^2 + dy^2 + dz^2 - dt^2$ thus becomes Euclidean, $ds^2 = dx^2 + dy^2 + dz^2 + d\tau^2$.

Formula (7.19) results from adding the contributions of all solutions with 1, 2, 3, ... instantons. It is clearly not the simplest way to obtain (7.20) unless one wants to have the precise expression of Δ. Then the path integral method is not necessarily more complicated than alternative methods.

When dealing with a spin instead of a particle, the complication increases again, as will now be seen.

7.4   The path integral method applied to spins

Until now, in this chapter, the problem treated was that of a particle, more precisely an oscillator. Harmonic or not, an oscillator is described in quantum mechanics by creation and destruction operators which obey the Bose commutation rules. Their eigenvalues go from $-\infty$ to ∞. However, in this book, we are interested in spin operators, which have a finite number of eigenvalues.

We shall follow the articles of Garg (1993) and Schilling (1995).

We shall consider the Hamiltonian (6.81) for $H_z = 0$. One wishes to write the 'Euclidean' action (i.e. for an imaginary time, as above) analogous to (7.15) for a classical spin whose polar angles θ and ϕ are defined by the spin components $S_x = s\cos\theta$ along the hard axis, $S_z = s\sin\theta\cos\phi$ along the easy axis, and $S_y = s\sin\theta\sin\phi$ along the intermediate axis. The motion between times $-t_1/2$ and $t_1/2$ in classical mechanics corresponds to the minimum of the action. An expression which has this property can be shown to be

$$S_{\mathrm{E}} = \int_{-t_1/2}^{t_1/2} d\tau [-i\hbar s(1-\cos\theta)\dot{\phi}(\tau) + \mathcal{H}(\theta(\tau), \phi(\tau))]. \tag{7.21}$$

This expression turns out to be correct, and may be justified. However, the justification which has been given here is not sufficient, because one might add any constant to S_{E} without modifying the property that its minimization gives the classical equation of motion. For instance, as noticed by Schilling (1995), the following (wrong) expression might be used (with a real time).

$$S = \int_0^t dt[\hbar s\dot{\phi}(t)\cos\theta(t) - \mathcal{H}(\theta(t), \phi(t))]. \tag{7.22}$$

The essential difference between (7.21) and (7.22) is the expression $s\hbar \int dt\dot{\phi}(t) = s\hbar[\phi_f - \phi_i]$, where ϕ_i and ϕ_f are the initial and final values of ϕ, i.e. the values at equilibrium. This expression is independent of the path which joins the equilibrium points. It is therefore a constant. It turns out that (7.21) is correct while (7.22) would lead to wrong results, e.g. it would violate Kramers' theorem.

In the case of particle tunnelling, there was only (with an imaginary time) one classical path from one equilibrium point to the other. This is the instanton (7.14). In the case of a spin, we have the surprise to find two paths. In vanishing field, they both correspond to $\theta = \pi/2$, but to two different sign of $\sin\phi$ (Fig. 7.1). This is related to the topology of the sphere, because the end of the **S** vector moves on a sphere.

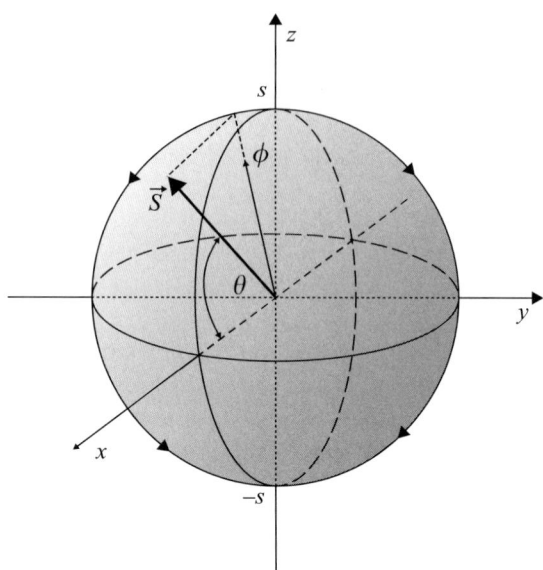

FIG. 7.1. The vector **S** which represents a classical spin lies on a sphere. It makes
an angle θ with the hard axis x, and its projection on the yz plane makes
an angle ϕ with the easy axis z. The equilibrium points in zero field are the
top $(S_z = s)$ and the bottom $(S_z = -s)$ of the sphere. Those points are
linked by two imaginary time trajectories which are the two halves (marked
by arrows) of the section $S_x = s \cos \theta = 0$ by the plane defined by the easy
axis z and the intermediate axis y. Kramers' degeneracy can be interpreted as
an interference between both trajectories. The diabolic points which appear
in a non-vanishing field can be obtained by an analogous argument with more
complicated trajectories.

It will be seen that the two paths can interfere. Destructive interference leads
to a vanishing tunnel splitting. Thus, the diabolic points of the previous chapter
are obtained in a different way, which establishes an unexpected analogy with
other quantum phenomena such as the Bohm–Aharonov effect, or a SQUID. In
the case of a half-integer spin in vanishing magnetic field, the interference gives
rise to Kramers degeneracy.

The interference effect appears via formula (7.2). It is dominated by the two
classical paths defined above in vanishing field. In the argument of the hyperbolic
sine in (7.19), the action is therefore replaced by the sum of the actions $S_{\mathrm{E}}^{(1)}$ and
$S_{\mathrm{E}}^{(2)}$ which correspond to the two paths. They are given by (7.21), where the
second term has the same value for both paths. In the first term, $\cos \theta$ vanishes,
and what remains is $-i\hbar s \int d\phi$, with an integration from 0 to π for one path
and from 0 to $-\pi$ for the other path. Therefore, the actions $S_{\mathrm{E}}^{(1)}$ and $S_{\mathrm{E}}^{(2)}$ differ
by a factor $\exp(2\pi i s)$. Since this factor is -1 when s is even, Kramers' theorem

is recovered. The same argument using (7.22) instead of (7.21), would yield a result in contradiction with Kramers' theorem, therefore (7.22) is wrong.

In the case of an integer spin, similar interferences appear in a field parallel to the hard axis. The calculation is a little more complicated than in vanishing field, but since alternative methods are also quite complicated, path integrals are really useful.

It is of interest to look more carefully at the term $-is\hbar[\phi_f - \phi_i]$ which appears in (7.21). This term is independent of the Hamiltonian. Its nature is 'geometrical'. It is of no interest in the purely classical case $s = \infty$, when there is no tunnelling. If one looks at formula (7.2), in real time this term can be seen to be related to the phase of the wavefunction. Instead of considering an interference between two paths, one can consider the closed path constituted by the two paths. The variation of the phase of the wavefunction along such a closed path is a famous problem treated by Berry (1984). For that reason, the first term of (7.21) is often called the 'Berry phase'. However, the paths considered by Berry were in the parameter (H_x, H_z, H_y, B), with a very slow ('reversible') variation of these parameters with time.

In this very simplified and incomplete presentation, the difficulties related to the commutation rules between spin operators have been ignored. The detailed calculation can only be performed if this problem is solved, and this is not very easy. There are two possibilities (Schilling 1995). The first one is to use 'coherent spin states', which are quantum states which mimic the classical states. An alternative method is the transformation of the spin operators into boson operators b, b^+. Among these 'bosonization' procedures, the best known is the Holstein–Primakoff representation $S^z = s - b^+ b$. However, it can only be used when the spin has a single favoured direction, not if the spin tunnels between two positions. A transformation which is appropriate to this case has been described and applied by Schilling (1995).

The extension of the path integral method to Hamiltonians more complicated than (6.81) requires additional complications. In the presence of quartic terms, Keçecioğlu and Garg (2003) have argued that the classical path should be replaced by a discontinuous path.

TUNNELLING IN A TIME-DEPENDENT MAGNETIC FIELD AT LOW TEMPERATURE

8.1 Advantages of a time-dependent magnetic field

In Chapter 5 the evolution of a molecular magnet in a constant magnetic field was studied. One can get more information, especially at low temperature, by measuring its magnetization $M(t)$ in a time-dependent magnetic field. In this chapter, the field will be assumed parallel to the easy magnetization axis z and linear in time,

$$H_z(t) = H_1 - \alpha t. \tag{8.1}$$

At low temperatures, the magnetization curve exhibits a succession of steep parts separated by regions where dM/dt is much smaller. This is displayed by Fig. 8.1 in the case of Fe$_8$ and was already seen from Fig. 4.26 in the case of Mn$_{12}$ac. It turns out that the steep parts correspond to the level crossings found in Chapter 6, when tunnelling is possible. The magnetization variation dM/dt is accelerated in those regions because the relaxation is faster. Thus, the physical content of Fig. 8.1 is the same as that of Fig. 6.7. Both figures show that at low temperature the evolution toward equilibrium is easier if it does not require excitation to the top of a barrier, but just tunnelling through the barrier. One can also mention that Fig. 8.1 exhibits hysteresis. The determination of the magnetization curve allows the detection of resonances by a single experiment. It has other advantages, as will be seen.

At low temperatures, as seen in Chapter 6, there are sharp maxima of the relaxation rate (or 'resonances') at particular values $H^{(mm')}$ of H_z. It is not surprising to find (Fig. 8.1) that the variation dM/dt is steeper near resonances. Thus, the method allows the detection of resonances by a single experiment. It has other advantages, as will be seen.

In the remainder of this chapter, the effect of a time-dependent field on a single spin at low temperature is considered. The spin is described by a wavefunction $|\Phi(t)\rangle$ which satisfies the time-dependent Schrödinger equation

$$i\hbar \frac{d}{dt} |\Phi(t)\rangle = \mathcal{H}(t) |\Phi(t)\rangle \tag{8.2}$$

where

$$\mathcal{H}(t) = \mathcal{H}_0 + g\mu_B H_z(t) S_z. \tag{8.3}$$

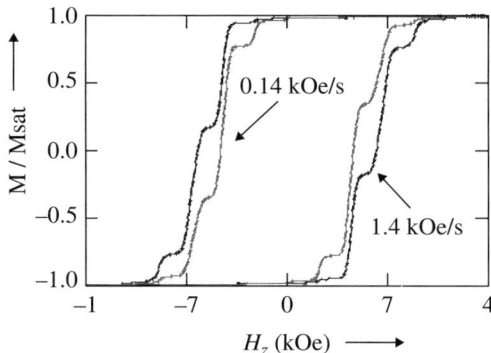

FIG. 8.1. Magnetization as a function of the field H for Fe_8 at two different values of dH/dt and $T = 80$ mK. Courteously provided by W. Wernsdorfer

The differential equation (8.2) should be complemented by an initial condition to specify the value of $|\Phi(t)\rangle$ at the initial time, which can be chosen as $t=0$. At $t=0$ the system will be assumed to be at equilibrium, and since the temperature is low, $|\Phi(0)\rangle$ is, with a very good approximation, the lowest state of one of the wells. If $H_1 > 0$, then $|\Phi(0)\rangle = |-s^*\rangle$. However, for the sake of generality, $|\Phi(0)\rangle$ will be assumed to be any eigenvector $|m^*\rangle$ of $\mathcal{H}(0)$, with $m < 0$. It will be assumed that tunnelling is negligible at $t=0$. Then the field evolves according to (8.1) and crosses a region where tunnelling to a state $|m'^*\rangle$ is possible, with $m' > 0$. The experiment is stopped at a time t_f where tunnelling is again impossible. No other tunnelling process to a state $|m''^*\rangle$ is possible between 0 and t_f. The value m' depends on H_1. The field region in which tunnelling is important has a width $2\delta H^{(mm')} \approx 2\hbar\omega_T^{(mm')}/(g\mu_B|m' - m|)$ as follows from (6.28).

Equation (8.2) is a satisfactory description of tunnelling near the zero-field resonance, i.e. for $m' = -m = s$. For other resonances, when at least one of the two tunnelling states is an excited one, one has to take into account the possibility of deexcitation with emission of phonons. That case will be investigated in Section 10.6.

In the present chapter, on the contrary, the spin is assumed to be alone, only subject to its anisotropy Hamiltonian and to the time-dependent magnetic field. In Chapter 5, it was stressed that this is a very poor approximation. In the present chapter it will be a good approximation for $|m| = |m'| = s$, at least if the sweeping velocity α is large enough. At low sweeping velocity, the spins have time to interact in a dynamic way between themselves and with nuclear spins, as will be seen in Chapter 9. But if α is large enough, this interaction may be viewed as a static field. Thus, each spin feels a local field H_{loc} which is the sum of the external field and the field produced by the other spins (electronic and nuclear). Therefore, the quantity H_1 in (8.1) is different for different spins.

However, if α is large enough, it can be assumed to be independent of time during the experiment, i.e. for $0 < t < t_f$. In that case, the wavefunction at t_f is the same for all spins apart from phase factors which have no physical significance. This can be seen from the equations of the next sections, or directly from Schrödinger's equation (8.2). The only difference between different spins is that they begin to tunnel at different times, stop tunnelling at different times, but the duration of tunnelling is the same and the net evolution is the same.

If the local field changes during the experiment, it may be expected that its variation has a weaker effect than that of the external field, provided the sweeping velocity α is large enough.

In other words, the measurement of the variation δM of the magnetization across a resonance gives access to the properties of an isolated spin subject to its crystal field. In the absence of crystal defects, as will be seen in Section 8.3, it gives access to the tunnel splitting $\hbar\omega_T$ which corresponds to each resonance.

The method described above has been applied by Wernsdorfer $et\ al.$ (2000a) with α between 0.001 and 1 T/s. It has also been used by Del Barco $et\ al.$ (2002).

8.2   Fast sweeping and adiabatic limit

From textbooks we learn that, if a quantum mechanical system with discrete energy levels subject to a time-dependent Hamiltonian is at a particular time in its ground state, it remains in its (time-dependent!) ground state provided the Hamiltonian varies very slowly. The precise condition is that the energy $\Delta E(t)$ of the first excited level (counted from the ground state) satisfies the relation

$$\hbar \frac{d}{dt} \frac{1}{\Delta E(t)} \ll 1. \tag{8.4}$$

When this condition is satisfied, the evolution is said to be $adiabatic$. This adiabaticity introduced by Ehrenfest is different from, though related to, adiabaticity in thermodynamics. The latter meaning was relevant in chapter 3, where the susceptibility and the specific heat were studied. This concept can be extended to excited states, but it is of greatest experimental interest for the ground state.

In the case of the experiments described in this chapter, condition (8.4) is satisfied, except near a level crossing, especially when the field passes through the value $H = 0$. Then the ground state crosses over from a left-localized state $|\,g(t)\rangle$ to a right-localized state $|\,d(t)\rangle$ (Fig. 8.2). Since the evolution is not adiabatic, the actual wavefunction, which satisfies (8.2), is a linear combination

$$|\,\Phi(t)\rangle = x(t)\,|\,g(t)\rangle + y(t)\,|\,d(t)\rangle. \tag{8.5}$$

If $y(0) = 0$ at the initial time $t = 0$, the reversal probability at time t_f is

$$\delta P = |y(t_f)|^2 \tag{8.6}$$

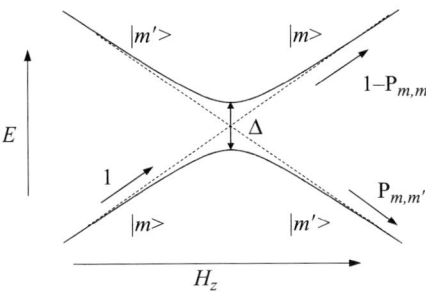

FIG. 8.2. Effect of a time-dependent magnetic field when two levels cross. The system goes from state $|\, m^* \rangle$ to state $|\, m'^* \rangle$ (on the figure the stars are omitted) with a probability $P_{mm'}$ which is equal to 1 in the adiabatic limit (slow velocity).

and does not appreciably depend on t_f if the field at t_f is far from any resonance. The state vectors $|\, g(t) \rangle$ and $|\, d(t) \rangle$ will be assumed to be orthogonal and normalized.

It should be stressed that the evolution of the spin in the experiments of interest (Wernsdorfer *et al.* 2000b) is always adiabatic with respect to the states which do not cross. Apart from $|\, g(t) \rangle$ and $|\, d(t) \rangle$, other states play no part and can be ignored. For instance, if the system is at equilibrium at the beginning of the experiment, and if the zero-field resonance is swept, then the system remains in the space of the lowest states $|\, -s^* \rangle$ and $|\, s^* \rangle$ of both wells. In that space, the evolution can be adiabatic or non-adiabatic, depending on the sweeping velocity α defined by formula (8.1). There is an evolution from the adiabatic limit ($\alpha \to 0$) to the opposite limit which will be called 'rapid sweeping' ($\alpha \to \infty$).

8.3   Calculation of the reversal probability

8.3.1   *Equations of motion*

Inserting the wavefunctions (8.5) into the Schrödinger equation (8.2), one obtains

$$x(t)\,|\, \dot{g}(t) \rangle + \dot{x}(t)\,|\, g(t) \rangle + y(t)\,|\, \dot{d}(t) \rangle + \dot{y}(t)\,|\, d(t) \rangle$$
$$= \frac{1}{i\hbar} x(t) \mathcal{H} \,|\, g(t) \rangle + \frac{1}{i\hbar} y(t) \mathcal{H} \,|\, d(t) \rangle. \tag{8.7}$$

Multiplying by $\langle g(t)\,|$, then by $\langle d(t)\,|$, one obtains the system of two equations

$$\begin{cases} x(t)\,\langle g(t)\,|\, \dot{g}(t) \rangle + \dot{x}(t) + y(t)\langle g(t)\,|\, \dot{d}(t) \rangle \\ \quad = \frac{1}{i\hbar} x(t)\,\langle g(t)\,|\, \mathcal{H}(t)\,|\, g(t) \rangle + \frac{1}{i\hbar} y(t)\,\langle g(t)\,|\, \mathcal{H}(t)\,|\, d(t) \rangle \\[2mm] x(t)\,\langle d(t)\,|\, \dot{g}(t) \rangle + y(t)\langle d(t)\,|\, \dot{d}(t) \rangle \langle + \dot{y}(t) \\ \quad = \frac{1}{i\hbar} x(t)\,\langle d(t)\,|\, \mathcal{H}(t)\,|\, g(t) \rangle + \frac{1}{i\hbar} y(t)\,\langle d(t)\,|\, \mathcal{H}(t)\,|\, d(t) \rangle. \end{cases} \tag{8.8}$$

Far from a resonance, $|g(t)\rangle$ and $|d(t)\rangle$ separately satisfy the Schrödinger equation, and $\dot{x}(t) = \dot{y}(t) = 0$.

On the contrary, near a level crossing between $|m^*\rangle$ and $|m'^*\rangle$, the components $x(t)$ and $y(t)$ vary abruptly and $\dot{x}(t)$ and $\dot{y}(t)$ are so large that $|\dot{g}(t)\rangle$ and $\left|\dot{d}(t)\right\rangle$ can be neglected. The localized state vectors $|g(t)\rangle = |m^*\rangle$ and $|d(t)\rangle = |m'^*\rangle$ are almost independent of time t.

The system (8.8) now reads

$$
\begin{cases}
\dot{x}(t) = \frac{1}{i\hbar}x(t)\langle m^* | \mathcal{H}(t) | m^*\rangle - i\tilde{\omega}_{\mathrm{T}}^{(mm')}y(t) \\[2mm]
\dot{y}(t) = -i\tilde{\omega}_{\mathrm{T}}^{(mm')}x(t) + \frac{1}{i\hbar}y(t)\langle m'^* | \mathcal{H}(t) | m'^*\rangle
\end{cases}
\tag{8.9}
$$

where $\tilde{\omega}_{\mathrm{T}}^{(mm')}$ is given by

$$
\hbar\tilde{\omega}_{\mathrm{T}}^{(mm')} = \langle m^* | \mathcal{H} | m'^*\rangle = \langle m'^* | \mathcal{H} | m^*\rangle^*.
$$

This quantity is practically independent of time although $\mathcal{H} = \mathcal{H}(t)$ contains a time-dependent field parallel to z. This can be seen for instance by perturbation theory. The tunnel frequency $\omega_{\mathrm{T}}^{(mm')}$ defined by (6.24) is the modulus of $\tilde{\omega}_{\mathrm{T}}^{(mm')}$, which will be assumed real to make the equations slightly simpler. Extension to complex values would create no difficulty.

The system (8.9) can be simplified by introducing new variables $X(t)$ and $Y(t)$ defined by

$$
\begin{cases}
x(t) = \exp[-iu(t)]X(t) \\[2mm]
y(t) = \exp[-iw(t)]Y(t)
\end{cases}
\tag{8.10}
$$

where $u(t)$ and $w(t)$ are defined by

$$
\begin{cases}
u(t) = \frac{1}{\hbar}\int_{t_0}^{t} dt' \langle m^* | \mathcal{H}(t') | m^*\rangle \\[2mm]
w(t) = \frac{1}{\hbar}\int_{t_0}^{t} dt' \langle m'^* | \mathcal{H}(t') | m'^*\rangle
\end{cases}
\tag{8.11}
$$

where the time t_0 may be arbitrarily chosen. The system (8.9) now takes the simple form

$$
\begin{cases}
\dot{X}(t) = -i\tilde{\omega}_{\mathrm{T}}^{(mm')}Y(t)e^{iU(t)} & \text{(a)} \\[2mm]
\dot{Y}(t) = -i\tilde{\omega}_{\mathrm{T}}^{(mm')}X(t)e^{-iU(t)} & \text{(b)}
\end{cases}
\tag{8.12}
$$

where

$$
U(t) = u(t) - w(t) = \frac{1}{\hbar}\int_{t_0}^{t} dt' \left[\langle m^* | \mathcal{H}(t') | m^*\rangle - \langle m'^* | \mathcal{H}(t') | m'^*\rangle\right]. \tag{8.13}
$$

This expression simplifies because (8.3) can be written as

$$
\mathcal{H}(t) = \mathcal{H}_0' + g\mu_{\mathrm{B}}\delta H_z(t)S_z
$$

where $\mathcal{H}'_0$ is independent of time and $\delta H_z(t) = H_z(t) - H^{(mm')}$ is the distance from the field value at which levels cross. Expression (8.13) reads

$$U(t) = \frac{g\mu_{\mathrm{B}}}{\hbar}(\tilde{m} - \tilde{m}')\int_{t_0}^{t} dt' \delta H_z(t') \qquad (8.14)$$

where $\tilde{m} = \langle m^* \,|\, S_z \,|\, m^* \rangle$ will generally be approximated by m.

The time t_0 can be chosen such that $\delta H_z(t_0) = 0$. Then (8.1) implies $\delta H_z(t) = \alpha(t - t_0)$, and integration of (8.14) yields

$$U(t) = \frac{v_{mm'}}{2\hbar}(t - t_0)^2 \qquad (8.15)$$

where

$$v_{mm'} = g\mu_{\mathrm{B}}(m' - m)\alpha = g\mu_{\mathrm{B}}(m - m')dH_z/dt. \qquad (8.16)$$

8.3.2 The solution of Landau, Zener, and Stückelberg

The solution of (8.12) when $U(t)$ is given by (8.15) has been given independently by Landau (1932), Zener (1932), and Stückelberg (1932). The problem is to solve the time-dependent Schrödinger equation in a two-dimensional Hilbert space, and this looks simple. Actually it is a pretty hard algebraic exercise (Grifoni and Hänggi 1998). The clearest derivation is that of Kayanuma (1984), which is reproduced in Appendix H. Only the result will be given here. The spin-flip probability when crossing a resonance is

$$\delta P = 1 - \exp\left(-\frac{\pi \Delta_{mm'}^2}{2\hbar v_{mm'}}\right) \qquad (8.17)$$

where the tunnel splitting $\Delta_{mm'} = 2\hbar|\omega_{\mathrm{T}}^{(mm')}|$ takes very different values at the various resonances, as seen from Section 6.4. For a given value of $\alpha = dH_z/dt$, $v_{mm'}$ has a relatively weaker dependence on m and m', given by (8.16).

Formula (8.17) interpolates between both extreme limits which are the adiabatic case ($\delta P = 1$ if v is small) and the fast sweeping case ($\delta P = 0$ if $v = \infty$).

The experiments of Wernsdorfer et al. (2000a) on Fe$_8$ are in agreement with (8.17) for a sweeping rate $dH/dt > 0.001$ T/s (Fig. 8.3) and yield $2\hbar\omega_{\mathrm{T}}/k_{\mathrm{B}} \simeq 10^{-7}$K for the ground doublet in zero field ($m = -m' = 10$).

For a slower sweeping rate, experimental results are not in agreement with (8.17). At least two explanations are possible. The first one is that a too slow sweeping velocity enables the spin to interact with other (nuclear and electronic) spins. Thus, the Landau–Zener–Stückelberg theory, valid for a single spin, cannot be applied. Another possible explanation is that the crystal contains defects. In that case, there is a distribution of tunnel frequencies (Chudnovsky and Garanin 2001, Mertes et al. 2001), and the experimental result is an average over the tunnel frequencies. For a high velocity v, (8.17) reduces (omitting the

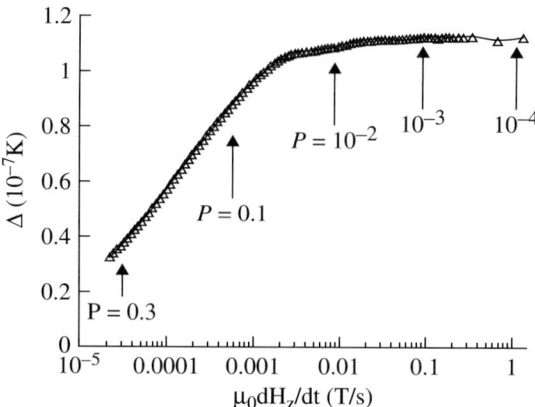

FIG. 8.3. The quantity $\Delta = \Delta_{s,-s}$ in Fe$_8$, as deduced from (8.17) and the experimental data (Wernsdorfer *et al.* 2000a). Identification with the tunnel splitting is only possible if the result is independent of the sweeping velocity v. It is only so at sufficiently high velocity. The arrows show particular values of the spin-flip probability δP (8.17), here denoted P. Copyright (2000) American Institut of Physics.

indices m, m') to

$$\delta P = \frac{2\pi\hbar}{v}\omega_{\mathrm{T}}^2.\tag{8.18}$$

The average over the tunnel frequencies is

$$\delta P = \frac{2\pi\hbar}{v}\left\langle\omega_{\mathrm{T}}^2\right\rangle.\tag{8.19}$$

With or without averaging, δP is proportional to $1/v$. The experimentalist who makes that observation may be tempted to claim an agreement with the Landau-Zener–Stückelberg theory. However, there is no agreement at lower velocity, because averaging destroys the form of (8.17).

8.3.3 *Fast sweeping*

In this section, the derivation of formula (8.18), valid in the fast sweeping limit, will be given. It is in fact easy while the derivation of the exact formula (8.17) is more difficult and therefore given only in Appendix H.

It results from (8.15) that the exponential $e^{-iU(t)}$ in (8.12) oscillates rapidly with time if the sweeping velocity v is fast. In that case the integral

$$Y(t_f) = \int_0^{t_f}\dot{Y}(t)\,dt\tag{8.20}$$

is small, and the flipping probability $\delta P = |Y(t_f)|^2$ is weak. The calculation easily follows from (8.12b) when $X(t)$ is replaced by $X(0) = 1$ and $U(t)$ by its expression

(8.15). One obtains

$$Y(t_f) = -i\omega_T \int_0^{t_f} dt \exp\left[\frac{-iv}{2\hbar}(t - t_0)^2\right]. \tag{8.21}$$

The exponential has a real part and an imaginary part which both fluctuate around 0. For $t < 0$ and $t > t_f$, the fluctuations are so fast that the exponential can be replaced by its mean value 0, in agreement with the statements made at the beginning of this chapter. Thus, the integration bounds may be replaced by $-\infty$ and ∞, and a definite integral is obtained, which is easily found in books or in software:

$$Y(t_f) = -i\omega_T \int_{-\infty}^{\infty} dt \exp\left[\frac{-iv}{2\hbar}(t - t_0)^2\right] = -(1 - i)\omega_T \sqrt{\frac{\pi\hbar}{v}}. \tag{8.22}$$

Therefore

$$\delta P = |Y(t_f)|^2 = \frac{2\pi}{v}\hbar\omega_T^2$$

which coincides with (8.18).

As stressed at the beginning of this chapter, the above calculation is only correct when sweeping the zero-field resonance. For the other resonances, the spin tunnels from the lowest state of the left-hand well to an excited state of the other well, whose lifetime is rather short. The spin de-excites rapidly from that state to lower energy states, so that it is not possible to consider only two states $|m^*\rangle$ and $|m'^*\rangle$. It turns out that in that case too, formula (8.18) still applies. This will be shown in Section 10.6.

8.3.4 *Sweeping back and forth through the resonances*

As seen in Section 8.3.2, the basic formula (8.17) is based on the hypothesis of an isolated spin, which is only justified for a fast sweeping velocity $\alpha = dH/dt$. But if α is large, the reversal probability δP is weak. However, it is easy to enhance the sensitivity of the experiment by sweeping back and forth several times through the same resonance instead of increasing the field at a uniform velocity. The observed resonances are between $|m^*\rangle$ and $|m'^*\rangle$, but in practice m is equal to $-s$ in a real experiment, so that each resonance is characterized by $N = |m + m'| = |m' - s|$. At any sweep a fraction δP of the molecules which are still in the metastable well escape from that well, so that the fraction R_N remaining in the metastable well is multiplied by $(1 - \delta P)$. After n cycles around the $(m - m')$ resonance (Fig. 8.4) the fraction of magnetization that has not yet relaxed is given by

$$R_N^{(n)} = R_N^{(0)}(1 - \delta P)^{2n} = \exp\left(\frac{-n\pi\Delta_{mm'}^2}{\hbar g\alpha\mu_B|m - m'|}\right). \tag{8.23}$$

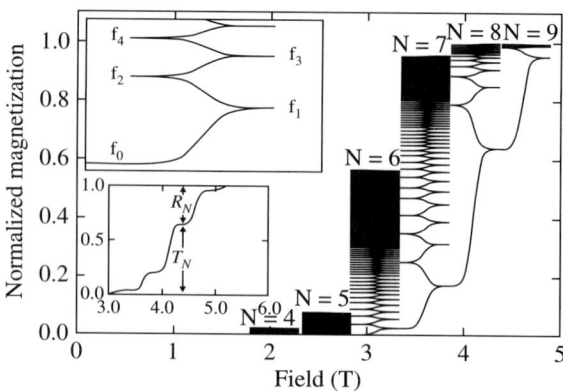

FIG. 8.4. An example of experimental application of a time-dependent longitudinal field to extract the tunnel splitting from the LZS model. The main
picture shows the variation of the magnetization when the magnetic field is
swept back and forth around a tunnelling resonance. In the upper inset is
an enlargement of the main picture. The lower inset helps to define the fraction of the magnetization that has tunnelled, T_N, and that associated to the
molecules that are still in the metastable well, R_N. Modified from Mertes
et al. 2003. Copyright (2003) American Institut of Physics.

The factor 2 in the exponent $2n$ accounts for the fact that at any cycle
the resonance condition is met twice. It is interesting to note that $R_N^{(n)}$ should
decay exponentially with the number n of cycles. Deviations from this behaviour have been encountered and attributed to a distribution of tunnel splitting, in
agreement with the hypothesis made in Section 8.3.2. This point will be discussed
in more detail in Chapter 12.

9

INTERACTION OF A SPIN WITH THE EXTERNAL WORLD AT LOW TEMPERATURE

9.1 Coherence, incoherence, and relaxation

9.1.1 *Low temperatures*

The present chapter is devoted to the study of a *real* molecular spin $\mathbf{S}$ at low temperature, when no phonons are available but tunnelling is possible. A real spin interacts with its environment, especially the other electronic spins, and also the nuclear spins. The properties of this spin are very different from those of the isolated spin considered in Chapter 6. Of course the latter has no real existence. However, its study was an unavoidable preliminary to that of real spins.

The case of a constant external field will first be considered. The tunnel motion of an ideal, isolated spin is then characterized by a periodic oscillation between the left-hand and the right-hand well, as described by formula (6.47). The phase $t\delta E/\hbar$, or at resonance $2\omega_T t$, is well defined. This tunnelling with a precise phase is called *coherent*. It is quite different from *relaxation*, which is a stochastic evolution toward equilibrium.

In contrast, when the spin interacts with its environment (as it always does!) the wavefunction loses the memory of its phase. This phenomenon is called loss of coherence or *decoherence* (Zurek 2003). This implies that the wavefunction is no longer an appropriate description of the spin. Even at low temperature, a spin in contact with the external world can usually not be described by a wavefunction, but only by its probability to be in state $|-s^*\rangle$ or in state $|s^*\rangle$. This probability goes to its equilibrium value, and this is relaxation.

At the low temperatures which are of interest in the present chapter, phonon absorption is negligible since no phonons are available. However, phonon *emission* is still possible and is responsible for the finite lifetime of the excited states of each well. A typical situation is shown in Fig. 6.2, when the field is such that tunnelling can take place between the lowest state of one well, here $|-s^*\rangle$, and an excited state of the other well, which is $|(s-1)^*\rangle$ in Fig. 6.2. The ideal, isolated spin of Section 6.5 would oscillate between $|-s^*\rangle$ and $|(s-1)^*\rangle$. For the real spin, these oscillations are damped because the excited state $|(s-1)^*\rangle$ deexcites to the ground state $|s^*\rangle$ with emission of phonons. This implies decoherence and relaxation. The inverse lifetime $\gamma = 1/\tau_1$ of the excited state is given at low temperature by (5.41) with $m = s - 1$, $m' = s$ and $\beta = \infty$.

The case of weak fields and low temperatures is more difficult. In that case, tunnelling occurs between $|-s^*\rangle$ and $|s^*\rangle$, and no phonons can be emitted because $|s^*\rangle$ is the ground state and no energy is available. Of course, no phonons can be absorbed either because the temperature is low and no phonons are available. Thus, phonons have no effect on the molecular spin. Decoherence and relaxation are mainly the result of the interaction of each spin with other spins, molecular or nuclear.

9.1.2   The window mechanism

This interaction may be represented by a local, time-dependent field $\mathbf{H}_{\mathrm{loc}}(t)$ acting on each spin. If there is an external field $\mathbf{H}$, the total local field is $\mathbf{H}'_{\mathrm{loc}}(t) = \mathbf{H} + \mathbf{H}_{\mathrm{loc}}(t)$ while $\mathbf{H}_{\mathrm{loc}}(t)$ can be called the *internal local field*. As in Chapter 6, tunnelling is only possible when two levels cross, but the local field has to be taken into account in the level-crossing condition.

More precisely, it was seen in Chapter 6 that tunnelling is possible for an isolated spin when the external field H_z in the z direction satisfies $H^{(mm')} - \delta H_0^{(mm')} < H_z < H^{(mm')} + \delta H_0^{(mm')}$, where $g\mu_{\mathrm{B}}\delta H_0^{(mm')} s \simeq \hbar\omega_{\mathrm{T}}^{(mm')}$ according to (6.28), while $H^{(mm')}$ is the field for which the levels $|m^*\rangle$ and $|m'^*\rangle$ cross. This field is given by (6.4) if the anisotropy Hamiltonian has the form (2.10). Thus, tunnelling is possible if the external field lies in a window of width $\hbar\omega_{\mathrm{T}}^{(mm')}$ centred about one of the level-crossing fields $H^{(mm')}$.

For a real spin, the possibility of tunnelling requires the same condition, except that the external field should be replaced by the total local field $H_z + H_{\mathrm{loc}}^z$. Thus tunnelling is possible if

$$H^{(mm')} - \delta H_0^{(mm')} < H_z + H_{\mathrm{loc}}^z < H^{(mm')} + \delta H_0^{(mm')}. \tag{9.1}$$

This 'window condition' gives a very special character to the relaxation mechanism. It was called 'degeneracy blocking' by Prokofev and Stamp (1996).

A reasonable assumption is that the local field H_{loc}^z is a random variable, whose distribution is characterized by its width δH. Then, (9.1) can be satisfied, and tunnelling is possible, if

$$H_0^{(mm')} - \langle H_{\mathrm{loc}}^z \rangle - \delta H < H_z < H_0^{(mm')} - \langle H_{\mathrm{loc}}^z \rangle + \delta H. \tag{9.2}$$

Thus, there is a resonance width of the order of δH, which may be much larger than the natural width $2\hbar\omega_{\mathrm{T}}^{(mm')}$.

In the relaxation mechanism, the time dependence of the local field $H_{\mathrm{loc}}^z(t)$ is essential. If H_{loc}^z does not depend on time, the few spins which satisfy (9.2) do tunnel, and most of the spins do not move (Prokofev and Stamp 1996). Thus, a complete theory of relaxation has to worry about why and how the local field depends on t. This question will be addressed in Section 9.2.6, but it can already be said that it is not yet really solved.

Another important feature to be taken into account is the decoherence of the tunnel process. Coherent tunnelling, as studied in Section 6.5, would indeed imply a periodic motion rather than relaxation toward the equilibrium state (Prokofev and Stamp 1996). Decoherence can arise from the time dependence of

$H_{\text{loc}}^z(t)$, and this will generally be assumed to happen, thus following Fernández (2003). However, faster decoherence processes can also take place as will be seen in Section 9.4.2.

Consequences of the above remarks are the following:

- Relaxation cannot arise from 'frozen' impurities or defects, e.g. motionless dislocations.
- If relaxation arises from interactions with nuclear spins (the 'hyperfine' interactions introduced in Chapter 2) the relaxation time τ of the molecular spins depends on the relaxation time τ_ν of the nuclear spins, and goes to infinity when τ_ν goes to infinity.
- If relaxation arises from interactions with other molecular spins, it is a complicated many-body problem. This mechanism will be seen to lead to *non-exponential* relaxation.

As a matter of fact, magnetic relaxation arises from both interactions with nuclear spins, and interactions with other molecular spins. However, to simplify the analysis, the two mechanisms will be studied separately.

9.2 Hyperfine interactions

9.2.1 *The hyperfine field and its order of magnitude*

As seen in Section 2.1, 'hyperfine' interactions between nuclear spins and electronic spins include 'contact' and dipole interactions. Contact interactions are mainly important between a nucleus and electrons of the same ion. They are essential for $Mn_{12}ac$ because all Mn nuclei are ^{55}Mn, with nuclear spin 5/2 and magnetic moment

$$\mu_{\text{Mn}} = 3.47\mu_{\text{n}} = 1.89 \times 10^{-3}\mu_{\text{B}} \tag{9.3}$$

where the nuclear magneton μ_{n} is about $1/1837$ times μ_{B}.

The dipole interaction energy between two magnetic moments $\mathbf{m}$ and $\mathbf{m}'$ is given by formula (2.50):

$$W = \frac{\mu_0}{4\pi} \sum_{\alpha\gamma} \frac{1}{r^3} \left[\delta_{\alpha\gamma} - \frac{3}{r^2}r^\alpha r^\gamma \right] m_\alpha m'_\gamma$$

where $\mu_0/(4\pi) = 10^{-7}$ Henry/m.

The magnetically active nuclei which do not carry the unpaired electrons, like the hydrogen atoms in $Mn_{12}ac$, do give a contribution to the hyperfine field, also called super- or transfer-hyperfine, with contact (through bonds) and dipolar (through space) terms. The first one is proportional to the contribution of the s orbitals of atoms with magnetic nuclei to the molecular orbitals carrying the unpaired electron.

In $Mn_{12}ac$, the most significant contribution to the hyperfine field is provided by the manganese and the hydrogen nuclei. Indeed, natural carbon contains

98.9% of ^{12}C isotope which has spin 0, and natural oxygen contains 99.76 % of ^{16}O isotope which also has spin 0. In contrast, natural manganese contains 100% of ^{55}Mn isotope which has spin 5/2. The actual formula

$$Mn_{12}O_{12}(CH_3COO)_{16}(H_2O)_4].2CH_3COOH.4H_2O$$

indicates that there are 72 hydrogen nuclei. Protons have a spin 1/2 and a magnetic moment

$$\mu_H = 2.79\mu_n = 1.5 \times 10^{-3}\mu_B \tag{9.4}$$

while deuterons have a spin 1 and a magnetic moment

$$\mu_D = 0.86\mu_n = 0.47 \times 10^{-3}\mu_B. \tag{9.5}$$

The order of magnitude of the magnetostatic energy between a Mn ion of magnetic moment $4\mu_B$ and a proton at distance 0.3 nm (a typical value) is thus

$$|W| \approx 10^{-7}\frac{1}{r^3}mm' = -10^{-7} \times 0.04 \times 10^{30} \times 6 \times 10^{-3}\mu_B^2 \approx 0.24 \times 10^{-26} \text{ J.} \tag{9.6}$$

By an oversimplification of the discussion presented in Section 3.3.2, the corresponding field H_{hf} felt by the manganese moment can be obtained by identifying (9.6) with $g\mu_B H_{hf}$, namely

$$H_{hf} \approx 10^{-3} \text{ tesla} \tag{9.7}$$

where the factor 0.24 has been dropped, because several protons contribute significantly to the hyperfine field. For that reason, the actual typical value of the hyperfine field of dipolar origin is a few mT. Contact interactions provide an approximately 10 times higher value, which can be evaluated with satisfactory accuracy, from nuclear magnetic resonance data in different materials where the Mn ions have a similar environment and electronic state.

Thus the typical hyperfine field (energy divided by $g\mu_B s$) can be evaluated in Mn$_{12}$ac, and turns out to be between 20 and 40 mT. This is of the same order of magnitude as the dipole interactions between molecular (electronic) spins. This is an exciting peculiarity of molecular magnetism.

In Fe$_8$ the most abundant iron isotope ^{56}Fe has spin $I = 0$, while the only magnetic isotope is ^{57}Fe, with spin $I = 1/2$, and an abundance of 2.12%. Only one molecule in about six contains a ^{57}Fe nucleus. The main contribution to the hyperfine field comes from other nuclei, mainly H and N while ^{78}Br and ^{81}Br can be neglected. The hyperfine field is therefore expected to be much smaller than in Mn$_{12}$ac. Actually, it is about 10 times as small.

The energy (9.6) corresponds to $W/k_B \approx 1$ mK. Even though it must be multiplied by a factor of about 10 in the case of a contact interaction, nuclear spins are not oriented by the neighbouring molecular spins at the temperatures reached in usual experiments.

9.2.2 *Experimental evidence of hyperfine interactions*

The role of the nuclear magnetic moment in promoting tunnelling has been experimentally confirmed by isotopically modifying the Fe_8 clusters (Wernsdorfer *et al.* 2000c). In particular both ^{57}Fe and D enriched samples of Fe_8 have been synthesized. In the former the hyperfine field is larger than in the natural abundance species while a weaker field is operative in the latter. As seen in the previous subsection, ^2H has nuclear spin $I = 1$ which is larger than that of the proton ($I = 1/2$) but has a significantly smaller gyromagnetic factor, thus reducing the nuclear magnetic moment. The two isotopically substituted samples showed the same oscillations of the tunnelling rate with the transverse field, as displayed in Fig. 6.7 for natural Fe_8. The position of the quenching of the tunnelling rate remains unchanged in the three samples confirming that the magnetic anisotropy is not influenced by the isotopic enrichment. On the contrary, the relaxation time of the magnetization is significantly changed. In Fig. 9.1 is reported the temperature dependence of the time needed to relax 1% of the saturation magnetization. In the tunnelling regime of low temperature the ^{57}Fe enriched sample shows the fastest relaxation while the slowest relaxation is observed for the deuterium enriched crystal. This marked isotopic effect does not depend on the mass, which is increased in both isotopically enriched samples, but rather on the nuclear magnetism.

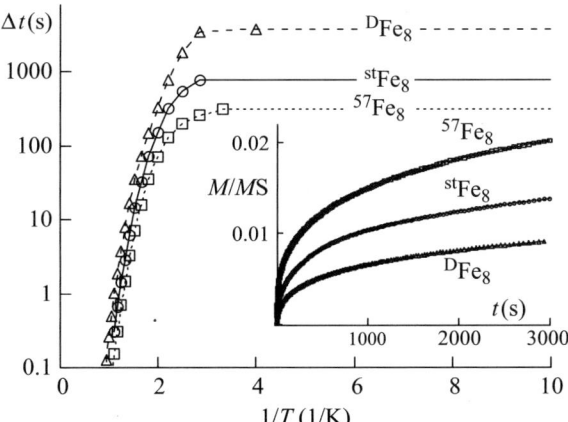

FIG. 9.1. Time needed to relax 1% of the saturation magnetization for the three isotopic Fe_8 crystals versus the inverse of temperature. The inset shows the time dependence of the magnetization of the three crystals measured after cooling in zero field and then applying a longitudinal field of 42 mT in the pure tunnelling regime ($T = 40$ mK). From Wernsdorfer *et al.* (2000c). Copyright (2000) by American Physical Society.

It is interesting to verify experimentally if the increased relaxation corresponds also to an increase of the linewidth of the resonance and particularly that
is the effect of the hyperfine field. This has been possible thanks to an experimental procedure developed by Wernsdorfer and coworkers that exploits the
possibility to have access not to the entire sample experiencing a distribution of
local field but to address a fraction of it by digging, or burning, a hole in the
distribution (Wernsdorfer *et al.* 1999). In view of the relevance of the technique
for the following discussion, it will be briefly presented here.

The hole digging method consists of three steps as schematized in Fig. 9.2

(1) Preparing the initial state. A well-defined initial magnetization state of
the crystal of molecular clusters can be achieved by cooling the sample from
high to low temperatures in a magnetic field H_0 parallel to the easy direction
z. For definiteness, this field will be assumed strong, so that the saturated
magnetization state is reached.

(2) Modifying the initial state–hole digging. After preparing the initial state,
a field H_{dig} is applied during a time t_{dig}, called the 'digging field' and 'digging
time' respectively. For definiteness, this field will be assumed to be close to 0,
so that a fraction of the molecular spins feel a vanishing local field and relax to

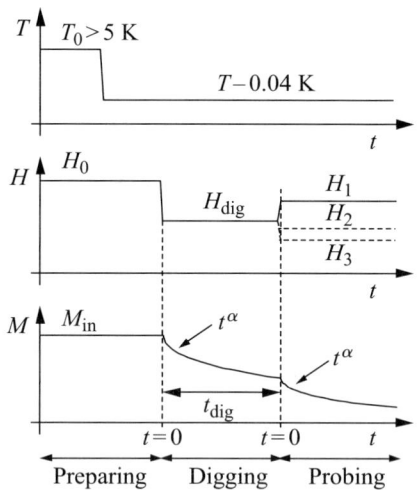

FIG. 9.2. Schematic representation of the temperature (top), applied field
(middle), and magnetization (bottom) variation during the three steps of the
hole-digging procedure. The temperature of the sample is rapidly quenched
from high temperature to the investigated temperature (top). The magnetic
field is kept at H_0 during the cooling process then a field H_{dig} is applied for
a time period t_{dig} and after the measuring field H_1 is set. The procedure is
repeated for different measuring fields H_2, H_3, etc.

their equilibrium magnetization by tunnelling. Thus a hole opens in the curve representing the number of unrelaxed spins as a function of the local field. The field sweeping rate from H_0 to H_{dig} should be fast enough to minimize the change of the initial state.

(3) Probing the final state. To see the hole, a tunable field H_z is applied to measure the short-time relaxation as a function of H_z. The velocity of that relaxation gives information about the number of spins which have not yet relaxed. Technically, as will be seen later in this chapter, the relaxed fraction is proportional to $t^{1/2}$ and the proportionality coefficient $\Gamma_{sqrt}^{1/2}(H_z, H_{dig}, t_{dig})$ is the quantity to measure. It is displayed as a function of H_z in Fig. 9.3 for $t_{dig} = 0$, while Fig. 9.4 shows the hole which opens for a non-vanishing value of t_{dig}.

In practice, the initial state is not necessarily the saturated one. If not, the quench should be fast (of the order of 1 s) in order to avoid partial relaxation, which in some cases might qualitatively perturb the internal field distribution. In Fig. 9.3 the field dependence of $\Gamma_{sqrt}(H_z)$ for three different values of the initial magnetization are reported. The narrowest distribution is observed for the almost saturated sample which had initial magnetization $M_{in} = -0.998M_s$. A proof of the power of the method in providing information on the bias field distribution is given by the remarkable structure seen for $M_{in} = -0.870M_s$. The peak

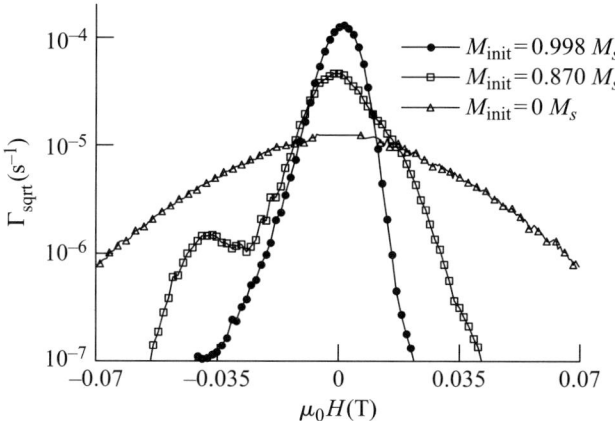

FIG. 9.3. Dependence of the short-time tunnelling rate as a function of the longitudinal field for three different initial states characterized by the initial magnetization $M_{in} = -0.998M_s$ (circles), $M_{in} = -0.870M_s$ (squares), and equal to 0, $M_{in} = 0$ (triangles). The structure observed at intermediate magnetization is due to clusters which experience the dipolar environment generated by the reversal of one neighbouring cluster along the a (-0.04 T), b (0.035T), and c (0.025T) crystallographic directions. From Wernsdorfer et al. (1999). Copyright (1999) by the American Physical Society.

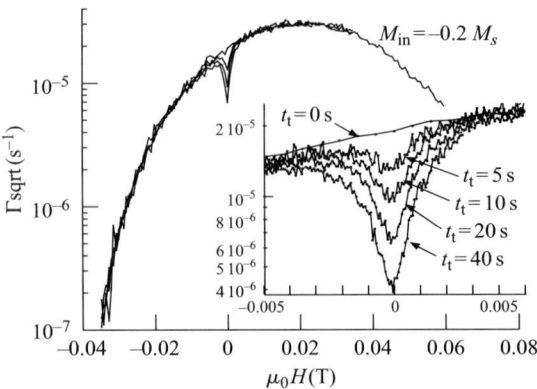

FIG. 9.4. The hole observed by Wernsdorfer *et al.* in the function $\Gamma_{\text{sqrt}}(H)$ at very low temperature for Fe$_8$. Since relaxation is only possible when the external field compensates the local field it can be assumed that $\Gamma(H)$ is proportional to the probability that the local field has the value $-H$. From Wernsdorfer *et al.* (1999).

at -0.04 T as well as the shoulder at $+0.02$ T and $+0.04$ T originate from the clusters which have one nearest neighbour cluster with reversed magnetization. Taking into account the variation of the dipolar field due to the flipping of a nearest-neighbour spin it is possible to attribute the peak at negative field to the reversal of the neighbouring cluster along the crystallographic a axis, this one being almost coincident with the easy axis of magnetization z. The two shoulders at positive field are due to the reversal of the spin of a nearest-neighbour along b or c, the one at smaller field corresponding to the reversal of the spin of the nearest-neighbour cluster at larger distance and therefore along c. These results are in good agreement with simulations (Ohm *et al.* 1998; Cuccoli *et al.* 1999). The experimental results suggest that when a hole is dug in a saturated sample, it is dominated by the change of intermolecular dipolar fields during the digging. However, in special conditions, it is possible to reduce the change of intermolecular dipolar fields in order to be sensitive to local field fluctuations coming from nuclear spins. This can be done for instance by digging a hole into the tail of the dipolar distribution of a demagnetized sample as shown in Fig. 9.5.

In these conditions almost all molecules are out of resonance and only a very small fraction ($<10^{-4}$) might be brought into resonance by the hyperfine field fluctuations. Therefore, for short digging times, the variation of the intermolecular field is negligible and the hole width directly reflects the hyperfine field fluctuations (Wernsdorfer *et al.* 2000c). The described procedure allows us to get rid of the dipolar contribution to the linewidth and thus to show the contribution of the hyperfine field to the broadening of the intrinsic linewidth. The hole-digging method previously described has been used to compare the intrinsic linewidth of the tunnelling resonance for the three isotopic samples. The conditions that minimize the dipolar broadening have been selected so that the observed linewidth is essentially dominated by the hyperfine contribution. The

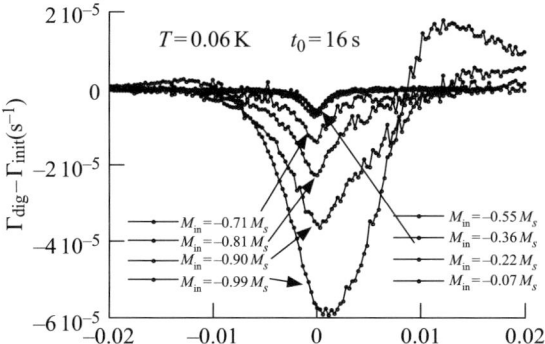

FIG. 9.5. Hole-digging at 0.06 K for $t_{\text{dig}} = 16$ s starting from different initial magnetization. For initial magnetization $|M_{\text{in}}| < 0.5M_s$ the linewidth of the hole becomes constant, showing that the linewidth is not dependent on the initial magnetization. From Wernsdorfer *et al.* (1999). Copyright (1999) by the American Physical Society.

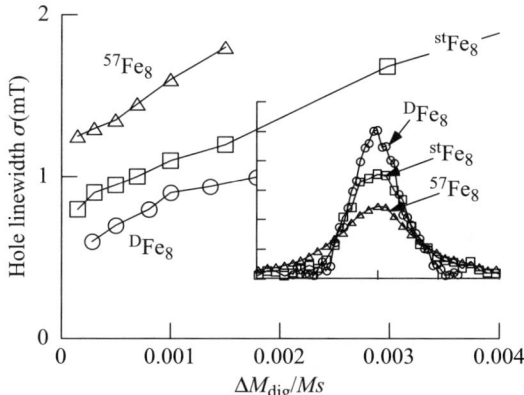

FIG. 9.6. Linewidth of the hole burnt in the distribution of tunnel rate for the three isotopic crystals of Fe_8 as a function of the fraction of magnetization reversed during the hole-digging procedure. The insert shows the experimental hole line-shape. The experimental conditions of $H_z = 42$ mT and $H_x = 200$ mT have been selected as those for which the narrowest hole is obtained. From Wernsdorfer *et al.* (2000c). Copyright (2000) by the American Physical Society.

linewidth of the hole depends on the amount ΔM_{dig} of magnetization that has been reversed. By extrapolating the linewidth to $\Delta M_{\text{dig}} = 0$ it is possible to get the intrinsic linewidth. These results are 0.6 ± 0.1 mT, 0.8 ± 0.1 mT, and 1.2 ± 0.1 mT for the deuterated, natural, and ^{57}Fe enriched samples, as reported in Fig. 9.6.

The hyperfine interaction between a nuclear spin $\mathbf{I}_i$ and the spin σ_r of a *single* Fe ion can be expressed as in (2.19). Summing over the eight Fe ions yields

$$\mathcal{H}_{\text{hf}} = \sum_i \sum_{r=1}^8 \sigma_r \cdot A_{ir} \cdot \mathbf{I}_i. \tag{9.8}$$

One would like to have a formula involving the total spin $\mathbf{S} = \sum \sigma_r$ rather than the eight spins σ_r. This is possible because, at the low temperatures of interest, only the lowest exchange multiplet $s = 10$ is populated. Inside this multiplet, σ_r may be replaced by $c_r \mathbf{S}$, where c_r is a numerical constant. This is the projection theorem, mentioned in Chapter 2 and derived in textbooks (Cohen-Tannoudji *et al.* 1986). An equivalent statement is

$$\mathcal{P}\sigma_r \mathcal{P} = c_r \mathcal{P}\mathbf{S}\mathcal{P} \tag{9.9}$$

where $\mathcal{P}$ is the projection operator onto the lowest exchange multiplet $s = 10$. This property is a special case of the more general Wigner–Eckart theorem mentioned in Chapter 2. Insertion of (9.9) into (9.8) yields

$$\mathcal{H}_{\text{hf}} = \sum_i \mathbf{S} \cdot A_i \cdot \mathbf{I}_i \tag{9.10}$$

with

$$A_i = \sum_{r=1}^8 c_r A_{ir}. \tag{9.11}$$

The projection coefficients c_j can be roughly evaluated as follows. The ground state of Fe$_8$ is approximately given (with the notation of Fig. 3.36) by $\sigma_1^z = \sigma_2^z = -\sigma_3^z = -\sigma_4^z = \sigma_5^z = \sigma_6^z = \sigma_7^z = \sigma_8^z = 5/2$. This is often called the 'Néel ground state'. In that state, $S^z = 10$ and therefore $c_r = -1/4$ for $r = 3$ or 4, and $c_r = 1/4$ for other values. A quantum treatment, using a better ground state, can be performed by using recurrently the projection technique for coupled angular momenta as explained in Chapter 2. One finds $c_3 = c_4 = -5/22$ and $c_1 = c_2 = c_5 = \cdots = c_8 = 8/33$.

9.2.3 *Linewidth of hyperfine origin*

An important effect of nuclear spins is to give to the resonance a width which is appreciably larger than the natural width, which is of the order of $\hbar\omega_{\text{T}}$ as seen from (6.28). Indeed, tunnel occurs between *hyperfine* states $|m^*, \nu\rangle$ and $|m'^*, \nu'\rangle$, where $|m^*\rangle$ and $|m'^*\rangle$ are local states of the molecular spin, as before, while $|\nu\rangle$ and $|\nu'\rangle$ are states of the nuclear spins. The states $|m^*\rangle$ and $|m'^*\rangle$ of an isolated molecular spin have a well-defined energy. In contrast, since there are many nuclear spins which interact with that molecular spin, there are many hyperfine states (Fig. 9.7) which correspond to $|m^*\rangle$ and $|m'^*\rangle$. If the distance between hyperfine levels is smaller than $\hbar\omega_{\text{T}}$, the energy levels can be regarded as forming a continuous band.

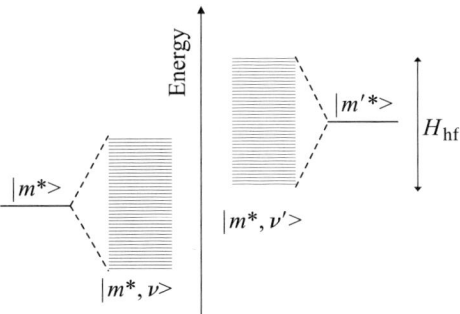

F<small>IG</small>. 9.7. Hyperfine splitting of electronic levels. In the absence of hyperfine inter-
actions, the localized states $|m^*\rangle$ and $|m'^*\rangle$ have a well-defined energy (if
tunnelling is neglected). They give rise to hyperfine states which practically
form two bands. These bands are much broader than the natural width $\hbar\omega_T$.
Tunnelling can occur if the bands have at least a partial overlap.

The exploitation of (9.11) requires knowledge of the single-spin hyperfine
coupling constant A_{ir}. It contains both through-space (dipolar) and through-
bond (contact) contributions. While the former are easily calculated using the
point-dipole approximation, the latter can only be roughly estimated. The density functional approach can been used to evaluate the contact term of the
hyperfine coupling on the proton of the bridging OH groups on the nitrogen and
proton of the NH group of the ligands by using a model species (Wernsdorfer $et\ al.$
2000c; Sessoli $et\ al.$ 2001). The results, even though semiquantitative, reveal that
the contact term has the same order of magnitude as the dipolar one, as confirmed
by ^{1}H NMR experiments performed at low temperature (Furukawa $et\ al.$ 2001b).
    A more quantitative analysis can be performed in the case of a completely
substituted [57]Fe$_8$ cluster, because in this case the interaction A_{rr} between the
nuclear and electronic spin of the same iron atom is dominated by the the contact term. It can be measured in other materials and is of the order of 1 mT
(McGarvey 1966). The NMR data of Fig. 3.37 confirm this value.
    Using (9.10), the hyperfine field resulting from [57]Fe is found to be a (roughly
Gaussian) random variable whose linewidth σ_{57} is about 0.9 mT. This contribution is to be added to the contribution σ_0 of the other magnetic nuclei, already
present in natural Fe$_8$. The widths of the two independent random variables add
to give a total width

$$\sigma_{\text{tot}} = \sqrt{\sigma_0^2 + \sigma_{57}^2}. \tag{9.12}$$

Formula (9.12) yields $\sigma_{\text{tot}} \simeq 1.1 \pm 0.1$ mT, in good agreement with the
experimental value of 1.2 ± 0.1 mT.
    As stated above, the hyperfine states can be approximated by a continuous
band. Let this statement be quantitatively justified. The width of this band,

expressed in terms of the hyperfine field H_{hf} seen by a molecular spin, is of the order of one millitesla in Fe_8 (see Fig. 9.6) and about 10 times more in $Mn_{12}ac$. The corresponding energy is $w_{hf} = g\mu_B s H_{hf}$, so that the distribution of w_{hf}/k_B in Fe_8 (for instance) has a width of 0.01 K. Since the tunnel splitting, as seen in Section 8.3.2, is 10^{-7} K, the ratio is $10^5 \approx 2^{17}$, so that an interaction with 17 spins $1/2$ is sufficient to justify the continuum approximation. The number of nuclear spins is higher than 17, but a proper discussion should take into account their distance to the molecular spin. This calculation is left as an exercise to the reader. Even if the distance between hyperfine levels is not so small, the fluctuations of the magnetic field are probably larger than this distance, so that it is legitimate to treat the hyperfine field as a continuous variable.

Since the hyperfine energy distribution in Fe_8 is 10^5 times broader than $\hbar\omega_T$, this suggests that the resonance width arising from hyperfine interactions is 10^5 times larger than the natural width $2\hbar\omega_T$. This is correct, but it should be mentioned that the conditions for tunnelling between two hyperfine states are more complicated than might be expected from the above argument. This argument uses states of the molecular spin localized in the $S^z > 0$ region or in the $S^z < 0$ region, and takes into account their anisotropy energy, their Zeeman energy and their hyperfine energy. Tunnelling between two such localized states, and thus delocalization, can only be possible if their energy is the same with a precision of $\hbar\omega_T$. However, this is a necessary, but not sufficient condition! Tunnelling is only possible if the matrix element of the Hamiltonian between the two hyperfine states is large enough. This point is discussed in Appendix I and the above assertion is justified. The resonance width arising from hyperfine interactions is 10^5 times larger than $\hbar\omega_T$ in Fe_8. In $Mn_{12}ac$ the ratio is even larger.

9.2.4 Relaxation of hyperfine origin

While the evaluation of the resonance width due to nuclear spins is simple, the relaxation is a more difficult problem. What is the type of relaxation (exponential or not)? What is the relaxation time τ? The hyperfine field is generally replaced by a randomly fluctuating field of root mean square amplitude δH and relaxation time τ_H (Tupitsyn and Barbara 2002; Fernández and Alonso 2000,2002).

In this section, for definiteness, the external field will be assumed to be 0 and the temperature will be assumed so low that the component S_z of any spin is close to either $-s$ or s. Initially, it may be assumed to be s.

The probabilities $P^+(t) = |X(t)|^2$ and $P^-(t) = |Y(t)|^2$ that a spin is up or down can in principle be obtained from equations (8.12) which determine $X(t)$ and $Y(t)$. The initial condition will be assumed to be $P^+(0) = 1$ and $P^-(0) = 0$. For instance, (8.12b) reads $\dot{Y}(t) = -i\omega_T X(t)e^{-iU(t)}$, where $U(t)$ is given by (8.13) and (8.3) or, in zero external field,

$$U(t) = \frac{2g\mu_B s}{\hbar} \int_{t_0}^{t} dt_1 H_z(t_1). \qquad (9.13)$$

Here, $H_z(t)$ is the hyperfine field. It will be assumed to undergo random fluctuations characterized by their mean square δH^2 and their relaxation time τ_H. The average value is $\langle H_z \rangle = 0$.

It is convenient to investigate first short times during which $X(t)$ is almost equal to $X(0) = 1$. Thus

$$\dot{Y}(t) = -i\omega_T e^{-iU(t)}. \tag{9.14}$$

Since $U(t)$ is real, $\dot{Y}(t)$ fluctuates between its initial value $-i\omega_T$ and $i\omega_T$. It is important to calculate the time τ_1 it requires to change sign. Roughly speaking, this time is the time required for $U(t)$ to reach the value π. It follows from (9.13) that $U(t)$ is the sum of t/τ_H random, uncorrelated quantities δU_n which have average value 0 and variance $\delta U^2 \approx (2g\mu_B s/\hbar)^2 \delta H^2 \tau_H^2$. The time τ_1 is thus approximated by $(\tau_1/\tau_H)\delta U^2 = \pi^2$ or

$$\tau_1 = \frac{\hbar^2}{(2g\mu_B s)^2 \delta H^2 \tau_H}. \tag{9.15}$$

It is now possible to study spin relaxation. It is determined by the behaviour of $Y(t)$. Roughly speaking, this quantity is the sum of (t/τ_1) random, uncorrelated quantities of the order of $i\omega_T \tau_1$. Therefore

$$\langle |Y(t)|^2 \rangle \simeq \omega_T^2 t \tau_1 \tag{9.16}$$

The relaxation time τ is determined by $\langle |Y(\tau)|^2 \rangle \simeq 1/2$. Hence

$$1/\tau \simeq 2\omega_T^2 \tau_1 \simeq \frac{2\hbar^2 \omega_T^2}{(2g\mu_B s)^2 \delta H^2 \tau_H} \tag{9.17}$$

This formula should not be understood as a recipe which can be systematically applied. Its derivation just gives a flavour of the type of methods which can be applied when the relaxation process results from a sequence of uncorrelated stochastic processes, each of which is a small perturbation. In that case the relaxation is exponential, $P^\pm(t) = ((1/2)[1 \pm \exp(-t/\tau)]$.

The validity of (9.17) is actually very limited. It relies on (9.15) and (9.16). These relations are valid if $\tau_H \ll \tau_1 \ll \tau$. This implies $\omega_T \tau_H \ll \omega_T \tau_1 \ll 1$ or

$$\frac{\hbar^2 \omega_T}{(2g\mu_B s)^2 \delta H^2} \ll \tau_H \ll \frac{\hbar}{2g\mu_B s \delta H}. \tag{9.18}$$

At low temperature, τ_H may become very long, as discussed in Section 9.2.6, and the second condition (9.18) may fail to be satisfied. On the other hand, if the first condition (9.18) is not satisfied, the coherence time τ_1 given by (9.15) can become very long, and other sources of decoherence can become important as will be seen in Section 9.4.2.

9.2.5 Effect of hyperfine interactions in the case of time-dependent fields

As already said, the tunnel frequency ω_T is a property of the isolated spin, which is just a virtual concept, so that ω_T can only be measured indirectly. When, as reported in Chapter 8, Wernsdorfer had the idea to apply a longitudinal, time-dependent field to a magnetized sample, his hope was that the demagnetizing effect of this field should be independent of the hyperfine field, and more generally of the local field. Indeed, the external field at which partial demagnetization occurs depends on the external field, but in all cases it takes place by the same amount (except if complex interaction mechanism are considered).

On the other hand, the hyperfine field also has a transverse component, which can contribute to tunnelling. The best way to assess the effect of this component is to measure the effect of an external transverse field of the same magnitude. The results of Wernsdorfer and Sessoli (1999a) clearly show that this effect is negligible (see Fig. 6.8).

However, Wernsdorfer *et al.* (2000d) measured the tunnel frequency by the Landau–Zener method on several Fe_8 samples with different isotopic compositions. These samples were the same as in the experiments mentioned in Section 9.2.2 but now the field was time-dependent. The demagnetization was found to depend very much on the isotopic composition. Essentially, the tunnel splitting is larger when the nuclear magnetic moment is larger. Three samples were studied. (i) The natural material, where Fe is ^{56}Fe (with no nuclear spin) at a concentration of 98% while hydrogen is mainly ^1H. (ii) A sample enriched in ^{57}Fe (nuclear spin 1/2). (iii) A sample enriched in deuterium, which has a weaker nuclear magnetism than ^1H. Magnetic relaxation is ten time faster in sample (ii) than in sample (i). Relaxation in standard Fe_8, sample (i), is intermediate.

The effect of a time-dependent field (Landau–Zener method) was also investigated (Wernsdorfer *et al.* 2000b). The effect of isotopic substitution on tunnelling is found to be quite strong. This result is disturbing since the motivation of the Landau–Zener method was to get rid of the local field. A lazy explanation would be that the quality of the samples is different because the chemists are unable to make crystals of identical quality for different isotopic concentrations. The effect of crystal defects will be addressed in Chapter 12.

9.2.6 *How do nuclear spins relax?*

If the local field is attributed to hyperfine interactions, its relaxation time τ_H (which is a very important quantity as seen above) should be the relaxation time of nuclear spins. This is a familiar object in nuclear magnetic resonance.

In the literature on NMR, nuclear spins are usually characterized by two relaxation times T_1 and T_2. The former corresponds to the longitudinal component, i.e. parallel to the external field (which is usually high in NMR). The longitudinal relaxation usually results from interactions with phonons. The relaxation time T_2 corresponds to the transverse component and is related to interactions between nuclear spins. However, interactions between nuclear spins can also flip the longitudinal components I_1^z and I_2^z of a pair of nuclear spins if they have the same Zeeman energy. This is spin diffusion. Its characteristic time is, in the simplest cases, of the order of T_2.

For molecular nanomagnets, at the low temperatures of interest in this chapter, the effect of phonons is very small (Prokofev and Stamp 1996; Morello *et al.* 2004). Consequently, the relaxation of the hyperfine interactions, characterized by the time τ_H, is attributed to spin diffusion.

However, this picture leads to difficulties. Each nuclear spin is mainly subject to the dipole (or contact) field $\mathbf{H}_{\mathrm{int}}$ created by the nearest electronic spins. Even if an external field is applied, it does not dominate $\mathbf{H}_{\mathrm{int}}$ except, perhaps, for a small proportion of nuclear spins for which $\mathbf{H}_{\mathrm{int}}$ turns out to vanish. For the sake of simplicity, the external field will be assumed to vanish. The interaction with molecular spins creates for each nuclear spin $\mathbf{I}$ a local field $\mathbf{H}_{\mathrm{int}}$ in some direction Z (which is not the same for all nuclear spins). The component I_Z can be flipped by phonons when phonons are available. The corresponding relaxation time may be called T_1, and it is extremely long at low temperature. The transverse components, with respect to Z, have a relaxation time T_2 which is not particularly long.

Can this time be identified with the relaxation time τ_H of the z-component of the hyperfine field seen by the nuclear spins? This identification is acceptable if spin diffusion is easy, i.e. if the longitudinal components I_1^Z and I_2^Z of a pair of nuclear spins can flip. This double flip is only possible if the local fields $\mathbf{H}_{\mathrm{int}}^{(1)}$ and $\mathbf{H}_{\mathrm{int}}^{(2)}$ on the two nuclei have the same modulus, and this is a severe restriction.

More precisely, in the absence of an external field, the interaction between a molecular spin $\mathbf{S}$ and a nuclear spin $\mathbf{I}_k$ is

$$\mathcal{H}_k = -g_k^{zZ} S_z I_k^Z - \sum_{\alpha=xy} \sum_{\gamma=XYZ} g_k^{\alpha\gamma} S_\alpha I_k^\gamma \qquad (9.19)$$

where g_k takes into account the dipole and contact interactions and the gyromagnetic ratios of the two spins This formula is easily extended to the case of a non-vanishing external field, see formula (I.7) of Appendix I. In (I.7) the dominant term is the first one, where S_z plays the part of the external field in NMR. If other terms are neglected, the eigenstates are those of I_k^Z, and their spacing sg_k^{zZ} is generally fairly large in comparison with interactions between nuclear spins. These are therefore not expected to produce transitions between the levels, and spin diffusion should be hindered.

The actual situation is probably complicated. In $\mathrm{Mn}_{12}\mathrm{ac}$, for instance, it might be necessary to clarify the respective role of Mn nuclei, which have a strong interaction with a single molecular spin $\mathbf{S}$, and the other nuclei which interact more weakly with several $\mathrm{Mn}_{12}\mathrm{ac}$ clusters.

An article of Morello *et al.* (2004) gives precise experimental information on the puzzle to solve in $\mathrm{Mn}_{12}\mathrm{ac}$, and some theoretical suggestions about the way to solve this puzzle. They confirm that 'extrapolating the observed high-T nuclear spin–lattice relaxation to the millikelvin range would lead to astronomically long relaxation times.' Nevertheless, they 'observe that, upon cooling down to 20 mK, the nuclear spin–lattice relaxation saturates to a roughly temperature-independent plateau'. Moreover, the nuclear spin temperature remains equal to the lattice temperature. Morello *et al.* suggest that the contact with the lattice is ensured by the interaction with particular molecules which can more easily flip, while nuclear spin dynamics results from nuclear 'spin diffusion'. Spin diffusion would be an effect of the interactions between nuclear spins, which, as seen above, raises certain questions.

A last remark about extremely low temperatures will close this section. At such temperatures, nuclear spins would order and T_2 would become infinite, but such temperatures cannot be reached experimentally. Ordered phases of nuclear spins can be obtained in certain materials as a transient phenomenon, by imposing a nuclear spin temperature which is temporarily much lower than the temperature of the other degrees of freedom, such as phonons or electrons.

9.3 Relaxation by dipole interactions between molecular spins

In this section, the relaxation of the magnetization at low temperature in the presence of dipole interactions between molecular spins will be considered. As before, 'low temperature' means that only the lowest level of each well can be populated. However, the stable equilibrium state is assumed to be paramagnetic.

The contribution of nuclear spins to the local field will be assumed to be small with respect to that of the dipole interactions between molecular spins. This assumption seems correct for Fe_8, as seen before.

The spins will be treated as classical Ising spins with a flipping probability per unit time which is 0 unless the local field H_{loc} satisfies $|H_{loc}| < \delta H_0$, where δH_0, in the absence of hyperfine interactions, is the natural half-width $2\hbar\omega_T$. The flipping probability per unit time for $|H_{loc}| < \delta H_0$ is assumed to have a constant value α. This scheme gives rise to a very peculiar, non-exponential relaxation discovered by Prokofev and Stamp (1998a,b).

In a first stage, the following assumptions will be made: (i) all spins are in the $-z$ direction at time $t = 0$; (ii) constant and uniform external field; (iii) ellipsoidal sample with an axis parallel to the easy direction z. Thus, the total local field H_{loc} is uniform at time $t = 0$.

If H_{loc} is small enough to satisfy $|H_{loc}| < \delta H_0$ the spins start flipping. Note that H_{loc} designates the total local field previously called H'_{loc} in Section 9.1.

The novel feature is that this reversal of the spins causes a modification of the dipole field, so that the local field no longer satisfies the flipping condition $|H_{loc}| < \delta H_0$ for all spins. However, the flipping condition can be satisfied for a few spins. The question is: how many?

The local field at time t is the sum of the initial field $H_{loc}(0)$ and the variation of the dipole field which results from the reversal of a certain number $\delta N(t)$ of spins in time t. Since $H_{loc}(0) \simeq 0$, the local field is twice the dipole field created by the reversed spins. This field has a distribution centred at $H_{loc} = 0$, with a width $\delta H_{loc}(t)$ whose order of magnitude is the typical order of magnitude of the dipole field. Therefore, it is proportional to the reciprocal of the cube of the typical distance $\ell(t)$ between reversed spins. Introducing the unit cell volume a^3, one obtains (assuming a cubic symmetry for simplicity)

$$\delta H_{\text{loc}}(t) = \text{Const} \times a^3/\ell^3(t). \tag{9.20}$$

That is very nice, because $a^3/\ell^3(t)$ is (apart from a geometrical factor of order unity) the proportion $\delta N(t)/N$ of reversed spins, which is equal, apart from a factor -2, to the relative variation $\delta M(t)/M(0)$ of the magnetization. Indeed, $\delta M(t)/M(0) = -2\delta N(t)/N = -a^3/\ell^3(t)$. Thus, (9.20) reads

$$\delta H_{\text{loc}}(t) = \text{Const} \times \delta M(t). \tag{9.21}$$

The local field distribution has a width $\delta H_{\text{loc}}(t)$ and a spin can flip if its local field is smaller than δH_0. This quantity is small, so that, after a small time, $\delta H_{\text{loc}}(t)$ becomes larger than δH_0 (Fig. 9.8). The proportion of spins which can flip is then $\delta H_0/\delta H_{\text{loc}}(t)$. For those spins, the flipping probability per unit time is α. The proportion of spins which flip in time dt is therefore $\alpha \delta H_0 dt/\delta H_{\text{loc}}(t)$. But it is also equal to half the relative magnetization variation $-(1/M)dM/dt$. For short times, M may be replaced by its initial value and $dM/dt = \text{Const} \times \delta H_0/\delta H_{\text{loc}}(t)$ or, according to (9.21),

$$\frac{dM}{dt} = \text{Const}/\delta M(t) \tag{9.22}$$

where the constant depends on α and δH_0. Multiplying both sides of (9.22) by $\delta M(t)$ and noticing that $dM/dt = d(\delta M)/dt$, one obtains $d(\delta M)^2/dt = \text{Const}$ and therefore (Prokofev and Stamp 1998a,b)

$$|\delta M(t)| = \sqrt{\Gamma_{\text{sqrt}}t} \tag{9.23}$$

where Γ_{sqrt} is a constant.

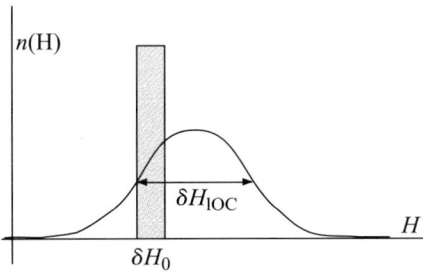

FIG. 9.8. Proportion of spins which feel a local field H. The width δH_{loc} of the distribution increases with time according to formula (9.20). On the other hand, a spin can flip by tunnelling only if its local field is in the grey window, whose width is given by (6.28).

Of course, $\delta M(t)$ cannot be infinite, and (9.23) applies only to short enough times. On the other hand, it does not apply to too short times, when (9.21) is smaller than δH_0. For those very short times, $\delta M(t)$ is proportional to t. For long times, the relaxation has either an exponential form (Tupitsyn and Barbara 2002), or that of a stretched exponential (Sangregorio *et al.* 1997). The latter can be reproduced by simulations (Cuccoli *et al.* 1999).

The hypotheses made in the above argument are very restrictive. As a first extension, let the sample have a non-ellipsoidal shape. Even if the initial magnetization is uniform, the local field is not. However, if the external field is in the appropriate range, there is a region ($\mathcal{R}$) where the local field vanishes and relaxation can begin. One can redo the preceding result, replacing the sample by the region ($\mathcal{R}$). For short times, the evolution still has the form (9.23). For longer times, simulations of Cuccoli *et al.* (1999) predict a relaxation which depends on the shape of the sample.

The representation of the system by Ising spins with a flipping probability is an extension of a time-dependent Ising model devised by Glauber (1963). The new feature with respect to Glauber's model is the crucial dependence of the flipping probability on the local field. The justification of this model is not easy. However, the theoretical result (9.23) is in nice agreement with experiment as will now be seen.

Formula (9.23) describes quite well the experimental results of Ohm *et al.* (1998) on Fe$_8$, displayed in (Fig. 9.9). It also fits the results of Wernsdorfer *et al.* (1999) who, however, used an experimental procedure which is rather different from that described in the beginning of this section. They quench the system in vanishing external field, so that the magnetization is 0, and then they suddenly apply a field along the easy z axis. At short times, the magnetization

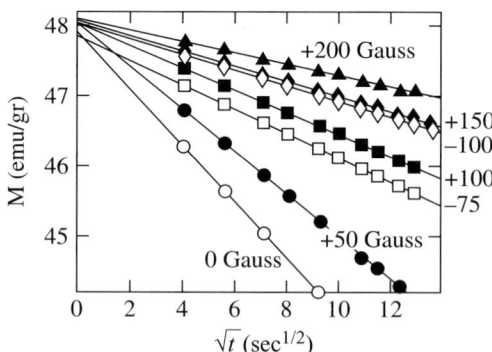

FIG. 9.9. Experimentally observed magnetic relaxation at short times and low temperature for various values of the final magnetic field in Fe$_8$ according to Ohm *et al.* (1998b). There is a good agreement with the $\sqrt{t}$ law of Prokofev and Stamp.

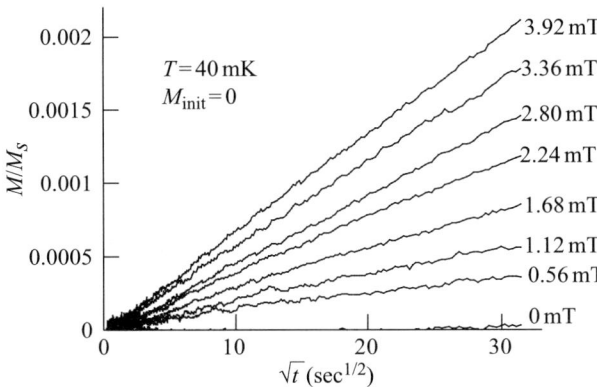

FIG. 9.10. Experimentally observed magnetic relaxation from zero magnetization at short times and low temperature in Fe$_8$ according to Wernsdorfer *et al.* (1999). Copyright (1999) by the American Physical Society.

$\delta M(t)$ (Fig. 9.10) is well described by the $\sqrt{t}$ law (9.23). In contrast with the assumptions made at the beginning of this section, the initial magnetization is not saturated. Is initial saturation essential in the argument of Prokofev and Stamp? How universal is the square-root behaviour? Prokofev and Stamp (1998a) already pointed out the possibility of an exponential relaxation, and gave a precise expression (their formula 13) in certain cases. Experimentally, the $\sqrt{t}$ behaviour is often observed (e.g. Thomas *et al.* 1999; Tupitsyn and Barbara 2002) in cases where a theoretical explanation is not easy to find. Using simulations, Fernández and Alonso (2003, 2004) obtain the $\sqrt{t}$ behaviour even with an unsaturated initial state, but find that it is neither general nor universal. Their results were criticized by Tupitsyn and Stamp (2004). Another contribution to the theoretical debate was that of Chudnovsky (2000), also criticized by Prokofev and Stamp (2000).

In Fe$_8$, the hyperfine field is but a weak part (10%) of the local field. Therefore the $\sqrt{t}$ behaviour is not expected to arise from hyperfine interactions except in diluted samples, when the hyperfine field dominates the dipole field.

As expected from (9.2), the measured relaxation rate depends on the field. In Fe$_8$, if one increases the field from the value 0, the relaxation becomes very slow for $H > 0.04$ tesla, which is twice the resonance half-width (Wernsdorfer *et al.* 1999).

Square-root relaxation was also observed in Mn$_{12}$ac by Bokacheva *et al.* (2000), who applied a transverse magnetic field to accelerate the dynamics.

As explained before in this section, the $\sqrt{t}$ law is not applicable to very short times. At these times, the only spins which can relax are those whose local dipole field is exactly compensated by the external field plus the hyperfine field. Therefore, as seen in Section 9.2, a hole should appear in the distribution $P(H_{dip})$ of the dipole fields, and the width of the hole is the typical hyperfine field.

9.4   Other theoretical approches

9.4.1   *The theory of Caldeira and Leggett*

Caldeira and Leggett (1983) elaborated the first theory of tunnelling with friction. Their calculation, which is detailed and rather long, remains a basic reference for the topic. However, the problem is different from that addressed in this book, and their results are different. They indeed considered a particle in a single potential well, e.g.

$$V(x) = Ax^2 - Bx^3$$

so that the tunnelling particle escapes and never comes back. Tunnelling takes place between localized states, which are quantized, and delocalized states, which form a continuum. There is no level-crossing condition to satisfy. There is no resonance. Instead, if friction is introduced, for instance through the interaction with phonons, the escape rate is always reduced. This result is easy to remember since friction decreases the velocity when there is no tunnelling. It is not surprising that the result is similar when tunnelling is active.

9.4.2   *Hyperfine interactions according to Prokofev and Stamp*

Prokofev and Stamp (1996) have investigated in great detail the 'quantum relaxation of magnetization in magnetic particles'. Their result in the case of a 'giant spin' interacting with a large number N of nuclear spins will be reproduced here. They assume 'the hyperfine couplings to be tightly clustered' (with a spread Γ_μ) 'around a principal value $\omega_0 \ll \Delta_0$', where $\Delta_0 = \hbar\omega_T$ in our notation. This assumption is not appropriate to a molecular spin interacting with nuclear spins, but it is worth giving a flavour of the results obtained by Prokofev and Stamp. The probability $P(t)$ that the giant spin is in state $|s^*\rangle$ at time t if it was in state $|s^*\rangle$ at time 0 is given by formula (4.44) of Prokofev and Stamp (1996)

$$P(t) - \frac{1}{2} \approx \sum_p C_N^{(N+p-p_H)/2} \frac{\exp[-\beta\omega_0(p - p_H)/2]}{2^N Z(\beta)}$$

$$\times \left\{ 1 - \exp\left[-\frac{p^2}{\lambda e^2} \left(\frac{\pi^2 \Gamma_\mu}{2\Delta_0^2 t} \right)^{1/p} \right] \right\} \tag{9.24}$$

where λ characterizes the relative modulation of the tunnel frequency by nuclear spins. The factor $Z(\beta)$ (with $\beta = 1/(k_B T)$) ensures that $P(0) = 1$. The summation is over the number p of nuclear spins which flip together with the giant spin during tunnelling. The quantity p_H is proportional to the external field.

When deriving (9.24), Prokofev and Stamp assume that nuclear spin dynamics, at low temperature, results from nuclear spin diffusion, i.e. the simultaneous flip of two nuclear spins, and is fast. The characteristic time for nuclear spin diffusion, called τ_H in Section 9.2, is identified with the transverse coherence time T_2. This identification is not always correct.

It is remarkable that the time T_2 does not appear in (9.24). According to (9.24) the parameters which rule the dynamics are Γ_μ and λ. At time t, the spins which have relaxed correspond to p values such that

$$\frac{p^2}{\lambda e^2} \left(\frac{\pi^2 \Gamma_\mu}{2\Delta_0^2 t} \right)^{1/p} < 1$$

or

$$\frac{p^2}{\lambda e^2} < \left(\frac{2\Delta_0^2 t}{\pi^2 \Gamma_\mu} \right)^{1/p},$$

or

$$\ln \frac{p^2}{\lambda e^2} < (1/p) \ln \frac{2\Delta_0^2 t}{\pi^2 \Gamma_\mu},$$

or finally

$$p < \ln \frac{2\Delta_0^2 t}{\pi^2 \Gamma_\mu} \bigg/ \ln \frac{p^2}{\lambda e^2} \qquad (9.25)$$

Thus, for short times, (9.24) can be approximated by a linear function of $\ln t$. This behaviour results from a summation over many values of p.

The theory of Prokofev and Stamp (1996) differs from the scheme outlined in Section 9.2.4 in (at least) two respects. First, it takes into account the combined flip of nuclear and electronic spins. Second, the cause of decoherence is the coupling between transverse components of the nuclear and electronic spins, rather than the variation of the longitudinal hyperfine field. Thus, the time T_2 or τ_H, which characterizes the velocity of this variation, does not appear in (9.24), in contrast with (9.17). The former difference may correspond to a difference between the physical problems. Combined flips are probably less important for molecular nanomagnets than for superparamagnetic grains investigated by Prokofev and Stamp (1996). The latter difference probably corresponds to different values of the hyperfine relaxation time.

9.4.3 Hyperfine interactions as a random walk

In the previous sections it has been assumed, in agreement with (9.18), that the relaxation of nuclear spins is relatively fast in comparison with tunnelling. This is not necessarily a good approximation at low temperature since the nuclear relaxation time becomes very long. In this paragraph, the opposite limit will be considered. It will be assumed that a spin which starts relaxing at $t = 0$ has totally relaxed to its equilibrium value at time t if its local field has taken the value 0 between times 0 and t. The argument can be extended to any field which allows tunnelling, but in the next argument the case value $m = s, m' = -s$ is considered. The external field will be assumed to vanish, and dipole interactions with other spins will be neglected.

The magnetization at time t thus depends on the probability $p(h, t)$ that the hyperfine field $H^z(t')$ has taken the value 0 for $0 < t' < t$ if the initial field was $H^z(0) = h$. The field H^z is the sum of contributions H_i^z of many nuclei. These nuclei will be assumed to flip by random, uncorrelated jumps. Thus, they are similar to the steps of a random walker, which may be forward or backward, and the evolution of the total field $H^z(t)$ is equivalent to a random walk. This is a classical problem. The quantity which is generally calculated is the probability $\rho(h, h', t)$ that the random walker is at h' at time t if he started from h at time 0. In the simplest case, the random walker has the same probability to go forward or backward, and $\rho(h, h', t)$ obeys the diffusion equation

$$\frac{\partial}{\partial t}\rho(h, h', t) = \alpha \frac{\partial^2}{\partial h'^2}\rho(h, h', t) \tag{9.26}$$

where α is related to the relaxation time of nuclear spins. The solution is

$$\rho(h, h', t) = \frac{1}{2\sqrt{\pi \alpha t}} \exp \frac{-(h - h')^2}{4\alpha t}. \tag{9.27}$$

This suggests that, during the time t, the hyperfine field has fluctuated by an amount of the order of $\sqrt{\alpha t}$. Therefore, it has taken the value 0 if $|h|$ is smaller than $\sqrt{\alpha t}$, and the proportion of spins which has relaxed is proportional to $\sqrt{\alpha t}$. Thus the magnetization $M_z(t)$ varies at short times as

$$M_z(t) = M_z(0) + [M_z(\infty) - M_z(0)]A\sqrt{t} \tag{9.28}$$

where A is a constant. The result $\delta M_z(t) \sim \sqrt{t}$ is identical to that obtained in Section 9.3 but the physical mechanism is very different.

Miyashita and Saito (2001) derived for the first time the $\sqrt{t}$ behaviour from a random walk mechanism. However, they did not notice that the argument is only correct if nuclear spin relaxation is very slow. They assumed the initial local field $H_z(0)$ to be tuned at a level crossing, exactly at resonance. Then, it goes away from resonance and comes back and forth, so that the resonance value is crossed $n(t)$ times, on the average, in time t. At each crossing event, the spin has a weak probability to flip, so that the magnetization change $\delta M_z(t)$ at time t is proportional to $n(t)$, which is proportional to $\sqrt{t}$. However, in practice, the initial local field $H_z(0)$ does not correspond to resonance, and the probability $p(t)$ that the resonance condition has been satisfied between times 0 and t is proportional to $\sqrt{t}$. In a realistic description, $\delta M_z(t)$ is proportional to the product $n(t)p(t)$ which is proportional to t for short time. The assumption that the spin has a weak probability to flip at each resonance event is correct if nuclear spin relaxation is sufficiently fast. In the opposite case, the electronic, molecular spin follows adiabatically the hyperfine field and $\delta M_z(t)$ is proportional to $n(t)$, and therefore to $\sqrt{t}$.

The above argument is oversimplified. As a matter of fact, the number of nuclei which contribute appreciably to the hyperfine field is finite. Therefore the hyperfine field $|H^z|$ cannot increase beyond a bound δH. Alternatively, the random walker is kept near the origin by a force $f(h')$ which, for short time, may be replaced by $f(h)$ and imposes a drift velocity proportional to $f(h)$. Therefore, a term proportional to $tf(h)$ has to be added to the right-hand side of (9.28).

However for small t, this term is negligible with respect to $\sqrt{t}$ and (9.28) is still valid.

9.4.4 *Tunnelling as an effect of hyperfine interactions*

In this chapter, spin tunnelling has always been assumed to be mostly an effect of single ion anisotropy, and hyperfine interactions can only modulate this effect. In principle, hyperfine interactions alone can produce tunnelling. This possibility has been investigated by Garanin *et al.* (2000).

10

TUNNELLING BETWEEN EXCITED STATES

10.1 The three relaxation regimes

In the preceding chapters two relaxation regimes have been encountered (Fig. 10.1).

- A 'high'-temperature regime (T larger than 7 or 8 K in Mn$_{12}$ac) in which the spin climbs to the top of the barrier by absorbing phonons (Fig. 5.2) then goes down emitting phonons. This regime was addressed in Chapter 5 and corresponds to the dashed arrows of Fig. 10.1.

- A low-temperature regime (T lower than 2 K in Mn$_{12}$ac) in which the spin crosses the barrier without exchanging energy with the external world and without going through an excited state. This regime was addressed in Chapters 6 and 9. It corresponds to the dotted arrow of Fig. 10.1.

It was experimentally demonstrated by Friedman *et al.* (1996) that in Mn$_{12}$ac, at intermediate temperatures, another regime is possible. As displayed by the dot-dashed arrows of Fig. 10.1, the spin absorbs energy from phonons, however not enough to reach the top of the barrier, but enough to reach an approximate eigenstate $|m^*\rangle$ (see Sections 5.3 and 6.1.7) where tunnelling into a state $|m'^*\rangle$ is possible in a reasonably short time $1/\omega_{\mathrm{T}}^{(mm')}$. Indeed, even if $\omega_{\mathrm{T}}^{s,-s}$ is very small as it is in Mn$_{12}$ac, $\omega_{\mathrm{T}}^{(mm')}$ can be much larger for lower values of $|m|$ and $|m'|$. This can be seen by applying the perturbative treatment of Section 6.4. For instance, if tunnelling is the result of a transverse field H_x, $\omega_{\mathrm{T}}^{(mm')}$ is obtained at order $|m'-m|$, through a formula analogous to (6.42) where the exponent is now $|m'-m|$ instead of $2s$. Thus, for $H_z = 0$,

$$\hbar\omega_{\mathrm{T}}^{(mm')} = K(s, m, m')|D| \left(\frac{g\mu_{\mathrm{B}}H_x}{2|D|}\right)^{|m'-m|} \tag{10.1}$$

where the calculation of the coefficient $K(s, m, m')$ is left as an exercise to the reader. Thus, in the limit $g\mu_{\mathrm{B}}H_x \ll |D|$, the tunnel frequency $\omega_{\mathrm{T}}^{(mm')}$ decreases exponentially when $|m'-m|$ increases. In Mn$_{12}$ac, it is negligible for $-m = m' = s = 10$, but may be appreciable for $-m = m' = 4$.

10.2 Tunnelling at resonance

As seen in Chapters 6 and 8, tunnelling between $|m^*\rangle$ and $|m'^*\rangle$ is only possible near a particular value $H_0^{mm'}$ of the magnetic field H_z in the easy direction.

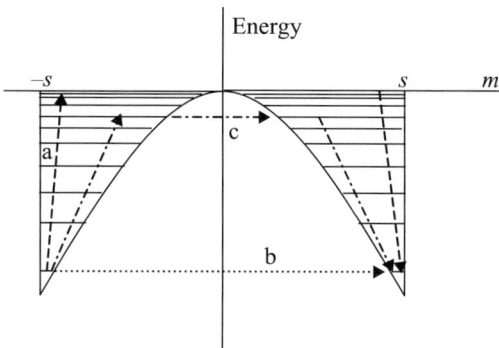

FIG. 10.1. The three possible relaxation processes in vanishing field. (a) The spin, initially in the left-hand well, absorbs phonon energy and goes to the highest level which is delocalized, so that the spin can deexcite into the right-hand well (dashed arrows). (b) The spin tunnels without change of energy (dotted arrows). (c) The spin absorbs phonon energy, goes to an excited level and then tunnels into the right-hand well (dot-dashed arrows).

This value is determined by the equality $E_m^{(0)} = E_{m'}^{(0)}$, which is the level crossing condition (6.3). More precisely, in agreement with (6.28), tunnelling is negligible if $g\mu_B|(m'-m)(H_0^{mm'} - H_z)|$ is much larger than $\hbar\omega_T^{(mm')}$. In this expression, H_z designates the local field, which has to include the interaction with other spins (electronic and nuclear). When the external field varies, the domains where tunnelling is important ('resonances') may be broader than $\hbar\omega_T^{(mm')}$ but are still pretty narrow as can be seen from Chapter 9.

Since tunnelling is only possible near level crossings, the relaxation time has a succession of minima when H_z is varied. The observation of these minima (or, equivalently, of steps in the hysteresis curve) by Friedman *et al.* (1996) and Thomas *et al.* (1996) is the experimental proof of the intermediate regime (Fig. 6.7).

If, instead of changing the field at constant temperature, one changes the temperature at constant field, one expects the relaxation time to follow the Arrhenius formula (5.2). However, in contrast with Chapter 5, the activation energy $k_B T_0$ is now the energy difference between the ground state and the state where tunnelling is possible (dash-dotted arrow in Fig. 10.1). Therefore, in vanishing field, $k_B T_0$ can be appreciably lower than its value $|D|s^2$ predicted by (5.5). More generally, it can be appreciably lower than (5.10) for all values of H_z which correspond to level crossings. In conclusion, $k_B T_0(H_z)$ has to show minima at level crossings. Experimentally, this has actually been observed by Novak and Sessoli (1995).

The next section is devoted to magnetic tunnelling in a constant magnetic field, when one of the localized states is an excited one. Since the other localized state may be the lowest state in its well, the theory also applies to resonances in non-vanishing field at very low temperature.

10.3 Tunnel probability into an excited state

To take tunnelling into account, an extension of the master equation (5.12) is necessary. The additional ingredient to be introduced into the master equation is intuitive. One needs a transition probability per unit time $\Gamma_m^{m'}$ when the levels m and m' cross (see formulae 6.3 and 6.4). These new coefficients are not given by the golden rule and do not satisfy the selection rule $\delta m = \pm 1$ or ± 2, which results from second-order perturbation theory. Tunnelling requires a high order in perturbation theory. In the present section the new transition probability $\Gamma_m^{m'}$ will be calculated without using high-order perturbation theory, but just the following ingredients:

(1) the lifetimes τ_m and $\tau_{m'}$ calculated without tunnelling;

(2) the tunnel frequency $\omega_{\mathrm{T}}^{(mm')}$ calculated for an isolated spin, i.e. for $\tau_m = \tau_{m'} = \infty$. As seen in Section 6.4, $\omega_{\mathrm{T}}^{(mm')}$ can be calculated using high-order perturbation theory.

General, approximate formulae for $\Gamma_m^{m'}$ have been derived by Garanin and Chudnovsky (1997), Fort *et al.* (1997), and Villain *et al.* (1997). Before writing them, it is of interest to examine a few particular cases which shed some light on this problem. In all these cases, the unperturbed energies $E_m^{(0)}$ and $E_{m'}^{(0)}$ of two particular levels are assumed to be exactly equal to each other. The spin is initially assumed to be in a state $\mid m^* \rangle$ localized in the left-hand well. It can deexcite into this well with probability $(1-\alpha)$ or into the other well with probability α. We wish to calculate the total reversal probability α and the reversal probability per unit time Γ as functions of the lifetimes τ_m and $\tau_{m'}$ and of the tunnel frequency $\omega_{\mathrm{T}}^{(mm')}$ (sometimes abbreviated as ω_{T} in the following) of the ideal, isolated spin.

First example

$$\omega_{\mathrm{T}}^{(mm')}\tau_m < \omega_{\mathrm{T}}^{(mm')}\tau_{m'} < 1.$$

The reader will perhaps be surprised that the tunnel period may be longer than the lifetime. But it should be kept in mind that ω_{T} is the tunnel frequency that the spin would have if it were isolated. The case studied in this example is that of 'overdamped' tunnel oscillations, which are not oscillations at all. Similarly, elastic waves (phonons) are sometimes overdamped, and it is of interest to attribute them a frequency although they are not really waves.

For $t < \tau_m$, equations (6.50) can be used to get a rough approximation of

$$\dot{y}(t) = -i\omega_{\mathrm{T}}x(t) \simeq -i\omega_{\mathrm{T}}x(0) = -i\omega_{\mathrm{T}} \tag{10.2}$$

so that $|y(t)| \simeq \omega_{\mathrm{T}}t$. This formula, as well as (10.2), only holds approximately for $t < \tau_m$ at resonance. At $t \approx \tau_m$ the reversal probability is $\alpha \approx |y(\tau_m)|^2 \simeq \omega_{\mathrm{T}}^2\tau_m^2$.

This is the *total* reversal probability. The reversal probability per unit time at resonance is $\Gamma \approx \alpha/\tau_m \approx \omega_T^2 \tau_m$.

Second example

$$\omega_T^{(mm')} \tau_{m'} < \omega_T^{(mm')} \tau_m < 1.$$

Again, equations (6.50) can be used, but only for t shorter than the smaller lifetime, i.e. $t < \tau_{m'}$. On the other hand, the spin is still able to flip for $\tau_{m'} < t < \tau_m$. In order to represent the evolution for such times, one can add a phenomenological damping term into (10.2):

$$\dot{y}(t) \simeq -i\omega_T - y(t)/\tau_{m'}. \tag{10.3}$$

After some time, y saturates at the value $y = -i\omega_T\tau_{m'}$. The first equation (6.50) reads $\dot{x}(t) = \frac{1}{i\hbar}(E_0 + \Delta)x(t) - i\omega_T^2 y$. The first term vanishes at resonance if the origin of the energies is suitably chosen, and one obtains $\dot{x}(t) = -\omega_T^2 \tau_{m'}$ and $x(t) = 1 - \omega_T^2 \tau_{m'} t$. The probability to be in state $|m\rangle$ at time t is $|x(t)|^2 = 1 - 2\omega_T^2 \tau_{m'} t$. Since the possibility of deexcitation in the left-hand well is ignored in the first equation (6.50), this corresponds to a spin reversal probability per unit time $\Gamma \approx 2\omega_T^2 \tau_{m'}$, at resonance. The total reversal probability is $\alpha \approx \Gamma\tau_m \approx 2\omega_T^2 \tau_m \tau_{m'}$.

Third example

$\omega_T^{(mm')} \tau_m$ and $\omega_T^{(mm')} \tau_{m'}$ are much greater than 1.

The spin, which is initially in state $|m\rangle$, has enough time to perform several oscillations from one side to the other. It spends the same time on both sides but the deexcitation probability per unit time is $1/\tau_m$ when it is on the left-hand side and $1/\tau_{m'}$ when it is on the right-hand side. Therefore, at resonance,

$$\alpha = \frac{1/\tau_{m'}}{1/\tau_m + 1/\tau_{m'}} = \frac{\tau_m}{\tau_m + \tau_{m'}} \tag{10.4}$$

which is independent of $\omega_T^{(mm')}$. It is more difficult to define a reversal probability per unit time Γ. However, if the purpose is to calculate the relaxation time τ, this does not matter because τ depends only on $\tau_{m'}$, not on $\omega_T^{(mm')}$ and Γ. Indeed, before it deexcites, the spin spends half of its time in each well, regardless of the value of Γ. Thus any sufficiently large value of Γ (more precisely, $\Gamma \gg 1/\tau_{m'} \gg 1$) is acceptable.

The general formula for the reversal probability per unit time proposed by Garanin and Chudnovsky (1997), and Villain *et al.* (1997), Leuenberger and Loss (1999) is

$$\Gamma_m^{m'} = \frac{2\omega_T^2 \tau_{mm'}}{1 + \tau_{mm'}^2 (E_m - E_{m'})^2/\hbar^2} \frac{2}{1 + \exp[\beta(E_{m'} - E_m)]} \tag{10.5}$$

where

$$\frac{1}{\tau_{mm'}} = \frac{1}{\tau_m} + \frac{1}{\tau_{m'}}. \tag{10.6}$$

The second factor in (10.5) accounts for detailed balance. As will now be shown it is not important in a relaxation process, so that (10.5) takes the simplified form

$$\Gamma_m^{m'} = \frac{2\omega_T^2 \tau_{mm'}}{1 + \tau_{mm'}^2 (E_m - E_{m'})^2/\hbar^2}. \tag{10.7}$$

Indeed, if $(E_m - E_{m'})$ is small (smaller than $k_B T$), (10.5) and (10.7) coincide to a good approximation. If $(E_m - E_{m'})$ is not small (larger than $\hbar/\tau_{mm'}$), both expressions are small and $\Gamma_m^{m'} \approx 0$. Problems could only arise in the case $k_B T < |E_m - E_{m'}| < \hbar/\tau_{mm'}$, a condition which would require an unphysically short lifetime $\tau_{mm'} < \beta\hbar$, i.e. $\tau_{mm'} < 10^{-11}$ s if $T > 1$ K.

Formula (10.5) has been criticized because, for large values of ω_T, the resulting value of $\Gamma_m^{m'}$ can be larger than ω_T. This is clearly absurd, but leads to a correct relaxation time as explained above (for $E_m = E_{m'}$) when discussing the third example. A way to avoid this difficulty (Luis et al. 1998) is to work with the true eigenvectors of the Hamiltonian of the isolated spin, which are delocalized at resonance. Thus, one has to calculate the effect of phonons on these delocalized states. This approach is the best one for strongly excited states, when tunnelling is not a weak effect with respect to damping by phonons.

The general result for the total reversal probability α from the state $|m^*\rangle$ can be deduced from (10.7) and is (Villain et al. 1997)

$$\alpha = \frac{\tau_{mm'}}{2\tau_{m'}} \left\{ 1 + \frac{1}{2\omega_T^2 \tau_m \tau_{m'}} [1 + \tau_{mm'}^2 (E_m - E_{m'})^2/\hbar^2] \right\}^{-1}. \tag{10.8}$$

The reader will easily check that the three particular cases considered at the beginning of this section are in agreement with the general formulae (10.5) and (10.8). These three cases correspond to $E_m = E_{m'}$, so that (10.8) reads $\alpha = \omega_T^2 \tau_m \tau_{mm'}$ in examples (1) and (2), and $\alpha = \tau_{mm'}/2\tau_{m'}$ in example (3).

The approximation used in formula (10.3), where a damping of the wavefunction was introduced, is not recommended. It does not allow us, for instance, to describe the possibility for the spin to deexcite from state $|m'^*\rangle$ and then to reexcite to that state. The correct formulation of the problem involves writing a master equation for the density matrix. This method will be seen in Section 11.2.

The use of the density matrix makes the analysis more complicated. If the problem of interest is the long-time behaviour, it is simpler to use the ordinary master equation of Section 5.3, just introducing new transition probabilities which allow tunnelling. They are given by (10.5) and therefore they are only important near a level crossing (formula 6.4). Their effect is that the temperature T_0 which appears in the Arrhenius relation (5.2) is lower than expression (5.5) near a level crossing.

10.4   A remark about the dynamic susceptibility

While it can be neglected in the relaxation process, the second factor of (10.5) is important, for instance, in the calculation of the dynamic susceptibility (Fernández, unpublished). To show that, it is sufficient to consider two states $|m^*\rangle$ or state $|m'^*\rangle$ which are decoupled from the other ones (as occurs for $m = -s$ and $m' = s$ at very low temperature). Their probabilities $p_m(t) = p(t)$ and $p_{m'}(t) = 1 - p(t)$ satisfy the equations $\dot{p}_m(t) = \Gamma^m_{m'} p_{m'}(t) - \Gamma^{m'}_m p_m(t)$ and $\dot{p}_{m'}(t) = \Gamma^{m'}_m p_m(t) - \Gamma^m_{m'} p_{m'}(t)$ which reduce to

$$\dot{p}(t) = \Gamma^m_{m'} - [\Gamma^{m'}_m + \Gamma^m_{m'}]p(t). \tag{10.9}$$

If the spin fluctuates around equilibrium under the influence of a field $H_z(t) = \bar{H} + \delta h(t)$, where $\delta h(t) = \delta h_0 \cos \omega t$, one can write $\Gamma^m_{m'} = \bar{\Gamma}^m_{m'} + (\partial/\partial h)\Gamma^m_{m'}\delta h(t)$ and a similar formula for $\Gamma^{m'}_m$. Then $p(t) = \bar{p} + \delta p(t)$ where $\bar{p}$ can be obtained by writing that the terms of order 0 in (10.9) vanish. The terms of order 1 yield

$$\dot{p}(t) = (\partial/\partial h)\Gamma^m_{m'}\delta h(t) - [\bar{\Gamma}^{m'}_m + \bar{\Gamma}^m_{m'}]\delta p(t) - \bar{p}[\partial/\partial h)\Gamma^{m'}_m + \partial/\partial h)\Gamma^m_{m'}]\delta h(t). \tag{10.10}$$

The derivatives $(\partial/\partial h)\Gamma^m_{m'}$ and $\partial/\partial h)\Gamma^{m'}_m$ are easily deduced from (10.5). In particular, at resonance, the last term of the right-hand side of (10.10) vanishes, and the second factor in (10.5) is essential. Formula (10.7) would yield no term linear in (10.10), and a vanishing susceptibility.

10.5   Two different types of relaxation

It is of interest to note the difference between the novel type of relaxation observed in vanishing field and very low temperature, and the classical relaxation which takes place in other cases as high or moderate temperatures and resonances in non-vanishing field.

We have seen that an isolated spin would be able to oscillate periodically between two localized states. When interactions with its environment are taken into account, these oscillations are destroyed, and this is decoherence. When spin tunnelling occurs between the lowest two states of the two potential wells, decoherence gives rise to unusual phenomena, e.g. the $\sqrt{t}$ behaviour of formula (9.23), discovered by Prokofev and Stamp (1998a,b).

If at least one of the tunnelling states is an excited state, the situation is different, and simpler. In that case, decoherence results from deexcitation, which mainly results from the emission of phonons. The theory (given in Section 5.7) is well established and not very complicated. Moreover, as far as tunnelling is concerned, a single result of the theory is required, namely the lifetime τ_m of the excited state or states. It is related to the coefficients γ^q_m of Section 5.7 by $1/\tau_m = \sum_q \gamma^q_m$. This lifetime and the coefficients γ^q_m do not take tunnelling into account.

10.6 Effect of a time-dependent field at $T \neq 0$

Up to now, we used the master equation (5.12) for a time-independent Hamiltonian, i.e. a constant field. This is correct, as seen in Section 5.3, for times larger than a microscopic time τ_{col}, necessary for the establishment of local equilibrium. Equation (5.12) can also be applied in a variable field provided the variation is slow enough. The solution of the master equation is not very simple except at very low temperature, when only the lowest states in both wells are populated. In this case, tunnelling occurs when the initially populated state (which will be assumed to be $|-s^*\rangle$) crosses a level $|(s-k)^*\rangle$ of the opposite well, with $k > 0$. If $k = 0$, coherence is essential, (5.12) cannot be applied, and the Landau–Zener–Stückelberg theory of Chapter 8 should be used. For $k \neq 0$, at low temperature, in the vicinity of the level crossing, (5.12) implies the very simple equation (Leuenberger and Loss 1999; Chudnovsky and Garanin 2001)

$$\frac{d}{dt}p_{-s}(t) = -\Gamma^{s-k}_{-s}p_{-s}(t) \tag{10.11}$$

or, using (10.5), where the second factor is neglected:

$$\frac{d}{dt}\ln p_{-s}(t) = -\frac{2\omega_T^2 \tau_k}{1 + \tau_k^2(E_{-s} - E_{s-k})^2/\hbar^2}. \tag{10.12}$$

Assuming $(E_{-s} - E_{s-k}) = v_k t$ as in Chapter 8 one obtains the ratio of the initial and final probabilities $p_{-s}(t_i)$ and $p_{-s}(t_f)$:

$$\ln\frac{p_{-s}(t_f)}{p_{-s}(t_i)} = \int_{-\infty}^{\infty}\frac{-2\omega_T^2 \tau_k dt}{1 + \tau_k^2(E_{-s} - E_{s-k})^2/\hbar^2} = \frac{-2\hbar\omega_T^2}{v_k}\int_{-\infty}^{\infty}\frac{du}{1+u^2}. \tag{10.13}$$

The extension of the integration from $-\infty$ to ∞ is correct if the (natural) width of the resonance is much smaller than the distance between resonances, as in the Landau–Zener–Stückelberg theory. Replacing the integral by its value, (10.13) reads

$$\frac{p_{-s}(t_f)}{p_{-s}(t_i)} = \exp\frac{-2\pi\hbar\omega_T^2}{v_k}. \tag{10.14}$$

This formula coincides with (8.17) although the derivation and the physical context are quite different. Formula (10.14) holds if the relaxation time τ_k of the state $|s-k\rangle$ to the state $|s\rangle$ is very short. If this is not so, the problem is still easy to solve, and left as an exercise to the reader.

The experimental curve of Fig. 8.1 indeed shows very steep parts for field values which correspond to level crossings.

10.7 Role of excited spin states

Until now, in Chapters 5–9, the exchange interaction inside each molecular group has been assumed so strong that, for a given molecule only the states with lowest exchange energy can be reached and the magnetic properties of the molecule

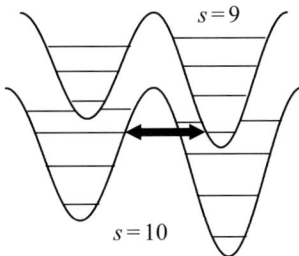

FIG. 10.2. A possible additional tunnelling pathway involving an excited spin state.

are those of a large spin – a spin $s = 10$ in the case of $Mn_{12}ac$ and Fe_8. At very low temperature (below 3 or 4 K) this is certainly a correct approximation. At high temperatures, when tunnelling plays no part, this is certainly not true, but the effect of the excited states with higher exchange energies is just a quantitative correction, and generates no characteristic property. At intermediate temperature, it may be different, if tunnelling can occur between a 'high'-energy state of the exchange multiplet $s = 10$ and a 'low'-energy state of the first excited exchange multiplet $s = 9$, if both have approximately the same energy, as schematized in Fig. 10.2. Low-symmetry components of the magnetic and exchange anisotropy can promote tunnelling, as $(\mathbf{S})^2$ is not a good quantum number.

This mechanism might explain the maxima observed in specific heat (Gaudin *et al.* 2002) and ac susceptibility of Fe_8 (Anfuso *et al.* 2004), at fields where no level crossing is expected, but clear evidence of the presence of a low-lying $s = 9$ state has not yet been provided.

10.8   Magnetic specific heat in the presence of spin tunnelling

In Section 3.2.3 the measurement of the specific heat in an alternating current was described. This method is especially adequate for molecular magnets in view of the small size of the available crystals. Moreover, the specific heat of these materials has the interesting property of depending on the frequency ω and thus, to provide information on the magnetic relaxation time τ by a method which may be cheaper and easier than others.

If $\omega\tau \ll 1$ the specific heat $C(\omega)$ measured by the method of Section 3.2.3 is the equilibrium specific heat C_{eq}. Each spin has time to explore all states of both potential wells (see Fig. 5.2). On the other hand, if $\omega\tau \gg 1$, during a period, the spin has only time to explore a single well. The specific heat $C(\omega)$ is that of a particle trapped in a single well C_{uni}, which can be called the 'unilateral' specific heat. In vanishing field, $C_{uni} = C_{eq}$. In a magnetic field, however, both specific heats have different values as seen in Appendix J, and $C_{uni} < C_{eq}$. In that appendix it is also shown that (Fernández *et al.* 1998, Fominaya *et al.* 1999)

$$|C(\omega)| = \sqrt{C_{\text{uni}}^2 + \frac{C_{\text{eq}}^2 - C_{\text{uni}}^2}{1 + \omega^2 \tau^2}}. \tag{10.15}$$

This formula interpolates between the obvious limits $C(0) = C_{\text{eq}}$ and $C(\omega) = C_{\text{uni}}$ for $\omega\tau \gg 1$. The frequency is always assumed low with respect to the reciprocal equilibration time within each well.

From the measurement of $|C(\omega)|$ it is possible to deduce the magnetic relaxation time τ by relation (10.15) which can more explicitly be written as

$$\omega^2 \tau^2 = \frac{C_{\text{eq}}^2 - C_{\text{uni}}^2}{C^2(\omega) - C_{\text{uni}}^2} - 1. \tag{10.16}$$

Thus, the sharp minima of τ at the field values which correspond to level crossings, $E_m(H_z) = E_{m'}(H_z)$, can give rise to sharp maxima of $|C(\omega)|$ for appropriate values of the frequency ω and of the temperature. This occurs if $\omega\tau$ is higher than 1 far from resonances, and smaller than 1 near resonances. The maxima of the specific heat have been observed in Mn_{12}ac by Fominaya *et al.* (1997b, 1999).

At the frequency of about 25 s^{-1} used by Fominaya *et al.*, $C(\omega)$ coincides with C_{uni} below 4 K, and coincides with C_{eq} above 5.5 K, so that there are no sharp peaks and no frequency dependence. Only between these temperatures can the relaxation time be evaluated from specific heat data, because it is of the order of the period of the heating modulation.

Typical results (after subtraction of the phonon specific heat, which should be suitably evaluated) are displayed at various temperatures in Fig. 10.3. The resonances (sharp peaks) are as impressive as in magnetic measurements, but the

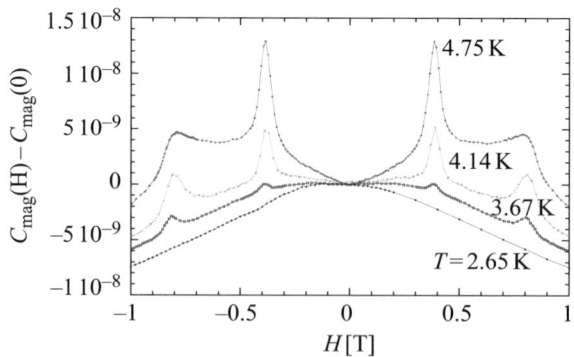

FIG. 10.3. Experimental specific heat measured on a single crystal of Mn_{12}ac at various temperatures and at fixed frequency (25.3 s^{-1}) with the magnetic field applied along the easy axis. From Fominaya *et al.* (1999).

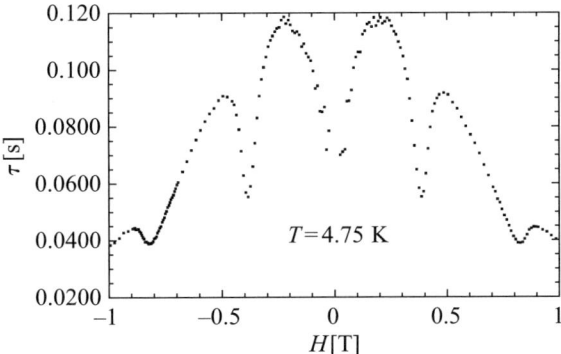

FIG. 10.4. Relaxation time τ of $Mn_{12}ac$ as a function of the longitudinal magnetic field. From Fominaya *et al.* (1999).

applicability of the method is limited because of the experimental limitations on the frequency.

The peaks of the specific heat can be related to maxima of the relaxation rate. $1/\tau$, which can then be deduced from (10.15). The result is displayed by Fig. 10.4.

10.9 Tunnelling out of resonance

The sharp peaks displayed by so many figures in this book, e.g. Fig. 6.7, are spectacular. However, between the peaks, relaxation takes place although it is much slower. Indeed, if a spin is in the down position in a field H_z which favours the up position, this spin can tunnel to the up position with emission of a phonon which ensures energy conservation (Hartmann-Boutron *et al.* 1996). The probability of this process per unit time is given by the golden rule.

Actually, this golden rule is just the same formula (5.32) as in Chapter 5. It should just be kept in mind that the vector $|\, m, n_q\rangle$ which appears in this formula should be built from the eigenvectors $|\, m^*\rangle$ of the spin Hamiltonian, not from the eigenvectors $|\, m\rangle$ of S^z. The vectors $|\, m^*\rangle$ are only approximately localized on the right or on the left, and the matrix element

$$\langle m, n_{q\rho}\,|\,\mathcal{H}_{\text{s-ph}}\,|\,m', n_q \pm 1\rangle$$

which appears in (5.32) does not strictly vanish even if $|\, m^*\rangle$ and $|\, m'^*\rangle$ are (approximately) localized on different sides. It is expected to be proportional to the tunnel frequency, times a quantity which depends on the spin–phonon interaction. Since the *square* of this matrix element appears in the transition probability, tunnelling out of resonance is expected to be much weaker than resonant tunnelling.

Information about the numerical calculation of the relaxation rate are available in Appendix F.2.

11

COHERENCE AND DECOHERENCE

11.1 The mystery of Schrödinger's cat

The study of coherence destruction ('decoherence') sheds some light on uncertainty, which is the most mysterious feature of quantum mechanics. A recent review is that of Zurek (2003).

While Heisenberg's relations give a mathematical form to uncertainty, Schrödinger formulated it as the striking paradox of a cat whose wavefunction can be a linear combination of the living state and the dead state. Obviously, this idea contradicts our daily experience that a cat can be alive or it can be dead, not both. In the early developments of quantum mechanics, this contradiction was explained by the interaction of the system (cat or whatever) with the laboratory instrument during the measurement process.

More recently an alternative explanation was proposed (Zurek 1991; Brune *et al.* 1996). If the system has a well-defined Hamiltonian with two possible eigenstates $|1\rangle$ and $|2\rangle$ (e.g. $|\,\text{alive}\rangle$ and $|\,\text{dead}\rangle$) then it can be in an 'entangled' state such as $[|1\rangle + |2\rangle]/\sqrt{2}$. However, there is generally an interaction with the external world, which destroys entanglement. This is decoherence. Decoherence is a very general phenomenon, which goes far beyond molecular magnetism. If it can be avoided, this will open the way to quantum computers. In these computers, the information will not be contained in ordinary *bits* as a binary variable ($+$ or $-$) but in quantum bits (or *q-bits*) as a continuous phase of a wavefunction ($+$ or $-$). Because of uncertainty relations, a q-bit does not carry more information than a bit, but quantum computers should provide a convenient way to do parallel computing.

As far as we understand, the new interpretation of quantum uncertainty does not replace the old one based on the measurement process. In some cases the old explanation may be right, in other cases the new one applies. In particular, it generally applies to a macroscopic system. Rather than a cat, which is a complicated object, one can imagine a billiard ball which can be in either one of two bowls and is initially prepared in an entangled state which is a superposition of the two possible states.

11.2 The density matrix

An isolated quantum system is described by a wavefunction. If it interacts with the external world, this is no longer possible. The spin considered in Chapters 5 and 6 is a good example. In Chapter 6 it was described by a wavefunction.

In Chapter 5 it was described by a probability. One can imagine that at a particular time it has a well-defined wavefunction, but after some time, the interaction with the external world will make this description impossible. Even at very low temperature, we are almost sure that the spin is in its ground state, but the phase of the wavefunction is no longer known after some time. This is decoherence.

How can decoherence be described? How can a quantum system be described at intermediate times, when the wavefunction has already partly, but not completely lost the memory of its phase? An appropriate tool is the density matrix (Zurek 1991).

The density matrix $\rho_{nn'}(t)$ is defined in textbooks of statistical mechanics by the property that the average value of a real observable quantity a at time t is $\langle A(t) \rangle = \mathrm{Tr}(\rho(t)A)$, where A is the Hermitian operator associated to a. For any set of orthonormal basis vectors $|n\rangle$ this can be written as

$$\langle A(t) \rangle = \sum_{nn'} \rho_{nn'}(t) \langle n' \,|\, A \,|\, n \rangle \tag{11.1}$$

where A is the operator associated to a.

The density matrix has to be Hermitian. This can be seen by choosing an operator A which has only two non-vanishing elements $\langle n' \,|\, A \,|\, n \rangle = \langle n \,|\, A \,|\, n' \rangle^* = e^{i\varphi}$. Then $\langle A(t) \rangle = \rho_{nn'}(t)e^{i\varphi} + \rho_{n'n}(t)e^{-i\varphi}$. This should be real for any value of the real quantity φ, therefore $\rho_{nn'}(t) = \rho_{n'n}^*(t)$.

For an isolated system subject to a Hamiltonian $\mathcal{H}(t)$, the density matrix is

$$\rho_{nn'}(t) = \int_0^t dt' \exp(-it'\mathcal{H}(t')/\hbar) \,|\, n' \rangle \langle n \,|\, \int_0^t dt'' \exp(it''\mathcal{H}(t'')/\hbar)$$

$$= \exp[it(E_{n'} - E_n)/\hbar] \,|\, n' \rangle \langle n \,| \tag{11.2}$$

and satisfies the equation

$$i\hbar \frac{d}{dt}\rho(t) = [\mathcal{H}, \rho] \tag{11.3}$$

which is equivalent to the Schrödinger equation. This equation can for instance represent the evolution of an isolated spin subject to an anisotropy Hamiltonian $\mathcal{H} = -|D|S_z^2$.

Equation (11.3) includes neither decoherence nor energy dissipation. Those two effects result from the interaction with the external world. The interaction we have in mind in this chapter is the spin–phonon Hamiltonian as in Chapter 5. Interactions with other spins will be ignored for the sake of simplicity.

If a spin (or another system) is initially prepared in some state (for instance $|-s\rangle$) and is then left in contact with a thermal bath, it evolves to equilibrium. In this equilibrium state, if the basis vectors $|n\rangle$ are those which diagonalize the Hamiltonian, the density matrix is diagonal, $\rho_{nn'} = \rho_n \delta_{nn'}$, and $\rho_n = \exp(\beta E_n)/Z$ is the Boltzmann distribution.

11.3 Master equation for the density matrix

The spin–phonon interaction will be taken into account by adding phenomeno-logical terms into (11.3). These terms will be assumed linear, for simplicity and in analogy with the classical master equation (5.12). The new master equation reads

$$\hbar \frac{d}{dt}\rho(t) = -i[\mathcal{H}, \rho(t)] - \Lambda\rho(t). \tag{11.4}$$

For a spin s, the density matrix $\rho(t)$ is a $(2s+1) \times (2s+1)$ matrix, and the 'master matrix' Λ has $(2s+1)^4$ components $\lambda_{mm'}^{nn'}$.

In the following, the Hamiltonian will be assumed to be independent of time. Then, using the basis of vectors $|n\rangle$ which diagonalize the spin Hamiltonian $\mathcal{H}$, equation (11.4) reads

$$\frac{d}{dt}\rho_{nn'}(t) = -\sum_{pp'} \lambda_{nn'}^{pp'}\rho_{pp'}(t) - \frac{i}{\hbar}(E_n - E_{n'})\rho_{nn'}(t). \tag{11.5}$$

In this basis, the diagonal elements ρ_{nn} of the density matrix are the prob-abilities that the system is in state $|n\rangle$. They coincide with the probabilities ρ_n of Chapter 5. Thus, (11.5) should coincide with (5.12) if off-diagonal elements $\rho_{nn'}$ are ignored. For $n \neq n'$, one concludes that $\lambda_{nn}^{n'n'} = -\gamma_n^{n'}$, where $\gamma_n^{n'}$ are the coefficients which appear in (5.12). Since they are real, equation (11.4) violates time reversal invariance and, just as (5.12), it can only be valid for t larger than a microscopic time τ_{col}, introduced in Section 5.3. Thus, equation (11.4) provides a satisfactory description of decoherence only if decoherence is slow, and develops in a time longer than τ_{col}. A more general formalism has been given by Würger (1998).

A non-vanishing off-diagonal element $\rho_{nn'}$ indicates that the system has some probability to be in a linear combination $|n^*\rangle \cos\varphi + |n'^*\rangle \sin\varphi \exp(i\theta)$, *with a non-uniform probability distribution of the phase θ.*

For instance, if the system is in state $(|m^*\rangle + |m'^*\rangle)/\sqrt{2}$ or in state $(|m^*\rangle - |m'^*\rangle)/\sqrt{2}$, it is easily seen that $\rho_{mm'}$ has a non-vanishing value which is different in the two cases (and respectively equal to $1/2$ and $-1/2$). At equi-librium, all values of θ have the same probability since θ has no effect on the energy, and $\rho_{nn'} = \rho_n\delta_{nn'}$ is diagonal. However, a diagonal density matrix does not imply thermal equilibrium, but a less restrictive property which is the absence of information on the phases (e.g. the wavefunction θ in the above example). This absence of information is called quantum incoherence, while the knowledge of all phases is called quantum coherence, in analogy for instance with coherent light, characterized by the knowledge of the phases of all Fourier components.

Of course, uncertainty is still present in the diagonal density matrix, but it is a classical uncertainty which is not contrary to our intuition, in contrast with the quantum uncertainty of Schrödinger's cat.

11.4 Properties of the master matrix Λ

As seen in the previous section the elements $\lambda_{nn}^{n'n'}$ in equation (11.4) are (apart from the sign if $n \neq n'$) identical to the coefficients $\gamma_{n'}^{n}$ of (5.12). They were evaluated in Chapter 5. It will now be seen that many of the other matrix elements of Λ vanish, if one works in the basis of the eigenvectors $|\,n\,\rangle$ of $\mathcal{H}$.

Assume indeed that, at $t = 0$, all off-diagonal elements $\rho_{pp'}(0)$ vanish except for a pair of values, e.g. $p, p' = 1, 2$. If $\lambda_{nn'}^{pp'} \neq 0$ for $n, n' \neq 1, 2$, this implies that coherence, which was absent for $t = 0$, appears at $t \neq 0$. This is contrary to our intuition that coherence should decrease with increasing time. Therefore, if one of the four indices p, p', n, n' is not equal to another one, then

$$\lambda_{nn'}^{pp'} = 0. \tag{11.6}$$

More precisely it was shown by Garanin and Chudnovsky (1997) within certain approximations, and it is proven in Appendix K, that for $n \neq n'$

$$\lambda_{nn'}^{nn'} = (1/2)\lambda_{nn}^{nn} + (1/2)\lambda_{n'n'}^{n'n'} = \frac{1}{2}\left(\frac{1}{\tau_n} + \frac{1}{\tau_{n'}}\right) \tag{11.7}$$

where τ_n is the lifetime of the state $|\,n\,\rangle$, namely

$$\frac{1}{\tau_n} = \lambda_{nn}^{nn} = \sum_{p \neq n} |\lambda_{pp}^{nn}| = \sum_{p} \gamma_{n}^{p}. \tag{11.8}$$

Formula (11.7) expresses a relation between the loss of coherence between states $|\,n\,\rangle$ and $|\,n'\,\rangle$, and the decay of these states into other states. However, this is only an approximate result, a consequence of our decision to neglect multiphonon processes briefly addressed in Section 5.8. In reality, while relaxation implies decoherence, the opposite is not true. This can be seen as follows. Let the spin Hamiltonian have the simple form (2.10), $\mathcal{H}_{an} = DS_z^2$, and let the spin–phonon interaction have the form

$$\mathcal{H}_{\text{s-ph}} = \sum_{\alpha,\gamma} \tilde{\Lambda}_{\alpha,\gamma,\xi,\zeta}\, \partial_\alpha u_\gamma(\mathbf{r})\, S_z^2 \tag{11.9}$$

which corresponds to formula (5.27) with $\xi = \zeta = z$. This Hamiltonian is unable to produce transitions and therefore it is unable to produce relaxation. But it does produce decoherence since the phonons modulate the energy difference between the eigenstates of the spin Hamiltonian.

Using (11.5) and (11.8), one finds for $n \neq n'$

$$\rho_{nn'}(t) = \rho_{nn'}(0) \exp\frac{it(E_n - E_{n'})}{\hbar} \exp\left[-t(\tau_n^{-1} + \tau_{n'}^{-1})/2\right]. \tag{11.10}$$

11.5 Coherence and muon spectroscopy

It is first appropriate to give a microscopic formula for the density matrix, using the total Hamiltonian $\mathcal{H}_{\text{tot}}$, e.g. $\mathcal{H}_{\text{tot}} = \mathcal{H} + \mathcal{H}_{\text{phon}} + \mathcal{H}_{\text{int}}$, where $\mathcal{H}_{\text{phon}}$ is the phonon Hamiltonian and $\mathcal{H}_{\text{int}}$ is the spin–phonon interaction. These three Hamiltonians will be assumed independent of time. One can choose basis vectors $|\,nk\rangle$ which are direct products of basis vectors $|\,n\rangle$ of the spin (or more generally of the small system) of basis vectors $|\,k\rangle$ of the phonons (more generally, the thermal bath). To exploit (11.1), one can write

$$\langle A(t)\rangle = \frac{1}{Z_\theta} \sum_{nn'k} \rho_{nn'}(0)\exp(-\beta\mathcal{E}_k)\,\langle n'k\,|\exp(it\mathcal{H}_{\text{tot}})A\exp(-it\mathcal{H}_{\text{tot}})\,|\,nk\rangle$$

(11.11)

where the thermal bath is assumed to be at thermal equilibrium, while the small system is not necessarily at equilibrium. In formula (11.11) it is assumed that the small system perturbs the thermal bath and its eigenenergies $\mathcal{E}_k$ in a negligible way. The partition function $Z_\theta = \sum \exp(-\beta\mathcal{E}_k)$ of the thermal bath has been introduced. Formula (11.11) can be rewritten as

$$\langle A(t)\rangle = \frac{1}{Z_\theta} \sum_{nn'k} \rho_{nn'}(0) \sum_{mm'k'} \exp(-\beta\mathcal{E}_k)$$

$$\langle n'k\,|\exp(it\mathcal{H}_{\text{tot}})\,|\,mk'\rangle\,\langle mk'\,|\,A\,|\,m'k'\rangle\,\langle m'k'\,|\exp(-it\mathcal{H}_{\text{tot}})\,|\,nk\rangle\,.$$

(11.12)

Comparison with (11.1) yields

$$\rho_{m'm}(t) = \sum_{nn'} \rho_{nn'}(0)Q_{nn'}^{m'm}(t)$$

(11.13)

where

$$Q_{nn'}^{m'm}(t) = \frac{1}{Z_\theta} \sum_{kk'} \exp(-\beta\mathcal{E}_k)\,\langle n'k\,|\exp(it\mathcal{H}_{\text{tot}})\,|\,mk'\rangle\,\langle m'k'\,|\exp(-it\mathcal{H}_{\text{tot}})\,|\,nk\rangle\,.$$

(11.14)

Comparison with (11.6) shows that, if $n \neq n'$, expression (11.14) differs from 0 only if $n' = m$ and $n = m'$. It results from (11.10) that, if $n \neq n'$,

$$Q_{nn'}^{m'm}(t) = \delta_{n'm}\delta_{nm'}\exp\frac{it(E_n - E_{n'})}{\hbar}\exp\left[-t(\tau_n^{-1} + \tau_{n'}^{-1})/2\right].$$

(11.15)

In Chapter 3, when discussing muon spectroscopy, the correlation function $\langle S^-(t)S^+(0)\rangle$ was encountered. It can be written as

$$\langle S^-(t)S^+(0)\rangle = \frac{1}{ZZ_\theta} \sum_{nn'k} \sum_{mm'k'} \exp(-\beta E_n)\exp(-\beta\mathcal{E}_k)$$

$$\langle nk\,|\exp(it\mathcal{H}_{\text{tot}})\,|\,mk'\rangle\,\langle m\,|\,S^-\,|\,m'\rangle$$

$$\langle m'k'\,|\exp(-it\mathcal{H}_{\text{tot}})\,|\,n'k\rangle\,\langle n'\,|\,S^+\,|\,n\rangle$$

(11.16)

or

$$\langle S^-(t)S^+(0)\rangle = \frac{1}{Z}\sum_{nn'}\sum_{mm'}\exp(-\beta E_n)Q_{n'n}^{m'm}(t)\,\langle m\,|\,S^-\,|\,m'\rangle\,\langle n'\,|\,S^+\,|\,n\rangle\,.$$

$$(11.17)$$

If the spin Hamiltonian $\mathcal{H}$ commutes with S^z (e.g. $\mathcal{H} = DS_z^2$) the only non-vanishing terms are those with $m' = m+1$ and $n' = n+1$, and (11.15) implies $n = m$. Thus, (11.17) reads

$$\langle S^-(t)S^+(0)\rangle = \frac{1}{Z}\sum_{m}\exp(-\beta E_m)Q_{m+1,m}^{m+1,m}(t)\,\langle m\,|\,S^-\,|\,m+1\rangle\,\langle m+1\,|\,S^+\,|\,m\rangle$$

or, according to (11.15),

$$\langle S^-(t)S^+(0)\rangle = \frac{1}{Z}\sum_{m}\exp(-\beta E_m)\,|\,\langle m\,|\,S^-\,|\,m+1\rangle\,|^2$$

$$\exp\frac{it(E_{m+1} - E_m)}{\hbar}\exp\left[-t(\tau_m^{-1} + \tau_{m'}^{-1})/2\right]. \qquad (11.18)$$

This formula can be inserted into formula (3.59) which gives the muon spin relaxation time. The result is in agreement with that of Lancaster *et al.* (2004) apart from the factor $1/Z$ (forgotten by these authors) and a slight modification of the last factor.

11.6 Case of spin tunnelling

The case of a spin in a double well is an instructive illustration of the density matrix formalism. Far from a resonance, there is no major difficulty (Luis *et al.* 1998). The eigenvectors $|n\rangle$ of the spin Hamiltonian $\mathcal{H}$ may be identified with the localized states $|m^*\rangle$ introduced in the earlier chapters. At $t = 0$ the matrix elements $\rho_{mm'}(0)$ may be assumed to vanish except for $m = m' = -s$. It follows from (11.5) and (11.6) that all off-diagonal elements remain equal to 0 at all times t. Therefore, (11.5) reduces to (5.12). However, the selection rules (5.28) are not strictly valid since $|m^*\rangle$ is not exactly an eigenvector of S^z. This is not a consequence of the use of the density matrix.

The master matrix Λ is a $(2s+1)^2 \times (2s+1)^2$ Hermitian matrix, which therefore has $(2s+1)^2$ eigenvectors and eigenvalues. Since its use in the absence of tunnelling is equivalent to the method of Chapter 5, this implies that there are $2s(2s+1)$ eigenvectors and eigenvalues which play no part.

The situation is more complicated at a resonance. Let the case of a vanishing field be considered, $H_x = H_y = H_z = 0$. For the sake of simplicity, only the lowest two eigenvectors,

$$|\,\mathrm{sym}\rangle = \frac{1}{\sqrt{2}}\,(|\,s\rangle + |-s\rangle)$$

and

$$| \text{ anti} \rangle = \frac{1}{\sqrt{2}} \left(| s \rangle - | -s \rangle \right)$$

will be considered. The non-vanishing matrix elements of S^z are

$$\langle \text{sym} \, | \, S^z \, | \, \text{anti} \rangle = \langle \text{anti} \, | \, S^z \, | \, \text{sym} \rangle = s \qquad (11.19)$$

while $\langle \text{sym} \, | \, S^z \, | \, \text{sym} \rangle = \langle \text{anti} \, | \, S^z \, | \, \text{anti} \rangle = 0$. Thus

$$\langle S^z(t) \rangle = \frac{1}{\sqrt{2}} \langle \text{sym} \, | \, \rho(t) \, | \, \text{anti} \rangle + \frac{1}{\sqrt{2}} \langle \text{anti} \, | \, \rho(t) \, | \, \text{sym} \rangle \, . \qquad (11.20)$$

If one wishes to study the relaxation of S^z, what should now be done? In Chapter 5, the relaxation rate was the lowest of the $2s$ non-vanishing eigenvalues of the 'master matrix'. Now the master matrix has $(2s + 1)^2$ eigenvalues. In the absence of tunnelling, many of them have no physical relevance. In the presence of tunnelling, the method does not look reliable.

A safer approach is to give a non-vanishing value to the off-diagonal elements of the density matrix, e.g.

$$\langle \text{sym} \, | \, \rho(0) \, | \, \text{anti} \rangle = \langle \text{anti} \, | \, \rho(0) \, | \, \text{sym} \rangle = 1/2. \qquad (11.21)$$

The relaxation time τ is defined at low temperature (when only $| \text{ sym} \rangle$ and $| \text{ anti} \rangle$ are appreciably populated) by

$$\langle \text{sym} \, | \, \rho(\tau) \, | \, \text{anti} \rangle = 1/(2e) \qquad (11.22)$$

and the extension to higher temperatures is not difficult.

It is now absolutely necessary to take into account the second term of the right-hand side of (11.5) which is responsible for tunnelling, even though the Λ matrix also takes tunnelling partially into account a phonon-assisted tunnelling which, at resonance, is smaller than direct tunnelling.

11.7 Spin tunnelling between localized states

At resonance, it may be more transparent to work in the basis of the localized, quasi-eigenstates $| m^* \rangle$ defined in Section 6.1.7. For an exact, numerical calculation, this is not very convenient. But this method is suitable to make approximations.

The initial condition is now $\rho_{-s,-s} = 1$ and τ is defined at low temperature by

$$\langle -s \, | \, \rho(\tau) \, | \, -s \rangle = 1/(e). \qquad (11.23)$$

These conditions only involve diagonal elements of the density matrix. However, off-diagonal elements do appear in the master equation. Indeed, additional terms should be introduced because the Hamiltonian $\mathcal{H}$ has off-diagonal matrix elements $\hbar\omega_{mm''}$. Therefore (11.5) is replaced by

$$\frac{d}{dt}\rho_{mm'}(t) = \sum_{pp'}\Gamma^{pp'}_{mm'}\rho_{pp'}(t) - \frac{i}{\hbar}(E_m - E_{m'})\rho_{mm'}(t)$$

$$+ i\sum_{m''}[\omega_{mm''}\rho_{m''m'}(t) - \omega_{m''m'}\rho_{mm''}(t)]. \qquad (11.24)$$

The third term has the main responsibility in the generation of off-diagonal matrix elements $\rho_{mm'}(t)$. Assuming these elements to be small, they can be neglected in the third term of the right-hand side of (11.24) which can be rewritten as

$$\frac{d}{dt}\rho_{mm'}(t) = \sum_{pp'}\Gamma^{pp'}_{mm'}\rho_{pp'}(t) - \frac{i}{\hbar}(E_m - E_{m'})\rho_{mm'}(t)$$

$$+ i\omega_{mm'}[\rho_{m'm'}(t) - \rho_{mm}(t)]. \qquad (11.25)$$

An additional approximation is possible. Indeed the off-diagonal elements $\rho_{mm'}(t)$ which are created oscillate with a frequency $(E_m - E_{m'})/\hbar$ because of the second term of the right-hand side. Therefore they remain small except near a resonance. Thus, the third term can be omitted except if $(E_m - E_{m'})$ is small.

Note that in the basis of the localized states $|m^*\rangle$, the equilibrium density matrix $\rho(\infty)$ is not diagonal. If one defines a pseudospin $1/2$ operator σ_x by $\langle\text{sym}|\sigma_x|\text{sym}\rangle = \langle\text{anti}|\sigma_x|\text{anti}\rangle = 1/2$ and $\langle\text{sym}|\sigma_x|\text{anti}\rangle = \langle\text{anti}|\sigma_x|\text{sym}\rangle = 0$, the Hamiltonian can be written at low temperature in the ground doublet as $\mathcal{H} = -\hbar\omega_T\sigma_x$ plus a constant. Therefore $\langle\sigma_x\rangle = \langle\text{sym}|\rho(\infty)|\text{anti}\rangle + \langle\text{anti}|\rho(\infty)|\text{sym}\rangle$ does not vanish. However, it is extremely small at all temperatures which can be reached.

11.7.1 Decoherence by nuclear spins in zero field

In the case of tunnelling between the lowest two states of the double well in very weak field, decoherence is a many-body problem which cannot be treated by writing a phenomenological master equation for the density matrix.

A possible approach starts from equation (8.12b), which yields the probability $|Y(t)|^2$ that the spin has tunnelled at time t. Equation (8.12b) reads $\dot{Y}(t) = -i\omega_T X(t)e^{-iU(t)}$. Integration is easy for times much shorter than both τ and $1/\omega_T$. Since $Y(0) = 0$, one obtains

$$Y(t) = -i\omega_T X(0)\int_0^t e^{-iU(t')}dt'. \qquad (11.26)$$

It is appropriate to write this expression as

$$Y(t) = -i\omega_T X(0)\exp[-iU(0)]\int_0^t \exp\{-i[U(t') - U(0)]\}dt'. \qquad (11.27)$$

The average value of this quantity for a given value of $X(0)$ and $U(0)$ is

$$\langle Y(t) \rangle = -i\omega_T X(0) \exp[-iU(0)] \int_0^t \langle \exp\{-i[U(t') - U(0)]\}\rangle dt'. \quad (11.28)$$

The small fluctuations of the small quantity $[U(t') - U(0)]$ around its mean value 0 are responsible for decoherence. The average value at the right-hand side of (11.28) will be approximated by its first cumulant

$$\langle Y(t) \rangle = -i\omega_T X(0) \exp[-iU(0)] \int_0^t dt' \exp\left\{-(1/2)\langle[U(t') - U(0)]^2\rangle\right\} \quad (11.29)$$

Using the definition (9.13) of $U(t')$,

$$U(t) = \frac{2g\mu_B s}{\hbar} \int_{t_0}^t dt_1 H_z(t_1)$$

expression (11.29) reads

$$\langle Y(t) \rangle = -i\omega_T X(0) \exp[-iU(0)] \int_0^t dt'$$

$$\exp\left[-2\left(\frac{g\mu_B s}{\hbar}\right)^2 \int_0^{t'} dt_1 \int_0^{t'} dt_2 \langle H_z(t_1) H_z(t_2)\rangle\right] \quad (11.30)$$

Assuming $\tau \gg t \gg \tau_H$ where τ_H has been introduced in Section 9.2.4, (11.30) can be approximated as

$$\langle Y(t) \rangle = -i\omega_T X(0) \exp[-iU(0)] \int_0^t dt' \exp\left[-\left(2\frac{g\mu_B s}{\hbar}\right)^2 \int_0^{t'} dt_1 \tau_H \langle H_z^2\rangle\right]$$

$$= -i\omega_T X(0) \exp[-iU(0)] \int_0^t dt' \exp\left[-2\left(\frac{g\mu_B s}{\hbar}\right)^2 \tau_H \langle H_z^2\rangle t'\right]$$

or finally, integrating the exponential,

$$\langle Y(t) \rangle = -\frac{i\hbar^2 \omega_T X(0)}{2g^2 \mu_B^2 s^2 \tau_H \langle H_z^2\rangle} \exp[-iU(0)]\left\{1 - \exp\left[-2\left(\frac{g\mu_B s}{\hbar}\right)^2 \tau_H \langle H_z^2\rangle t'\right]\right\}. \quad (11.31)$$

This expression increases linearly at short times, exactly as it would in the absence of fluctuations, i.e. in the absence of decoherence. Decoherence sets in when the expression between curly brackets saturates to the value 1. This occurs at a time of the order of τ_{coh} defined by

$$\frac{1}{\tau_{\text{coh}}} = 2\left(\frac{g\mu_B s}{\hbar}\right)^2 \tau_H \langle H_z^2\rangle. \quad (11.32)$$

11.8 Potential applications of quantum coherence: quantum computing

If decoherence can be sufficiently reduced, then one can build quantum computers. One day, perhaps, this will be possible.

What is a quantum computer? What has it to do with molecular magnetism? Before giving an oversimplified answer, it is convenient to recall that, in a classical computer (the one we all have on our desks), numbers are written in the binary system, so that the digits are 0 or 1, and they are represented by the state of a system, e.g. a MOSFET transistor. Each bit may be in state 0 or in state 1, not both. In a quantum computer, bits are replaced by quantum bits ('qubits' or 'q-bits') which can be in a linear combination of the two states. A spin $1/2$ is a good candidate to the function of a q-bit since it can be in state $|+\rangle$, in state $|-\rangle$, or in a linear combination $|+\rangle \cos\varphi + |-\rangle \sin\varphi e^{i\theta}$. Although this state depends on two continuous parameters, the information accessible from it is restricted by quantum uncertainty, and a q-bit cannot yield more information than a classical bit. The advantage of a quantum computer is that it operates on several numbers at the same time. For instance, integers from 0 to eight can be represented by the states of three spins $1/2$, namely $|+++\rangle$, $|++-\rangle$, $|+-+\rangle, \ldots, |---\rangle$. If the system is initially in a combination of these eight states, operations on this state involve simultaneous operations (additions, multiplications, for instance) on the eight integers. A quantum computer makes something like parallel computing.

The advantage of parallel computing is illustrated by the following example taken from Leuenberger and Loss (2001). Suppose you want to find the name of the citizen who has, in your country, the social security number 11 547 672. Your classical computer will write this number in base 2, and requires 23 figures to do this. Then the computer needs 23 operations to give the answer. A parallel computer can do that in a single operation, doing all 23 classical ones at the same time, in a parallel way. And this spares computer time.

However a quantum computer does not make really parallel computing since, at the end, a single result can be read and not the results of all parallel calculations. But a quantum computer can be very efficient in certain cases, and a convincing example is the factorization of a large number. Suppose you want to find the prime factors of a number of 130 digits. If it turns out to be the product of two prime numbers of 65 figures, at the beginning of the 21st century an array of supercomputers (which of course had to be classical) required several months to find them! An algorithm to do that in a much shorter time with a quantum computer was proposed by Shor (1994) and investigated in detail by Ekert and Jozsa (1996). There was just one problem: a quantum computer did not exist! The first implementation of Shor's algorithm (Vandersypen *et al.* 2001) was done by researchers from IBM and Stanford University directed by Isaac Chuang and published in *Nature*. Unfortunately, their 'quantum computer' was a mere toy of seven q-bits, and it was not able to factorize a number larger than 15, with the expected, but uninteresting result that $15 = 3 \times 5$. The seven q-bits were the spins of seven magnetic nuclei in a solution, and were triggered by a magnetic resonance system. One works with an ensemble of quantum computers rather than a single one.

The factorization problem is just one of those which are difficult for classical computers. Quantum computers might be useful for other issues, e.g. (i) the search of the ground state of a glass or a spin-glass; (ii) deciphering of encrypted messages without knowing the key; (iii) chess. The last two points deserve some comments. While quantum computing may be a weapon against cryptography, quantum physics can also help cryptography, and quantum cryptography might soon become operational. Regarding chess, already in the year 1997 the classical computer Deep Blue from IBM was able to defeat the world champion Kasparov. Quantum computers might put an end to chess championships by proving either that white can always win, or that black can always win, or that a draw is unavoidable if both players play as well as possible.

Is the quantum toy computer constructed in the USA in 2001 the prototype of an operational one? This is not at all sure. Many bits, i.e. many spins, are required, and the simultaneous manipulation of all these spins, without losing quantum coherence, will not be easy. The 2001 experiment demonstrates that spins in molecules may at least in principle be used for quantum computing. However, those spins are nuclear ones, their interaction is weak, and we are far from the subject of this book! The use of single-molecule magnets for quantum computation was proposed by Leuenberger and Loss (2001). They considered a single molecule of spin 10 and assumed one can populate the lowest levels of a single well, using appropriate electromagnetic radiation. Technically, the real-ization of such a computer would probably be more difficult than with nuclear spins since the interactions are stronger and decoherence is faster. As these lines are written, in the year 2005, an experimental proof of the possibility of using electronic spins for quantum computing is still lacking, but there is active spec-ulative research on the subject. For instance Troiani *et al.* (2005a) propose using antiferromagnetic rings as quantum gates. The following statement of Troiani *et al.* (2005b) sheds light on the time-scales in a future quantum computer based on electronic spins: 'Our simulations of the single q-bit gates provide negligible values for the leakage ... even for gating times of the order of 100 picoseconds, i.e. well below the tens of nanoseconds estimated for the spin decoherence time.'

12

DISORDER AND MAGNETIC TUNNELLING

12.1 Experimental evidence of disorder

In the tetragonal material $Mn_{12}ac$, in the absence of impurities, tunnelling should satisfy the selection rule (6.20). Experimentally, this selection rule is not satisfied, and all minima of the relaxation time corresponding to level crossings show up. In this chapter, it will be seen that experimental results are consistent with the existence in the spin Hamiltonian of a pretty large anisotropy term of the form $E(S_x^2 - S_y^2)$. Such a term can only be due to disorder.[1] Moreover, Mertes *et al.* (2003), employing the experimental procedure described in Section 8.4 based on sweeping back and forth the longitudinal field across a tunnel resonance, have shown (Fig. 12.1) that the plot of the logarithm of the magnetization of oscillation cycles is not linear as would be expected from (8.23) for a pure material.

At least two types of disorder have been invoked. Chudnovsky and Garanin (2001) have suggested that dislocations, which are present in almost all crystals (apart from Si crystals used in microelectronics) might play the dominant part. On the other hand, as explained in some detail in Section 4.7.1, more peculiar microscopic defects are always present in $Mn_{12}ac$, as shown by low-temperature X-ray diffraction data (Cornia *et al.* 2002a). In the remainder of this chapter, the effect of disorder will be theoretically analysed.

12.2 Landau–Zener–Stückelberg experiment with a distribution of tunnel frequencies

Assume all spins are initially polarized in the $-z$ direction, as an effect of a strong magnetic field along the z axis. The temperature will be assumed so low that all spins are in state $|-s^*\rangle$. Then, the magnetic field is lowered with a constant velocity and changes sign. When the field goes through the level-crossing value between $m = -s$ and $p = s - k$, the reversal probability p_k of a spin is given by formula (8.17), which will be written as

$$p_k = 1 - \exp\left(\frac{-2\pi\hbar\omega_{Tk}^2}{v_k}\right) \tag{12.1}$$

where $2\hbar\omega_{Tk}$ is the tunnel splitting and

$$v_k = g\mu_B(2s - k)dH_z/dt. \tag{12.2}$$

---

[1] Remember that, in this chapter, E is the anisotropy parameter, not the energy!

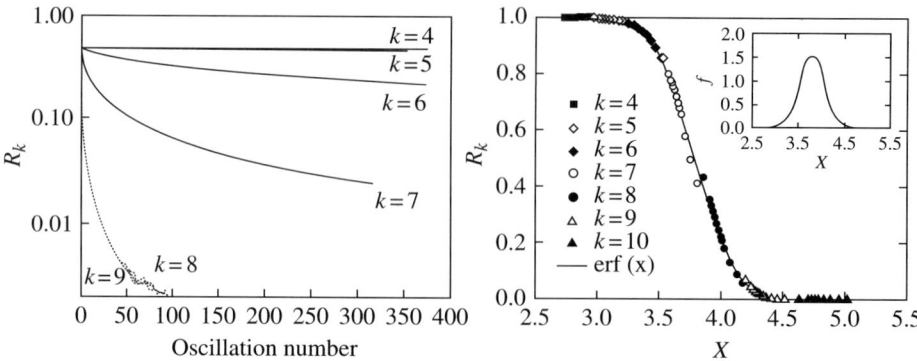

FIG. 12.1. Left: fraction of molecules of Mn$_{12}$ac that remain in the metastable well as a function of the number of field sweepings across the tunnel resonance k. Right: the same quantity measured for different sweeping rates and plotted versus the scaling variable $X = (s - k/2)^{-1} \ln(g_k \sqrt{v_k})$; From Mertes *et al.* 2001 and 2003.

Following Chudnovsky and Garanin, we shall consider the case of a broad distribution of tunnel frequencies. Spins which have a large enough tunnel frequency flip, and the other spins do not. The value $\omega_{\mathrm{T}k} = \omega_k^{\min}(k, v_k)$ which separates the two behaviours depends on k and v. As a first approximation, these authors assume that the only term of the Hamiltonian which does not commute with S_z, and therefore is responsible for tunnelling, is the term $E(S_x^2 - S_y^2)$ of formula (2.5).

In this model, tunnelling occurs only if k is even, and the relation between E and $\omega_{\mathrm{T}k}$ is given according to perturbation theory by formula (G.11) which can be written as

$$2\hbar\omega_{\mathrm{T}k} = g_k \left(\frac{E}{2|D|} \right)^{s - k/2} \tag{12.3}$$

where

$$g_k = \frac{2|D|}{[(2s - k - 2)!!]^2} \sqrt{\frac{(2s - k)!(2s)!}{k!}}$$

$$= \frac{2|D|}{2^{2s-k-2}[(s - k/2 - 1)!]^2} \sqrt{\frac{(2s - k)!(2s)!}{k!}} \tag{12.4}$$

where $n!! = n(n - 2)(n - 4)(n - 6).....$, so that $(2n)!! = 2^n n!$ and $(2n + 1)!! = (2n + 1)!/(2^n n!)$.

According to (12.1), $\omega_k^{\min}$ is given by $2\pi\hbar\omega_{\mathrm{T}k}^2 = v_k$, so that

$$\omega_k^{\min} = \sqrt{\frac{v_k}{2\pi\hbar}}. \tag{12.5}$$

Formulae (12.5) and (12.3) yield the order of magnitude of the lowest value of $|E|$ which allows for an observable tunnel effect, namely

$$E_{\min} \approx |D| \left(\frac{v_k}{g_k}\right)^{1/(s-k/2)}. \tag{12.6}$$

12.3   The scaling law of Chudnovsky and Garanin

Chudnovsky and Garanin (2001) have proposed an approximation based on the fact that, when $|E|$ increases for a given value of v_k, the reversal probability (8.17) changes fairly abruptly from 0 to 1. Therefore, for a distribution of the anisotropy parameter E, they approximate the reversal probability p by

$$p = 2 \int_{E_{\min}}^{\infty} \mathcal{P}(E)dE \tag{12.7}$$

where $\mathcal{P}(E)$ is the probability that E has a particular value. It has been assumed that $\mathcal{P}(-E) = \mathcal{P}(E)$ and the factor 2 accounts for negative values of E.

The function $\mathcal{P}(E)$ is not known. The interest in (12.7) is that it states that p is a function of $E_{\min}$. Therefore, according to (12.6), p_k is a function of $(\sqrt{v_k}/g_k)^{1/(s-k/2)}$ only. It may be preferable to take the logarithm of this quantity and to state that

$$p_k \text{ is a function of } (s - k/2)^{-1} \ln(\sqrt{v_k}/g_k) \text{ only}$$

This property will be called the 'Chudnovsky–Garanin scaling law'. The function can depend on the sample, because the disorder is generally irreproducible in a solid, but it is independent of k and v_k (see also Garanin and Chudnovsky 2002).

Experimental results of Mertes et al. (2001) on Mn_{12}ac are in good agreement with the Chudnovsky–Garanin scaling law. They have monitored the magnetization at different sweeping rates for the longitudinal field. They have then tested if all the curves, detected for different resonances k, scale on the same curve, as predicted by the Chudnovsky–Garanin scaling law. The results, plotted in Fig. 12.1, show good agreement with the theoretical expectations. This strongly suggests that the assumptions made to derive this law are correct, namely: (i) tunnelling is mainly produced by a quadratic term $E(S_x^2 - S_y^2)$ in the Hamiltonian; (ii) the distribution of E is broad. The domination of the quadratic term, which results from disorder, on the quartic term, is a bit surprising since other experimental data (e.g. NMR or X-ray diffractometry) suggest that the crystals are of good quality. The surprise is increased when one remembers that a relatively small high-order anisotropy can produce a relatively strong tunnelling as seen in Section 6.4. Chudnovsky and Garanin (2001) suggested that the strong effect of disorder might be a result of long-range disturbances which enhance the effect of a rather low concentration of defects. Actually, a crystal defect necessarily produces a disturbance of the elastic deformation, which is long-ranged.

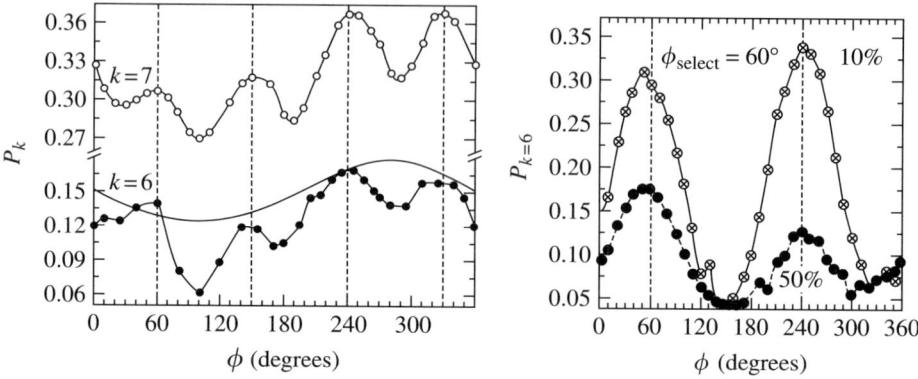

FIG. 12.2. Left: tunnel probability for the resonances $k = 6$ and $k = 7$ as a function of the angle ϕ of the transverse field $H_\perp = 0.3$ T with respect to one of the faces of the crystal. The material is $Mn_{12}ac$ and the temperature is $T = 0.6$ K. The solid line corresponds to the trend due to a small misalignment of the crystal. Right: a similar experiment where the investigated molecules have been selected by applying a longitudinal field at $\phi = 60°$. Two different subsets of the population, corresponding to the 50% and 10% fraction with the largest tunnel splitting, respectively, have been investigated. The two-fold symmetry is evident. (After del Barco *et al.* 2003.)

As an example, the strain produced by an infinite, straight dislocation decreases as

$$\epsilon \sim 1/r \tag{12.8}$$

while a dislocation loop of radius R produces at distance $r \gg R$ a strain

$$\epsilon \sim R^2/r^3. \tag{12.9}$$

The theory of Chudnovsky and Garanin (2001) is attractive although it encounters several difficulties. First, the elastic strain generates interactions between defects, which tend to decrease their effect at long distance. For instance, in a pair of straight, parallel dislocations, the state with opposite signs is favoured. Moreover, the Chudnovsky–Garanin theory predicts that the distribution $P(E)$ of the anisotropy E should have a maximum at $E = 0$, in disagreement with experiment, as will be seen. The maximum at $E = 0$ predicted by Chudnovsky and Garanin is an extension of the central limit theorem which applies to a random variable $X = \sum x_i$ which is the sum of many independent random variables x_i with the same distribution $p(x_i)$ centred at $x_i = 0$) (but not necessarily maximum at $x_i = 0$). Obviously, the most probable value of X is its average value, namely 0. In the present case, $E = \sum E_r$ is the sum of the effects of defects at various distances r, and the distribution $p_r(E_r)$ depends on r. However, if the decrease with r is slow as in formula (12.8), a maximum at $E = 0$ is expected as indeed found in the theory of Chudnovsky and Garanin (2001).

The experimental data of Mertes et $al.$ (2001) allow also an evaluation of the distribution of the transverse anisotropy E. This distribution, shown in the inset of Fig. 12.1, is well approximated by two Gaussians centred at $E = \pm 0.03D$, and not at 0 as predicted by Chudnovsky and Garanin. This value is much larger than those calculated in Table 4.3 taking into account the different isomers present in Mn$_{12}$ac. It therefore suggests that other sources of tunnelling are present. Certainly the quartic anisotropy term $(C/2)(S_x^4 + S_y^4)$ must be taken into account, although it violates the Chudnovsky–Garanin scaling law.

Del Barco et $al.$ (2002) have repeated the previous experiment and they have found that, comparing the data for resonance with $k = 6$ and $k = 8$, the latter has a narrower distribution. The resonance $k = 8$ is in fact permitted in four-fold symmetry and the quadratic term plays a minor role. The same authors have ingeniously designed a more sophisticated experiment in order to high-light the presence of a local quadratic anisotropy. If a transverse anisotropy is present, P_k should be affected by a transverse field, promoting tunnelling if applied along the intermediate axis or depressing it if applied along the hard axis. A standard experiment with $H_\perp = 0.3$ T gives the result repor-ted on the left of Fig. 12.2. An overall $\frac{\pi}{2}$ periodicity is observed even though, according to the model suggested by Cornia et $al.$ (2002a) and described in Section 4.7.1, most molecules experience a two-fold symmetry. This is not sur-prising because positive and negative E values are equally probable, so that on the average the four-fold symmetry of the crystal is preserved. However, if the sample is prepared by applying a positive longitudinal field together with a transverse field, only those molecules that experience the largest tun-nel splitting would be selectively magnetized. Those are characterized by the transverse field applied parallel to their intermediate axis. After having pre-pared the sample with a transverse field of 0.6 T applied at 60° from a crystal face, they repeated the first experiment monitoring the reversal of the mag-netization on application of a negative field. This time the largest contribution comes from those molecules that have been previously selected, and their tun-nel probability P_k now shows π periodicity, as displayed in the right part of Fig. 12.2. The angle at which a maximum in P_k is observed agrees with those calculated in Table 4.3. These results strongly support the hypothesis that the tunnel distribution in Mn$_{12}$ac is primarily due to the presence of different iso-mers, even if a slightly larger quadratic transverse anisotropy, E, up to 10 mK, is necessary to reproduce the experimental data. The data of Mertes et $al.$ (2001) can be reconciled with those of Del Barco et $al.$ (2003) if a large con-centration of defects was present in the crystal, thus overwhelming the intrinsic disorder due to the presence of different isomers. HF-EPR experiments have fur-ther confirmed this picture and have demonstrated the role of mismatching in the principal axes of transverse two-fold and four-fold anisotropies (Del Barco et $al.$ 2005).

Although the previous analysis explains many experimental features, some refinements are necessary in view of minor details. As noticed in Section 12.2, formula (12.3) predicts tunnelling only if k is even. However, Mertes *et al.* observed it for odd values too. They could explain that by the random transverse fields induced by disorder. This implies an appropriate redefinition of g_k.

12.4 Other distortion isomers in $Mn_{12}ac$

As seen at the end of Section 4.7.1, any sample of $Mn_{12}ac$ contains a small fraction of molecules which exhibit a much faster relaxation. In those molecules the barrier is reduced to 35 K or even 15 K, yielding much faster tunnelling dynamics.

This type of disorder is quite different from that described by Cornia *et al.* (2002a,b). It affects a limited proportion of molecules (5% have a barrier of 35 K and 0.5% have a barrier of 15 K) but the affected molecules are strongly perturbed since the barrier is reduced by a factor of the order of 2 and 4, respectively. Geometrically, the anisotropy axes might be rotated by an angle of the order of 10° (Wernsdorfer, private communication). In contrast, the type of disorder described by Cornia *et al.* (2002a,b) affects all the molecules which have the anisotropy axis nearly along the crystallographic four-fold axis.

According to Morello *et al.* (2004), this type of disorder might play an essential part in nuclear spin relaxation at low temperature. It should also contribute to destroy the four-fold symmetry.

12.5 Spin glass phases?

Structurally, a 'spin glass' (Binder and Young 1986) is the magnetic equivalent of a glass. In a glass, atoms are disordered as in a liquid, but immobile as in a crystal. In a spin glass, the magnetic moments are disordered as in a paramagnet, but hardly flip. A magnetic moment which is up at time $t = 0$ has a large probability to be up a long time later. In contrast with ordinary glasses which result from a rapid cooling which do not allow the atoms to find the right place, spin glasses arise from atomic disorder, e.g. impurities. Another condition is necessary, namely competing interactions. This property is called *frustration*. A simple example of a frustrated system is a triangle of magnetic atoms with three equal, antiferromagnetic pair interactions. Another example is a triangular lattice of Ising spins with equal antiferromagnetic interactions between nearest neighbours. It is easy to see that this system is disordered at any temperature. However, if impurities are present, it can order at $T = 0$. The magnetic structure depends on the position of the impurities and is not periodic. The most common examples of spin glasses (Binder and Young 1986) are non-magnetic metals with magnetic impurities interacting through Rudermann–Kittel interactions. In molecular magnets, the interactions between molecular spins are presumably dipolar, and at low-temperature the spins are 'Ising' spins, i.e. they have two possible states $m = 10$ and $m = -10$. Frustration does exist in an array of Ising spins with dipolar interactions. Thus, the low temperature phase of Fe_8 and

Mn$_{12}$ac might well be a spin glass. If it is a spin glass, this has consequences for the spin dynamics, e.g. the relaxation of the magnetization should be very slow. In practice, since a slow relaxation may have various causes, it is difficult to say whether the very slow relaxation observed at low temperature in these systems is related to a spin glass structure. It is of interest to note that particularly slow relaxation has been observed in frozen dilute solutions of Mn$_{12}$ac, as seen in Section 4.7.1.

The kind of spin glass order that might appear in Fe$_8$ will be briefly discussed. This material may be roughly described as an array of parallel chains of clusters which are almost parallel to the easy axis z. The distance between chains is appreciably larger (50%) than the distance between molecules on the chain. An isolated chain would be ferromagnetically ordered under the effect of dipole interactions. If the chains were exactly parallel to the easy axis, and if the spin density were uniformly distributed along each chain, then the interaction between chains at distance x would vanish since it would be equal, apart from a constant factor, to

$$\int_{-\infty}^{\infty} \frac{dz}{(x^2 + z^2)^{3/2}} - \int_{-\infty}^{\infty} \frac{3z^2 dz}{(x^2 + z^2)^{5/2}} \tag{12.10}$$

which vanishes, as can be seen by integrating the second term by parts. This vanishing interchain interaction is a manifestation of frustration. The frustrated system hesitates between ferro- and antiferromagnetism. If defects are present, they can have a decisive influence. For instance, if the chains are interrupted, the interaction between the pieces of chains can become non-vanishing.

12.6 Conclusion

Three types of disorder have been considered to account for the properties of Mn$_{12}$ac: (i) dislocations; (ii) intrinsic, localized defects which are weak perturbations; (iii) dilute defects which produce strong local perturbations. In the light of the above discussion, dislocations do not quite fit the experimental properties well. Del Barco *et al.* (2003) have found an 'excellent accord with the isomer model of Cornia *et al.*' (2002b) and claim that their observations are 'not consistent with the dislocation model'. In view of the very small perturbations related to model (ii), it might be of interest to include also model (iii).

The picture of single-molecule magnets as non-interacting identical molecules, adopted in Chapters 5 and 6, is now seen to be strongly oversimplified. In Chapter 9, interactions between molecules were seen to be important. In the present chapter, disorder was found to be essential. Nevertheless, it is gratifying to see that the above-discussed experiments of del Barco *et al.* make it possible to distinguish the type of disorder, as well as to exploit it to select a subset of molecules by a kind of magnetic distillation. We will see in the following chapter that also intercluster interactions have been shown to be a precious tool for the investigation of the tunnelling of the magnetization.

13

MORE EXPERIMENTS ON SINGLE-MOLECULE MAGNETS

SMMs have been experimentally investigated using many different techniques and only a small fraction of these experiments have been discussed in the previous chapters. An exhaustive overview of the wide literature is not the goal of this book, nevertheless there are a few other experiments that we want to discuss here briefly, while more details can be retrieved from the original literature. They have been selected mainly for their relevance to the theoretical background provided in the previous chapters.

13.1   The advantages of complexity

For the sake of simplicity we have often limited our analysis of the resonant quantum tunnelling of the magnetization to the second-order spin Hamiltonian (2.5). However in Chapter 4 we have already seen that both $Mn_{12}ac$ and Fe_8 are better described if fourth-order axial and transverse anisotropy terms are included in the spin Hamiltonian. The present section is focused on the axial term $B_4^0 O_4^0$ defined in Appendix A.5, which can be replaced by $35B_4^0 S_z^4$ if B_4^0 is much smaller than D. This term modifies the energies of the $|m\rangle$ and $|m'\rangle$ levels, which now cross for

$$g\mu_B H_z = |D|\,(m + m') + 35B_4^0 (m^3 + m^2 m' + mm'^2 + m'^3) \qquad (13.1)$$

as shown in Fig. 13.1, and the level-crossing field does not depend only on $n = m + m'$ as was the case in Fig. 6.1.

This complication, which might look undesirable, has, however been, successfully exploited to investigate which is the most relevant channel for the mechanism of thermally activated resonant tunnelling, schematized in Fig. 10.1.

The hysteresis loop of a $Mn_{12}ac$ crystal at different temperatures has been carefully investigated, using a micro-Hall probe (Bokacheva *et al.* 2000; Kent *et al.* 2000) or using a cantilever torquemeter (Chiorescu *et al.* 2000b). Through the derivative of these curves the authors were able to better quantify the field at which the fastest relaxation was observed. In Fig. 13.2a the derivative of M *vs.* H curves taken at several temperatures is reported. For each resonance n the peak shows an up-field shift on lowering the temperature and develops a complex structure below 1 K. The effect is more evident when the fields of the maxima are reported *vs.* temperature as in Fig. 13.2b and c. These curves, being extracted from the hysteresis loop, necessarily correspond to a variable magnetization state of the sample.

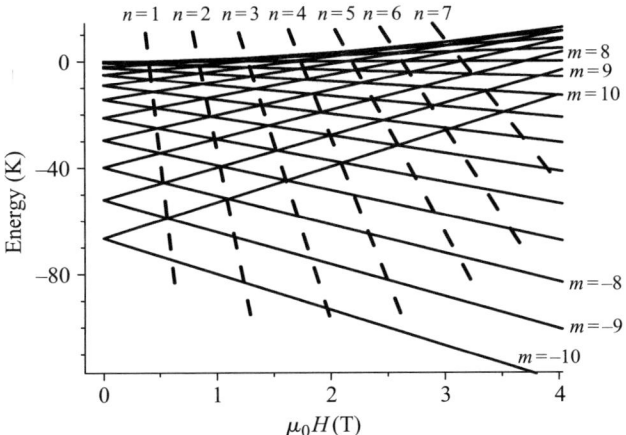

FIG. 13.1. Longitudinal field Zeeman level splitting of the $S = 10$ state of Mn$_{12}$ac assuming a fourth-order spin Hamiltonian for the spin anisotropy. The parameters are those of Table 4.1 extracted from INS data.

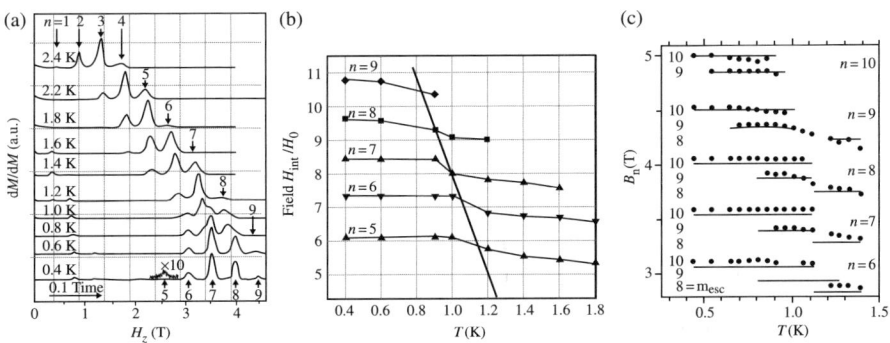

FIG. 13.2. (a) Derivative of the magnetization of a crystal of Mn$_{12}$ac with respect to applied longitudinal field at temperatures from 2.4 K to 0.4 K. (b) The same data plotted as the position of the resonant peak versus temperature. The solid line shows the approximate temperature below which peak positions are temperature independent. From Kent *et al.* (2000). (c) Similar results up to the tenth resonance. From Chiorescu *et al.* (2000b). The observed data are compared with the calculated fields assuming that the tunnelling involves the metastable state $m_{esc} = 10$, 9, or 8. With permission of EDP Sciences and the American Physical Society.

As already emphasized in Chapters 8 and 9, the effective field experienced by the molecule differs from the applied one because of the internal field generated by the other spins in the crystals and the hyperfine field. If the hyperfine field is ignored, the largest variation of the internal field cannot exceed $8\pi|M_s|$, where $|M_s|$ is the saturation magnetization per unit volume, which is about 0.064 T for $Mn_{12}ac$. This value is significantly smaller than the observed shift of Fig. 13.2. This observed shift also overwhelms the expected distribution in resonance fields due to the presence of the different isomers described in Section 4.7.1. The up-field shift of the peaks on decreasing the temperature is due to the fact that tunnelling pathways involving lower and lower energy states in the double well become more efficient when the population of the highest levels decreases (Bokacheva $et\ al.$ 2000; Chiorescu $et\ al.$ 2000b; Kent $et\ al.$ 2000). For $Mn_{12}ac$, B_4^0 is negative, and the lower levels cross at higher field, as shown in Fig. 13.1. In Fig. 13.2 are also reported the crossing fields for different pairs of levels involved for resonances going from $n = 5$ to $n = 9$ for micro-Hall experiments, and from $n = 6$ to $n = 10$ for torque measurements. Within the limit of the estimation of the internal field, the results at low temperatures agree with the calculated fields for the resonances where the escape from the metastable well occurs from the lowest state, $m_{esc} = 10$. It is thus possible to monitor the transition from a thermally activated tunnelling to pure tunnelling. This transition occurs gradually, and the temperature of this crossover depends on the investigated transition, decreasing on increasing n because the levels of the metastable well get closer and closer on increasing the field. The resonance in zero longitudinal field cannot be directly investigated in the pure tunnelling regime, because the process is too slow; however from an extrapolation of the crossing temperature of the resonance for higher n Chiorescu $et\ al.$ (2000b) estimated a cross-over temperature of 1.7 K in zero field.

13.2 Intercluster interactions in Mn_4 clusters

The class of compounds nicknamed Mn_4 and briefly described in Section 4.7.5 has been demonstrated to be a spectacular playground for the investigation of the interplay of intercluster interactions and quantum tunnelling of the magnetization. A key role is surely played by the large variety of Mn_4 clusters that have been synthesized, thus allowing a fine tuning of the magnetic anisotropy and of the intercluster interactions. Another important feature is the well isolated $S = 9/2$ ground state that makes the 'giant spin' approach a very good approximation.

13.2.1 The effects of intercluster interactions on magnetic tunnelling

We have already seen in Chapter 9 that the other clusters of the crystal generate an internal dipolar field which evolves in time as the magnetization relaxes, thus being responsible for the square-root time decay of the magnetization. Here we want to summarize the effects of an exchange interaction that couples two 'giant spins'. The practical realization of this model has been found in the Mn_4 cluster of formula $[Mn_4O_3Cl_4(O_2CEt)_3(py)_3]_2$, where $EtCO_2^-$ is the anion of propionic acid and py stands for pyridine. The unusual feature of this compound is the H-bond that connects $C-H$ groups to the chlorine atoms that belong to

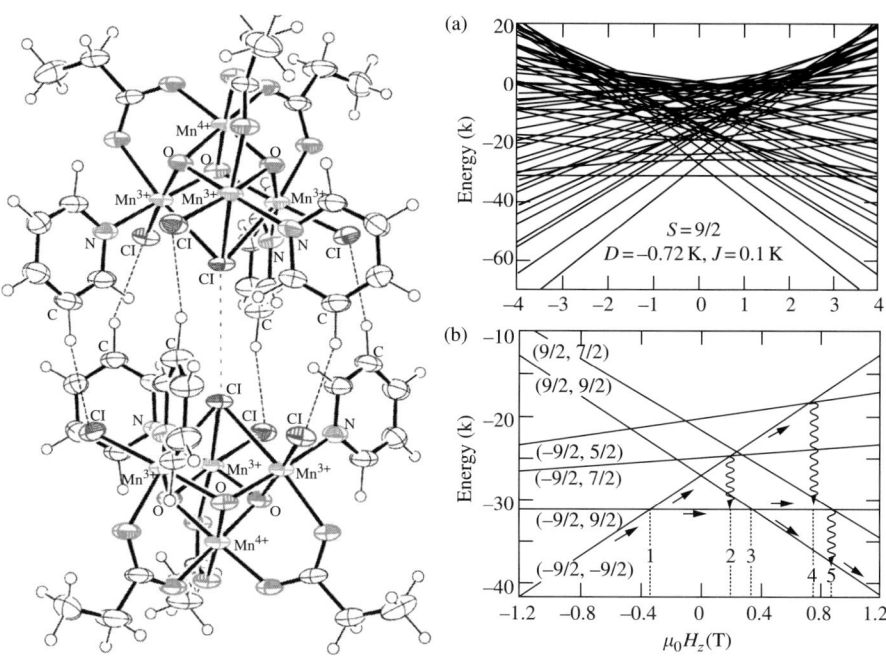

FIG. 13.3. Left: view of the structure of the [Mn$_4$O$_3$Cl$_4$(O$_2$CEt)$_3$(py)$_3$]$_2$ dimer. The dashed lines represent the C − H ⋯ Cl hydrogen bonds that connect the two Mn$_4$ moieties. Also the short Cl ⋯ Cl is shown. Right: Zeeman splitting of the $|m_1, m_2\rangle$ states arising from the coupling of the two $S = 9/2$ of the [Mn$_4$]$_2$ dimer. A weak antiferromagnetic interaction is added to the axial anisotropy of each spin. (a) The entire energy spectrum is reported. (b) An enlargement of the low-lying levels in the field range -1.2 up to 1.2 T is shown. The dotted lines labelled from 1 to 5 indicate the strongest tunnel resonances which involve the change of sign of only one m. The curled arrows show the tunnel resonances that involve excited states of the stable well. The process leading to the reversal of the magnetization, i.e. from $|-9/2, -9/2\rangle$ to $|9/2, 9/2\rangle$, is shown by light arrows. From Wernsdorfer *et al.* (2002). Reprinted with permission of Nature (http://www.nature.com).

another Mn$_4$ unit, thus forming the dimeric [Mn$_4$]$_2$ structure reported in Fig 13.3 (left).

While the intradimer interaction has been found to be antiferromagnetic and of the order of 0.1 K the interdimer interaction is negligible, especially in one pseudopolymorph that contains two molecules of hexane of crystallization in the crystal lattice.

The spin Hamiltonian for the pair of spins can be written as:

$$\mathcal{H} = \mathcal{H}_0 + \mathcal{H}' + \mathcal{H}'' \tag{13.2}$$

where, taking into account only the second-order magnetic anisotropy

$$H_0 = D(S_{1z}^2 + S_{2z}^2) + g\mu_B H_z(S_{1z} + S_{2z}) + J_z(S_{1z}S_{2z}) \tag{13.3}$$

$$H' = E(S_{1x}^2 + S_{2x}^2 - S_{1y}^2 - S_{2y}^2) + g\mu_B H_x(S_{1x} + S_{2x}) + g\mu_B H_y(S_{1y} + S_{2y}) \tag{13.4}$$

$$H'' = J_{xy}(S_{1x}S_{2x} + S_{1y}S_{2y}). \tag{13.5}$$

The spin Hamiltonian matrix is now written on the basis of the $|m_1, m_2\rangle$ states and has dimension $(2S+1)^2 = 100$. In the case of $J_z = J_{x,y} \gg D$ the states can be expressed in the coupled basis of the total spin $|S_T, M\rangle$, as already shown in Section 2.1.3, but this condition is not met in the present case. In the Hamiltonian matrix of (13.2) the terms of $\mathcal{H}_0$ are diagonal while those of $\mathcal{H}'$ couple elements of the basis that have $\delta m_1 = \pm 1, \pm 2$ provided that $\delta m_2 = 0$ (and vice versa).

Let us start from the case where $\mathcal{H}' \ll \mathcal{H}_0$ and $\mathcal{H}'' = 0$. The $|m_1, m_2\rangle$ states are eigenstates of the system, except close to an anticrossing where admixing occurs. However pairs of levels, namely $|\mathcal{M}, \mathcal{M}'\rangle$ and $|\mathcal{M}', \mathcal{M}\rangle$ remain degenerate. Even in this oversimplified scheme important effects can be expected, and have indeed been investigated by Wernsdorfer et al. (2002). In Fig. 13.3 (right) is reported the axial Zeeman splitting of the $|m_1, m_2\rangle$ states calculated for $[Mn_4]_2$. The lowest energy state in zero field is characterized by $M_T = m_1 + m_2 = 0$, as expected given the antiferromagnetic intradimer interaction. However, the exchange is weak compared to the magnetic anisotropy and many level crossings occur at low field. It is particularly important to notice that all level crossings in zero field involve pairs of levels where both m_1 and m_2 change sign. The zero-field anticrossing corresponds to a mechanism that implies cotunnelling of the two Mn_4 clusters, and therefore a much lower amplitude is expected than that involving pairs of level where only m_1 or m_2 changes sign.

A confirmation of this picture comes from the comparison of the hysteresis loops of $[Mn_4]_2$ with that of isolated Mn_4 clusters. In the latter a significant step at $H = 0$ is observed even if forbidden for a half-integer spin state. We have already seen that a transverse field of various origin (i.e. dipolar or hyperfine) relaxes this selection rule. On the contrary, when $[Mn_4]_2$ is investigated the step in zero field is found to be completely suppressed, because a transverse field does not remove the requirement of cotunnelling of the two giant spins of $[Mn_4]_2$.

Within the scheme of level crossings of Fig. 13.3 it has been possible to justify the field position and temperature dependence of the many steps observed in the hysteresis loop (Wernsdorfer et al. 2002). Beyond the fundamental interest of these results it is worth stressing their relevance for potential application. The switching of a weak intercluster interaction has suppressed the magnetic relaxation in zero field, a crucial point in magnetic data storage where a remnant magnetization stable in time is required

Equally interesting is the case with $\mathcal{H}'' \neq 0$. The pairs of levels with $|\mathcal{M}, \mathcal{M}'\rangle$ and $|\mathcal{M}', \mathcal{M}\rangle$ are now admixed and the eigenstates are now symmetric and anti-symmetric combinations of the pure states. The degeneracy is lifted and the splitting depends on the strength of the transverse exchange J_{xy}. The value of the transverse exchange interaction has been determined independently by Tiron *et al.* (2003) and Hill *et al.* (2003). In the first case the authors have performed a detailed investigation of the hysteresis loop, exploiting also the hole digging method already described in Section 9.2.2. They have been able to distinguish between the tunnelling resonances involving the symmetric and antisymmetric combinations of the $|-9/2, 7/2\rangle$ and $|7/2, -9/2\rangle$ states. The two steps in the hysteresis loops are separated by about 0.1 T. Hill *et al.* (2003) have used HF-EPR to investigate both field position and intensity of the many transitions that arise from the coupling of the two $S = 9/2$ spins, also providing some interesting hints for the potential use of dimers of SMMs for quantum computation. Both techniques have provided a value of J_{xy} very close to $J_z \approx 0.1$ K. An almost isotropic exchange interaction is expected given the small orbital contribution and the g value very close to that of the free electron for these clusters comprising Mn^{III} and Mn^{IV}, which both have an orbitally non-degenerate ground state.

In conclusion, two important observations have been originated by the study of weakly coupled Mn$_4$ clusters. The first is the possibility to significantly reduce the tunnelling in zero field as cotunnelling of the two spins of the dimer is required. Tunnelling in zero field is directly related to the loss of magnetic information in nanomagnets and the possibility of controlling it could have great potential interest for magnetic memory applications. The second point concerns the observation of entaglement of the two spins, thus providing some interesting hints for the potential use of dimers of SMMs for quantum computation (Hill *et al.* 2003).

13.2.2 *The effects of magnetic tunnelling on long-range magnetic order*

In Chapter 4 we have already shown that the intercluster interactions are small because of the organic shell provided by the terminal ligands. Dipolar interactions are, however, present, unless the molecules are diluted in a non-magnetic matrix. If r is the distance between magnetic groups in Å the order of magnitude of dipolar interactions, in K, is $\sim (gs)^2/r^3$. Thus, if $g = 2$, $s = 10$, and $r = 20$ Å, the ordering transition is expected to be of the order of 50 mK, or possibly lower because of frustration. The magnetic properties have been investigated down to the millikelvin temperature region where a magnetic ordered state should be attained. Magnetic ordering has indeed been observed in a few cases, for the Fe$_{19}$ cluster already reported in Chapter 4 (Affronte *et al.* 2002a), in a Mn$_6$ cluster (Morello *et al.* 2003), and in a Fe$_6$ cluster (Affronte *et al.* 2004a). In all these cases the dynamics of the magnetization of the cluster is relatively fast and the equilibrium state corresponding to magnetic order can be reached. On the contrary, Mn$_{12}$ac and Fe$_8$ are characterized by such a long relaxation time at low temperature and zero field that the magnetization freezes before reaching the ordering temperature. It would be interesting to address the question of the nature of this frozen state, that could in some respects resemble that of spin glasses (Martinez-Hidalgo *et al.* 2001), as seen in Section 12.5. Here, however, we want to describe an interesting study performed on two Mn$_4$ clusters of the family discussed in the previous section (Evangelisti *et al.* 2004).

These authors have measured the specific heat of two Mn$_4$ clusters of formula [Mn$_4$O$_3$Cl(O$_2$C-CH$_3$)$_3$(dbm)$_3$] and [Mn$_4$O$_3$(O$_2$C-C$_6$H$_4$-p-CH$_3$)$_4$(dbm)$_3$] respectively. The first one, abbreviated to Mn$_4$Cl hereafter, has the three-fold symmetry typical of these clusters and already discussed in Section 4.7.5. In the second case the apical bridging Cl$^-$ ion is replaced by the bridging anion of the p-methyl-benzoic acid, and the compound is thus referred to as Mn$_4$Me. In the case of Mn$_4$Me the three-fold symmetry is removed (see Fig. 4.32) and this gives rise to a much stronger transverse anisotropy.

The $|E/D|$ ratio is 0.21 for Mn$_4$Me, almost five times larger than for Mn$_4$Cl, while the axial anisotropy remains substantially unchanged. The larger transverse anisotropy results in a faster tunnelling in the low-temperature regime. Mn$_4$Me is thus expected to attain in a shorter time the equilibrium state, that at very low temperature could be an ordered state due to the weak intercluster interactions. However, the phase transition becomes observable only if the time required to attain equilibrium is comparable, or smaller, than the time window of the experimental technique.

Evangelisti *et al.* (2004) have measured the low-temperature specific heat using three different experimental time-scales, $\tau_e = 2$ s, 8 s, and 300 s, in the thermal relaxation method described in Section 3.2.3. The results for the two clusters are compared in Fig. 13.4. For Mn$_4$Cl a non-magnetic transition, probably a structural one, is observed at $T \approx 7$ K. The specific heat between 1 K and 7 K is well reproduced taking into account the Schottky anomaly due to the zero-field splitting of the $S = 9/2$ ground state. The increase of the specific heat at low temperature has been attributed to the hyperfine contribution. It is worth noting that the specific heat measured for the shortest experimental time is lower than that calculated for the nuclear contribution.

This observation suggests that the nuclear spin relaxation is strongly connected to the electron spin–lattice relaxation. In the specific heat of Mn$_4$Cl there is no evidence of a λ-anomaly due to magnetic ordering, which, on the contrary, is visible at $T_c = 0.21$ K in the case of Mn$_4$Me. The magnetic origin of this anomaly has been confirmed by its disappearance when a magnetic filed is applied, as shown by the empty circles in Fig. 13.4. It is also possible to evaluate the entropy change as a function of temperature. The magnetic contribution of the specific heat is obtained by a subtraction of the lattice part. However, two types of magnetic degree of freedom, the electronic and the nuclear ones, contribute to the specific heat. Evangelisti *et al.* (2004) evaluated the entropy of the electronic spins:

$$\Delta S(T) = \int_0^T \frac{C_M(T') - C_{nucl}(T')}{T'} dT' \tag{13.6}$$

where C_M is the magnetic contribution to the specific heat and C_{nucl} is the magnetic contribution responsible of the upturn observed at low temperature.

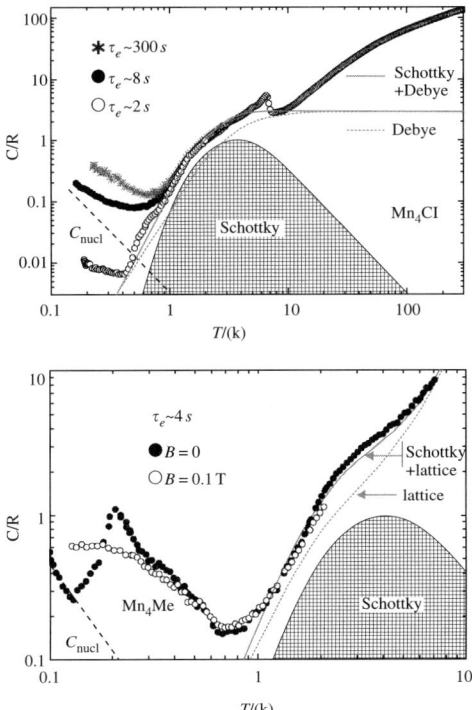

FIG. 13.4. Specific heat measured for [Mn$_4$O$_3$Cl(O$_2$C-CH$_3$)$_3$(dbm)$_3$] (top) and [Mn$_4$O$_3$(O$_2$C-C$_6$H$_4$-p-CH$_3$)$_4$(dbm)$_3$] (bottom). The experimental time τ_e employed in the relaxation method is indicated in the legend. The estimated Schottky contribution is shown as a grey bell-shaped curve, while the lattice and lattice plus Schottky contributions are reported as dotted and solid lines, respectively. The calculated nuclear specific heat is shown as a dashed line. Reprinted with permission from Evangelisti *et al.* (2004). Copyright (2004) of the American Physical Society.

Interestingly the electronic magnetic entropy tends to the value $R\ln(2S+1)$, with $S = 9/2$, at high temperature, where the 10 levels of the ground spin multiplet are all populated. Higher spin multiplets for this type of molecule are in fact more than 200 K higher in energy. On lowering the temperature ΔS reaches the value of $2R$ expected for a two-level system, only the $m = \pm 9/2$ pair being populated. This entropy is suddenly lost at T_c, due to the magnetic ordering.

The fastest tunnelling relaxation of Mn$_4$Me has allowed us to observe the ordering phenomena in the time-scale of the experiment. The nuclear contribution to the specific heat also agrees with the calculated one, confirming that the nuclei are now rapidly relaxing thanks to the faster relaxation time of the electronic spins.

These results clearly show how the possibility to observe the magnetic order of these non-diluted magnetic materials is strongly connected to the dynamics of the electron spins. The only efficient mechanism of relaxation at very low temperature is through incoherent tunnelling in the ground doublet. The tunnelling rate is, in turn, dominated by the transverse magnetic anisotropy and thus directly related to the structural symmetry of the molecule.

13.3   Tunnelling and electromagnetic radiation

The interplay between the electromagnetic radiation and the magnetic moment is at the basis of the magnetic resonance techniques discussed in Section 3.3, while its role in the dynamics of the magnetization has been discussed Section 5.9. Here we want to describe a type of experiment where the relaxation of the magnetization under applied electromagnetic radiation is monitored.

The principle of the experiments can be schematized as in Fig. 13.5. At very low temperature and zero field the only populated states are those with $m = \pm S$. If electromagnetic radiation of the appropriate frequency that matches the energy gap with the first excited states with $m' = \pm (S - 1)$ is applied, it can act as a pump thus populating these states. The spin temperature becomes different from the temperature of the thermal bath. As tunnelling involving states higher and higher in the barrier is more and more efficient, an acceleration of the overall relaxation rate is expected under irradiation. It is well known from EPR spectroscopy that the use of circular polarized radiation allows one to distinguish the sign of Δm of the promoted transition. In Fig. 13.5 it is shown that right polarization induces transitions with $\Delta m = -1$, thus is only efficient to pump the population in the right well, and vice versa.

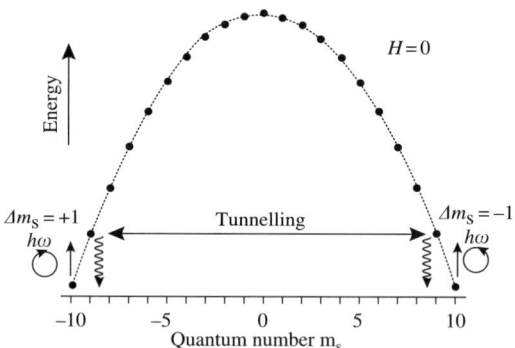

FIG. 13.5. Schematic diagram showing the double well potential in zero field of a SMM with $S = 10$. The electromagnetic radiation with matching frequency induces a transfer of population from the $m = \pm 10$ to ± 9 states where tunnelling is more efficient. Circularly polarized radiation selects the sign of Δm of the transition. From Sorace *et al.* (2003). Copyright (2003) of the American Physical Society.

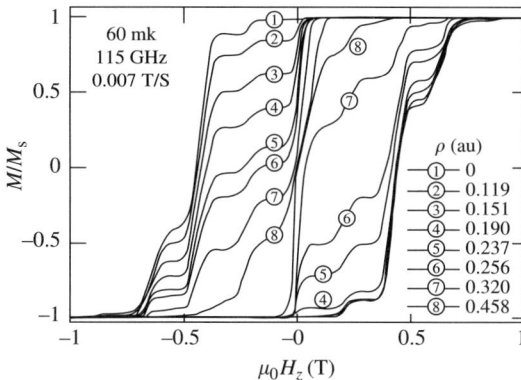

FIG. 13.6. Hysteresis loop of a single crystal of Fe_8 measured with a micro-Hall probe under irradiation at 115 GHz. The numbered curves represent the loop recorded with increasing power. Arbitrary units are used as the power has been measured at the Gunn-diode source but not at the sample. From Sorace *et al.* (2003). Copyright (2003) of the American Physical Society.

In the experiment of Sorace *et al.* (2003) a microcrystal of Fe_8 was mounted on a micro-Hall probe and inserted in a dilution refrigerator equipped with a superconducting magnet. Thanks to an oversized waveguide the sample is irradiated with the radiation emitted by a Gunn-diode source. This type of source is not tuneable but the gap in zero field between $m = \pm 10$ and $m' = \pm 9$ nicely matches the 115 GHz of a rather common source used in HF-EPR. In Fig. 13.6 the magnetic hysteresis loops under decreasing attenuation of the Gunn diode are shown. The typical stepped curves are observed, but, on increasing the power of the radiation, the zero-field step becomes more and more pronounced and the loops become narrower. In principle these effects are also encountered on heating the sample. Therefore, it can be hard to distinguish between an overall heating of the sample or a selective pump effect of the radiation. In Fig. 13.6 it is, however, clear that the loops become strongly asymmetric, as only the descending branch of the hysteresis loop is significantly affected by the increasing power of the radiation. This is a direct consequence of the use of highly polarized (97%) radiation, and nicely confirms the expected resonant effect of the radiation.

The combined use of magnetometry and microwave radiation is, at the time of writing, one of the hottest topics in molecular nanomagnetism. While in the previous example the electromagnetic radiation has been used to promote transitions within the same well, a lower frequency can be employed to monitor the effect of the radiation on the tunnelling rate at an avoided level crossing (Del Barco *et al.* 2004). In order to do so, these authors have employed a sample, a tetranuclear Ni(II) cluster of formula [Ni(hpm)(t-Bu-EtOH)Cl]$_4$ (Yang *et al.* 2003), which has a significantly larger splitting, further increased by the application of a transverse field.

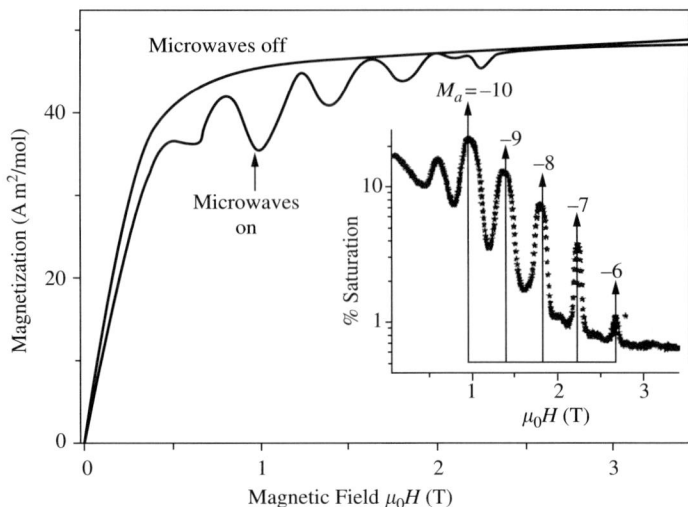

FIG. 13.7. Magnetization variation induced by the application of microwave radiation (141 GHz) versus applied longitudinal field measured on a single crystal of Fe$_8$ in the thermally activated tunnelling regime (T = 4 K). When the frequency of the radiation coincides with the energy splitting of $|m\rangle$ and $|m+1\rangle$ states of the ground $S = 10$ multiplet the radiation is absorbed and the spin temperature is modified with a significant decrease of the measured magnetization. Reprinted with permission from Cage *et al.* (2005). Copyright (2005) American Institut of Physics.

The magnetization state of these quantum objects can therefore be manipulated thanks to electromagnetic radiation, thus opening new horizons in the field of molecular nanomagnetism. The outcome of this lively research goes even beyond the results shown above. In fact the change in magnetization under the effect of electromagnetic radiation represents a new type of EPR detection as shown in Fig. 13.7 (Cage *et al.* 2005). When this technique is associated with micro-Hall probes, extraordinary sensitivity can be achieved (Bay *et al.* 2005; Petukhov *et al.* 2005).

14

OTHER MAGNETIC MOLECULES

Beyond molecules with slow relaxation of the magnetization at low temperature, several other classes of molecules have attracted widespread interest. In particular we will focus our attention on rings, grids, and large antiferromagnetic clusters, showing how they can provide interesting magnetic properties even if they do not show slow relaxation of the magnetization.

Rings were initially devised as theoretical tools (not to say toys) to help understand the magnetic properties and the spin dynamics of one-dimensional materials. However, at the end of the twentieth century, real rings were synthesized, and it became possible to test theories on real objects. Often these new materials are labelled Ferric wheels, after the first widely investigated system, namely a decanuclear iron(III) derivative of formula $[\mathrm{Fe(OCH_3)_2(O_2CCH_2Cl)}]_{10}$ whose structure is shown in Fig. 14.1 (Taft *et al.* 1994). The name is associated with the well-known Ferris wheel, also depicted in Fig. 14.1, which was the attraction of the 1893 exhibition in Chicago, the American answer to the Eiffel

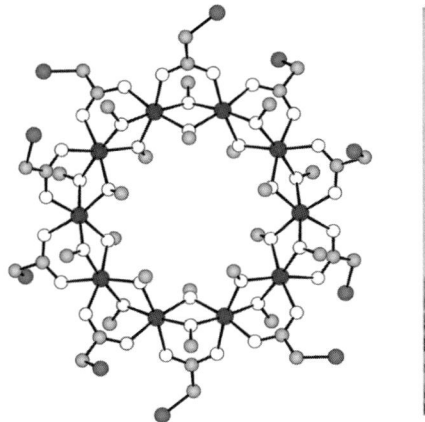
Ferric wheel

Ferris wheel

FIG. 14.1. The Ferric wheel, left, with the iron atoms drawn as large black spheres, oxygen white, and chlorine as large grey spheres. On the right, a picture of Ferris wheel.

tower built for the Paris exhibition in 1889. The wheel was created by the bridge builder George W. Ferris.

The Ferric wheel is the archetype of antiferromagnetic rings, exactly as $Mn_{12}ac$ is the archetype of SMM. The interest in antiferromagnetic rings increased when it was suggested that it should be possible to observe quantum coherent oscillations (Meier and Loss 2001). These points will be worked out in Section 14.1.

The largest wheel reported so far, in the form of a torus of 84 manganese ions (Tasiopoulos et al. 2004), was described by Christou and co-workers and its structure was shown in Fig. 1.5. The compound behaves as a SMM at low temperature.

While the rings are models for one-dimensional materials, grids are models for two-dimensional magnetic materials (Lehn 1995; Thompson et al. 2003). The schematic structure of a grid is shown below.

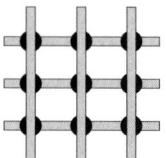

The bars indicate suitable ligands which can bind to three different metal ions, while the circles represent the metal ions. Clearly the 3×3 grid is an excellent model of a square lattice.

Grids, however, are not the only type of two-dimensional clusters. Fe_8, which has been widely described above (Wieghardt et al. 1984), after all is a planar molecule comprising a layer of iron ions. Larger clusters comprising more than 15 ions arranged as in the brucite structure have also been reported (Heath and Powell 1992), and the structure of one of them has been discussed in Section 4.2. The Brucite structure is essentially one layer of close packed metal ions separated from other layers. A more regular structure was observed in a Mn_{19} cluster (Pohl et al. 2001). The sketch in Fig. 14.2 shows how the metal ions are arranged. First we can imagine having a ring of six ions around a central one. The expansion of this shell requires 12 additional ions. This defines a set of 'magic numbers', 1, 7, and 19. The next step would be to add another ring of 18 ions, yielding a magic number 37, but so far no such cluster has been reported.

The metal ions in these typical structures are connected by suitable bridges, for instance oxygen atoms, which form a close packed layer even if they are connected to organic moieties (Caneschi et al. 1995).

Also, more complex structures can be observed, which correspond to three-dimensional arrangements of the magnetic ions. These spin topologies are more frequently met in polyoxometallates (Coronado and Gomez-García 1998; Müller et al. 1998), including oxovanadium(IV) ions, spin $S = \frac{1}{2}$, and

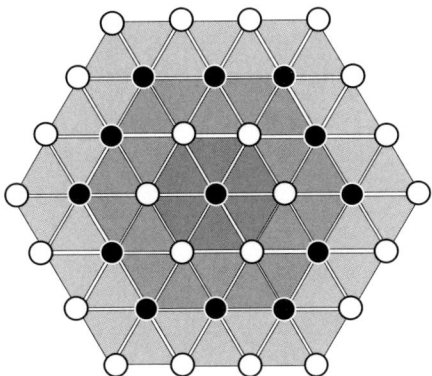

FIG. 14.2. Sketch of the metal ion layers in a brucite structure. The first, second, and third shells are drawn in a grey scale and provide finite structure with 7, 19, and 37 spins, respectively.

iron(III), spin $S = 5/2$. These systems will be treated in Section 14.3. Among them perhaps the most fascinating from the esthetical point of view is $[Mo_{72}Fe_{30}O_{252}(Mo_2O_7)(H_2O)_2(Mo_2O_8H_2(H_2O))(CH_3COO)_{12}(H_2O)_{91}]\cdot150H_2O$, $Mo_{72}Fe_{30}$, whose structure was described in Chapter 4 (Müller $et\ al.$ 1999). The molybdenum ions can be considered to be diamagnetic, while the 30 iron(III) ions have $S = 5/2$.

14.1 Magnetic wheels

The most numerous class of antiferromagnetic wheels reported so far contain iron(III) ions, but other metal ions, like copper(II) and chromium(III), have also been described. Rings with a number of magnetic centres $N = 6, 8, 10, 12$, and 18 have been reported and their properties analysed in some detail (Caneschi $et\ al.$ 1996; Caneschi $et\ al.$ 1999; Saalfrank $et\ al.$ 1997; Taft $et\ al.$ 1994; Watton $et\ al.$ 1997; Van Slageren $et\ al.$ 2002; Waldmann 2002a; Waldmann $et\ al.$ 2003). Beyond being of interest as models for low-dimensional magnets antiferromagnetic rings are attracting interest for the hypothesis that they may provide good opportunities for observing quantum coherence in the oscillation of the Néel vector. A representation of this last one is shown in Fig. 14.3 for a six-member ring. Some relevant references are available (Meier and Loss 2001; Honecker $et\ al.$ 2002; Waldmann 2002b) and only a sketchy description will be given below.

The Néel vector operator is defined as:

$$\mathbf{n} = \frac{1}{Ns} \sum_{i=1}^{N} (-1)^i \mathbf{s}_i. \qquad (14.1)$$

There are several advantages of antiferromagnetic rings compared to ferro- or ferrimagnetic clusters. The main advantage is associated to the fact that in principle the tunnel frequency between almost degenerate states in AF rings,

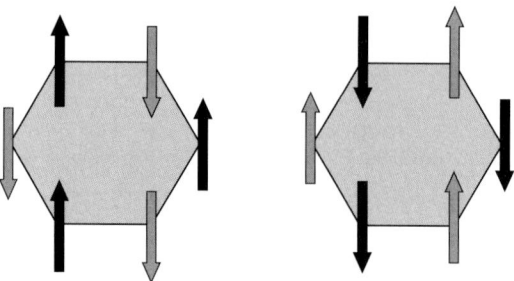

FIG. 14.3. The Néel vector in a six-member AF ring. In classical terms there are two degenerate configurations, left and right, and the system can tunnel between them.

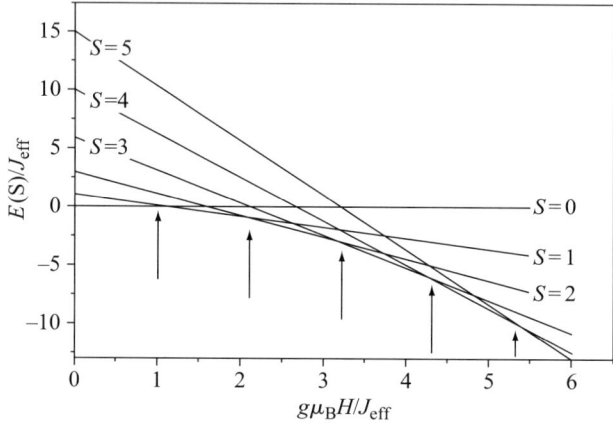

FIG. 14.4. Field effect on the energy levels of a ring with an even number of antiferromagnetically coupled spins.

Δ, is much larger than in ferro- or ferrimagnetic clusters. In order to observe coherence the decoherence frequency, Γ, must be significantly smaller than the tunnel frequency, therefore a larger tunnel splitting makes AF clusters better suited than the FM, in this respect.

In fact the gap between levels for AF rings is large, being determined by isotropic exchange. The external field can be used to tune the tunnel splitting. This phenomenon has already been described in Chapter 3. Figure 14.4 shows the levels of a ring of antiferromagnetically coupled spins, and the effect of a magnetic field applied parallel to the Z axis, this last corresponding to the unique axis normal to the ring.

At zero field the ground state is $S = 0$, and then several cross-overs can be observed (Taft *et al.* 1994). The situation is analogous to that described in

Section 3.1.1 for antiferromagnetic dimers. It will be seen in Section 14.1.1 that the energies $E(S)$ of the zero-field levels above the ground state scale according to the formula

$$E(S) = -\frac{J_{\text{eff}}}{2}S(S+1) \qquad (14.2)$$

where J_{eff} is an effective coupling constant (Caneschi *et al.* 1996). Formula (14.2) turns out to coincide with the Landé interval rule valid for atoms. It implies that the cross-overs from the S to $S+1$ ground state occur with identical spacing:

$$H_{S \to S+1} = -\frac{J_{\text{eff}}}{g\mu_{\text{B}}}(S+1). \qquad (14.3)$$

In the neighbourhood of the cross-over at sufficiently low temperature the system behaves as a two-level system, an ideal one to observe quantum tunnelling related phenomena provided that the tunnelling splitting is not too small.

Level crossing can also be induced by transverse fields applied in the ring plane, and it is this type of configuration which is under investigation for observing the tunnelling of the Néel vector. The ideal conditions would consist in choosing a field such that the AF ring can be described by a two-level state, $|0\rangle, |1\rangle$. The system must be prepared in a superposition of energy eigenstates of the Hamiltonian, $\Psi = \frac{1}{\sqrt{2}}[|0\rangle + |1\rangle]$. This corresponds to a Néel vector oriented in the $+Z$ direction. The state has a temporal evolution such that after every odd number of half-cycles it evolves into the degenerate state $\Psi' = \frac{1}{\sqrt{2}}[|0\rangle - |1\rangle]$, with the Néel vector oriented in the $-Z$ direction.

14.1.1   *Iron rings*

This is so far the most complete series of AF rings. The Ferric wheel was the first to be characterized, in particular for its typical stepped magnetization. The origin of this behaviour has been discussed above. The experimental data were obtained at low temperature, both in a static, up to 20 T, and in a dynamic, up to 50 T, magnetic field and they are shown in Fig. 14.5 (Taft *et al.* 1994).

The steps in the static field correspond to the cross-overs $S = 0 \to 1, 1 \to 2, 2 \to 3$, while those in the right-hand side show also the transitions up to $S = 9$. The energies of the excited states were fitted using (14.3) to give $J_{\text{eff}} = -13.8$ K.

Very detailed studies were performed on six-member rings, taking advantage of the smaller number of states, which allow the diagonalization of the Hamiltonian matrix in a relatively simple way. Perhaps one of the most thoroughly investigated systems is [NaFe$_6$(OMe)$_{12}$(pmdbm)$_6$]ClO$_4$, Hpmdbm= 1,3-bis(4-methoxyphenyl)-1,3-propanedione, whose structure is shown in Fig. 14.6 (Caneschi *et al.* 1996). The ring of six iron(III) ions has a sodium ion in the centre. In fact the centre of the ring comprises six oxygen atoms which define an octahedral coordination site for an additional ion. The compound crystallizes in the R$\bar{3}$ space group, with the cluster in an S$_6$ symmetry site, which makes the iron ions all equivalent to each other.

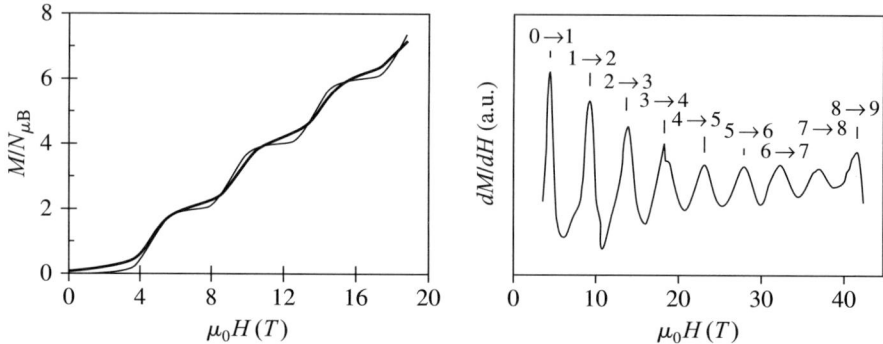

FIG. 14.5. Left: magnetization of Fe_{10} at 0.7 K as a function of the applied field in a static experiment; right derivative of the magnetization recorded in a pulsed field. From Taft *et al.* (1994). Copyright (1996) American Chemical Society.

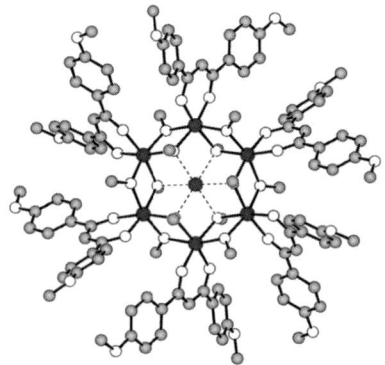

FIG. 14.6. Structure of $[NaFe_6(OMe)_{12}(pmdbm)_6]ClO_4$. The grey scheme of Fig. 14.1 has been employed. From Caneschi *et al.* (1996) with permission of Wiley-VCH.

The temperature dependence of the magnetic susceptibility was satisfactorily fitted with $J = -28.6$ K. From the analysis of the high-field magnetization of $[NaFe_6(OMe)_{12}(pmdbm)_6]ClO_4$ at low temperature the coupling constant J_{eff} defined by (14.3) was found to be equal to -33.0 K (Cornia *et al.* 1999).

The origin of J_{eff} is associated with an approximate treatment of the low-lying levels. For classical spins, the ground state is the so-called Néel state, all odd spins are parallel and all even spins are antiparallel to the odd spins. Thus $S_{2i+1}^z = s$ and $S_{2i}^z = -s$, where z is an arbitrary direction. The total spin S is thus 0. To increase S by a given amount with a minimal energy increase, the recipe is to keep the odd spins parallel and the even spins parallel too, but to tilt the direction of the odd spins with respect to the even spins by an angle 2θ,

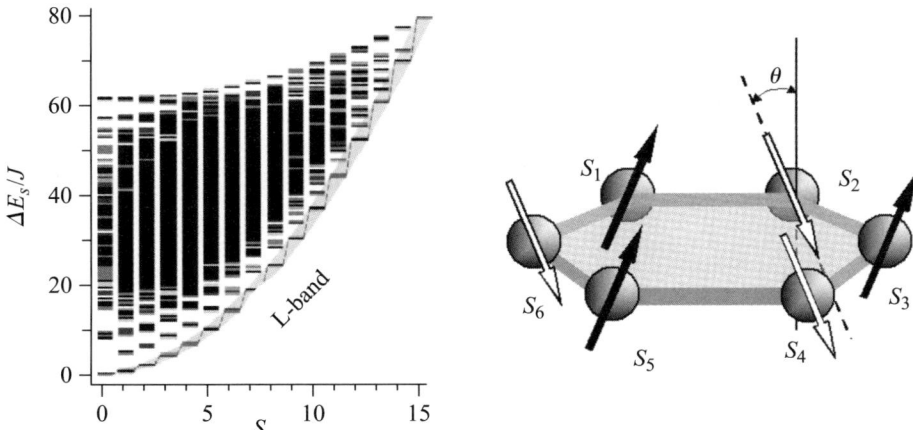

FIG. 14.7. Left: the energy spectrum of the spin levels for an antiferromagnetic ring of six $s = 5/2$, where the first rotational band, or L-band, is shown in grey. Right: a schematization of how the L-band is generated by the relative tilting of the two sublattices (of an angle 2θ in the classical picture), the spins of one sublattice remaining however parallel to each other.

as shown in Fig. 14.7. Thus, the energy is increased by

$$E = |J|Ns^2(1 - \cos 2\theta) \tag{14.4}$$

while

$$S = Ns\sin\theta = Ns\sqrt{\frac{1}{2}(1 - \cos 2\theta)}. \tag{14.5}$$

Introducing (14.4) in (14.5) yields $E = 2|J|S^2/N$. This coincides with (14.2), with

$$J_{\text{eff}} = \frac{4J}{N}. \tag{14.6}$$

The same formula for J_{eff} is obtained using the ITO formalism associated with a partition of the individual quantum spin in odd and even sites. The total spins of the two sublattices are given as

$$\mathbf{S_A} = \sum_{i=0}^{N/2-1} \mathbf{S_{2i+1}} \text{ and } \mathbf{S_B} = \sum_{1}^{N/2} \mathbf{S_{2i}} \tag{14.7}$$

and the lowest lying states correspond well to those arising from the coupling of these two intermediate spins:

$$\mathcal{H} = -J_{\text{eff}}\mathbf{S_A} \cdot \mathbf{S_B} \tag{14.8}$$

where $S_A = S_B = (N/2)s$ correspond to the parallel alignment of all the spins of a sublattice. The validity of the approximate treatment leading to (14.6) increases on increasing the s value of the individual spins. Moreover, it decreases on increasing the size of the ring, N. In fact these low-energy excitations, schematized as a coherent rotation of the spins of the two sublattices, are typical of finite systems. In the infinite chain the lowest excitations are described by spin waves that delocalize the excitations due to flipped spins. A discussion of the limitation of this approach is also available (Santini *et al.* 2005).

The low-lying levels calculated for a ring of six $s = 5/2$ ions is shown in Fig. 14.7. It is apparent that there are regularities in the low-lying levels. Looking at the lowest levels of each S value we note a parabolic behaviour, as suggested by (14.2). The sequence of levels is also called a rotational band, as the energy of the spin levels scales as those of a rigid rotor. Similar behaviour is observed if we look at the second lowest lying levels of each S value. The lowest band, also called the L-band, corresponds to $S_A = S_B = (N/2)s$, the first excited band to $S_A = [(N/2) - 1]s, S_B = (N/2)s$, and $S_B = [(N/2) - 1]s, S_A = (N/2)s$, and so on. This observation has been extended to other spin topologies and will be discussed in more detail below.

Information on the lowest lying levels of $[NaFe_6(OMe)_{12}(pmdbm)_6]ClO_4$ was obtained using several different techniques (Abbati *et al.* 2001), like torque magnetometry and neutron measurements. In particular it was possible to measure the zero-field splitting parameters of the first excited $S = 1$ state as $D = 6.22(3)$ K from Inelastic Neutron Scattering experiments and $D = 6.58(3)$ K from torque magnetometry. Attempts to reproduce this value using the dipolar approximation were unsuccessful, showing that single-ion contributions may be operative.

In order to obtain first-hand information on the single-ion anisotropy the analogous and isomorphous diamagnetic gallium(III) derivative was synthesized. By doping the gallium derivative with iron(III) ions mixed species with stoichiometry $Ga_{N-n}Fe_n$ are obtained ($n \leq N$) whose relative abundance can be calculated if a statistical distribution of the two metal ions is assumed. In fact it can be shown that the probability of a given species, P_{Nn}, is given by a binomial distribution:

$$P_{Nn} = \frac{N!}{n!(N-n)!} P^n (1 - P)^{N-m} \tag{14.9}$$

where P is the iron mole fraction. By choosing the mole fraction of iron it was possible to obtain rings where the predominant species is Ga_5Fe_1 and the magnetic data, in particular high-field magnetic torque measurements, and high-field EPR spectra, provided direct information on the single-ion anisotropy. The zero-field splitting tensor was found to be rhombic with $D = 0.62(2)$ K, $E/D = 0.15$. The Z axis of the tensor is not parallel to the unique axis of the Fe_6 cluster, z, but makes an angle of $79.2(4)°$ with it.

Using the formalism developed in Chapter 2, the zero-field splitting component of the single-ion anisotropy parallel to the Z axis is given by:

$$D_{S=1,ZZ} = \sum_{i=1}^{6} d_i D_{ZZ} \qquad (14.10)$$

where $\mathbf{D}_{S=1}$ is the tensor for the first excited $S = 1$ state, $\mathbf{D}$ is the single-ion tensor and d_i is a coefficient which depends on the nature of the $S = 1$ spin state. Assuming that it can be described as $|S_A S_B S\rangle$, where $S_A = S_B = 15/2, S = 1$ the d_i coefficients are calculated as $-12/5$ using iteratively (2.27), as shown in Section 2.5.2. Since the individual $\mathbf{D}$ tensors of the iron ions have their local anisotropy axis, z, essentially perpendicular to the unique Z axis of the ring, and D is positive, a negative component is projected parallel to Z. Using the coefficients given above and the experimental values the single-ion contribution is calculated as $D_{S=1} = 5.2(1)$ K. It must be remarked that the positive D value derives from a negative single-ion contribution along the unique Z axis and a negative projection on the total spin S. Adding, with the same procedure, the calculated values of the dipolar contributions, $D_{S=1}$ is calculated as 6.8(1) K in excellent agreement with the experimental data (Abbati *et al.* 2001).

One of the open problems is bound to the phenomenon of the crossing of the levels. In fact as already seen in the previous chapters two quantum levels can either simply cross each other or go through an anti- (or avoided-) crossing depending of whether there is a matrix element connecting them or not. Clearly this point is relevant for the observation of quantum effects and tunnelling in this type of material.

Evidence of anticrossing of the low-lying levels in a field has been achieved for $[NaFe_6(OMe)_{12}(pmdbm)_6]ClO_4$ and for $[LiFe_6(OMe)_{12}(dbm)_6]$ $B(C_6H_5)_4 \cdot 5CH_2Cl_2$, where Hdbm=di-benzoylmethane. For the latter the best evidence came from specific heat measurements performed at 0.78 K (Affronte *et al.* 2002b), shown in Fig. 14.8.

Two pronounced peaks are observed at 10.4 and 13.0 T with a relative minimum at the crossing field between the ground $S = 0$ and the first excited $S = 1$ state, $B_{c1} = 11.7$ T. A single broad peak is observed at $B_{c2} = 22.4$ T. The observed behaviour is associated with a Schottky-type anomaly which occurs because two levels are populated in the crossing regions. The corresponding expression for the specific heat C_S is given by:

$$C_S = \left(\frac{\Delta}{k_B T}\right)^2 \frac{\exp(\Delta/k_B T)}{[1 + \exp(\Delta/k_B T)]^2} \qquad (14.11)$$

where Δ is the separation between the pair of populated levels, which is field dependent. Close to the cross-over fields, $B_{cn}\Delta$ is given by $\Delta(B) = g\mu_B|B_{cn} - B|$. The specific heat must show two peaks at $B \approx B_{cn} \pm k_B T/(g\mu_B)$ and go to zero at B_{cn} if the levels cross. In fact at B_{cn} the two levels are degenerate and the contribution to the specific heat is zero. The experimental data were fitted with

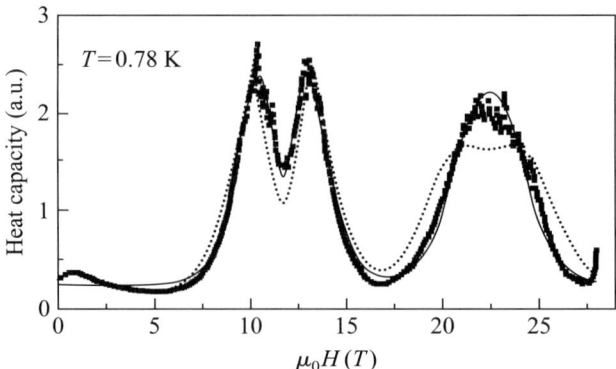

FIG. 14.8. Specific heat measurements of $[\text{LiFe}_6(\text{OMe})_{12}(\text{dbm})_6]\text{B}(\text{C}_6\text{H}_5)_4 \cdot 5\text{CH}_2\text{Cl}_2$ at 0.78 K. The solid and dotted curves are calculated with the models described in the text. From Affronte *et al.* (2002b). Copyright (2002) of the American Physical Society.

two different models, one taking into account the presence of a distribution of cross-over fields due to sample inhomogeneity (shown as dotted lines) and the other introducing avoided crossing of the levels, shown as a solid line. For the latter, which gives the best fit, the field dependence of the level separation is given as:

$$\Delta(B) = \{[g\mu_B(B_{cn} - B)]^2 + 4K_n^2\}^{1/2} \qquad (14.12)$$

where K_n is the matrix element connecting the two states. The best-fit parameters are: $B_{c1} = 11.81(1)$ T, $K_1 = 0.43$ K; $B_{c2} = 22.43(1)$ T; $K_2 = 1.18$ K. The second peak does not show any double structure because the tunnel splitting is larger than the thermal energy, thus the Schottky contribution to the specific heat reaches its maximum at B_{c2}.

The physical origin of the coupling responsible of the avoided crossing is not yet clarified. In fact a perturbation which is capable of coupling states of different parity is needed, because S and $S + 1$ states are involved. The obvious candidate for such behaviour is the antisymmetric exchange, but it requires a non-centrosymmetric structure. The crystal structure of the compounds does not fulfil this requirement, so it has been postulated that distortions are operative at low temperature (Nakano and Miyashita 2001). This is a feature which perhaps occurs more often than anticipated. For instance in the interpretation of the neutron data and of the EPR spectra of a Cr_8 ring, which at room temperature has tetragonal symmetry, it was necessary to include a rhombic term (Carretta *et al.* 2003a; Van Slageren *et al.* 2002). Low-temperature structural data showed that indeed at low T the structure is no longer tetragonal.

14.2 Grids

The synthesis of grids is a beautiful example of self-assembly (Baxter *et al.* 1994; Lehn 1995; Thompson 2002), which gives the required cluster with high yield. The structure of the largest magnetic grid reported so far (Zhao *et al.* 2000a,b), the $[\text{Mn}_9(2\text{-POAP-2H})_6]^{6+}$ cation, was shown in Fig. 4.7.

The temperature dependence of the magnetic susceptibility clearly shows that the predominant coupling is antiferromagnetic. Low-temperature magnetization suggests a ground $S = 5/2$ state, with low-lying excited states with higher spin multiplicity. Using qualitative considerations it is rather easy to give a schematic representation of the ground state, as shown in Fig. 14.9. Full-scale calculations confirmed this view (Carretta *et al.* 2003b; Guidi *et al.* 2004).

An important piece of information also came from INS experiments (Guidi *et al.* 2004). These showed that the first excited state is a $S = 7/2$ state, while the second is a $S = 3/2$. The zero-field splitting factor for the ground S = 5/2 was found to be $D_{5/2} = -0.48$ K, while that of the first excited state is $D_{7/2} = -2.9 \times 10^{-2}$ K. A convenient form for the Hamiltonian was found to be:

$$H = -J_R \left(\sum_{i=1}^{7} \mathbf{S_i} \cdot \mathbf{S_{i+1}} + \mathbf{S_8} \cdot \mathbf{S_1} \right) - J_C \left(\mathbf{S_2} + \mathbf{S_4} + \mathbf{S_6} + \mathbf{S_8} \right) \cdot \mathbf{S_9}$$

$$+ D_R \sum_{i=1}^{8} S_{i,z}^2 + D_C S_{9,z}^2 + g\mu_{\text{B}} \mathbf{S} \cdot \mathbf{H} \qquad (14.13)$$

where R refers to spin in the external ring of eight ions and C to the central ion. In order to describe the low-temperature behaviour the simplified effective spin Hamiltonian shown below was used:

$$H_{\text{eff}} = -J_R^{\text{eff}} \mathbf{S_A} \cdot \mathbf{S_B} + D_R^{\text{eff}} (S_{A,z}^2 + S_{B,z}^2) - J_C \mathbf{S_R} \cdot \mathbf{S_9} + D_C S_{9,z}^2 + g\mu_{\text{B}} \mathbf{S} \cdot \mathbf{H}$$
$$(14.14)$$

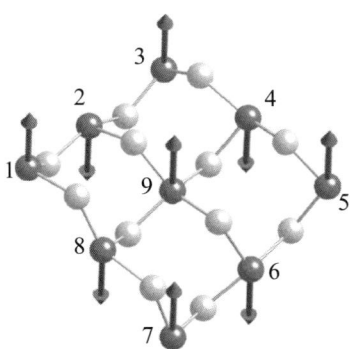

FIG. 14.9. Schematic structure of the magnetic core structure of $[\text{Mn}_9(2\text{-POAP-}2\text{H})_6]^{6+}$ with the labelling of the magnetic sites and the spin orientation in the ground state.

where

$$\mathbf{S}_A = \mathbf{S}_1 + \mathbf{S}_3 + \mathbf{S}_5 + \mathbf{S}_7; \mathbf{S}_B = \mathbf{S}_2 + \mathbf{S}_4 + \mathbf{S}_6 + \mathbf{S}_8; \mathbf{S}_R = \mathbf{S}_A + \mathbf{S}_B; \mathbf{S} = \mathbf{S}_R + \mathbf{S}_9.$$
$$(14.15)$$

The lowest lying S_R levels can be obtained by antiferromagnetically coupling $S_A = S_B = 10$. The S_R levels approximately follow a Landé interval rule. In the assumption $J_R^{\mathrm{eff}} > J_C$ the order of the low-lying levels can be easily approximated assuming that the energies of the S states can be obtained by combining the $S_R = S_A + S_B$ states with S_9. Using spin projection techniques it turns out that for the lowest lying states $J_R^{\mathrm{eff}} = 0.526 \, J_R$ and $D_R^{\mathrm{eff}} = 0.197 \, D_R$.

An interesting feature of Mn$_9$ is the observation of peaks in the torque curves at 0.4 K (Carretta et $al.$ 2003; Waldmann et $al.$ 2004), as shown in Fig. 14.10.

The torque rapidly increases at low field, reaching a maximum at about 2 T. This is due to the zero-field splitting of the ground $S = 5/2$ state. At higher fields oscillations are observed. The torque decreases on increasing field and eventually changes sign. The change in sign was initially associated to a change in magnetic anisotropy on passing from the ground $S = 5/2$ state in zero field to the first excited state $S = 7/2$. These features are only observed in the torque and not in the magnetization. Actually the INS data described above showed that there is an admixture of the S states with the higher lying $S = 9/2$ state, which has different sign in the anisotropy and determines the observed change in the torque. The oscillations are determined by the admixing of states according to the model outlined below.

In the level-crossing region only two states $|S, m = -S\rangle$ and $|S + 1, m = -S - 1\rangle$ are thermally populated and the energy spectrum can be described by

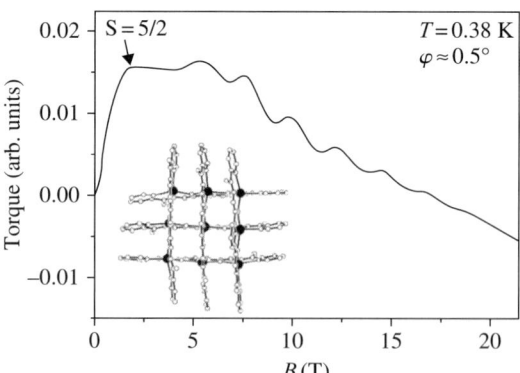

FIG. 14.10. Torque versus magnetic field for a single crystal of Mn$_9$. The applied field is in the grid plane. From Waldmann et $al.$ (2004). Copyright (2004) of the American Physical Society.

the following Hamiltonian matrix:

$$\begin{bmatrix} \varepsilon_s & \Delta/2 \\ \Delta/2 & \varepsilon_{s+1} \end{bmatrix} \tag{14.16}$$

where $\varepsilon_S(B,\varphi)$ and $\varepsilon_{S+1}(B,\varphi)$ describe the field and angular dependence of the two levels taking into account the magnetic anisotropy (zero-field splitting) but excluding level mixing at the cross-over field. The level mixing is parametrized by Δ.

14.3 Three-dimensional clusters

14.3.1 *Spherical antiferromagnets*

As anticipated above the most symmetrical large cluster reported so far is $Mo_{72}Fe_{30}$. The geometry of the 30 iron(III) ions is particularly interesting because it defines an icosidocahedron, as sketched in Fig. 14.11.

In a sense it can be considered as a half-fullerene. The iron ions are on the vertices of 12 pentagons, like in fullerene, but each pentagon is directly connected with five nearest-neighbour pentagons, giving rise to 20 triangles on the whole. It is apparent that the structure is extremely spin frustrated, due to the number of antiferromagnetic interactions in rings with an odd number of members. Consequently it must be expected that the ground state is characterized by a high degeneracy.

The temperature dependence of the magnetic susceptibility has been measured, showing that indeed a weak antiferromagnetic coupling is operative between nearest neighbour iron ions. Approximate calculations, due to the impossibility of using exact approaches for the 6^{30} states, suggested a coupling

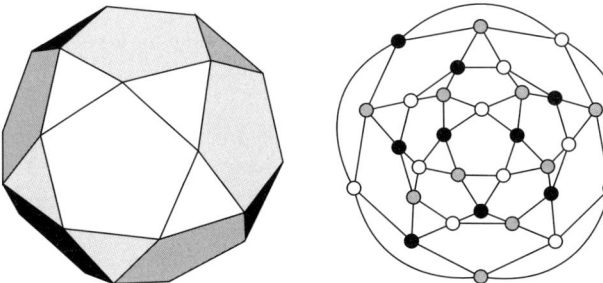

FIG. 14.11. Left: sketch of the icosidodecahedral arrangement of iron(III) ions in $Mo_{72}Fe_{30}$. Right: planar projection. Classically each spin of a triangle forms an angle of 120° with the two neighbours, thus defining three magnetic sublattices that are highlighted with different shades. From Axenovich and Luban (2001). The icosidodecahedron is a convex quasiregular polyhedron (i.e. all corners are identical) with 32 faces, 20 of which ($\varepsilon\iota\kappa o\sigma\iota$) are triangles while 12 ($\delta o\delta\varepsilon\kappa\alpha$) are pentagons.

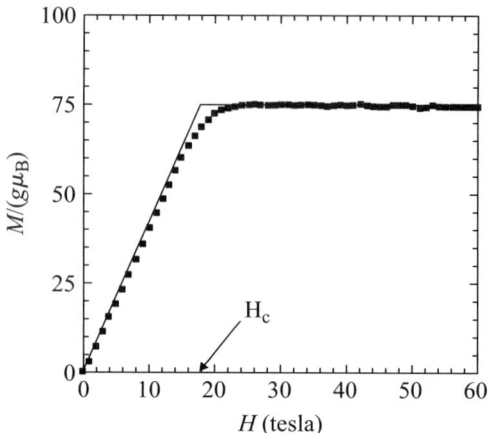

FIG. 14.12. Field dependence of the magnetization of $Mo_{72}Fe_{30}$ in a pulsed field. The nominal temperature is 0.46 K, but the spin temperature has been estimated to be close to 4 K. From Axenovich and Luban (2001); original data in Müller *et al.* (2001b).

constant $J = -1.57$ K (Müller *et al.* 2001b). These were based on classical vectors, including only nearest-neighbour coupling constants.

Beyond the fit of the temperature dependence of the susceptibility, the most interesting feature of $Mo_{72}Fe_{30}$ is that down to 100 mK the magnetization is not expected to show any evidence of a gap between the $S = 0$ ground state and the first excited $S = 1$ state. Pulsed magnetization measurements seem to support this view, although it was impossible to control accurately the spin temperature in the pulsed field measurements reported in Fig. 14.12 (Müller *et al.* 2001b).

The theoretical model used to describe the low-lying levels is a generalization of the treatment used above for the antiferromagnetic rings. We recall that the low-lying levels of the rings can be described by dividing the N spins into two subsets, the odd and even sites. The conjecture of Luban (Schnack and Luban 2001) is that this approach can be extended to all the systems of N identical spins s which can be divided into n_{sl} sublattices according to a symmetry transformation. In the case of the icosidodecahedron there are three sublattices, defined by the white, grey, and black shading in Fig. 14.11. The spins in each sublattice must be coupled to their maximal values $S_{sl} = Ns/n_{sl}$, and the energies of the lowest lying states for each S spin are given by:

$$E_{S,\min} \approx -\frac{J}{2}\left[\frac{A(N, s)}{N}\right] S(S + 1) + E_a \qquad (14.17)$$

where $A(N, s)$ is a coefficient, to be discussed below. E_a is a normalization factor which guarantees that (14.17) gives the right energy for the ferromagnetic state,

$S = Ns$. The low-lying levels are indicated as rotational bands because the energies of the levels vary quadratically with the total spin like in a rigid rotor. For the icosidodecahedron the coefficient $A(N, s)$ can be obtained either from a comparison with the diagonalization of the Hamiltonian matrices or from first principles following an approach similar to that outlined in Section 14.1 for rings. For $Mo_{72}Fe_{30}$ the lowest rotational band takes the form:

$$E_{S,\min} = -\frac{J}{10}S(S+1) + 30Js\left(s + \frac{1}{10}\right) \tag{14.18}$$

which can be rewritten in an equivalent way to express the energies also of the excited rotational bands, as:

$$E_S = -\frac{J}{10}\left[S(S+1) - S_A(S_A+1) - S_B(S_B+1) - S_C(S_C+1)\right] \tag{14.19}$$

where S_A, S_B, and S_C are the three sublattice spin vectors. For the ground band $S_A = S_B = S_C = 25$; for the first excited band $S_A = 24$; $S_B = S_C = 25$, and all the required permutations, and so on.

The calculated magnetization at 0 K yields a staircase with 75 steps which terminates at the critical field $B_c = 15|J|/(g\mu_B)$. Above B_c the magnetization saturates and all the spins are parallel to each other. Introducing the above value of J the critical field is calculated as 17.7 T, in excellent agreement with the experimental data of Fig. 14.12.

14.3.2   *Vanadium cluster*

Another three-dimensional cluster containing a high symmetry distribution of magnetic ions is provided by $[V_{15}^{IV}As_6O_{42}(H_2O)]$, V_{15}. The oxovanadium(IV) ions, $s = \frac{1}{2}$, are arranged in three layers, containing six, three, and six ions, respectively, as sketched in Fig. 14.13.

The coupling in the hexagonal layers is strongly antiferromagnetic, in such a way that at relatively high temperature the spins in the hexagons freeze into the ground $S_h = 0$ state, as shown in Fig. 14.13. The coupling between the spins in the middle triangular layer is very weak, while they are coupled to the spins in the hexagons. As a consequence the two $S_t = \frac{1}{2}$ and the $S_t = 3/2$ levels for the triangle are thermally populated down to low temperature. The quartet state was reported to lie 3.8 K above the doublets. INS measurements in the presence and in the absence of an applied magnetic field showed that the ground pair of doublets is actually split by about 0.31(4) K (Chaboussant *et al.* 2004).

From the structural point of view the triangle is equilateral, so that elementary theory predicts that the two ground $S_t = 1/2$ states are degenerate. However, the presence of degenerate ground levels always gives rise to some instability, because perturbations tend to break it. The possible perturbations are spin–orbit coupling and phonon coupling. In general one of the two is dominant and quenches the effect of the other. If phonon coupling dominates the system will lower its symmetry, transforming the equilateral triangle into an isosceles or

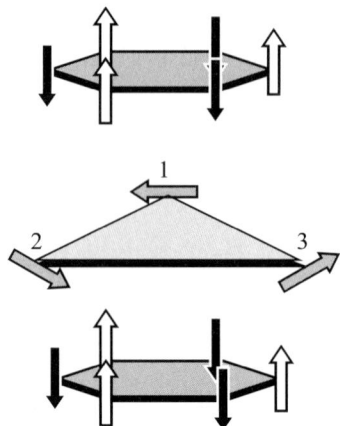

FIG. 14.13. Spin alignment in the three layers of V_{15}. The spins in the middle triangle are frustrated.

scalene one. In the isosceles case two different coupling constants must be taken into consideration. If we assume that the coupling between S_1 and S_2 is different from the other two, then we will have $J = J_{13} = J_{23}$ and $J' = J_{12}$. As a consequence of symmetry breaking the double degenerate ground state is split into two $S_t = \frac{1}{2}$ levels separated by $(J - J')$.

Spin–orbit coupling operates through the antisymmetric exchange term of Chapter 2.

$$\mathcal{H} = \Sigma_{i<j}\mathbf{d}_{ij} \cdot \mathbf{S}_i \times \mathbf{S}_j. \tag{14.20}$$

The symmetry requirements of antisymmetric exchange impose that the only components of $\mathbf{d}_{ij}$ different from zero are those parallel to the trigonal axis, which can be assumed to be the Z axis of the cluster. Also in this case the ground state is split into two Kramers doublets, separated by the energy $2d$, where $d = d_{12}^Z = d_{13}^Z = d_{23}^Z$. Detailed treatments of triangular systems have been made by Tsukerblat *et al.* (1987), and thorough investigations of simple trinuclear species have been performed. The analytical expression for the energies of the two Kramers doublets, in the simplifying assumption of isotropic Zeeman interaction, is given by

$$E_{1(2)}^{\pm} = \pm\sqrt{d^2 + \left(\frac{g\mu_B H}{2}\right)^2} \pm dg\mu_B H \cos\theta \tag{14.21}$$

where H is the applied magnetic field and θ is the angle between the field and the trigonal axis of the cluster. 1 and 2 identify the Kramers doublets and $\pm$ the spin components. If the field is parallel to the trigonal axis the two Kramers doublets are split exactly like in the absence of antisymmetric exchange, and four levels are observed. If the field is orthogonal to the trigonal axis the Zeeman splitting goes to zero and the two levels become degenerate. This is due to the competition between the antisymmetric exchange which tends to orient the spins parallel to the trigonal axis, and the field which orients it perpendicular. The application of high fields again generate the four split sublevels. The consequence of (14.21) is that the magnetization of the system becomes anisotropic at low field:

$$\chi_\perp = \frac{N(g\mu_B)^2}{6\sqrt{4d^2 + (g\mu_B H)^2}} \tanh \frac{\sqrt{4d^2 + (g\mu_B H)^2}}{2k_B T}. \tag{14.22}$$

This kind of behaviour was experimentally observed in a trinuclear copper(II) derivative (Tsukerblat *et al.* 1987) and in a vanadyl polyoxomolybdate (Gatteschi *et al.* 1996).

Formula (14.21) shows that the splitting of the two low-lying spin doublets is field dependent when the field is not applied parallel to Z. Since no field effects can be observed in the INS spectra of polycrystalline powder of V_{15} the antisymmetric exchange can be considered to be not influential in that case. The analysis of the spectra showed that the data can indeed be interpreted with a scalene deformation of the triangle occurring at low temperature (Chaboussant *et al.* 2004).

14.3.3 *Mixed-valence systems*

Most of the compounds we have described so far have localized valences, i.e. the oxidation state of each metal ion is well described assuming a given number of electrons. However, it is possible to imagine situations in which the charges are not rigidly localized on the individual ions, but they can hop from one site to the other. This is indeed a very interesting case, because the corresponding clusters can be considered as tiny models of magnetic conductors, while the localized charge clusters are examples of magnetic insulators. Perhaps the most versatile mixed-valence clusters are those of general formula $[V_{18}O_{42}]^{n-}$, whose structure is sketched in Fig. 14.14. There are two forms which can be obtained by rotation of 45° along the tetragonal axis. The cage can contain also solvent, indicating the possible operation of a template effect in the synthesis of the clusters (Müller *et al.* 1997).

According to the synthetic conditions the negative charge n can be 4, 6, and 12. The case with $n = 12$ corresponds to a localized charge configuration, with all the vanadium ions formally having a +4 charge. In the $n = 6$ case six vanadium(IV) are oxidised to vanadium(V), non-magnetic, and in the $n = 4$ there are eight vanadium(V) and 10 vanadium(IV). From the analysis of the X-ray diffraction data the charges are delocalized, in the sense that it is impossible to indicate which vanadium is +5 and which is +4. Therefore the clusters can be considered as examples of class III of the Robin and Day classification of mixed-valence systems.

The magnetic data for the completely reduced V_{18}^{IV} and for the two partially oxidized species are shown in Fig. 14.15. The data are reported as χT

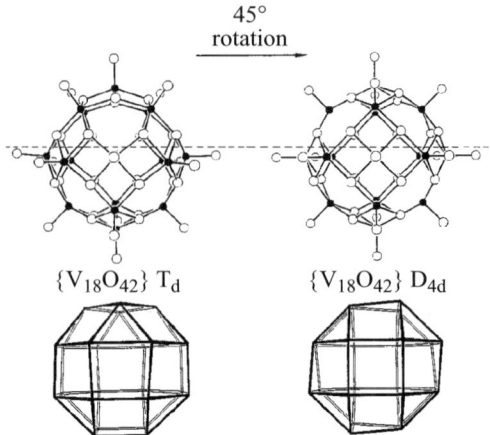

FIG. 14.14. Sketch of the structure of $[V_{18}O_{42}]^{n-}$.

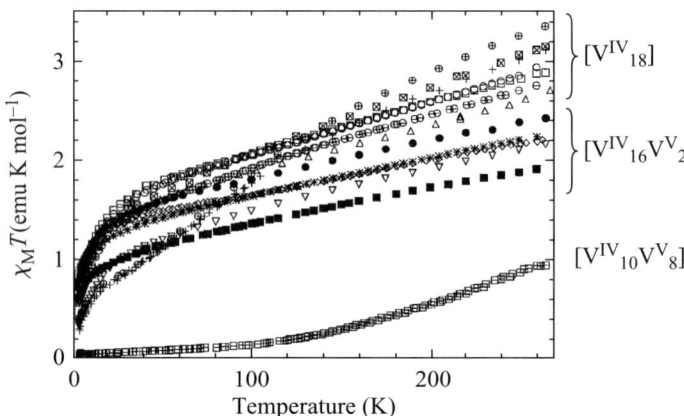

FIG. 14.15. Temperature dependence of the χT product of $[V_{18}O_{42}]^{n-}$ clusters. From Müller *et al.* (1997). Copyright (1997) American Chemical Society.

per mole of cluster and it is immediately apparent that the system with more unpaired electrons is less antiferromagnetically coupled than the systems with fewer unpaired electrons. At first this may seem to be counterintuitive, because with more formally non-magnetic ions one would expect that the average distance between unpaired electrons should be greater, thus reducing the coupling. Several qualitative arguments have been suggested.

One possibility is the increase of the exchange due to change of the global charge on the cluster. A second one is related to spin frustration effects which hamper the antiparallel orientation of the spins. These should be comparatively stronger in the clusters with higher numbers of magnetic centres. Finally a third possibility is the combination of magnetic exchange and electron transfer, which is present in the partially oxidized systems. Recent calculations performed at the *ab initio* level suggest that indeed it is the increase of the interaction through the delocalization which increases the effective interactions (Gaita, 2004).

15

EMERGING TRENDS IN MOLECULAR NANOMAGNETISM

At the time of writing, the field of SMMs is experiencing fast evolution. As more and more accurate control of the synthetic process is achieved, new systems with unprecedented and interesting magnetic phenomena are being produced. An interesting example is the evolution of the AF rings discussed in the previous chapter. Thanks to the template effect discussed in Section 4.5 large odd-member antiferromagnetic rings have recently become available. It is rather intuitive that these rings represent ideal models to investigate frustration effects at the nano-scale (Cador *et al.* 2004), as all the antiferromagnetic interactions cannot be satisfied at the same time. It has also been suggested that the spin structure of such rings has some analogies to the Möbius strip, as shown in Fig. 15.1.

Up to now only odd-member rings containing one metal ion with a different charge on the remaining ones have been investigated, as the templating technique described in Section 4.5 requires the presence of an unbalanced charge on the ring. Equally interesting is the selective substitution of just one metal centre of an even-member ring, with an ion carrying a different spin. The ground state of the ring is therefore no longer zero. The dynamics of this uncompensated spin seems

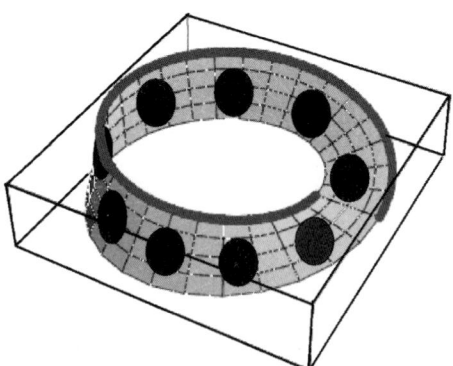

FIG. 15.1. The spin frustration in an odd-member antiferromagnetic ring is reminiscent of the Möbius strip. In fact, if the antiferromagnetic interactions are satisfied in a region of the ring a 'knot' is found where the two ends of the string meet. In ideal systems the 'knot' is delocalized on the entire ring.

particularly appealing for its relevance in quantum computing (Troiani *et al.* 2005a,b) and for low-temperature magnetocaloric effects (Affronte *et al.* 2004b).

As far as slow relaxing magnetic clusters are concerned the competition to obtain higher spin ground states and stronger anisotropy is expected to continue. However, even if molecular nanomagnets with blocking temperatures suitable for applications were successfully obtained, this would not be enough to fully exploit the potential of these materials. The magnetization of each molecule must become addressable to reach information storage at the molecular level. The organization of SMMs on surfaces represents a first step in this direction, as scanning probe techniques used in surface science are the only ones that, at the moment, can reach such a high spatial resolution. Moreover, scanning probe techniques, and in particular scanning tunnel microscopy, are in principle able to modify the properties of an individual molecule (Park *et al.* 2000).

The very first attempts to organize SMMs were based on the inclusion of Mn_{12} derivatives in Langmuir–Blodgett films (Clemente-Leon *et al.* 1998). Some successful attempts to obtain thin films or monolayers of SMMs have recently been presented. We can divide them into two categories based on the strength of the interaction of the SMM with the surface, which is significantly weaker when simple physical adsorption is occurring, compared to the case where a chemical bond is formed. This last case is mainly based on the strong affinity of gold towards metal surfaces (Cornia *et al.* 2003; Abdi *et al.* 2004; Zobbi *et al.* 2005) or on the reactivity of unsaturated C=C bonds with a native silicon surface (Condorelli *et al.* 2004; Fleury *et al.* 2005). A more relevant difference for the aims of this book is the distinction between those approaches that employ non-modified SMMs and those that are based on the specific functionalization of SMMs with linker groups that are able to graft the surface. The second examples concern Mn_{12} derivatives where the acetate has been substituted by carboxylic acid carrying a sulphur atom on the aliphatic or the aromatic residue. This type of encapsulation with functionalized ligands is quite common for a wide range of particles but in molecular nanomagnets it can reach 'surgical' precision. An interesting example of tailored on-demand molecule is shown here.

It is quite intuitive that, given the large magnetic anisotropy intrinsic of SMMs, their organization on surfaces requires controlling also the orientation of the molecules on the surface. A possible solution is the controlled substitution of only some ligands by playing on their different reactivity (Fleury *et al.* 2005), thus favouring anchorage in a preferred direction. A different strategy has been used that relies on the design of bicarboxylate ligands based on anthracene-1,8-dicarboxylic acid (Pacchioni *et al.* 2004). As shown in Fig. 15.2a the separation between the two −COO−groups ($\sim$5.2 Å) nicely fits the distance between the axially coordinated carboxylates in Mn_{12} clusters. Indeed two possible substitutions of the axial carboxylates can be expected, as depicted in Fig. 15.2b, but bridging mode I is observed in the cluster of formula $[Mn_{12}O_{12}(O_2CC_6H_5)_8(L)_4(H_2O)_4]\cdot8CH_2Cl_2$, with H_2L = 10-(4-acetylsulfanylmethyl-phenyl)-anthracene-1,8-dicarboxylic acid. Grafting of this

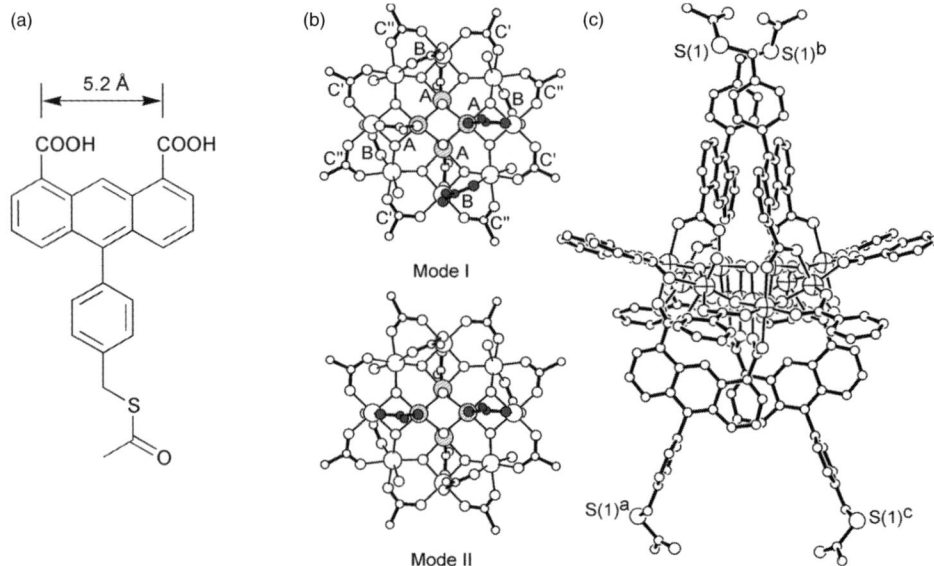

FIG. 15.2. (a) Structure of 10-(4-acetylsulfanylmethyl-phenyl)-anthracene-1,8-dicarboxylic acid. (b) Two possible coordination modes of the bicarboxylic acid replacing axial ligand of Mn_{12} clusters. (c) X-ray structure of $[Mn_{12}O_{12}(O_2CC_6H_5)_8(L)_4(H_2O)_4]\cdot 8CH_2Cl_2$. Adapted from Pacchioni *et al.* (2004) by permission of the Royal Society of Chemistry.

molecule is expected to favour the orientation of the easy axis of magnetization perpendicular to the surface.

The organization on surfaces of SMMs does not differ substantially from that of organic molecules but the discussion of these aspects goes far beyond the goals of this book. Many texts and reviews are available on the matter (Ulman 1991; Xia and Whitesides 1998). On the contrary, data on the magnetic properties of monolayers of SMMs are unfortunately not yet available, despite the great relevance that the interplay of magnetic quantum tunnelling and the tunnelling of conducting electrons through the anchored molecules could have for potential applications (Kim and Kim 2004).

We are therefore abandoning here this fascinating subject to dedicate the last few pages to other molecular systems exhibiting slow relaxation of the magnetization.

15.1   SMMs based on a single metal ion

The synthetic approaches to molecular nanomagnets described in this book have provided fascinating examples of giant clusters that approach the dimensions of small proteins or those of conventional magnetic nanoparticles. The opposite trend is, however, also observed: the search for smaller and smaller SMMs, down to the ultimate limit, a single magnetic centre that shows magnetic memory. The largest spin value that can be observed for a metal ion is $S = 7/2$ for Gd^{3+}. More promising is, however, the use of more anisotropic rare earth ions, which can reach a multiplicity as high as $J = 8$ for Ho^{3+} (Benelli and Gatteschi 2002). Rare earths have been introduced in molecular clusters containing also transition metal ions, with the aim to increase the magnetic anisotropy, and mixed clusters with nuclearity Cu_2Tb_2 (Osa *et al.* 2004), or Dy_6Mn_6 (Zaleski *et al.* 2004) have been obtained. A non-vanishing value of χ'' has been observed at low frequency, and this is a fingerprint of SMM behaviour as seen in Section 3.1.5. However, no real increase in the energy barrier, compared to Mn_{12} clusters, has been obtained up to now.

A separate discussion is necessary for the class of molecules investigated by Ishikawa and co-workers (2003a) comprising one lanthanide metal ion, e.g. Dy^{3+}, Ho^{3+}, or Tb^{3+}, in double decker compounds with phthalocyanine. Phthalocyanine are macrocyclic flat molecules that contain eight nitrogen atoms, but only four act as coordinating sites (see Fig. 15.3a). The donor atoms are in the five-member rings. Two *bis*-deprotonated phhtalocyanine molecules, abbreviated as Pc, coordinate a rare earth metal ion forming a mononegative anion of general formula $[(Pc)_2M]^-$ with the metal sandwiched between two ligands, as shown in Fig. 15.3b. The environment around the lanthanide is strictly tetragonal and the strong spin–orbit coupling induces a large splitting in zero field of the $J = 6$, $15/2$, and 8 for Tb^{3+}, Dy^{3+}, and Ho^{3+}, respectively. According to Ishikawa *et al.* (2003b), the ground doublet, which assumes the maximum $|m_J|$ value for Tb^{3+} and intermediate values for Dy^{3+} and Ho^{3+}, is separated from the first excited doublets by 627 K, 50 K, 21.6 K, for the three ions, respectively.

These compounds, even when diluted in a diamagnetic matrix constituted by $[(Pc)_2Y]^-TBA^+$ ($TBA^+ =$ tetrabutylammonium cation) show a non-zero χ''.

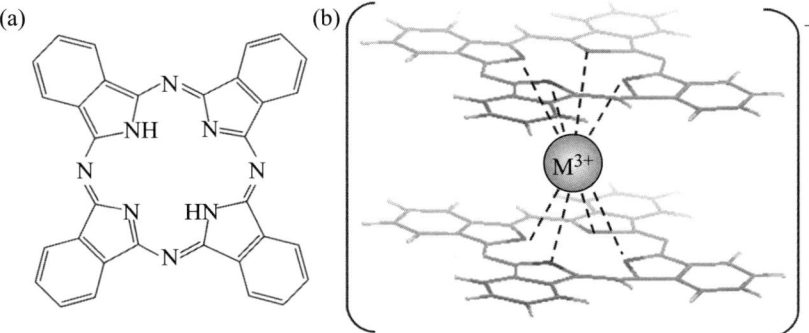

FIG. 15.3. (a) View of the structure of the phtalocyanine molecule. (b) Schematic structure of the double-decker $[(Pc)_2M]^-$ anion.

The fingerprint is observed at higher temperature than in SMM based on 3d metal ions (Ishikawa *et al.* 2004), hysteresis is only observed well below 1 K and is significantly narrower than that of SMMs. The hysteresis is characterized by a structure of equally spaced steps, with a field separation of about 20 mT for $[(\text{Ho})_2\text{Y}]^-\text{TBA}^+$ (Ishikawa *et al.* 2005).

The molecular nature of the material seems to play a minor role here, and indeed a similar stepped hysteresis was previously observed in the inorganic compound LiYF_4 doped with Ho^{3+} magnetic ions (Giraud *et al.* 2001). The origin of these close steps cannot be due to the level crossing of the states of ground doublet with the excited ones because these crossings occur at much higher fields. This suggested a different nature of the steps: the crossing or avoided crossing of the hyperfine sublevels. Let us take as an example the Ho^{3+} ion. The ^{163}Ho isotope has a natural abundance of 100% and is characterized by $I = 7/2$. The ground doublet is characterized by $m_J = \pm 5$, and each of these states is split into eight hyperfine levels, as shown in Fig. 15.4. The hyperfine splitting is described by

$$\mathcal{H}_{\text{hf}} = A_J \mathbf{J} \cdot \mathbf{I} \tag{15.1}$$

where A_J is of the order of 40 mK for Ho^{3+} (Abragam and Bleaney 1986). The level crossings thus show a separation of about 20 mT and can become

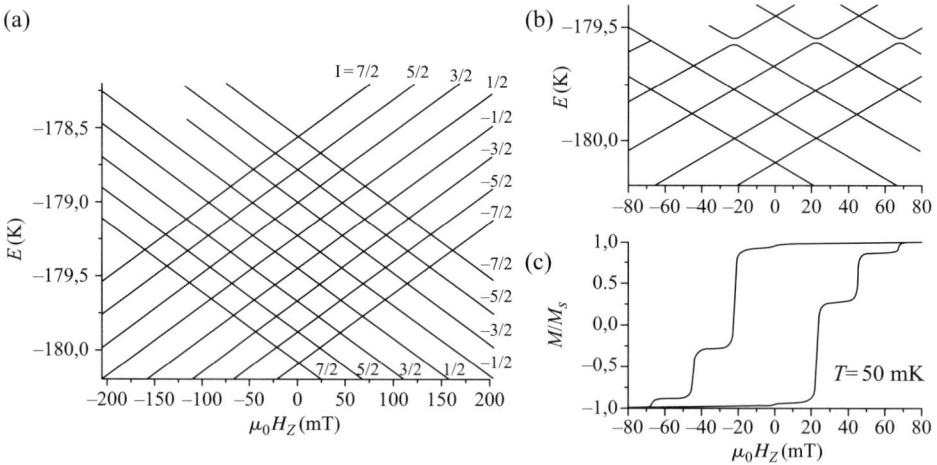

FIG. 15.4. (a) Calculated longitudinal field Zeeman splitting of the ground doublet of the $J = 8$ multiplet of Ho^{3+}. The ^{163}Ho isotope has $I = 7/2$ and 100% natural abundance. (b) Enlargement of the central part of the same graph to show the coincidence between the field value of the hyperfine level crossings and the steps in the hysteresis cycle (c) recorded with a micro-SQUID at 50 mK and with a sweeping rate of 0.11 mT/s. Courteously provided by B. Barbara.

avoided level crossings if the crystal field effects, allowed by tetragonal symmetry and described by the Stevens operators of Appendix A.5, are taken into account.

Acceleration of the reversal of the magnetization, and thus steps in the hysteresis cycle shown in Fig. 15.4, are observed at level crossings, giving rise to a phenomenon similar to the resonant quantum tunnelling described in Chapter 6 (Giraud *et al.* 2001; Ishikawa *et al.* 2005).

15.2 Single chain magnets

Despite much effort, which we hope has been efficiently highlighted in the previous pages of this book, the critical temperatures needed to observe SMM behaviour are still lower than 4 K. The route to increase the anisotropy barriers in zero-dimensional magnets, the clusters described so far, is a difficult one, therefore alternative roads deserve to be explored. A rather obvious idea is that of employing one-dimensional systems, because they may afford extended correlation lengths of the magnetization at relatively high temperature, especially if they are associated to Ising type magnetic anisotropy. Indeed a time-dependent version of the Ising model was proposed as early as 1963 by Glauber, who assumed that each spin can independently flip with a probability which depends on its environment and temperature, as will be discussed in detail below. The Glauber dynamics, predicting an exponential dependence of the relaxation time on the ratio between exchange interaction and temperature, was widely used by theorists in different fields even besides magnetism, but until the beginning of the twenty-first century there was no application to the original objects, namely to one-dimensional ferromagnets. This can be justified by the fact that at the low temperatures where slow Glauber dynamics becomes observable in real time or in ac susceptibility, most systems have already undergone a cross-over to three-dimensional magnetic order (Steiner *et al.* 1976).

In fact ideal Ising one-dimensional ferromagnets do not order above 0 K, but real systems do show transitions to three-dimensional order due to the combined effect of the long correlation developed at low temperatures and the weak residual interchain interactions (dipolar and/or exchange in nature). Therefore the recipe to observe Glauber dynamics is to assemble Ising-type magnetic building blocks in chains and isolate the chains as efficiently as possible. The latter requirement can be achieved by using molecular building blocks, where the magnetic active centre, for instance a metal ion, is embedded in a cover of an essentially magnetically inert organic moiety which allows having interchain metal distances longer than 1 nm.

The compound with the structure sketched in Fig. 15.5 was the first reported to show Glauber dynamics (Caneschi *et al.* 2001). It has formula [Co(hfac)$_2$(NITPhOMe)], CoPhOMe, where hfac = hexafluoroacetylacetonate, and NITPhOMe = 4'-methoxy-phenyl-4,4,5,5-tetramethylimidazoline-1-oxyl-3-oxide). This material does not show evidence of long-range order although the

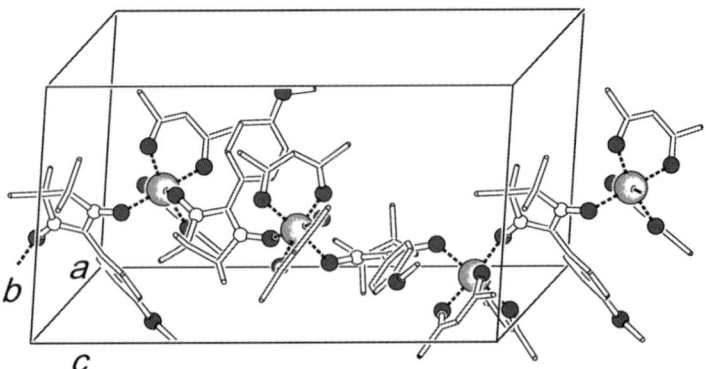

FIG. 15.5. Sketch of the structure of CoPhOMe. The cobalt ions are shown as large dark grey spheres, while oxygen atoms are drawn as smaller black spheres and nitrogen as white spheres. The helix winds up along the crystallographic c axis.

intrachain exchange interaction is as large as 200 K. It can be considered as an analogue of the ring comprising six manganese(II) and six radicals described in Chapter 1, whose structure was given in Fig. 1.17. An interesting feature of the crystals is that they either comprise right or left helices, with a spontaneous resolution of the enantiomers from solution. The chains have imposed crystal symmetry C_3 and they are formed by cobalt(II) ions bridged by the NITPhOMe radicals.

As previously described, cobalt(II) in octahedral symmetry has a ground $^4T_{1g}$ state which is split by crystal field and spin–orbit coupling effects to yield an anisotropic Kramers doublet which is the one populated at low temperature. In a spin Hamiltonian approach this corresponds to an effective anisotropic $S = \frac{1}{2}$ state, which couples to the isotropic $S = \frac{1}{2}$ spin of the radical. The overall coupling between the two spins is antiferromagnetic but, since the g values of the cobalt and of the radical are very different from each other, the system behaves like a one-dimensional ferrimagnet. Experimentally the out-of-phase ac magnetic susceptibility goes through a frequency-dependent maximum as shown in Fig. 15.6, showing that the magnetization relaxes slowly below 15 K. In fact the relaxation time of the magnetization, deduced from the frequency and temperature dependence of the position of the maximum of χ'', follows an Arrhenius law with $\Delta = 154(2)$ K and $\tau_0 = 3.0(2) \times 10^{-11}$ s. Compared to Mn$_{12}$ac the barrier is more than doubled, while the pre-exponential factor is faster by four orders of magnitude. Magnetization measurements on single crystals below 4 K show a hysteresis loop when the field is parallel to the trigonal axis and no hysteresis when it is applied perpendicular to that. The frequency dependence of the magnetic susceptibility rules out the transition to long-range magnetic order and is compatible with either 'superparamagnetic' or 'spin glass' behaviour. It must be recalled that

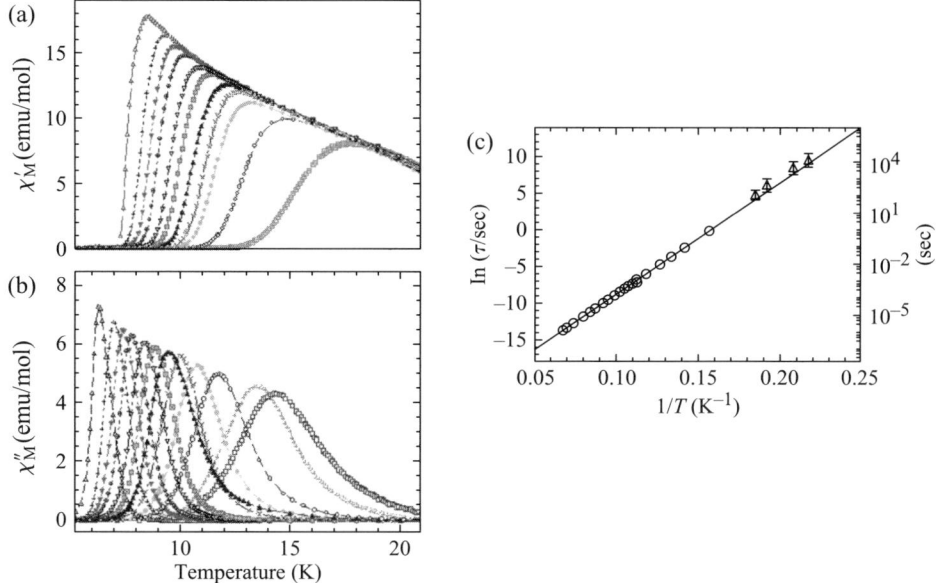

FIG. 15.6. Temperature dependence of the in-phase (a) and out-of-phase (b) ac magnetic susceptibility of CoPhOMe, in the frequency interval 0.18Hz–95 kHz. (c) The temperature dependence of the relaxation time. Triangles correspond to data extracted from the time decay of the remnant magnetization. Redrawn from Caneschi *et al.* (2001) and Caneschi *et al.* (2002). With the permission of Wiley-VCH and EDP Sciences.

spin glass behaviour had previously been associated with the slow relaxation of the magnetization of [MnTPP][TCNE], where TPP^{2-} is tetraphenylporphinate and TCNE^{-} is the radical anion of tetracyanoethylene, also described in Chapter 1 (Hibbs *et al.* 2001). On the other hand, spin glass behaviour is in general associated with very fast, sometimes unphysical, pre-exponential factors, suggesting that indeed CoPhOMe follows Glauber dynamics.

Shortly after the report of the unusual magnetic properties of CoPhOMe, a French–Japanese team reported another example, formed by the repetition in space of a binuclear manganese(III) species bridged by nickel(II) complex, as sketched in Fig. 15.7 (Clerac *et al.* 2002).

Also in this case the organic groups efficiently separate the chains one from the other, with the shortest interchain Mn-Ni distance of 10.39 Å. The dominant exchange interaction is the antiferromagnetic one between nickel and the two nearest-neighbour manganese ions, thus at low temperature the system behaves as a collection of $S = 3$ spins, which interact ferromagnetically thanks to the manganese–manganese ferromagnetic coupling (Miyasaka *et al.* 2004). At low temperature the uniaxial magnetic anisotropy of the $S = 3$ units warrants

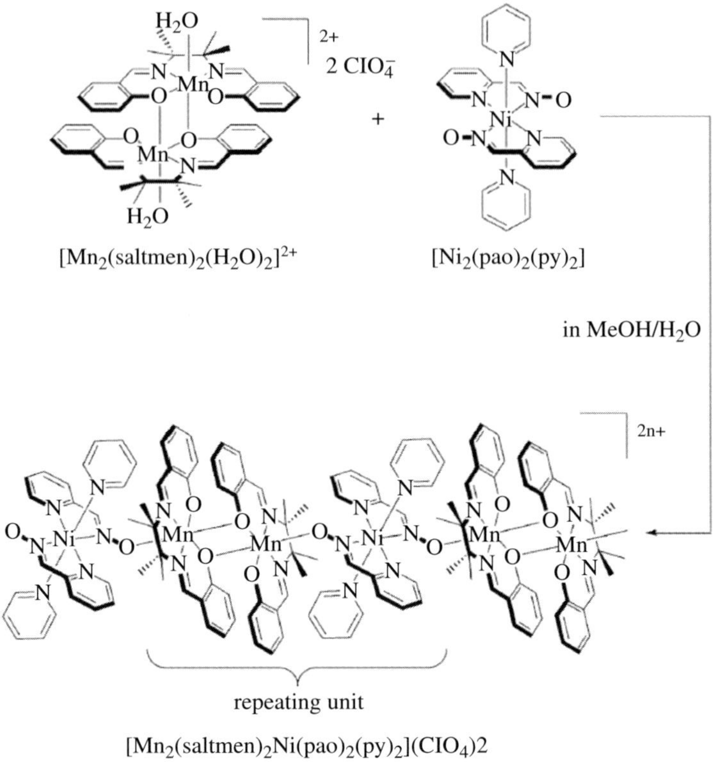

[Mn$_2$(saltmen)$_2$(H$_2$O)$_2$]$^{2+}$ [Ni$_2$(pao)$_2$(py)$_2$]

in MeOH/H$_2$O

repeating unit

[Mn$_2$(saltmen)$_2$Ni(pao)$_2$(py)$_2$](ClO$_4$)2

FIG. 15.7. Sketch of the structure of [Mn$_2$(saltmen)$_2$Ni(pao)$_2$(py)$_2$](ClO$_4$)$_2$. saltmen^{2-} = N, N'-(1,1,2,2-tetramethylethylene) *bis* (salicylideneiminate); pao$^-$ = pyridine 2-aldoxime; py = pyridine. From Clerac *et al.* (2002). Copyright (2002) American Chemical Society.

the Ising behaviour of the chain, which is characterized by slow relaxation of the magnetization and the opening of the hysteresis loop. The expression 'single chain magnets' was suggested in analogy to single molecule magnets (Clerac *et al.* 2002). The same comments on the real meaning of the expression which were made for SMMs at page 11 can be repeated here. The Arrhenius behaviour of the relaxation time was fitted with $\tau_0 = 5.5(1) \times 10^{-11}, \Delta = 72(1)$ K. Several other examples of SCM have been reported since and the list keeps growing (Lescouezec *et al.* 2003; Miyasaka *et al.* 2003; Liu *et al.* 2003; Chakov *et al.* 2004; Shaikh *et al.* 2004; Miyasaka *et al.* 2004; Costes *et al.* 2004; Pardo *et al.* 2004; Ferbinteanu *et al.* 2005).

The key idea of the Glauber dynamics is based on the consideration of the probability of reversing a spin in a chain where only nearest-neighbour interactions of the Ising type are operative according to the spin Hamiltonian:

$$\mathcal{H} = -J\Sigma_{k=1,N}\, \sigma_k\, \sigma_{k+1} \tag{15.2}$$

where $\sigma_k = \pm 1$. Indicating with α the intrinsic probability of reversal for an isolated spin, three types of transitions must be taken into account, depending on the orientations of the neighbouring spins $\sigma_{k\pm 1}$, as schematized below:

$$\omega_{\sigma_k \to -\sigma_k} = \frac{1}{2}\alpha(1-\gamma) \qquad \Uparrow \uparrow \Uparrow \qquad (15.3\text{a})$$

$$\omega_{\sigma_k \to -\sigma_k} = \frac{1}{2}\alpha \qquad \Uparrow \uparrow \Downarrow \qquad (15.3\text{b})$$

$$\omega_{\sigma_k \to -\sigma_k} = \frac{1}{2}\alpha(1+\gamma) \qquad \Downarrow \uparrow \Downarrow \qquad (15.3\text{c})$$

where γ is a factor which depends on the strength of the nearest-neighbour interaction:

$$\gamma = \tanh(2J/kT). \qquad (15.4)$$

For a ferromagnetic coupling, $J > 0$, γ tends to 1 when the temperature approaches zero, and the probability in (15.3a), the one involved in the relaxation of a saturated sample, goes to zero too.

The average time τ needed to completely reverse the magnetization starting from a saturated configuration can be obtained considering that there must be a sequence of events. At a given site k the magnetization is reversed with a cost of energy equivalent to the break of the two bonds with the two nearest neighbours, as shown in Fig. 15.8b. This, according to (15.2), corresponds to a barrier

$$\Delta = 4J. \qquad (15.5)$$

After the first reversal the successive steps cost zero, because the flip occurs at a site with a positive and a negative interaction, respectively.

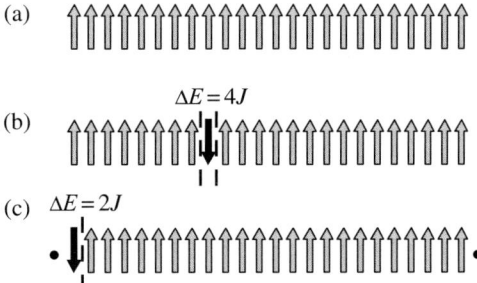

FIG. 15.8. Schematic view of the mechanism of relaxation in 1D Ising systems. The sample is prepared in the saturated state (a) and the relaxation starts with the reversal of one spin in the infinite chain with an energy cost $\Delta E = 4J$ (b). When the chain has a finite length the reversal of the spin at one edge has a halved energy cost (c).

The relaxation time is therefore given by:

$$\tau = \tau_0 \exp(4J/kT) \tag{15.6}$$

where τ_0 is directly related to $\alpha, \tau_0 = \alpha^{-1}$.

Comparing with the Arrhenius behaviour of SMM, Single Chain Magnet are in principle able to show a larger barrier than SMM because for SCM its height depends on the strength of the exchange interaction. In fact CoPhOMe has a much larger barrier than $Mn_{12}ac$, and also Mn_2Ni described above has a higher barrier. In both cases, however, the pre-exponential factor is much shorter than observed in SMM, thus the increase of the blocking temperature T_B, assuming that T_B corresponds to the temperature at which $\tau = 100$ s, is sizeable but not dramatic, as shown in Fig. 15.9 where the calculated temperature dependence of the relaxation times of $Mn_{12}ac$ and CoPhOMe is compared.

Many questions remain open in SCMs. One important point is that of reconciling the absence of long-range order and the Ising-type behaviour of the chains. In fact the exponential divergence required by one-dimensional Ising magnets is such that even very small dipolar interactions between chains might trigger the cross-over to three-dimensional magnetic order. A possible explanation of this might lie in the role of defects. It is well known that one-dimensional magnets are much more sensitive than three-dimensional magnets to the presence of defects. In fact if in a one-dimensional system a defect is present it breaks the chain of magnetic interactions and produces large finite size effects. For three-dimensional systems it is always possible to find alternative pathways in the presence of a concentration of defects that remains below the percolation limit. On the contrary, the infinite chain will be cut in a number of segments whose average length depends on the concentration of defects. For a system with concentration c of non-magnetic defects the probability P_L of having a segment of

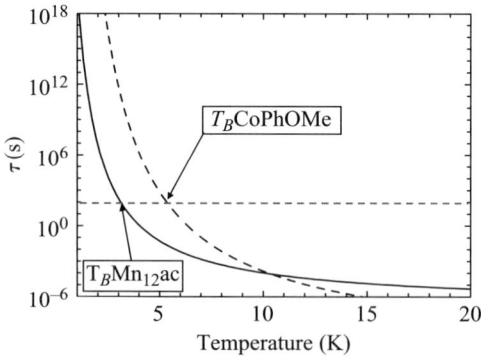

FIG. 15.9. Calculated temperature dependence of the relaxation time of the magnetization of $Mn_{12}ac$ (black solid line) and CoPhOMe (black broken line) assuming an Arrhenius behaviour. The estimated blocking temperatures, assumed to corresponds to $\tau = 100$ s, are also shown.

length L, expressed in lattice steps, is given by:

$$P_L = c^2(1 - c)^L. \tag{15.7}$$

The presence of defects has major effects on the dynamics of the magnetization, because the reversing of a spin next to a non-magnetic site halves the barrier to be overcome, being $\Delta = 2J$. This is clearly seen in Fig. 15.8c. The magnetization reversal is likely to occur at the end of the segment because this costs only the break of one interaction, however only a small fraction of spins is located next to a defect. Two regimes are therefore expected, depending on the temperature and the defect concentration. When the length of the segment, L, is much longer than the correlation length of the spins, ξ, the reversal of the spin occurs at random and the barrier is given by (15.5). This situation corresponds to the high-temperature limit. On the other hand, when ξ is much longer than L, the reversal occurs preferentially at the defective sites and the barrier is halved (Kamphorst Leal da Silva et al. 1995; Luscombe et al. 1996). For the process to be efficient, thus leading to magnetization reversal, the wall that is nucleated at one edge of the segment must reach the other end. The number of steps required is proportional to the length of the segment, providing a linear dependence on L of the relaxation time.

Investigations on doped systems where diamagnetic zinc(II) replaces randomly cobalt(II) confirm this view, and show that even in nominally pure compounds defects still play a role (Bogani et al. 2004). In real systems, however, a distribution of lengths is observed centred around the value $\bar{L} = 1/c$. It is therefore necessary to take into account this distribution but the results are not very dissimilar from those obtained assuming an average value, $\bar{L}$. This fact arises from the peculiar linear dependence of the relaxation time on the size of the systems that strongly differentiates SCMs from SMMs. These latter in fact show an exponential dependence on S^2. The same difference is encountered with single-domain magnetic particles, where the relaxation time scales exponentially with the volume of the particle. Therefore SCMs seem to offer the possibility of storing information on segments of interacting spins. SCMs are characterized by comparatively higher blocking temperatures and higher magnetization than SMMs and show, at the same time, a less dramatic sensitivity to the dimensions.

The field of single chain magnets merges with that of single molecule magnets in some cases. In fact examples have been reported of SCMs built up with building blocks that show, on their own, slow relaxation of the magnetization (Ferbinteanu et al. 2005). The relaxation time has been found to follow the Arrhenius law with a barrier which appears to be the sum of the Glauber barrier, due to the magnetic exchange along the chain, and the anisotropy barrier of the slow relaxing building blocks (Coulon et al. 2004). This composite effect is expected to be exploited in the future to sensibly increase the blocking temperature of molecular nanomagnets.

APPENDIX A

SYSTEMS OF UNITS, PHYSICAL CONSTANTS AND BASIC MATHEMATICAL TOOLS

A.1 International system of units, electromagnetic CGS and electrostatic CGS systems

In these three systems the Coulomb interaction energy between two electric charges q and q' at distance r in vacuum can be written as

$$W_{\text{cou}} = \frac{qq'}{\tilde{\epsilon}_0 r} \tag{A.1}$$

while the interaction energy between two magnetic dipoles $\boldsymbol{\mu}$ and $\boldsymbol{\mu}'$ at distance $\mathbf{r}$ in vacuum is

$$W_{\text{mag}} = -\tilde{\mu}_0 \sum_{\alpha\gamma} r^{-3} \left[\delta_{\alpha\gamma} - \frac{3}{r^2} r^\alpha r^\gamma \right] \mu_\alpha \mu_\gamma. \tag{A.2}$$

The usual constants ϵ_0 and μ_0 are related to $\tilde{\epsilon}_0$ and $\tilde{\mu}_0$ in the international system (SI) by

$$\tilde{\epsilon}_0 = 4\pi\epsilon_0, \quad \tilde{\mu}_0 = \frac{\mu_0}{4\pi} \tag{A.3}$$

while, in the electrostatic (es) and electromagnetic CGS systems, one has simply

$$\tilde{\epsilon}_0 = \epsilon_0, \quad \tilde{\mu}_0 = \mu_0. \tag{A.4}$$

In vacuum, the electric and magnetic fields and inductions $\mathbf{E}$, $\mathbf{H}$, $\mathbf{D}$, $\mathbf{B}$ are related by the relations

$$\mathbf{D} = \epsilon_0 \mathbf{E}, \quad \mathbf{B} = \mu_0 \mathbf{H} \tag{A.5}$$

and satisfy the Maxwell equations

$$\frac{\partial}{\partial t}\mathbf{B} = -\text{rot}\mathbf{E}, \quad \frac{\partial}{\partial t}\mathbf{D} = \text{rot}\mathbf{H} \tag{A.6}$$

and

$$\text{div}\mathbf{D} = \text{div}\mathbf{B} = 0. \tag{A.7}$$

From the above equations it is easy to deduce $\epsilon_0 \mu_0 \partial^2 \mathbf{H}/\partial t^2 = -\nabla^2 \mathbf{H}$ and therefore, in the three systems of units,

$$\epsilon_0 \mu_0 c^2 = \tilde{\epsilon}_0 \tilde{\mu}_0 c^2 = 1 \tag{A.8}$$

where c is the speed of light.

In the international system, $\mu_0/(4\pi) = \tilde{\mu}_0 = 10^{-7}$ henry/m and ϵ_0 can be deduced from (A.8). In the emCGS system, $\mu_0 = 1$ emCGS unit and $\tilde{\epsilon}_0$ can be deduced from (A.8). In the esCGS system, $\epsilon_0 = 1$ esCGS unit and μ_0 can be deduced from (A.8).

A.2 Gauss' system of units

In this system, also called symmetric CGS, (A.1) and (A.2) hold with $\mu_0 = \epsilon_0 = 1$, so that (A.8) is not valid. The Maxwell equations (A.6) are replaced by

$$\frac{\partial}{\partial t}\mathbf{B} = -c \text{ rot } \mathbf{E}, \quad \frac{\partial}{\partial t}\mathbf{D} = c \text{ rot } \mathbf{H} \tag{A.9}$$

where the value of c is equal to that of the speed of light.

For those who are afraid to make mistakes and want to be able to check that their formulae have the correct dimension, Gauss' system should be prohibited since it seems to imply that ϵ_0 and μ_0 have the same dimensions.

In this book, the magnetic induction B is usually called the field and designated by H.

A.3 Other common units

In aircraft and on the roads of the USA, miles and yards are still used as length units. Physicists do not use them, but do use the Ångström, which is 0.1 nm. More complicated are the energy units used by physicists in various areas. To measure an energy W in statistical physics, it is common to give the ratio W/k_B of W to Boltzmann's constant and very often a physicist will say that W is a certain number of kelvin, e.g. 2.3. This means that $W/k_B = 2.3$ K. A spectro-scopist will alternatively say that W is a certain number of cm^{-1}, e.g. 3. This means that k defined by $W = \hbar c k$ is 3. A specialist of electrons in solids will measure W in electron-volts (eV), and 1 eV is the energy $W = e$ of an electron of charge $-e$ subject to a potential of 1 volt. Neutron spectroscopists often use the milli-electron-volt (meV) and high-energy physicists use the kev, the MeV or the GeV etc. The conversion is easily performed if one knows the essential physical constants.

A.4 Physical constants

Velocity of light:

$$c = 2.998 \times 10^8 \text{ m/s}$$

Charge of an electron: $-e$, with

$$e = 1.602 \times 10^{-19} \text{ coulomb}$$

Mass of an electron:

$$m = 0.9109 \times 10^{-31} \text{ kilograms}$$

Planck constant:

$$\hbar = \frac{h}{2\pi} = 1.054 \times 10^{-34} \text{ Js} \tag{A.10}$$

Boltzmann constant:

$$k_B = 1.3806 \times 10^{-23} \text{ JK}^{-1} \tag{A.11}$$

Bohr magneton:

$$\mu_B = \frac{e\hbar}{2m} = 0.9274 \times 10^{-23} \text{ JT}^{-1} \tag{A.12}$$

Energy units:

$$1 \text{ eV} = 1.602 \times 10^{-19} \text{ J}$$

is equivalent to 1.16×10^4 kelvin and 0.807×10^4 cm^{-1}.

A.5 Stevens operators

The O_n^m operators are defined as:

$$O_2^0 = 3S_z^2 - s(s+1) \tag{A.13}$$

$$O_2^2 = \frac{1}{2}(S_+^2 + S_-^2) \tag{A.14}$$

$$O_4^0 = 35S_z^4 - [30s(s+1) - 25]S_z^2 + 3s^2(s+1)^2 - 6s(s+1) \tag{A.15}$$

$$O_4^2 = \frac{1}{4}[7S_z^2 - s(s+1) - 5](S_+^2 + S_-^2)$$
$$+ \frac{1}{4}(S_+^2 + S_-^2)[7S_z^2 - s(s+1) - 5] \tag{A.16}$$

$$O_4^3 = \frac{1}{4}S_z(S_+^3 + S_-^3) + \frac{1}{4}(S_+^3 + S_-^3)S_z \tag{A.17}$$

$$O_4^4 = \frac{1}{2}(S_+^4 + S_-^4). \tag{A.18}$$

A.6 3j- and 6j-symbols

The 3j-symbols and 6j-symbols arise when considering, respectively, two and three coupled angular momenta. They can be found in the textbook by Messiah (1965) or on the web sites http://mathworld.wolfram.com/Wigner3j-Symbol.html and http://mathworld.wolfram.com/Wigner6j-Symbol.html.

The 3j-symbols are given by the formula

$$\begin{pmatrix} S & J & S' \\ M & Q & P \end{pmatrix} = (-1)^{S-J-P}\sqrt{\Gamma(S,J,S')}\sum_t (-1)^t$$

$$\times \frac{[(S+M)!(S-M)!(J+Q)!(J-Q)!(S'+P)!(S'-P)!]^{1/2}}{t!(S'-J+t+M)!(S'-M+t-Q)!(S+J-S'-t)!(S-t-M)!(J-t+Q)!} \tag{A.19}$$

with $M + P + Q = 0$ and $|S - J| \le S' \le S + J$. The quantity

$$\Gamma(S, J, S') = \frac{(S + J - S')!(S' + J - S)!(S - J + S')!}{(S + J + S' + 1)!}$$

has been introduced. Finally, the sum is extended to all integers t for which all the factorials are non-negative. The number of terms is $\mu + 1$ where μ is the smallest of the nine numbers $S \pm M$, $J \pm Q$, $S' \pm P$, $S + J - S'$, $S' + J - S$ and $S' + S - J$.

The $6j$-symbols are defined as

$$\begin{Bmatrix} J_1 & J_2 & J_3 \\ J_4 & J_5 & J_6 \end{Bmatrix} = [\Gamma(J_1, J_2, J_3)\Gamma(J_1, J_5, J_6)\Gamma(J_4, J_2, J_6)\Gamma(J_4, J_5, J_3)]^{1/2} \sum_t (-1)^t$$

$$\times \frac{(t + 1)!}{(t - J_1 - J_2 - J_3)!(t - J_1 - J_5 - J_6)!(t - J_4 - J_2 - J_6)!(t - J_4 - J_5 - J_3)!}$$

$$\times \frac{1}{(J_1 + J_2 + J_4 + J_5 - t)!(J_2 + J_3 + J_5 + J_6 - t)!(J_3 + J_1 + J_6 + J_4 - t)!}$$

$$(A.20)$$

where the sum over t has the same meaning as for the $3j$ symbols. Now the number of terms is $\mu + 1$ where μ is the smallest of the 12 numbers $J_1 + J_2 - J_3$; $J_1 + J_5 - J_6$; $J_4 + J_2 - J_6$; $J_4 + J_5 - J_3$; $J_2 + J_3 - J_1$; $J_5 + J_6 - J_1$; $J_2 + J_6 - J_4$; $J_5 + J_3 - J_4$; $J_3 + J_1 - J_2$; $J_6 + J_1 - J_5$; $J_6 + J_4 - J_2$; $J_3 + J_4 - J_5$.

A.7 Different notation

Not all authors use the same notation. For some authors, the electron charge is $e < 0$, while in this book it is assumed to be $-e$.

Kinetic moments, called $\hbar\mathbf{L}$, $\hbar\mathbf{S}$, $\hbar\mathbf{J}$, in this book, may be called $\mathbf{L}$, $\mathbf{S}$, $\mathbf{J}$ by other authors.

A majority of authors agree to write the magnetic moment of an orbital moment $\hbar\mathbf{L}$ as

$$\mathbf{m} = -\mu_B \mathbf{L} \qquad (A.21)$$

but for a spin $\mathbf{S}$ some authors, for instance Ibach & Lüth (1999), use a $+$ sign and write

$$\mathbf{m} = g_0 \mu_B \mathbf{S}$$

with $g_0 \simeq 2$.

In this book, the $-$ sign is always used, so that the kinetic moment $\mathbf{J}$, has magnetic moment

$$\mathbf{m} = -g_J \mu_B \mathbf{J} \qquad (A.22)$$

with $\mathbf{J} = \mathbf{L} + \mathbf{S}$.

It follows that the magnetic energy of a kinetic moment in a magnetic field **H** is

$$W = g\mu_{\mathrm{B}}\mathbf{H} \cdot \mathbf{S} \qquad (A.23)$$

with a $+$ sign. As a matter of fact, a $-$ sign appears in the literature approximately as frequently! This is (generally!) not a mistake, but an indication that g is negative, $g \simeq -2$ (Mohr and Taylor 2003) or that μ_{B} is negative, $\mu_{\mathrm{B}} = -e\hbar/2m = 0.9274 \times 10^{-23}$ joules/tesla (Cohen-Tannoudji *et al.* 1986). Our choice to define $e > 0$, $\mu_{\mathrm{B}} > 0$, $g > 0$ seems to be most widely used (e.g. Cohen and Taylor 1987).

A detailed analysis of these discrepancies will be found in the review of Villain (2003).

APPENDIX B

THE MAGNETIC FIELD

B.1 A complicated vocabulary

In books and articles about magnetism several kinds of magnetic fields are encountered:

the external field;
the demagnetizing field (resulting from dipole interactions);
the local field (also resulting from dipole interactions);
the 'Maxwellian' field and induction (those which appear in the Maxwell equations).

This appendix is intended to reduce the unavoidable confusion which arises from these various concepts.

In this book, what is officially called 'external magnetic induction' has generally been called magnetic field and designated by the symbol $\mathbf{H}$ (however, measured in teslas, which are units of induction). This terminology is appropriate to a microscopic, statistical mechanical description, which allows fluctuations.

In this appendix, the legal terminology is used. The relation between the external field (which will be called H_0), the 'Maxwellian' field $\mathbf{H}$ and the 'Maxwellian induction' $\mathbf{B} = \mu_0[\mathbf{H}+\mathbf{M}]$ will be recalled. The international system (SI) of units will be used and $\mathbf{M}$ is the magnetization.

B.2 Demagnetizing field and local field

As seen in Chapter 2, the interaction between magnetic moments $\mathbf{m}_i$ localized at lattice sites i may be assumed to have the bilinear expression

$$W = -\frac{1}{2}\sum_{ij}\sum_{\alpha\gamma}\Gamma_{ij}^{\alpha\gamma}m_i^\alpha m_j^\gamma \tag{B.1}$$

where $\Gamma_{ij}^{\alpha\gamma}$ is the sum of the long-range dipole interaction (2.50) and of the short-range exchange interaction (2.53).

In (B.1), the sum of the terms which contain a particular moment $\mathbf{m}_i$ can be written as

$$W = -\mu_0\mathbf{H}_i \cdot \mathbf{m}_i \tag{B.2}$$

where

$$\mu_0 H_{i\alpha} = \sum_j\sum_\gamma\Gamma_{ij}^{\alpha\gamma}m_i^\alpha m_j^\gamma. \tag{B.3}$$

Expression (B.2) is the same as the energy of a spin subject to a field $\mathbf{H}_i$. However, $\mathbf{H}_i$ can fluctuate, in contrast with a magnetic field. In macroscopic magnetism, fluctuations are ignored and (B.3) can be identified with a field which is called the 'local field'. The evaluation of the *electrostatic* local field in *fluids* at thermal equilibrium is a classical problem investigated in various cases by many authors from O.F. Mossotti in the middle of the nineteenth century to L. Onsager in 1936, not forgetting H.A. Lorentz at the beginning of the twentieth century.

The contribution of the long-range dipole interaction in (B.3) is particularly important. It systematically tends to decrease the magnetization and for that reason it is called the demagnetizing field. Its expression can be obtained from (2.50). Assuming a continuous variation of the magnetization in space, it can be transformed into

$$H_{\text{dem}}^{\alpha} = \frac{1}{4\pi} \int d^3 r' \left[(r'_\alpha - r_\alpha)|\mathbf{r}' - \mathbf{r}|^{-3} \right] \text{div} \mathbf{M}(\mathbf{r}'). \qquad (B.4)$$

A particularly simple case is that of a uniformly magnetized ellipsoid in a field parallel to one of its axes. Then the demagnetizing field is uniform and equal to

$$\mathbf{H}_{\text{dem}} = -d\mathbf{M} \qquad (B.5)$$

where the 'demagnetizing factor' d depends on the sample shape and is different for the three axes. Formula (B.5) is also correct for an arbitrary field direction, but d is now a tensor. For a sphere $d = \frac{1}{3}$ in the international systems of units. For an infinitely long ellipsoids, $d = 0$ for a field parallel to the long axis, while $d = \frac{1}{2}$ for a field orthogonal to that. For an infinitely large disk $d = 1$ for a field parallel to the short axis, while $d = 0$ for a field orthogonal to that axis. If CGS units are used, these values must be multiplied by 4π.

The use of the Maxwellian field $\boldsymbol{H} = \boldsymbol{H}_0 - \boldsymbol{H}_{\text{dem}}$ is a great advantage in macroscopic magnetism since, for instance, the magnetization M is related to H by the relation $\boldsymbol{M} = \boldsymbol{\chi H}$. However, for a microscopic calculation, fluctuations have to be taken into account and the relevant magnetic field is the external field.

B.3   Free energy

It is therefore of interest to be able to evaluate the free energy as a function of the external field (as is done in quantum mechanics or statistical mechanics) or as a function of the Maxwellian field (as is generally done in magnetostatics, electrostatics, or electrodynamics). The link between both pictures is well explained by Landau and Lifshitz (1969) in their Section 31.

Let the point of view of statistical mechanics be recalled first. One starts from the energy or Hamiltonian of the system in an external field H^0, e.g.

$$\mathcal{H} = -\mu_0 \sum_i \mathbf{H}_i^0 \cdot \mathbf{m}_i - \frac{1}{2} \sum_{ij} \sum_{\alpha\gamma} \Gamma_{ij}^{\alpha\gamma} m_i^\alpha m_j^\gamma + \mathcal{H}_{\text{an}}. \qquad (B.6)$$

Then one deduces the partition function

$$Z = \text{Tr}\mathcal{H} \qquad (B.7)$$

and the free energy

$$\tilde{\mathcal{F}} = -k_{\mathrm{B}} T \ln Z. \tag{B.8}$$

The tilde corresponds to the notation of Landau and Lifshitz (1969).

It follows from formulae (B.6)–(B.8) that at thermal equilibrium

$$\langle m_i^\alpha \rangle = -\frac{d}{dB_{i\alpha}^0} \tilde{\mathcal{F}} \tag{B.9}$$

where the derivative is at constant temperature and volume, and $\mathbf{B}_i^0 = \mu_0 \mathbf{H}_i^0$ is the external induction (generally called $\mathbf{H}_i$ in this book). Comparison of (B.9) with formula (31.3) of Landau and Lifshitz (1969) shows that their free energy $\tilde{\mathcal{F}}$ coincides with the above-defined function. However, Landau and Lifshitz (1969) define $\tilde{\mathcal{F}}$ in a different way, namely in terms of the Maxwellian field rather than the external field. They state that in an infinitesimal transformation at constant volume and temperature, the variation of $\tilde{\mathcal{F}}$ is

$$d\tilde{\mathcal{F}} = -\int d^3 r \, \mathbf{B} \cdot d\mathbf{H}. \tag{B.10}$$

This establishes the relation between a description in terms of the external field and the Maxwellian field.

The variation of $\tilde{\mathcal{F}}$ in an isothermal, reversible transformation at constant uniform field is the work provided to the magnetic system. It is a minimum at equilibrium. It is appropriate to introduce another free energy $\tilde{\mathcal{F}}$:

$$\mathcal{F} = \tilde{\mathcal{F}} + \int d^3 r \, \mathbf{B} \cdot \mathbf{H}. \tag{B.11}$$

The variation of $\tilde{\mathcal{F}}$ when magnetizing a system is the work which should be provided. In the simple case of a paramagnet of susceptibility χ and volume V in a uniform field H_0, without demagnetizing field, $\tilde{\mathcal{F}} = -\chi V H_0^2$ and $\mathcal{F} = \chi V H_0^2$. If one wants to magnetize a system by a magnetic field, one has to provide energy to create the field, but in a constant magnetic field, the magnetization of a system provides energy.

APPENDIX C

HOW IRREVERSIBILITY COMES IN

Microscopic equations of motion are invariant under time reversal, at least in classical mechanics and without a magnetic field, while the master equation (5.12), which correctly describes the behaviour of a macroscopic system (at least for long time as will be seen) does not have this invariance. To shed some light on this phenomenon, a simple example will be considered.

In the present appendix, we consider the case of a binary mixture and follow the evolution of the concentration $\rho(\mathbf{r}, t)$ of one of the components. This concentration will be assumed to satisfy a diffusion equation with a diffusion constant Λ. In terms of the Fourier transform $\rho_q(t)$ the diffusion equation reads

$$\partial \rho_q(t)/\partial t = -\Lambda q^2 \rho_q(t). \tag{C.1}$$

This is a particularly simple type of master equation. The solution is

$$\rho_q(t) = \rho_q(0) \exp(-\Lambda q^2 t). \tag{C.2}$$

It is appropriate to introduce the correlation function $\langle \rho_q(0)\rho_q(t) \rangle$. It follows from (C.2) that

$$\langle \rho_q(0)\rho_q(t) \rangle = \langle \rho_q(0)\rho_q(0) \rangle \exp(-\Lambda q^2 t). \tag{C.3}$$

On the other hand, the microscopic equations of motion are invariant under time reversal (at least in the simplest cases, excluding for instance charged particles in a magnetic field). This implies

$$\langle \rho_q(0)\rho_q(t) \rangle = \langle \rho_q(0)\rho_q(-t) \rangle \tag{C.4}$$

in contradiction with (C.3). This contradiction is solved if one notices that the argument which leads to the master equation implies an increasing, not decreasing time t. Therefore, (C.3) has no reason to be valid for negative time (and can obviously not be valid!). If we believe in (C.3) for positive time, and if we believe in (C.4) too, we must replace t by its absolute value in (C.3), which now reads

$$\langle \rho_q(0)\rho_q(t) \rangle = \langle \rho_q(0)\rho_q(0) \rangle \exp(-\Lambda q^2 |t|). \tag{C.5}$$

But we now face a new problem. The equations of motion are analytic and therefore $\langle \rho_q(0)\rho_q(t) \rangle$ should be an analytic function of t. Expression (C.5) is not analytic for short time t. This means that it is not correct for short times.

Near $t = 0$, the Taylor expansion

$$\langle \rho_q(0)\rho_q(t)\rangle = \langle \rho_q(0)\rho_q(0)\rangle - (t^2/2)\langle \dot{\rho}_q(0)\dot{\rho}_q(0)\rangle \tag{C.6}$$

applies. The cross-over to (C.5) takes place for a time τ_{coll} which can be deduced from (C.5) and (C.6).

The argument can be extended to quantum mechanics but this requires various mathematical tricks; in particular, one has to introduce an imaginary time as in chapter 7. The quantum version of linear response theory has been given by Kubo (1957) and can be found in the lectures of Noelle Pottier on the web site http://www.lpthe.jussieu.fr/DEA/pottier.html.

APPENDIX D

BASIC PROPERTIES OF THE MASTER EQUATION

In this appendix, it is shown that all eigenvalues of the 'master matrix'

$$\Theta_{ji} = \gamma_{ji} - \delta_{ji} \sum_{\ell \neq i} \gamma_{i\ell} \qquad (\text{D.1})$$

are real and negative, except that which corresponds to equilibrium and is equal to 0. The indices i, j denote the eigenstates of the (time-independent) Hamiltonian and the Θ matrix appears in equation (5.12) which can be written as

$$\frac{d}{dt}\mathbf{p}(t) = \Theta \mathbf{p}(t). \qquad (\text{D.2})$$

The transition probabilities γ_{ji} are real and positive and satisfy the principle of detailed balance

$$\gamma_{ji} = \gamma_{ij} \exp[\beta(\epsilon_j - \epsilon_i)]. \qquad (\text{D.3})$$

It is appropriate to introduce the real, symmetric, and therefore Hermitian matrix

$$\Gamma_{ij} = \gamma_{ij} \exp[\beta(\epsilon_j - \epsilon_i)/2] = \Gamma_{ji}. \qquad (\text{D.4})$$

Let v_i be an eigenvector of Θ with the eigenvalue λ:

$$\sum_{j \neq i} \gamma_{ji} v_j - v_i \sum_{\ell \neq i} \gamma_{i\ell} = \lambda v_i. \qquad (\text{D.5})$$

Substituting (D.4) one obtains

$$\sum_{j \neq i} \Gamma_{ji} u_j - \alpha_i u_i = \lambda u_i \qquad (\text{D.6})$$

where

$$u_i = v_i \exp[\beta \epsilon_i/2] \qquad (\text{D.7})$$

and

$$\alpha_i = \sum_{\ell \neq i} \gamma_{i\ell} = \sum_{\ell \neq i} \Gamma_{i\ell} \exp[\beta(\epsilon_i - \epsilon_\ell)/2]. \qquad (\text{D.8})$$

Thus, any eigenvalue λ of Θ is also an eigenvalue of the matrix $\Gamma_{ji} - \delta_{ji}\alpha_i$, which is real, symmetric, and therefore Hermitian. This implies that λ is real.

To show that these eigenvalues are non-positive, it is sufficient to show that the quadratic form

$$\Phi = \sum_{ij} \Gamma_{ji} x_i x_j - \sum_i \alpha_i x_i^2 \qquad (D.9)$$

cannot be positive. Indeed, it is easily shown that (D.9) reads

$$\Phi = -(1/2) \sum_{ij} \Gamma_{ji} \{x_i \exp[\beta(\epsilon_i - \epsilon_j)/4] - x_j \exp[\beta(\epsilon_j - \epsilon_i)/4]\}^2. \qquad (D.10)$$

The only non-vanishing contributions are those of coefficients Γ_{ij} with $i \neq j$ which are positive equal to zero. Therefore, expression (D.10) is negative except for $x_i = \exp(-\beta\epsilon_i/2)$, which corresponds to thermal equilibrium.

The desired property is thus proven.

APPENDIX E

DERIVATION OF THE ARRHENIUS LAW

As seen in Section 5.4, the general solution of the master equation is a sum of exponentials of time t, and at low temperature this sum is dominated by a single exponential $\exp(-t/\tau_1)$, where $1/\tau_1$ is the smallest non-vanishing eigenvalue of the Hermitian matrix $\tilde{\Gamma}$ defined by (5.19). The corresponding eigenvector $\tilde{\varphi}_m^{(1)}$ should be orthogonal to the eigenvector $\tilde{\varphi}_m^{(0)}$ which corresponds to the eigenvalue 0, and is given, according to (5.14) and (5.21), by

$$\tilde{\varphi}_m^{(0)} = (1/Z)\exp(-\beta E_m/2) \tag{E.1}$$

where

$$Z = \sum_m \exp(-\beta E_m). \tag{E.2}$$

It is convenient to write

$$\tilde{\varphi}_m^{(1)} = \lambda_m \exp(-\beta E_m/2) \tag{E.3}$$

The advantage of this expression is that, as will be seen, the variation of λ_m with m is slow.

The orthogonality of both vectors implies

$$\sum_m \lambda_m \exp(-\beta E_m) = 0 \tag{E.4}$$

while normalization implies

$$\sum_m \lambda_m^2 \exp(-\beta E_m) = 1. \tag{E.5}$$

The coefficients λ_m are determined by a variational principle, which is a standard method in quantum mechanics, when $-\tilde{\Gamma}$ is a Hamiltonian. They should minimize

$$-\langle \tilde{\varphi}_{(1)} | \tilde{\Gamma} | \tilde{\varphi}_{(1)} \rangle = -\sum_{mq} \tilde{\varphi}_m^{(1)} \tilde{\Gamma}_q^m \tilde{\varphi}_q^{(1)} \tag{E.6}$$

and the minimum of this quantity is just $1/\tau_1$, the quantity one wants to calculate
  Inserting (E.3) in (E.6), the sum over q can be written as

$$\sum_q \tilde{\Gamma}_q^m \tilde{\varphi}_q^{(1)} = \sum_q \tilde{\Gamma}_q^m \lambda_q \exp(-\beta E_q/2) = \sum_q \tilde{\Gamma}_q^m (\lambda_q - \lambda_m)\exp(-\beta E_q/2).$$

In the last expression, the subtracted term containing λ_m is 0 according to (E.1). Coming back to the original master matrix, one obtains

$$\sum_q \tilde{\Gamma}_q^m \tilde{\varphi}_q^{(1)} = \exp(\beta E_m/2)\sum_q \Gamma_q^m (\lambda_q - \lambda_m)\exp(-\beta E_q).$$

Insertion into (E.6) yields

$$\langle \tilde{\varphi}_{(1)} \,|\, \tilde{\Gamma} \,|\, \tilde{\varphi}_{(1)} \rangle = \sum_{mq} \Gamma_q^m \lambda_m (\lambda_q - \lambda_m) \exp(-\beta E_q). \tag{E.7}$$

According to the detailed balance principle (5.14), this can also be written as

$$\langle \tilde{\varphi}_{(1)} \,|\, \tilde{\Gamma} \,|\, \tilde{\varphi}_{(1)} \rangle = \sum_{mq} \Gamma_m^q \lambda_m (\lambda_q - \lambda_m) \exp(-\beta E_m)$$

or interchanging m and q:

$$\langle \tilde{\varphi}_{(1)} \,|\, \tilde{\Gamma} \,|\, \tilde{\varphi}_{(1)} \rangle = \sum_{mq} \Gamma_q^m \lambda_q (\lambda_m - \lambda_q) \exp(-\beta E_q). \tag{E.8}$$

Combining (E.7) and (E.8), one obtains

$$-\langle \tilde{\varphi}_{(1)} \,|\, \tilde{\Gamma} \,|\, \tilde{\varphi}_{(1)} \rangle = \frac{1}{2} \sum_{mq} \Gamma_q^m (\lambda_q - \lambda_m)^2 \exp(-\beta E_q). \tag{E.9}$$

This should be minimized with respect to the λ_m with the constraints (E.4) and (E.5). The qualitative behaviour of the solution is clear: λ_m should be approximately constant except near the maximum m_0 of E_m. If λ_m varies from λ_{-s} to λ_s in an interval of width Δm around m_0, then in this interval, $\exp(-\beta E_q) \approx \exp(-\beta E_{m_0 \pm \Delta m})$ and $(\lambda_q - \lambda_m) \approx |\lambda_{-s} - \lambda_s|/\Delta m$. Since there are Δm terms, (E.9) reads

$$-\langle \tilde{\varphi}_{(1)} \,|\, \tilde{\Gamma} \,|\, \tilde{\varphi}_{(1)} \rangle \approx |\Gamma_{m_0}^{m_0}| |\lambda_{-s} - \lambda_s|^2 \exp(-\beta E_{m_0 \pm \Delta m})/\Delta m. \tag{E.10}$$

At sufficiently low temperature, in a sufficiently weak magnetic field, $|\lambda_{-s} - \lambda_s|^2 \approx 4|\lambda_s|^2 \approx 4|\lambda_{-s}|^2$, which, according to (E.5), should be of the order of $\exp(\beta E_s) \approx \exp(\beta E_{-s})$. If the temperature is still low, but the magnetic field is so strong that $\beta(E_{-s} - E_s) \gg 1$, then the orthogonality condition requires $|\lambda_{-s} - \lambda_s|^2 \approx |\lambda_{-s}|^2 \approx \exp(\beta E_{-s})$. In both cases

$$-\langle \tilde{\varphi}_{(1)} \,|\, \tilde{\Gamma} \,|\, \tilde{\varphi}_{(1)} \rangle \approx |\Gamma_{m_0}^{m_0}| \exp[-\beta(E_{m_0} - E_{-s})] \exp[\beta(E_{m_0} - E_{m_0 \pm \Delta m})]/\Delta m. \tag{E.11}$$

This can now be identified with $1/\tau_1$, provided it is minimized with respect to Δm. Assuming $E_{m_0} - E_{m_0 \pm \Delta m} \approx |D|\Delta m^2$ as in Chapter 5, and treating m as a continuous variable, one finds that the relaxation time is

$$1/\tau_1 \approx |\Gamma_{m_0}^{m_0}| \sqrt{\beta|D|} \exp[-\beta(E_{m_0} - E_{-s})]. \tag{E.12}$$

The exponential factor corresponds to the Arrhenius law but the pre-exponential one is not temperature independent and deviations from the Arrhenius law are therefore expected. These are not essential, as argued in Appendix F.2. In fact, the correction induced by the $\sqrt{\beta}$ term is overcompensated by another one, hidden in the first factor $|\Gamma_{m_0}^{m_0}|$. Actually the transition probability (5.41) is proportional to T if $\beta|E_m - E_{m'}| \ll 1$.

APPENDIX F

PHONONS AND HOW TO USE THEM

F.1 Memento of the basic formulae

Lattice vibrations involve the elastic displacement $\mathbf{u}$ and the associated momentum $\mathbf{p}$. There are several possible descriptions of these fields.

- Each atom i has a mass M_i, a displacement $\mathbf{u}_i$, a velocity $\mathbf{v}_i$, a momentum $\mathbf{p}_i = M_i \mathbf{v}_i$.
- If the high-frequency, optical modes are ignored, all atoms of each unit cell $\mathbf{R}$ have approximately, in classical mechanics, the same displacement $\mathbf{u}_i = \mathbf{u}_R$ and the same velocity $\mathbf{v}_R$. The momentum $\mathbf{p}_R$ can be defined in *classical* mechanics as

$$\mathbf{p}_R = \sum^{i(R)} \mathbf{p}_i = \sum^{i(R)} M_i \mathbf{v}_i = \sum^{i(R)} M_i \mathbf{v}_R = M \mathbf{v}_R \tag{F.1}$$

where $M = \sum^{i(R)} M_i$ is the mass of the unit cell and the symbol $\sum^{i(R)}$ denotes a sum over the atoms i of the cell centred at $\mathbf{R}$. The displacement operators may be assumed to satisfy the relation

$$\mathbf{u}_R = \sum^{i(R)} \lambda_i \mathbf{u}_i \tag{F.2}$$

where $\sum^{i(R)} \lambda_i = 1$.

- In classical elasticity, within the same assumption that $\mathbf{u}_i$ and $\mathbf{u}_R$ do not vary much from one lattice site to the next one, a continuous field $\mathbf{u}(\mathbf{r})$ can be defined at each point $\mathbf{r}$ of the space.

In a periodic crystal, it is appropriate to introduce Fourier transforms of these fields. For a crystal of N unit cells, they will be defined as

$$\mathbf{u}_q = N^{-1/2} \sum_R \mathbf{u}_R \exp(i\mathbf{q} \cdot \mathbf{R})$$

$$= N^{-1/2} \sum_i \lambda_i \mathbf{u}_i \exp(i\mathbf{q} \cdot \mathbf{R}_i)$$

$$= v^{-1} N^{-1/2} \int d^3 r\, \mathbf{u}(\mathbf{r}) \exp(i\mathbf{q} \cdot \mathbf{r}) \tag{F.3}$$

where $v = V/N$ is the volume of the unit cell. The vectors $\mathbf{q}$ form a discrete set of N elements within the reciprocal unit cell (or Brillouin zone). Similarly

$$\mathbf{p}_q = N^{-1/2} \sum_R \mathbf{p}_R \exp(i\mathbf{q} \cdot \mathbf{R}) = N^{-1/2} \sum_i \mathbf{p}_i \exp(i\mathbf{q} \cdot \mathbf{R}_i). \qquad \text{(F.4)}$$

The kinetic energy can be written as

$$\mathcal{H}_{\text{kin}} = \sum_i \frac{p_i^2}{2M_i} = \sum_i \frac{M_i v_i^2}{2} = \sum_R \frac{M v_R^2}{2} = \sum_R \frac{p_R^2}{2M} = \sum_q \frac{|p_q|^2}{2M}. \qquad \text{(F.5)}$$

The elastic energy or free energy has the form

$$\mathcal{H}_{\text{el}} = \sum_q \sum_{\alpha,\gamma} \sum_{\xi,\zeta} \Omega_{\alpha,\gamma}^{\xi,\zeta} q_\alpha q_\gamma u_q^\xi u_q^\zeta. \qquad \text{(F.6)}$$

where $\Omega_{\alpha,\gamma}^{\xi,\zeta}$ are elastic constants. Note that, if the summation is over the whole Brillouin zone, as will be assumed, each vector $\mathbf{q}$ appears twice. The alternative would be to sum over half a Brillouin zone.

In order to avoid the tedious process of diagonalization of the matrix Ω, one often assumes an 'isotropic' elastic medium characterized by its Lamé coefficients λ and μ. This is totally unrealistic but there are so many parameters that there is not much hope to really solve the full problem. The 'isotropic' elastic solid has, in real space, a free energy density

$$f_{\text{el}} = \frac{\lambda}{2}(\text{div}(\mathbf{u}))^2 + \mu \left[(\partial_x u_x)^2 + (\partial_y u_y)^2 + (\partial_z u_z)^2\right]$$
$$\frac{\mu}{2} \left[(\partial_x u_y + \partial_y u_x)^2 + (\partial_y u_z + \partial_z u_y)^2 + (\partial_z u_x + \partial_x u_z)^2\right]. \qquad \text{(F.7)}$$

The energy per unit cell is obtained by multiplying f_{el} by the volume v of the unit cell. The total elastic free energy is obtained by adding the contribution of all cells R. As a function of the Fourier transform $\mathbf{u_q}$, the elastic free energy is therefore

$$\mathcal{H}_{\text{el}} = \frac{\lambda v}{2} \sum_q |q_x u_q^x + q_y u_q^y + q_z u_q^z|^2 + \mu v \sum_q \left[|q_x u_q^x|^2 + |q_y u_q^y|^2 + |q_z u_q z|^2\right]$$

$$+ \mu v \sum_q \left[|q_x q_y u_q^x u_q^y| + |q_y q_z u_q^y u_q^z| + |q_z q_x u_q^z u_q^x|\right]$$

$$+ \frac{\mu}{2} \left[(q_y^2 + q_z^2)|u_q^x|^2 + (q_z^2 + q_x^2)|u_q^y|^2 + (q_x^2 + q_y^2)|u_q^z|^2\right]$$

or

$$\mathcal{H}_{\text{el}} = v \sum_q \left[\frac{\lambda + \mu}{2}|\mathbf{q} \cdot \mathbf{u}_q|^2 + \frac{\mu}{2}q^2|\mathbf{u}_q|^2\right]. \qquad \text{(F.8)}$$

The sum of (F.8) and (F.5) will be treated as a Hamiltonian,

$$\mathcal{H}_{\text{ph}} = (v/2) \sum_q \left\{q^2 \left[(2\mu + \lambda)q^2|\mathbf{u}_{q1}|^2 + \mu q^2 |\mathbf{u}_q^{\text{tr}}|^2\right] + (1/M)|p_q|^2\right\} \qquad \text{(F.9)}$$

where $\mathbf{u}_{q,1} = \mathbf{q}(\mathbf{q}\cdot\mathbf{u}_q)/|\mathbf{q}|$ is the longitudinal component of $\mathbf{u}_q$ and $\mathbf{u}_q^{\mathrm{tr}} = \mathbf{u}_q - \mathbf{u}_q^{\mathrm{lg}}$ is the transverse component. The longitudinal component $\mathbf{p}_{q,1}$ and the transverse component $\mathbf{p}_q^{\mathrm{tr}}$ of $\mathbf{p}_q$ can be defined in a similar way.

Quantum mechanics can now be introduced via the usual commutation rule

$$[p_i^\alpha, u_j^\gamma] = i\hbar\delta_{ij}\delta_{\alpha\gamma} \tag{F.10}$$

where $\alpha, \gamma = x, y, z$

It follows from (F.10), (F.1) and (F.2) that

$$[p_R^\alpha, u_{R'}^\gamma] = i\hbar\delta_{RR'}\delta_{\alpha\gamma}. \tag{F.11}$$

It follows from (F.3), (F.4), (F.1) and (F.2) that

$$[p_q^\alpha, u_{-q'}^\gamma] = i\hbar\delta_{\mathbf{q},\mathbf{q}'}\delta_{\alpha\gamma}. \tag{F.12}$$

The Hamiltonian (F.9) is the sum of Hamiltonians of independent harmonic oscillators which can be diagonalized as explained in textbooks on quantum mechanics. In order to avoid mistakes by factors of 2, it should be remembered that each Fourier component, e.g. u_{q1}, has a real part $u'_{q1} = u'_{-q,1}$ and an imaginary part $u''_{q1} = -u''_{-q,1}$. A typical term of the sum (F.9) is

$$\mathcal{H}_{q1} = (2\mu + \lambda)vq^2\left(u'_{q1}\right)^2 + (1/M)\left(p'_{q1}\right)^2. \tag{F.13}$$

where the factor 2 has disappeared because the terms q and $-q$ of (F.9) are both included in (F.13).

Following textbooks, the harmonic oscillator Hamiltonian (F.13) is put into the standard or 'canonical' form

$$\mathcal{H}_{q1} = \hbar\omega_{q1}b_{q1}^*b_{q1} \tag{F.14}$$

via the transformation

$$b_{q1}\sqrt{\hbar} = u'_{q1}[(2\mu + \lambda)Mvq^2]^{1/4} - ip'_{q1}[(2\mu + \lambda)Mvq^2]^{-1/4}. \tag{F.15}$$

The longitudinal phonon frequency is

$$\omega_{q1} = c_1 q \tag{F.16}$$

where the longitudinal sound velocity is

$$c_1 = 2[(2\mu + \lambda)v/M]^{1/2}. \tag{F.17}$$

Similarly, the frequency of the two transverse phonon modes is

$$\omega_{q2} = \omega_{q3} = c_2 q \tag{F.18}$$

where

$$c_2 = (2\mu v/M)^{1/2}. \tag{F.19}$$

It follows from (F.15) and (F.17) that

$$b_{q1}\sqrt{\hbar} = u'_{q1}(c_1 Mq/2)^{1/2} - ip'_{q1}(c_1 Mq/2)^{-1/2}. \tag{F.20}$$

From (F.20) it is easily deduced that

$$u_{q\rho} = (c_\rho M q/\hbar)^{-1/2}[b_{q\rho} + b_{q\rho}^*]. \tag{F.21}$$

A similar formula holds for the imaginary parts u_{q1}'' and p_{q1}''. One can for instance separate the Brillouin zone $\mathcal{B}$ into two parts $\mathcal{B}_1$ and $\mathcal{B}_2$, such that, if $\mathbf{q}$ is an element of $\mathcal{B}_1$, $-\mathbf{q}$ is an element of $\mathcal{B}_2$. Then decide that, for any element $\mathbf{q}$ of $\mathcal{B}_1$, (F.20) defines b_{q1} while $b_{-q,1}$ is defined by the same relation where real parts are replaced by imaginary parts.

The total phonon Hamiltonian is

$$\mathcal{H}_{\mathrm{ph}} = \sum_q \sum_{\rho=1}^{3} \hbar\omega_{q\rho} b_{q\rho}^* b_{q\rho} \tag{F.22}$$

where q (which, strictly speaking, should be written $\mathbf{q}$) denotes N vectors of the reciprocal space, which fill the unit cell (or 'Brillouin zone') with a uniform density $8\pi^3 V = 8\pi^3 vN$, N being the number of unit cells, v the volume of the unit cell, and V the volume of the crystal. Finally, the index ρ labels the three phonon modes of a three-dimensional crystal.

F.2   Numerical calculation of the relaxation rate

It can be useful to give the reader some suggestions about how to develop a simple routine to estimate the temperature or field dependence of the relaxation time of a spin system with known spin Hamiltonian parameters describing the magnetic anisotropy. It is convenient to use the eigenvectors $|\, m\rangle$ of S_z as a basis to write the spin Hamiltonian. If an anisotropy or a transverse field is present, the Hamiltonian matrix is obviously no longer diagonal and a diagonalization procedure must be used. If a transverse field with a component along y is applied in conjunction with a transverse anisotropy, the Hamiltonian matrix turns out to be complex (but of course Hermitian) and the appropriate diagonalization subroutine must be chosen, even though the eigenvalues are real. However, this case will not treated here. Such a routine can be found in any eigenpackage, available in libraries such as EISPACK, or proprietary implementations such as the IMSL and NAG libraries. But it can simply be the Jacobi subroutine any student has written in the first course of informatics. The routines we refer to are those provided in *Numerical Recipes* by Press *et al.* (1989), where the algorithms are also discussed in detail. In order to write the master matrix we need to evaluate the true eigenstates $|\, m^*\rangle$ of the spin Hamiltonian. Each of the $(2s+1)$ eigenvalues $E(i)$ (where $i = 1, 2, \ldots, 2s+1$) is associated to an eigenstate that is given by a linear combination equivalent to (5.11), except that the index $m = -s, -s+1, ..., s-1, s$ is replaced by the index i. Similarly, the coefficients $\varphi_m^{m'}$, for the eigenstates in (5.11), are replaced by matrix elements $O(i, j)$, i.e. the elements of the ith row of the matrix of the eigenvectors.

All standard routines also provide the eigenvector matrix, but special care must be paid to the accuracy of the calculation of this one. In particular this simple numerical calculation is not suited to evaluate the relaxation time when

a pair of levels are almost degenerate, first because the accuracy in the determination of the eigenvectors is too poor and second because tunnelling is not
correctly treated at the resonance. However, the application of a longitudinal
field as strong as that of the Earth, and therefore present in most experimental
set-ups, is often enough to get out of this dark zone.

In a simplified form the master matrix Γ is now written on the basis of the
eigenstates $|p^*\rangle$ and $|q^*\rangle$ of the spin Hamiltonian by taking into account terms
of (5.27) that couple states with $|m - m'| = 1$ or 2. Thus the element γ_q^p of the
master matrix is given for $p \neq q$ by:

$$
\gamma_q^p = \frac{3v}{\pi\hbar^4 M c_s^5} \frac{(E_p - E_q)^3}{\exp[\beta(E_p - E_q) - 1]}
$$

$$
\left\{ |\tilde{D}_a|^2 \left[|\langle p | S_+^2 | q \rangle|^2 + |\langle p | S_-^2 | q \rangle|^2 \right] \right.
$$

$$
\left. + |\tilde{D}_b|^2 \left[|\langle p | \{S_+, S_z\} | q \rangle|^2 + |\langle p | \{S_-, S_z\} | q \rangle|^2 \right] \right\} \tag{F.23}
$$

where E_p and E_q are the eigenvalues corresponding to $|p^*\rangle$ and $|q^*\rangle$, $\tilde{D}_a$ and
$\tilde{D}_b$ are spin–phonon coupling coefficients,

$$
\langle p | S_+^2 | q \rangle = \sum_{mm'} (\varphi_p^{m'})^* \varphi_q^m \langle m' | S_+^2 | m \rangle
$$

$$
= \sum_m (\varphi_p^{m+2})^* \varphi_q^m \sqrt{[s(s+1) - m(m+1)][s(s+1) - (m+2)(m+1)]}
$$

$$
= \langle q | S_-^2 | p \rangle^* \tag{F.24}
$$

and

$$
\langle p | \{S_+, S_z\} | q \rangle = \sum_{mm'} (\varphi_p^{m'})^* \varphi_q^m \langle m' | \{S_-, S_z\} | m \rangle
$$

$$
= \sum_m (2m+1)(\varphi_p^{m+1})^* \varphi_q^m \sqrt{[s(s+1) - m(m+1)]}
$$

$$
= \langle q | \{S_-, S_z\} | p \rangle^* . \tag{F.25}
$$

The diagonal terms are given by $\Gamma_p^p = \sum_k \gamma_k^p$ in agreement with (5.17).

The stars which designate the complex conjugate quantities in (F.24) and
(F.25) can be be omitted in the case addressed here since these quantities are
real.

This simplified form is also based on the assumption that the dynamics of
the phonon-bath is much faster than that of the magnetization. The occupation
number of each phonon state thus depends only on the temperature of the bath.

As said in Section 5.4 the master matrix is not symmetric and thus a different
diagonalization procedure must be used. Non-symmetric matrices are difficult to
handle and much more sensitive to rounding errors. This difficulty might be overcome by transforming the master matrix into a symmetric one as explained in
Section 5.4. Another method is, before diagonalization, to transform the matrix

in order to have rows and columns of about the same norm. This can be done with a routine called BALANC that requires N^2 iterations, N being the dimension of the matrix. The matrix is then reduced to a simpler form, called Hessenberg. An upper Hessenberg matrix has non-zero elements in the upper triangle, the diagonal, and the first subdiagonal. This can be done by the ELMHES routine that requires a number of iterations proportional to N^3. The matrices we are handling are, however, reasonably small and computation time is not at all a problem. The final step is the search of the eigenvalues done by using the so-called QR algorithm performed by the HQR routine also available in *Numerical Recipes*. The eigenvalues are then ordered by modulus. The smallest one is found equal to zero and neglected, while the first nonvanishing one, λ_1, is used to calculate the relaxation time as $\tau = -1/\lambda_1$. This approximation of a unique contribution to the description of the overall process is correct at low temperature as already discussed. We have not mentioned the role played by the spin–phonon coupling coefficients. Their value acts on the relaxation time as a multiplicative coefficient. If, for instance, we are interested in calculating the temperature dependence of the relaxation time because we want to analyse it in the simple terms of the Arrhenius law, the choice of this coefficients only affects τ_0 and not the T_0. It is therefore common use to leave these as adjustable parameters to reproduce the observed behaviour. This numerical method can be used to calculate the relaxation time for a spin $s = 10$ subject to Hamiltonian (2.10) with $D = -0.7$ K, as in $Mn_{12}ac$ and to a small longitudinal field $\mu_0 H_z = 0.1$ mT to localize the states in the wells. We want to compare the numerical result with the Arrhenius law (5.3). The relaxation rate is approximated by the smallest nonvanishing eigenvalue $|\lambda_1|$ of the master matrix Γ. The calculation is repeated at several temperatures and we plot in Fig. F.1 the logarithm of the calculated relaxation time *vs.* $1/T$.

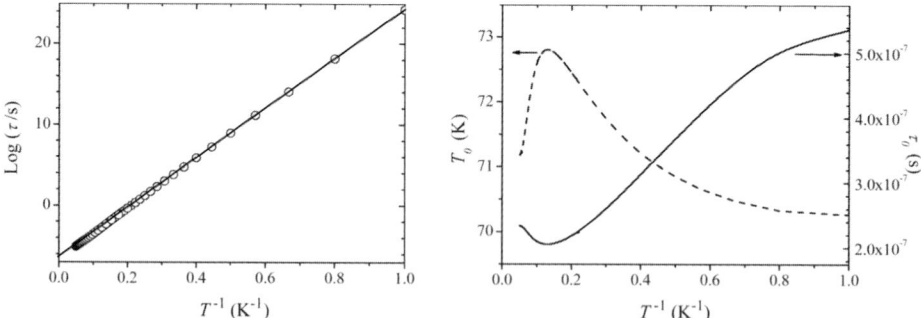

FIG. F.1. On the left, temperature dependence of the relaxation time (open symbols) reported in the Arrhenius plot. The solid line corresponds to the linear fit of the lowest temperature data. On the right, the temperature dependence of the parameters T_0 and τ_0. The relaxation time has been numerically calculated assuming $s = 10$ and $D = -0.7$ K.

The behaviour is well described by a straight line as predicted by the Arrhenius law. The relaxation time at high temperature tends to be slightly shorter than that predicted by extrapolating the linear behaviour observed at low temperature. This is in agreement with (E.12). The deviation is, however, significantly smaller than that observed experimentally in $Mn_{12}ac$ (Novak *et al.* 2004) and reported in Fig. 5.3. Changes in the efficiency of the spin-phonon coupling or the population of excited spin states could be responsible of the behaviour observed in $Mn_{12}ac$.

It is also interesting to quantify the deviations from the Arrhenius law already qualitatively discussed at the end of Appendix E and here evaluated from the numerical calculation of the relaxation time. To do that we plot on the right of Fig. F.1 the temperature dependence of both the barrier, T_0, defined as the derivative of $\ln(\tau)$ with respect to $1/T$, and of the pre-exponential factor τ_0 defined by $\ln(\tau_0) = \ln(\tau) - T_0/T$.

APPENDIX G

HIGH-ORDER PERTURBATION THEORY

The details of the calculation presented in Section 6.4 will first be given in the case of the perturbation (6.31). The non-vanishing elements of the matrices which appear in (6.39) and (6.40) are

$$\langle m \pm 1 \,|\, \mathcal{B} \,|\, m \rangle = \langle m \pm 1 \,|\, \delta\mathcal{H} \,|\, m \rangle = \langle m \pm 1 \,|\, g\mu_B H_x S_x \,|\, m \rangle \tag{G.1}$$

$$\langle p \pm 1 \,|\, \mathcal{B} \,|\, p \rangle = \langle p \pm 1 \,|\, \delta\mathcal{H} \,|\, p \rangle \tag{G.2}$$

$$\langle m' \pm 1 \,|\, \delta\mathcal{A} \,|\, m' \rangle = \langle m' \pm 1 \,|\, \delta\mathcal{H} \,|\, m' \rangle . \tag{G.3}$$

Inserting these relations into (6.39) and (6.40), one obtains at the lowest significant order, assuming $m < p$,

$$\langle m \,|\, \mathcal{B}^*(E - \mathcal{A})^{-1}\mathcal{B} \,|\, p \rangle = \langle m \,|\, \delta\mathcal{H} \,|\, m+1 \rangle \langle m+1 \,|\, (E - \mathcal{H}_0)^{-1} \,|\, m+1 \rangle$$
$$\langle m+1 \,|\, \delta\mathcal{H} \,|\, m+2 \rangle \langle m+2 \,|\, (E - \mathcal{H}_0)^{-1} \,|\, m+2 \rangle \ldots$$
$$\langle p-2 \,|\, \delta\mathcal{H} \,|\, p-1 \rangle \langle p-1 \,|\, (E - \mathcal{H}_0)^{-1} \,|\, p-1 \rangle$$
$$\times \langle p-1 \,|\, \delta\mathcal{H} \,|\, p \rangle . \tag{G.4}$$

At the lowest significant order, which is the order $(p-m)$, the quantity E can be replaced by the unperturbed value $E_m^{(0)}$ which is close to $E_p^{(0)}$ at resonance. This value is given by (5.6) if (6.30) holds. Thus, identifying the left-hand side with ω_T^{mp} in agreement with (6.39), (G.4) reduces to

$$\hbar\omega_T^{mp} = \left| \langle m \,|\, \delta\mathcal{H} \,|\, m+1 \rangle \frac{1}{E_m^{(0)} - E_{m+1}^{(0)}} \langle m+1 \,|\, \delta\mathcal{H} \,|\, m+2 \rangle \frac{1}{E_m^{(0)} - E_{m+2}^{(0)}} \ldots \right.$$
$$\left. \frac{1}{E_m^{(0)} - E_{p-2}^{(0)}} \langle p-2 \,|\, \delta\mathcal{H} \,|\, p-1 \rangle \frac{1}{E_m^{(0)} - E_{p-1}^{(0)}} \langle p-1 \,|\, \delta\mathcal{H} \,|\, p \rangle \right| . \tag{G.5}$$

In contrast with the off-diagonal elements of $\mathcal{B}^*(E - \mathcal{A})^{-1}\mathcal{B}$, the evaluation of the diagonal elements, which also enter in (6.38), would be much more difficult at the same order $(p - m)$. However, they are not important for the tunnelling frequency. They just shift the resonance as seen from Section 6.3.

If, instead of (6.31), $\delta\mathcal{H}$ is given by (6.32), the calculation is very similar and yields, if $(m - p)$ is a multiple of 2,

$$
\hbar\omega_{mp} = \left| \langle m \,|\, \delta\mathcal{H} \,|\, m+2 \rangle \, \frac{1}{E_m^{(0)} - E_{m+2}^{(0)}} \, \langle m+2 \,|\, \delta\mathcal{H} \,|\, m+4 \rangle \, \frac{1}{E_m^{(0)} - E_{m+4}^{(0)}} \cdots \right.
$$

$$
\left. \frac{1}{E_m^{(0)} - E_{p-4}^{(0)}} \, \langle p-4 \,|\, \delta\mathcal{H} \,|\, p-2 \rangle \, \frac{1}{E_m^{(0)} - E_{p-2}^{(0)}} \, \langle p-2 \,|\, \delta\mathcal{H} \,|\, p \rangle \right|. \tag{G.6}
$$

If $\delta\mathcal{H}$ is given by (6.33), a similar calculation yields, if $(m - p)$ is a multiple of 4,

$$
\hbar\omega_{mp} = \left| \langle m \,|\, \delta\mathcal{H} \,|\, m+4 \rangle \, \frac{1}{E_m^{(0)} - E_{m+4}^{(0)}} \, \langle m+4 \,|\, \delta\mathcal{H} \,|\, m+8 \rangle \, \frac{1}{E_m^{(0)} - E_{m+8}^{(0)}} \cdots \right.
$$

$$
\left. \frac{1}{E_m^{(0)} - E_{p-8}^{(0)}} \, \langle p-8 \,|\, \delta\mathcal{H} \,|\, p-4 \rangle \, \frac{1}{E_m^{(0)} - E_{p-4}^{(0)}} \, \langle p-4 \,|\, \delta\mathcal{H} \,|\, p \rangle \right|. \tag{G.7}
$$

Then, the case of a weak transverse field acting together with a quadratic anisotropy BS_x^2 will be considered. It is sufficient to treat it at the lowest possible order. For example, if $\delta\mathcal{H}$ is the sum of a strong term $\delta\mathcal{H}_2$ of the form (6.32) and a weak term $\delta\mathcal{H}_1$ of the form (6.31), the tunnelling frequency is given by (G.6) if $(p - m)$ is even, while for odd values,

$$
\hbar\omega_{mp} = \Big| \sum_{q=m}^{p-1} \langle m \,|\, \delta\mathcal{H}_2 \,|\, m+2 \rangle \, \frac{1}{E_m^{(0)} - E_{m+2}^{(0)}} \, \langle m+2 \,|\, \delta\mathcal{H}_2 \,|\, m+4 \rangle
$$

$$
\frac{1}{E_m^{(0)} - E_{m+4}^{(0)}} \cdots \frac{1}{E_m^{(0)} - E_{q-2}^{(0)}} \, \langle q-2 \,|\, \delta\mathcal{H}_2 \,|\, q \rangle \, \frac{1}{E_m^{(0)} - E_q^{(0)}}
$$

$$
\langle q \,|\, \delta\mathcal{H}_1 \,|\, q+1 \rangle \, \frac{1}{E_m^{(0)} - E_{q+1}^{(0)}} \, \langle q+1 \,|\, \delta\mathcal{H}_2 \,|\, q+3 \rangle
$$

$$
\frac{1}{E_m^{(0)} - E_{q+3}^{(0)}} \, \langle q+3 \,|\, \delta\mathcal{H}_2 \,|\, q+5 \rangle \cdots
$$

$$
\frac{1}{E_m^{(0)} - E_{p-4}^{(0)}} \, \langle p-4 \,|\, \delta\mathcal{H} \,|\, p-2 \rangle \, \frac{1}{E_m^{(0)} - E_{p-2}^{(0)}} \, \langle p-2 \,|\, \delta\mathcal{H} \,|\, p \rangle \Big|. \tag{G.8}
$$

Explicit forms of formulae (G.5)–(G.8) are easily obtained if $\mathcal{H}_0$ is given by (6.30). The level-crossing condition (6.3) imposes $g\mu_\mathrm{B} H_z = |D|(m + p)$, so that

$$
\begin{aligned}
E_m^{(0)} - E_q^{(0)} &= |D|(q^2 - m^2) + g\mu_\mathrm{B} H_z(m - q) \\
&= |D|(q^2 - m^2) + |D|(m + p)(m - q) \\
&= |D|(m - q)(p - q).
\end{aligned} \tag{G.9}
$$

For instance, insertion into (G.6) yields

$$\hbar\omega_{\mathrm{T}}^{mp} = |D| \left(\frac{B}{4|D|}\right)^{(p-m)/2} [(p-m-2)!!]^{-2} |\langle m | S_-^2 | m+2\rangle \langle m+2 | S_-^2 | m+4\rangle$$

$$\langle p-4 | S_-^2 | p-2\rangle \langle p-2 | S_-^2 | p\rangle|. \tag{G.10}$$

Since $\langle m | S_- | m+1\rangle = \sqrt{(s-m)(s+m+1)}$, this formula reads

$$\hbar\omega_{\mathrm{T}}^{mp} = |D| \left(\frac{B}{4|D|}\right)^{(p-m)/2} [(p-m-2)!!]^{-2} \sqrt{\frac{(s-m)!(s+p)!}{(s-p)!(s+m)!}}. \tag{G.11}$$

For instance, the ground state splitting $2\hbar\omega_{-s,s}$ in zero field is obtained for $p = -m = s$, namely

$$\hbar\omega_{\mathrm{T}}^{-s,s} = |D| \left(\frac{B}{4|D|}\right)^s \frac{(2s)!}{[(2s-2)!!]^2}$$

$$= 4s^2|D| \left(\frac{B}{4|D|}\right)^s \frac{(2s)!}{[(2s)!!]^2} = 4s^2|D| \left(\frac{B}{4|D|}\right)^s \frac{(2s)!}{[2^s(s)!]^2}.$$

or, using Stirling's formula $s! \simeq s^{s+1/2}e^{-s}\sqrt{2\pi}$ for large s,

$$\hbar\omega_{\mathrm{T}}^{-s,s} = 4s^2|D| \left(\frac{B}{4|D|}\right)^s \frac{(2s)^{2s+1/2}e^{-2s}\sqrt{2\pi}}{[2^s s^{s+1/2}e^{-s}\sqrt{2\pi}]^2} = \frac{4|D|}{\sqrt{\pi}} s^{3/2} \left(\frac{B}{4|D|}\right)^s. \tag{G.12}$$

APPENDIX H

PROOF OF THE LANDAU–ZENER–STÜCKELBERG FORMULA

The solution of system (8.12) which satisfies the initial condition $X(0) = 1$, $Y(0) = 0$ can be written as

$$
\begin{cases}
X(t) = 1 - i\omega_{\mathrm{T}} \int_0^t dt' \exp\left[iU(t')\right] Y(t') & \text{(a)} \\[2mm]
Y(t) = -i\omega_{\mathrm{T}} \int_0^t dt' \exp\left[-iU(t')\right] X(t') & \text{(b)}
\end{cases}
\tag{H.1}
$$

Eliminating $Y(t)$ in (H.1) one obtains

$$
X(t) = 1 - \omega_{\mathrm{T}}^2 \int_0^t dt' \exp\left[iU(t')\right] \int_0^{t'} dt'' \exp\left[-iF(t'')\right] U(t').
\tag{H.2}
$$

The lower integration bound can be replaced by $-\infty$ because there are no transitions between $t = -\infty$ and $t = 0$.

Equation (H.2) can be solved by iteration. One obtains a series which will only be written for $t = \infty$, namely

$$
X(\infty) = 1 + \sum_{n=1}^{\infty} (-\omega_{\mathrm{T}}^2)^n I_n
\tag{H.3}
$$

where

$$
I_n = \int_{-\infty}^{\infty} dt_{2n} \int_{-\infty}^{t_{2n}} dt_{2n-1} \cdots \int_{-\infty}^{t_2} dt_1 \, \exp f(\{t\})
\tag{H.4}
$$

with

$$
f(\{t\}) = i \sum_{j=1}^{n} \xi \left(t_{2j}^2 - t_{2j-1}^2 \right)
\tag{H.5}
$$

and $\xi = v/(2\hbar)$.

The following change of variables, introduced by Kayanuma (1984), will now be made:

$$
\begin{cases}
x_1 = t_1 & \text{(a)} \\
x_p = t_1 + \sum_{j=1}^{p-1} (t_{2j+1} - t_{2j}) & (2 \le p \le n) \quad \text{(b)} \\
y_p = t_{2p} - t_{2p-1} & (1 \le p \le n) \quad \text{(c)}
\end{cases}
\tag{H.6}
$$

The inverse transformation is

$$\begin{cases} t_{2p} = x_p + y_1 + y_2 + \cdots + y_p & \text{(a)} \\ t_{2p+1} = x_{p+1} + y_1 + y_2 + \cdots + y_p & \text{(b)} \end{cases} \tag{H.7}$$

It follows that

$$t_{2j}^2 - t_{2j-1}^2 = y_j(t_{2j} + t_{2j-1}) = 2y_j(x_j + y_1 + y_2 + \cdots + y_{j-1} + y_j/2)$$

and $f(\{t\}) = g(\{x\}, \{y\})$, with

$$g(\{x\}, \{y\}) = i\xi \left[2 \sum_{j=1}^{n} x_j y_j + \left(\sum_{j=1}^{n} y_j \right)^2 \right]. \tag{H.8}$$

Since $t_p \geq t_{p-1}$, the quantities x_j and y_j satisfy the relations

$$x_1 \leq x_2 \leq \cdots \leq x_j \leq x_{j-1} \leq \cdots \leq x_n \tag{H.9}$$

and

$$y_1, y_2, \ldots, y_j, \ldots, y_n \geq 0 \tag{H.10}$$

so that

$$I_n = \int_{-\infty}^{\infty} dx_1 \int_{x_1}^{\infty} dx_2 \cdots \int_{x_{n-1}}^{\infty} dx_n \int_0^{\infty} dy_1 \int_0^{\infty} dy_2 \cdots$$
$$\times \int_0^{\infty} dy_n \exp g(\{x\}, \{y\}). \tag{H.11}$$

The derivation of (H.11) makes use of the fact that the Jacobian of the transformation is equal to 1, as can be easily checked. This relation can also be written as

$$I_n = \int_{(D)} \mathcal{D}x \mathcal{D}y \exp g(\{x\}, \{y\}) \tag{H.12}$$

where the integration domain (D) is defined by (H.9) and (H.10).

It will now be shown that the variables x_j can be replaced in (H.9) by any perturbation (thus changing the domain (D)) without modifying the integral (H.12). Indeed, the integral is not modified by replacing the variables x_j and y_j since this is just changing the name of the variables. But the function (H.8) is symmetric, and does not change if the inverse permutation is done. This proves the proposition. For instance, in the case $n = 2$, the definition (H.12) is $I_2 = \int_{-\infty}^{\infty} dx_1 \int_{x_1}^{\infty} dx_2 \int_0^{\infty} dy_1 \int_0^{\infty} dy_2 \exp g(x_1, x_2, y_1, y_2,)$. Changing the name of the variables yields $I_2 = \int_{-\infty}^{\infty} dx_2 \int_{x_2}^{\infty} dx_1 \int_0^{\infty} dy_1 \int_0^{\infty} dy_2 \exp g(x_2, x_1, y_2, y_1)$, and since g is a symmetric function, this is equal to the expected expression

$$I_2 = \int_{-\infty}^{\infty} dx_2 \int_{x_2}^{\infty} dx_1 \int_0^{\infty} dy_1 \int_0^{\infty} dy_2 \exp g(x_1, x_2, y_1, y_2).$$

It follows that (H.11) can be written as

$$
I_n = \frac{1}{n!} \left(\prod_{j=1}^{n} \int_0^\infty dy_j \right) \exp\left[i\xi \left(\sum_{j=1}^{n} y_j \right) \right]^2 \left[\prod_{j=1}^{n} \int_{-\infty}^{\infty} dx_j \exp(2i\xi x_j y_j - \epsilon|y_j|) \right]
$$

(H.13)

where $\epsilon = 0$. Indeed that integral can be written as a sum of $n!$ integrals of the form (H.12), with $n!$ different definitions of the integration domain (D) corresponding to all permutations of the x_j, and all those integrals are equal.

In (H.13), the parameter $\epsilon = 0$ has been introduced to facilitate the calculation. Integrating over the x_j, one obtains

$$
I_n = \frac{1}{n!} \prod_{j=1}^{n} \left(\int_0^\infty dy_j \right) \exp\left[i\xi \left(\sum_{j=1}^{n} y_j \right) \right]^2 \prod_{j=1}^{n} \frac{2\epsilon}{\epsilon^2 + 4\xi^2 y_j^2}.
$$

(H.14)

In the limit $\epsilon \to 0$, the fraction becomes a delta function and the exponential may be replaced by 1. It follows that

$$
\begin{aligned}
I_n &= \frac{1}{n!} \prod_{j=1}^{n} \left(\int_0^\infty dy_j \frac{2\epsilon}{\epsilon^2 + 4\xi^2 y_j^2} \right) \\
&= \frac{1}{n!} \left(\int_0^\infty dy \frac{2\epsilon}{\epsilon^2 + 4\xi^2 y^2} \right)^n \\
&= \frac{1}{n!} \left((1/\xi) \int_0^\infty dy \frac{1}{1 + y^2} \right)^n \\
&= \frac{1}{n!} \left(\frac{\pi}{2\xi} \right)^n.
\end{aligned}
$$

(H.15)

Using this result in (H.3) and replacing ξ by $v/(2\hbar)$, one obtains

$$
X(\infty) = 1 + \sum_{n=1}^{\infty} \frac{1}{n!} \left(\frac{-\pi\hbar\omega_T^2}{v} \right)^n = \exp\left(\frac{-\pi\hbar\omega_T^2}{v} \right)
$$

(H.16)

The probability $1 - \delta P$ of still being in the left-hand well at $t = \infty$ is the square of this quantity. Formula (8.17) follows.

APPENDIX I

TUNNELLING BETWEEN HYPERFINE STATES

In the absence of hyperfine interactions, tunnelling takes place between two loc-
alized states, e.g. $|-s^*\rangle$ and $|-m^*\rangle$, which are approximate eigenvectors of the
spin Hamiltonian, which is the sum of an anisotropy term $\mathcal{H}_{an}$ and a Zeeman
term $\mathcal{H}_Z$. If hyperfine interactions are taken into account, one has to consider
the eigenvectors of the Hamiltonian

$$\mathcal{H} = \mathcal{H}_{an} + \mathcal{H}_Z + \mathcal{H}_{hf} + \mathcal{H}_\nu$$

where $\mathcal{H}_{hf} = -\sum_{\alpha\gamma=xyz} g_k^{\alpha\gamma} S_\alpha I_k^\gamma$ is the hyperfine interaction and $\mathcal{H}_\nu$ is the interaction between nuclear spins. The last two terms are small, and the eigenvectors
of $\mathcal{H}$ can be approximated by $|m^*,\nu\rangle$, where $|\nu\rangle$ designates nuclear spin states.
Generally, the nuclear spin states associated with different electronic states m
are different. Strictly speaking, nuclear spins mediate an interaction between the
molecular spins, but this effect will be ignored. A single molecular spin is taken
into account, and the other ones will be approximated by a field acting on the
nuclear spins.

Tunnelling takes place between *hyperfine* states $|g,\nu\rangle$ and $|d,\nu'\rangle$, where
$|g\rangle = |m^*\rangle$ and $|d\rangle = |m'^*\rangle$ are, respectively, left-localized and right-localized.
The calculation of Sections 8.3.2 and 8.3.3 will now be extended to this case.
If the initial, hyperfine state is $|g,\nu_0\rangle$, the state at time t is $X(t)|g,\nu_0\rangle +
\sum_\nu Y_\nu(t)|d,\nu\rangle$, with $X(t) \simeq X(0) \simeq 1$.

Equation (8.12) should now be replaced by a number of equations equal to the
number of hyperfine states. Assuming, as usual, that $|g\rangle$ and $|d\rangle$ are independent
of t, these equations are

$$\dot{Y}_\nu(t) = \frac{1}{i\hbar} X(t) \langle d,\nu \,|\, \mathcal{H} \,|\, g,\nu_0 \rangle \, e^{-i[u(t)-w(t)]}. \tag{I.1}$$

The Hamiltonian $\mathcal{H}$ will be approximated by its greatest term, the anisotropy
Hamiltonian which acts on the molecular spin. Thus

$$\langle d,\nu \,|\, \mathcal{H} \,|\, g,\nu_0 \rangle = \langle d \,|\, \mathcal{H} \,|\, g \rangle \langle \nu \,|\, \nu_0 \rangle = \omega_T^0 \langle \nu \,|\, \nu_0 \rangle \tag{I.2}$$

where ω_T^0 is the tunnel frequency in the absence of hyperfine interactions. The
reversal probability which replaces (8.18) is

$$\delta P = \sum_\nu \frac{2\pi\hbar}{v} \omega_T^2 |\langle \nu \,|\, \nu_0 \rangle|^2 \tag{I.3}$$

and this is equal to (8.18).

One can wonder what nuclear spins do during tunnelling: do they tunnel with the molecular spin or are they motionless? One can first consider the simple case when nuclear states $|\nu\rangle$ and $|\nu'\rangle$ associated to hyperfine states $|g,\nu\rangle$ and $|d,\nu'\rangle$ are the same. In that case, $|\langle\nu\,|\,\nu_0\rangle|^2 = \delta_{\nu,\nu_0}$. Then, (I.2) implies that nuclear spins are not modified. The interaction energy between nuclear spins and the molecular spin goes from the value $g\mu_B s H_{hf}$ to the opposite value $-g\mu_B s H_{hf}$. The Zeeman energy of the molecular spin goes from a value close to $g\mu_B H s$ to the opposite value $-g\mu_B H s$. Other contributions are small. Tunnelling is possible if the change in total energy is close to 0, i.e. if $H + H_{hf} \simeq 0$. Thus, the resonance width due to hyperfine interactions is of the order of $\langle H_{hf}\rangle$, as claimed in Section 9.2.

As a matter of fact, nuclear spins are not the same when the molecular spin is localized on the left or right hand side. Nevertheless, the factor $\langle\nu\,|\,\nu_0\rangle$ in (I.2) favours final nuclear spins which are close to the initial state. The resonance width should therefore be close to $\langle H_{hf}\rangle$.

The behaviour of nuclear spins during tunnelling has been discussed by Prokofev and Stamp (1995) and Tupitsyn *et al.* (1997). They conclude, in agreement with the above statement, that only a few nuclear spins flip.

The second part of this appendix regards the impossibility of relaxation in the absence of interactions between nuclear spins. At a given time, as seen in Section 9.2, most of the molecular spins cannot tunnel because the local field in the anisotropy direction z is too far from a level crossing, i.e. at a distance larger than the natural width $\hbar\omega_T$. Let this distance be smaller than the hyperfine width $g\mu_B H_{hf}$. The problem is whether the local field can change under the effect of the interaction between molecular and nuclear spins. The answer is 'no'. Only an approximate, simplified argument will be given, and a more elaborate proof will be found in the article of Prokofev and Stamp (1995). Neglecting interactions between nuclear spins, the Hamiltonian acting on $\mathbf{I}_k$ can be written as

$$\mathcal{H}_k = \gamma_k \mathbf{H}.\mathbf{I}_k - \sum_{\alpha\gamma=xyz} g_k^{\alpha\gamma} S_\alpha I_k^\gamma \qquad (I.4)$$

where γ_k is the gyromagnetic ratio of nucleus k. In the case of a dipole interaction between two localized spins at distance $\mathbf{r_k}$, $g_k^{\alpha\gamma}$ would be given by $g_k^{\alpha\gamma} = (C/r_k^3)(\delta_{\alpha\gamma} - r_k^\alpha r_k^\gamma/r_k^2)$ where C is a constant. Here, the story is more complicated because the molecular spin is the sum of single ion spins at different places.

Since the average value of the molecular spin $\mathbf{S}$ is in the easy direction z, it is convenient to separate the S_z component in (I.4):

$$\mathcal{H}_k = \gamma_k \mathbf{H}.\mathbf{I}_k - \sum_{\gamma=xyz} g_k^{z\gamma} S_z I_k^\gamma - \sum_{\alpha=yz}\sum_{\gamma=xyz} g_k^{\alpha\gamma} S_\alpha I_k^\gamma. \qquad (I.5)$$

The nuclear spin feels a local field of components

$$H^\gamma - g_k^{z\gamma}\langle S_z\rangle/\gamma_k. \qquad (I.6)$$

It is appropriate to define a local, k-dependent axis Z along this direction, and two other axes X, Y perpendicular to Z. Formula (I.5) can be approximately written as

$$\mathcal{H}_k = \left(\gamma_k H_Z - g_k^{zZ} S_z\right) I_k^Z - \sum_{\alpha=xy} \sum_{\gamma=XYZ} g_k^{\alpha\gamma} S_\alpha I_k^\gamma. \qquad (I.7)$$

Within an acceptable approximation, S_x and S_y can be replaced by their average value which is 0, so that the last term of the right-hand side can be ignored. Then the nuclear spin precesses around its local field. In this motion, the component I_k^Z is constant. Now, if (I.7) is regarded as acting on the molecular spin $\mathbf{S}$, this means that the z component of the local field is constant. Therefore, if the level-crossing condition is not satisfied at a given time, as is the case for most of the spins, it will never be satisfied. The precession of $\mathbf{I}$ does modify the second term of (I.7), thus producing a modulation of the tunnel frequency, but this has no effect on the relaxation time.

APPENDIX J

SPECIFIC HEAT

J.1 Specific heat at equilibrium and at high frequency

The equilibrium specific heat C_{eq} of a spin is given, as seen in Chapter 3, by

$$C_{eq} = \frac{\partial}{\partial T} \langle \mathcal{H} \rangle. \tag{J.1}$$

The quantity of interest is the magnetic specific heat, but the subscript 'mag' has been omitted because there is no ambiguity. The magnetic specific heat is assumed to be decoupled from the lattice, and the lattice specific heat is never considered in this appendix.

If the spin has $(2s+1)$ energy levels E_m, the mean energy $\langle \mathcal{H} \rangle$ is

$$\langle \mathcal{H} \rangle = \frac{\sum_{m=-s}^{s} E_m \exp[-\beta E_m]}{\sum_{m=-s}^{s} \exp[-\beta E_m]} \tag{J.2}$$

which is a generalization of the formulae seen in Chapter 3. The calculation yields

$$C_{eq} = \frac{1}{k_B T^2} \frac{1}{z} \sum_m E_m^2 \exp(-\beta E_m) - \frac{1}{k_B T^2} \left[\frac{1}{z} \sum_m E_m \exp(-\beta E_m) \right]^2 \tag{J.3}$$

where $z = \sum_m \exp(-\beta E_m)$.

As in Section 3.2.3 we now wish to consider a spin in a double potential well, and a high frequency $\omega \gg 1/\tau$, when the spin has almost no chance to go jump to the other part of the double well. Then one has to consider the specific heat C^+ (resp. C^-) of a spin confined in the right-hand (resp. left-hand) well. The eigenstates $| m \rangle$ will be assumed to be localized. In analogy with (J.3)

$$C^{\pm} = \frac{1}{k_B T^2} \frac{1}{z^{\pm}} \sum_m^{\pm} E_m^2 \exp(-\beta E_m) - \frac{1}{k_B T^2} \left[\frac{1}{z^{\pm}} \sum_m^{\pm} E_m \exp(-\beta E_m) \right]^2 \tag{J.4}$$

where $\sum_m^+$ and $\sum_m^-$, respectively, designate states localized in the right and left hand well, and $z = \sum_m^{\pm} \exp(-\beta E_m)$.

At equilibrium, the spin has probabilities z^+/z and z^-/z to be in the right or left-hand well, respectively. Therefore the high-frequency specific heat $C(\omega) = C_{uni}$ as defined in section 10.8 is

$$C_{uni} = \frac{z^+}{z} C^+ + \frac{z^-}{z} C^-$$

or according to (J.4)

$$C_{\text{uni}} = \frac{1}{k_B T^2} \frac{1}{z} \sum_m E_m^2 \exp(-\beta E_m)$$

$$- \frac{1}{z k_B T^2} \left\{ \frac{1}{z^+} \left[\sum_m^+ E_m \exp(-\beta E_m) \right]^2 + \frac{1}{z^-} \left[\sum_m^- E_m \exp(-\beta E_m) \right]^2 \right\}.$$

$$\text{(J.5)}$$

Subtraction of (J.5) from (J.3) yields

$$C_{\text{eq}} - C_{\text{uni}} = \frac{1}{z^2 k_B T^2} \left\{ \frac{z}{z^+} \Sigma_+^2 + \frac{z}{z^-} \Sigma_-^2 - [\Sigma_+ + \Sigma_-]^2 \right\} \qquad \text{(J.6)}$$

where

$$\Sigma_\pm = \sum_m^\pm E_m \exp(-\beta E_m). \qquad \text{(J.7)}$$

Since $z = z^+ + z^-$, (J.6) reads

$$C_{\text{eq}} - C_{\text{uni}} = \frac{1}{z^2 k_B T^2} \left\{ \frac{z^-}{z^+} \Sigma_+^2 + \frac{z^+}{z^-} \Sigma_-^2 - 2\Sigma_+ \Sigma_- \right\}$$

$$= \frac{1}{z^2 k_B T^2} \left\{ \sqrt{\frac{z^-}{z^+}} \Sigma_+ - \sqrt{\frac{z^+}{z^-}} \Sigma_- \right\}^2 \qquad \text{(J.8)}$$

which is positive.

J.2   Frequency-dependent specific heat

The next problem is to interpolate between the low-frequency specific heat $C(0) = C_{\text{eq}}$ and the high-frequency specific heat C_{uni}. It is appropriate to introduce a frequency-dependent specific heat $C(\omega)$ which should first be defined. The definition should be an extension of the equilibrium definition $C = \partial U / \partial T$ (at constant magnetic field, for instance). In the presence of a sinusoidal thermal excitation of frequency ω, the energy $U(t)$ and the temperature $T(t)$ are functions of time,

$$T(t) = T_0 + \delta T(t) = T_0 + \delta T_0 \cos(\omega t) = T_0 + Re[\delta T_0 \exp(i\omega t)] \qquad \text{(J.9)}$$

and

$$U(t) = U_0 + \delta U(t) = U_0 + \delta U_0 \cos(\omega t - \varphi) = U_0 + \Re \delta U_0 \exp[(-i\varphi) \exp(i\omega t)]. \qquad \text{(J.10)}$$

An appropriate definition of $C(\omega)$ is

$$C(\omega) = \frac{\delta U_0 \exp(-i\varphi)}{\delta T_0} \qquad \text{(J.11)}$$

which generalizes the equilibrium property $C = \partial U / \partial T$.

The next task is to calculate $\delta U(t)$ when $\delta T(t)$ is known. This requires knowledge of the probability $p_m(t)$ to be in state m at time t. Indeed

$$U(t) = \sum_{-s}^{s} p_m(t) E_m \tag{J.12}$$

where $p_m(t)$ is the probability to be in state m at time t. This probability depends on a single quantity $p^+(t)$, the probability to be in the right-hand well at time t. Indeed, inside each part of the double well, there is thermal equilibrium. Therefore, for all states $|\,m\rangle$ of the right-hand well,

$$p_m(t) = \frac{p^+(t)}{z^+} \exp(-\beta E_m) \tag{J.13}$$

while for all states of the left-hand well,

$$p_m(t) = \frac{p^-(t)}{z^-} \exp(-\beta E_m) \tag{J.14}$$

with $p^-(t) = 1 - p^+(t)$.

The only additional thing needed is an equation which determines the evolution of $p^+(t)$. This equation is

$$\frac{\partial}{\partial t} p^+(t) = -\frac{1}{\tau} \left[p^+(t) - p_T^+(t) \right] \tag{J.15}$$

where τ is the relaxation time already introduced before, and p_T^+ is the probability to be in the right-hand well when the system is at equilibrium at temperature T. It depends on t because T is given by (J.9). Of course, p_T^+ is readily written as the sum of the Boltzmann probabilities $z^{-1} \exp[-E_m/(k_B T)]$ on all states m of the right-hand well.

The above equations are easy to solve. Straightforward algebra (Fominaya et al. 1999) leads to (10.15).

APPENDIX K

MASTER EQUATION FOR THE DENSITY MATRIX

K.1 Basic hypotheses

The motion of a system described by a Schrödinger equation is invariant under time reversal. This is true indipendently on the size. However, if one considers a part of a large system, its motion becomes irreversible after a microscopic time τ_1. In a gas, the order of magnitude of τ_1 is the time between two collisions of a particular particle. In this book, the large system is a spin interacting with phonons. The small system is the spin. It is subject to a spin Hamiltonian $\mathcal{H}_{\rm sp}$, e.g.

$$\mathcal{H}_{\rm sp} = DS_z^2 + g\mu_B H S_z - C\left[S_+^4 + S_-^4\right] \tag{K.1}$$

the eigenvectors of which will be called $\mid m^*\rangle$. The probability for the spin to be in a particular state is given by a $2s \times 2s$ density matrix $\rho_{mn}(t)$. The basic hypothesis is that, for a time longer than a microscopic time τ_1, it satisfies the master equation (11.4). In the basis of the eigenvectors $\mid m^*\rangle$, this equation reads

$$\frac{d}{dt}\rho_{mn}(t) = i(\omega_n - \omega_m)\rho_{mn}(t) - \sum_{rr'} \Lambda_{mn}^{rr'}\rho_{rr'}(t). \tag{K.2}$$

For times much shorter than the relaxation time, but much longer than τ_1, this equation reads

$$\rho_{mn}(t) = \exp[it(\omega_n - \omega_m)]\rho_{mn}(0) - t\sum_{rr'} \Lambda_{mn}^{rr'}\rho_{rr'}(0). \tag{K.3}$$

This equation can be matched with a microscopic equation obtained by perturbation theory. This is the principle of the calculation of the coefficients $\Lambda_{mn}^{rr'}$, performed in this appendix.

The unperturbed Hamiltonian

$$\mathcal{H}_{\rm tot}^0 = \mathcal{H}_{\rm sp} + \mathcal{H}_{\rm ph} \tag{K.4}$$

is the sum of the spin Hamiltonian and the free phonon Hamiltonian (5.31).

The perturbation is the spin–phonon interaction defined by (5.34) and hereafter called $\mathcal{H}_1$. The notation is that of Section 5.6 except that the phonon modes will be labelled α rather than ρ to avoid confusion.

The total wavefunction is, at time t,

$$\mid\Psi(t)\rangle = \exp(-it\mathcal{H}_{\rm tot}/\hbar)\mid\Psi(0)\rangle \tag{K.5}$$

where $\mathcal{H}_{\rm tot} = \mathcal{H}_{\rm tot}^0 + \mathcal{H}_1$.

K.2 An expression for the density matrix of a spin system

It will first be assumed that the initial wavefunction $|\Psi(0)\rangle$ of the total system is well defined. It is the product

$$| \Psi(0)\rangle = | \psi(0), \Phi_Q\rangle \qquad \text{(K.6)}$$

of a spin function $| \psi(0)\rangle$ by a phonon function $| \Phi_Q\rangle$, which will be assumed to be an eigenfunction of the free phonon Hamiltonian $\mathcal{H}_{\text{ph}}$

The quantum average value of an operator $\tilde{\sigma}$ at time t is

$$\langle \Psi(t) \,|\, \tilde{\sigma} \,|\, \Psi(t)\rangle = \langle \Psi(0) \,|\, \exp(it\mathcal{H}_{\text{tot}}/\hbar)\tilde{\sigma}\exp(-it\mathcal{H}_{\text{tot}}/\hbar) \,|\, \Psi(0)\rangle. \qquad \text{(K.7)}$$

It will now be assumed that $\tilde{\sigma}$ is a spin operator. One wishes to calculate the average value of $\tilde{\sigma}$ at time t for a given initial spin wavefunction $| \psi(0)\rangle$, but assuming phonons to be in thermal equilibrium. To do that, $\Psi(0)$ can be replaced by (K.6) in (K.7), and multiplied by the Boltzmann factor ρ_Q which gives the probability of the phonon state Φ_Q. The sum over Q is then performed. The result is

$$\langle \tilde{\sigma}(t)\rangle = \sum_Q \rho_Q \langle \psi(0), \Phi_Q \,|\, \exp(it\mathcal{H}_{\text{tot}}/\hbar)\tilde{\sigma}\exp(-it\mathcal{H}_{\text{tot}}/\hbar) \,|\, \psi(0), \Phi_Q\rangle. \qquad \text{(K.8)}$$

On the other hand, the operator $\tilde{\sigma}$ may be written as

$$\tilde{\sigma} = \sum_{mnQ} | n^*, \Phi_Q\rangle \langle n^*, \Phi_Q \,|\, \tilde{\sigma} \,|\, m^*, \Phi_Q\rangle \langle m^*, \Phi_Q \,| \qquad \text{(K.9)}$$

so that (K.8) reads

$$\langle \tilde{\sigma}\rangle = \sum_{mnQ} \langle n^*, \Phi_Q \,|\, \tilde{\sigma} \,|\, m^*, \Phi_Q\rangle \, \rho_{mn}(t) \qquad \text{(K.10)}$$

where

$$\rho_{mn}(t) = \sum_{QQ'} \rho_Q \langle m^*, \Phi_{Q'} \,|\, \exp(-it\mathcal{H}_{\text{tot}}/\hbar) \,|\, \psi(0), \Phi_Q\rangle$$

$$\langle \psi(0), \Phi_Q \,|\, \exp(it\mathcal{H}_{\text{tot}}/\hbar) \,|\, n^*, \Phi_{Q'}\rangle \qquad \text{(K.11)}$$

is an element of the density matrix as defined in Section 11.2.

This expression must be identified with (K.3). Thus, neglecting terms or order t^2 or higher,

$$t \sum_{rr'} \Lambda^{rr'}_{mn}\rho_{rr'}(0) = \rho_{mn}(0)\exp[it(\omega_n - \omega_m)]$$

$$- \sum_Q \rho_Q \langle m^*, \Phi_{Q'} \,|\, \exp(-it\mathcal{H}_{\text{tot}}/\hbar) \,|\, \psi(0), \Phi_Q\rangle$$

$$\langle \psi(0), \Phi_Q \,|\, \exp(it\mathcal{H}_{\text{tot}}/\hbar) \,|\, n^*, \Phi_{Q'}\rangle \exp[it(\omega_m - \omega_n)]$$

or if $|\psi(0)\rangle = \sum_r \psi_r |r\rangle$

$$t\sum_{rr'}\Lambda^{rr'}_{mn}\psi_r\psi^*_{r'} = \psi_m\psi^*_n \exp[it(\omega_n - \omega_m)]$$

$$-\sum_{QQ'}\rho_Q \sum_{rr'}\psi_r\psi^*_{r'}\langle m^*|\exp(-it\mathcal{H}_{\text{tot}}/\hbar)|r^*,\Phi_Q\rangle$$

$$\langle r',\Phi_Q|\exp(it\mathcal{H}_{\text{tot}}/\hbar)|n^*\rangle \exp[it(\omega_m - \omega_n)]$$

or

$$t\Lambda^{rr'}_{mn} = \delta_{mr}\delta_{nr'}\exp[it(\omega_n - \omega_m)]$$

$$-\sum_{QQ'}\rho_Q\langle m^*,\Phi_{Q'}|\exp(-it\mathcal{H}_{\text{tot}}/\hbar)|r^*,\Phi_Q\rangle$$

$$\langle r',\Phi_Q|\exp(it\mathcal{H}_{\text{tot}}/\hbar)|n^*,\Phi_{Q'}\rangle \exp[it(\omega_m - \omega_n)] \qquad (\text{K.12})$$

K.3 Perturbation theory

K.3.1 *Diagrammatic expansion*

Within second-order perturbation theory

$$\exp(-it\mathcal{H}_{\text{tot}}/\hbar) = \exp(-it\mathcal{H}^0_{\text{tot}}/\hbar)$$

$$-\frac{i}{\hbar}\int_0^t dt' \exp\left[i(t'-t)\mathcal{H}^0_{\text{tot}}/\hbar\right]\mathcal{H}_1 \exp\left[-it'\mathcal{H}^0_{\text{tot}}/\hbar\right]$$

$$-\int_0^t \frac{dt'}{\hbar^2} \exp\left[i(t'-t)\mathcal{H}^0_{\text{tot}}/\hbar\right]\mathcal{H}_1$$

$$\times \int_0^t dt'' \exp\left[\frac{i(t''-t')}{\hbar}\mathcal{H}^0_{\text{tot}}\right]\mathcal{H}_1 \exp\left[\frac{-it''}{\hbar}\mathcal{H}^0_{\text{tot}}\right]. \qquad (\text{K.13})$$

Let this expression be inserted into the second term of the right-hand side of
(K.12). At order 0 combination with the first term yields 0. At order 1, the result
is also 0 because the average value of a phonon creation or destruction operator
$b^\pm_{q\alpha}$ is 0. The task is just to calculate the second-order contributions. They are
three, which correspond, respectively, to taking the first-order terms from both
exponentials, the second-order term of the first exponential and the second-order
term of the second exponential. Moreover each of the three contributions splits
into two, in which the phonon average values $\langle n_{q\alpha} + 1\rangle$ and $\langle n_{q\alpha}\rangle$, respectively,
appear. As first proposed by Feynman in the case of particle physics, the first
six terms of the perturbative expansion of (K.12) may be represented by six
diagrams as shown by Fig. K.1.

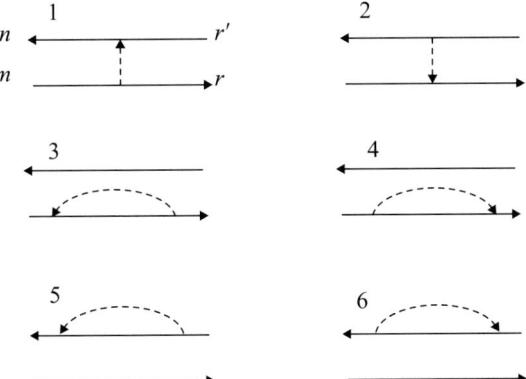

FIG. K.1. The six diagrams for second-order terms in formula (K.12).

K.3.2 First and second diagrams

The explicit value of the first diagram is

$$t[\Lambda_{mn}^{rr'}]_1 = -\frac{1}{N\hbar^2} \int_0^t dt' \int_0^t dt''$$

$$\sum_{q\alpha} \langle m^* | \exp\left[i(t'-t)\mathcal{H}_{\mathrm{sp}}/\hbar\right] U_{q\alpha}(S) \exp\left[-it'\mathcal{H}_{\mathrm{sp}}/\hbar\right] |r^*\rangle$$

$$\langle r'^* | \exp\left[it''\mathcal{H}_{\mathrm{sp}}/\hbar\right] U_{q\alpha}^*(S) \exp\left[i(t-t'')\mathcal{H}_{\mathrm{sp}}/\hbar\right] |n^*\rangle$$

$$\sum_{QQ'} \langle \Phi_Q | \exp\left[i(t'-t)\mathcal{H}_{\mathrm{ph}}/\hbar\right] b_{q\alpha} \exp\left[-it'\mathcal{H}_{\mathrm{ph}}/\hbar\right] | \Phi_{Q'}\rangle$$

$$\langle \Phi_{Q'} | \exp\left[it''\mathcal{H}_{\mathrm{ph}}/\hbar\right] b_{q\alpha}^+ \exp\left[i(t-t'')\mathcal{H}_{\mathrm{ph}}/\hbar\right] | \Phi_Q\rangle \rho_Q \qquad (K.14)$$

It is by no means clear that this formula yields a time-independent value of $[\Lambda_{mn}^{rr'}]_1$. If our hypotheses are correct this should be true *at long times*.

Introducing the energy E_Q^{ph} of the phonon state $| \Phi_Q\rangle$, (K.14) can be written as

$$t[\Lambda_{mn}^{rr'}]_1 = -\frac{1}{N\hbar^2} \int_0^t dt' \int_0^t dt''$$

$$\sum_{q\alpha} \langle m^* | \exp\left[i(t'-t)\mathcal{H}_{\mathrm{sp}}/\hbar\right] U_{q\alpha}(S) \exp\left[-it'\mathcal{H}_{\mathrm{sp}}/\hbar\right] |r^*\rangle$$

$$\langle r'^* | \exp\left[it''\mathcal{H}_{\mathrm{sp}}/\hbar\right] U_{q\alpha}^*(S) \exp\left[i(t-t'')\mathcal{H}_{\mathrm{sp}}/\hbar\right] |n^*\rangle$$

$$\sum_{QQ'} \langle \Phi_Q | b_{q\alpha} | \Phi_{Q'}\rangle \langle \Phi_{Q'} | b_{q\alpha}^+ | \Phi_Q\rangle \rho_Q \exp\left[i(t'-t'')(E_Q^{\mathrm{ph}} - E_{Q'}^{\mathrm{ph}})/\hbar\right]$$

Providentially, the difference $E_Q^{\text{ph}} - E_{Q'}^{\text{ph}} = -\hbar\omega_{q\alpha}$ is independent of the phonon states so that this formula reads

$$t[\Lambda_{mn}^{rr'}]_1 = -\frac{1}{N\hbar^2} \int_0^t dt' \int_0^t dt''$$

$$\sum_{q\alpha} \langle m^* | \exp\left[i(t'-t)\mathcal{H}_{\text{sp}}/\hbar\right] U_{q\alpha}(S) \exp\left[-it'\mathcal{H}_{\text{sp}}/\hbar\right] | r^* \rangle$$

$$\langle r'^* | \exp\left[it''\mathcal{H}_{\text{sp}}/\hbar\right] U_{q\alpha}^*(S) \exp\left[i(t-t'')\mathcal{H}_{\text{sp}}/\hbar\right] | n^* \rangle$$

$$\times \langle b_{q\alpha} b_{q\alpha}^+ \rangle \exp\left[i(t''-t')\omega_{q\alpha}\right] \tag{K.15}$$

Since the states $|m^*\rangle$, $|n^*\rangle$, $|r^*\rangle$ and $|r'^*\rangle$ are eigenstates of $\mathcal{H}_{\text{sp}}$ for the eigenvalues $\hbar\omega_m$, $\hbar\omega_n$, $\hbar\omega_r$ and $\hbar\omega_{r'}$, formula (K.15) reads

$$t[\Lambda_{mn}^{rr'}]_1 = -\frac{1}{N\hbar^2} \sum_{q\alpha} \langle m^* | U_{q\alpha}(S) | r^* \rangle \langle r'^* | U_{q\alpha}^*(S) | n^* \rangle \langle b_{q\alpha} b_{q\alpha}^+ \rangle$$

$$\int_0^t dt' \int_0^t dt'' \exp\left[i(t''-t')\omega_{q\alpha}\right] \exp\left[i(t'-t)\omega_m\right.$$

$$\left. -it'\omega_r + it''\omega_{r'} + i(t-t'')\omega_n\right] \tag{K.16}$$

It turns out to be convenient to introduce an additional variable ω through the formula

$$\frac{1}{N} \sum_{q\alpha} \rightarrow \frac{1}{N} \sum_{\alpha} \int_{-\infty}^{\infty} d\omega \sum_q \delta(\omega - \omega_{q\alpha}).$$

Formula (K.16) now becomes

$$t[\Lambda_{mn}^{rr'}]_1 = -\frac{1}{N\hbar^2} \sum_{\alpha} \int_{-\infty}^{\infty} d\omega \sum_q \delta(\omega - \omega_{q\alpha}) \langle m^* | U_{q\alpha}(S) | r^* \rangle \langle r'^* | U_{q\alpha}^*(S) | n^* \rangle$$

$$\langle b_{q\alpha} b_{q\alpha}^+ \rangle \int_0^t dt' \int_0^t dt'' \exp\left[i(t''-t')\omega_{q\alpha}\right] \exp\left[i(t'-t)\omega_m\right.$$

$$\left. -it'\omega_r + it''\omega_{r'} + i(t-t'')\omega_n\right] \tag{K.17}$$

This expression contains the function

$$F_{mrr'n\alpha}(\omega) = \frac{1}{N\hbar^2} \sum_q \langle m^* | U_{q\alpha} | r^* \rangle \langle r'^* | U_{q\alpha}^* | n^* \rangle \langle n_{q\alpha} + 1 \rangle \delta(\omega - \omega_{q\alpha}).$$

$$\tag{K.18}$$

From the absence of phonons at frequency 0 follows the property

$$F_{mpp'n\alpha}(0) = 0 \tag{K.19}$$

which will be useful later.

Inserting (K.18) into (K.17) one obtains

$$t[\Lambda_{mn}^{rr'}]_1 = -\sum_\alpha \int_{-\infty}^{\infty} d\omega\, F_{mrr'n\alpha}(\omega) \int_0^t dt' \int_0^t dt''$$

$$\exp\left[i(t'' - t')\omega\right] \exp\left[i(t' - t)\omega_m - it'\omega_r + it''\omega_{r'} + i(t - t'')\omega_n\right].$$

Integration over t'' yields

$$t[\Lambda_{mn}^{rr'}]_1 = -\sum_\alpha \int_{-\infty}^{\infty} d\omega\, F_{mrr'n\alpha}(\omega) \exp\left[it(\omega_n - \omega_m)\right]$$

$$\int_0^t dt' \exp\left[it'(\omega - \omega_r + \omega_m)\right] \frac{1 - \exp\left[it(\omega_{r'} - \omega_n - \omega)\right]}{i(\omega_n + \omega - \omega_{r'})}. \qquad (\text{K}.20)$$

One might integrate over t' too, but the form (K.20) is adequate. It contains the fraction

$$\frac{1 - \exp\left[it(\omega_{r'} - \omega_n - \omega)\right]}{i(\omega_n + \omega - \omega_{r'})} = 2\exp\left[it(\omega_{r'} - \omega_n - \omega)/2\right] \frac{\sin\left[t(\omega_n + \omega - \omega_{r'})/2\right]}{\omega_n + \omega - \omega_{r'}}. \qquad (\text{K}.21)$$

This fraction has a maximum at $\omega = \omega_{r'} - \omega_n$. As noticed at the beginning of this section, the region of interest is that of long times. Then the fraction is large and proportional to t in a band of width $1/t$ around the maximum, and oscillates around the value 0 outside this band. The fraction (K.21) can therefore be replaced by

$$2\frac{\sin\left[t(\omega_n + \omega - \omega_{r'})/2\right]}{\omega_n + \omega - \omega_{r'}} \rightarrow 2\delta(\omega_n + \omega - \omega_{r'}) \int_{-\infty}^{-\infty} dx \frac{\sin\left[tx/2\right]}{x}$$

$$= 2\pi\delta(\omega_n + \omega - \omega_{r'}). \qquad (\text{K}.22)$$

Making this replacement in (K.20) yields

$$t[\Lambda_{mn}^{rr'}]_1 = -2\pi \sum_\alpha F_{mrr'n\alpha}(\omega_{r'} - \omega_n) \exp\left[it(\omega_n - \omega_m)\right]$$

$$\int_0^t dt' \exp\left[it'(\omega_{r'} - \omega_n - \omega_r + \omega_m)\right]. \qquad (\text{K}.23)$$

The integral is generally finite for large t. This implies $[\Lambda_{mn}^{rr'}]_1 = 0$. The exception is when

$$\omega_{r'} - \omega_r + \omega_m - \omega_n = 0 \qquad (\text{K}.24)$$

This condition is satisfied in particular if

$$m = n, \quad r = r'. \qquad (\text{K}.25)$$

Condition (K.24) would also be satisfied for $m = r$ and $n = r'$, but then (K.23) vanishes because of (K.19).

One can wonder whether other solutions of (K.24) are possible. This is generally not so and it will be assumed that it is not. However, there may be other solutions of (K.24) for certain Hamiltonians, and an example is just, for instance, the case $\mathcal{H}_{\text{sp}} = DS_z^2$. Then (K.24) yields $r^2 + n^2 = r'^2 + m^2$. Apart from the trivial solutions (K.25), there are other solutions, namely $|m|, |n|, |r|, |r'| \leq 10$,

$m, r' = 0$ and ± 5, $n, r = \pm 3$ and ± 4
$m, r' = \pm 1$ and ± 7, $n, r = \pm 4$ and ± 8
$m, r' = \pm 1$ and ± 5, $n, r = \pm 5$ and ± 7
$m, r' = \pm 6$ and ± 7, $n, r = \pm 2$ and ± 9.

However, these combinations are no longer solutions of (K.24) if more anisotropy terms are added or if a magnetic field is applied. For an arbitrary anisotropy, there are non-trivial solutions of (K.24) for particular values of the field. This might be the cause of the unexpected relaxation maxima observed by Gaudin *et al.* (2002). However, the matrix elements of $U_{q\alpha}$ are expected to be small. For that reason, only the solution (K.25) will be retained.

Assuming (K.25) to hold and taking the limit $t \to \infty$ in (K.23), one finds

$$[\Lambda_{mn}^{rr'}]_1 = -2\pi\delta_{mn}\delta_{rr'}\sum_\alpha F_{mrrm\alpha}(\omega_r - \omega_n). \tag{K.26}$$

The second diagram of Fig. K.1 is given by a quite analogous formula

$$[\Lambda_{mn}^{rr'}]_2 = -2\pi\delta_{mn}\delta_{rr'}\sum_\alpha G_{mrrm\alpha}(\omega_r - \omega_n) \tag{K.27}$$

where

$$G_{mrr'n\alpha}(\omega) = \frac{1}{N\hbar^2}\sum_q \langle m^* | U_{q\alpha}|r^*\rangle\,\langle r'^* | U_{q\alpha}^*|n^*\rangle\langle n_{q\alpha}\rangle\delta(\omega + \omega_{q\alpha}). \tag{K.28}$$

K.3.3   *Third to sixth diagrams*

The contribution of the third and fourth diagrams of Fig. K.1 is, according to (K.12) and (K.13),

$$t[\Lambda_{mn}^{rr'}]_3 + t[\Lambda_{mn}^{rr'}]_4 = \frac{1}{2\hbar^2}\exp[it(\omega_m - \omega_n)]\int_0^t dt' \int_0^t dt'' \sum_{QQ'} \rho_Q\,\langle m^*, \Phi_{Q'}|$$

$$\exp\left[i(t'-t)\mathcal{H}_{\text{tot}}^0/\hbar\right]\mathcal{H}_1 \exp\left[i(t''-t')\mathcal{H}_{\text{tot}}^0/\hbar\right]\mathcal{H}_1$$

$$\exp\left[-it''\mathcal{H}_{\text{tot}}^0/\hbar\right]|r^*, \Phi_Q\rangle\,\langle r', \Phi_Q|\exp\left[it\mathcal{H}_{\text{tot}}^0/\hbar\right]|n^*, \Phi_{Q'}\rangle$$

or

$$t[\Lambda_{mn}^{rr'}]_3 + t[\Lambda_{mn}^{rr'}]_4 = \delta_{r'n}\frac{1}{2\hbar^2}\exp[it(\omega_m - \omega_n)]\int_0^t dt' \int_0^t dt'' \sum_Q \rho_Q\,\langle m^*, \Phi_Q|$$

$$\exp\left[i(t'-t)\mathcal{H}_{\text{tot}}^0/\hbar\right]\mathcal{H}_1 \exp\left[i(t''-t')\mathcal{H}_{\text{tot}}^0/\hbar\right]\mathcal{H}_1$$

$$\times \exp\left[-it''\mathcal{H}_{\text{tot}}^0/\hbar\right]|r^*, \Phi_Q\rangle\,\langle n^*, \Phi_Q|\exp\left[it\mathcal{H}_{\text{tot}}^0/\hbar\right]|n^*, \Phi_Q\rangle. \tag{K.29}$$

Insertion of (5.34) yields two contributions, one of which corresponds to the third diagram and reads

$$
t[\Lambda_{mn}^{rr'}]_3 = \delta_{r'n} \frac{1}{2N\hbar^2} \exp[it(\omega_m - \omega_n)] \int_0^t dt' \int_0^t dt'' \sum_{q\alpha} \sum_Q \rho_Q \langle m^*, \Phi_Q |
$$

$$
\exp\left[i(t'-t)\mathcal{H}_{\mathrm{tot}}^0/\hbar\right] U_{q\alpha}(\mathbf{S})b_{q\alpha} \exp\left[i(t''-t')\mathcal{H}_{\mathrm{tot}}^0/\hbar\right] U_{q\alpha}^*(\mathbf{S})b_{q\alpha}^+
$$

$$
\exp\left[-it''\mathcal{H}_{\mathrm{tot}}^0/\hbar\right] | r^*, \Phi_Q\rangle \langle n^*, \Phi_Q | \exp\left[it\mathcal{H}_{\mathrm{tot}}^0/\hbar\right] | n^*, \Phi_Q\rangle. \quad \text{(K.30)}
$$

Intercalating the quantities $\sum | p*, \Phi_{Q'}\rangle \langle p*, \Phi_{Q'} |$ and $\int_{-\infty}^{\infty} d\omega \delta(\omega - \omega_{q\alpha})$ which are both equal to 1, (K.30) reads

$$
t[\Lambda_{mn}^{rr'}]_3 = \delta_{r'n} \frac{1}{2N\hbar^2} \sum_p \int_{-\infty}^{\infty} d\omega \delta(\omega - \omega_{q\alpha}) \exp[it(\omega_m - \omega_n)] \int_0^t dt' \int_0^t dt'' \sum_{q\alpha} \sum_p
$$

$$
\sum_{QQ'} \rho_Q \langle m^*, \Phi_Q | \exp\left[i(t'-t)\mathcal{H}_{\mathrm{tot}}^0/\hbar\right] U_{q\alpha}(\mathbf{S})b_{q\alpha}
$$

$$
\exp\left[i(t''-t')\mathcal{H}_{\mathrm{tot}}^0/\hbar\right] | p^*, \Phi_{Q'}\rangle
$$

$$
\langle p^*, \Phi_{Q'} | U_{q\alpha}^*(\mathbf{S})b_{q\alpha}^+ \exp\left[-it''\mathcal{H}_{\mathrm{tot}}^0/\hbar\right] | r^*, \Phi_Q\rangle
$$

$$
\langle n^*, \Phi_Q | \exp\left[it\mathcal{H}_{\mathrm{tot}}^0/\hbar\right] | n^*, \Phi_Q\rangle. \quad \text{(K.31)}
$$

Treating this formula as (K.14), one obtains

$$
t[\Lambda_{mn}^{rr'}]_3 = \delta_{r'n} \frac{1}{2N\hbar^2} \sum_p \sum_{q\alpha} \int_{-\infty}^{\infty} d\omega \delta(\omega - \omega_{q\alpha}) \exp[it(\omega_m - \omega_n)] \int_0^t dt' \int_0^t dt''
$$

$$
\langle m^* | \exp\left[i(t'-t)\mathcal{H}_{\mathrm{sp}}/\hbar\right] U_{q\alpha}(\mathbf{S}) \exp\left[i(t''-t')\mathcal{H}_{\mathrm{sp}}/\hbar\right] | p^*\rangle
$$

$$
\langle p^* | U_{q\alpha}^*(\mathbf{S}) \exp\left[-it''\mathcal{H}_{\mathrm{sp}}/\hbar\right] | r^*\rangle \exp\left[i(t''-t')\omega\right] \langle b_{q\alpha}b_{q\alpha}^+\rangle. \quad \text{(K.32)}
$$

Introducing the function (K.18) one obtains

$$
t[\Lambda_{mn}^{rr'}]_3 = \frac{1}{2}\delta_{r'n} \sum_{\alpha} \sum_p \int_{-\infty}^{\infty} d\omega F_{mppr\alpha}(\omega) \int_0^t dt' \int_0^t dt''
$$

$$
\exp\left[it'(\omega_m - \omega_p) + it''(\omega_p - \omega_r)\right] \exp\left[i(t''-t')\omega\right] \quad \text{(K.33)}
$$

or, for long times, after integration over t'',

$$t[\Lambda_{mn}^{rr'}]_3 = \pi\delta_{r'n}\sum_\alpha\sum_p\int_{-\infty}^{\infty}d\omega\, F_{mppr\alpha}(\omega)\int_0^t dt'\exp\left[it'(\omega_m - \omega_r)\right].$$

If both sides are divided by t and if the long t limit is taken, one obtains

$$[\Lambda_{mn}^{rr'}]_3 = \pi\delta_{r'n}\delta_{rm}\sum_\alpha\sum_p F_{mppm\alpha}(\omega_p - \omega_r). \tag{K.34}$$

The fourth diagram of Fig. K.1 yields an analogous formula where $F_{mppr\alpha}(\omega)$ is replaced by $G_{mppr\alpha}(\omega)$ defined by (K.28). Finally the fifth and sixth diagrams are obtained from the third and fourth diagrams by permutation of m and n. Therefore, if $m \neq n$

$$\Lambda_{mn}^{rr'} = \pi\delta_{r'n}\delta_{rm}\sum_\alpha\sum_p\left[F_{mppm\alpha}(\omega_p - \omega_r) + G_{mppm\alpha}(\omega_p - \omega_r)\right.$$

$$\left. + F_{nppn\alpha}(\omega_p - \omega_r) + G_{nppn\alpha}(\omega_p - \omega_r)\right] \tag{K.35}$$

while, if $m \neq r$, according to (K.26) and (K.27)

$$\Lambda_{mn}^{rr'} = -2\pi\delta_{mn}\delta_{rr'}\sum_\alpha\left[F_{mrrm\alpha}(\omega_r - \omega_n) + G_{mrrm\alpha}(\omega_r - \omega_n)\right]. \tag{K.36}$$

which is negative, so that the transition probability $-\Lambda_{mm}^{rr}$ from $|m^*\rangle$ to $|m^*\rangle$ is positive.

K.3.4   *Summary of this section*

Thus the only non-vanishing coefficients are Λ_{mm}^{rr} and

$$\Lambda_{mn}^{mn} = -\frac{1}{2}\sum_p[\Lambda_{mm}^{pp} + \Lambda_{nn}^{pp}]. \tag{K.37}$$

As stressed in Chapter 11, this relation is not exact, but rather a consequence of the approximations. In particular, relation (K.19) has been deduced from the absence of phonons at frequency 0. It seems difficult to extend this argument to a spin–phonon interaction which would *not* be linear with respect to phonon operators.

Whatever the initial density matrix $\rho_{mn}(0)$, the density matrix $\rho_{mn}(t)$ ultimately assumes the form

$$\rho_{mn}(t) = \rho_m(t)\delta_{mn} \tag{K.38}$$

which testifies to the loss of coherence and in particular corresponds to the equilibrium distribution.

K.4   Tunnelling

When deriving the above formulae, and in particular when deducing (K.25) from (K.24), the possibility of exact or approximate degeneracy has been ignored. In the case of tunnelling, there is an approximate degeneracy. For instance, in zero field, the states $|\,m^*\rangle$ and $|-m^*\rangle$ are almost degenerate with a weak tunnel splitting $2\hbar\omega_{\mathrm{T}}^{(m)}$. If this splitting is neglected, the spin–phonon interaction gives a finite lifetime τ_m to the state $|\,m^*\rangle$. Formulae (K.35)–(K.37) are expected to hold if $\omega_{\mathrm{T}}^{(m)}\tau_m \ll 1$, provided the *localized* vectors $|\,m^*\rangle$ are used in the formulae. If $\omega_{\mathrm{T}}^{(m)}\tau_m \gg 1$, the *exact*, delocalized eigenvectors $|\,m^*\rangle\pm|-m^*\rangle$ should be used. For instance, let equation (K.2) be written at very low temperature, making the following assumptions. (1) Only the lowest four states are populated. (2) The tunnelling splitting is negligible in the ground doublet but has an appreciable value $\hbar\omega_{\mathrm{T}}$ in the excited doublet. (3) The excited doublet has a lifetime τ which satifies $\omega_{\mathrm{T}}\tau \ll 1$. It is therefore appropriate to use localized states $|\,s^*\rangle$, $|-s^*\rangle$, $|\,(s-1)^*\rangle$, $|-(s-1)^*\rangle$.

The density matrix has 16 elements and satisfies equation (K.2). However the situation simplifies if, at $t = 0$, it is assumed to have a single non-vanishing element, namely $\rho_{-s,-s}(0) = 1$. The second (dissipative) term of (K.2) generates a component $\rho_{-(s-1),-(s-1)}$. Then tunnelling, the first term of (K.2), generates all components $\rho_{-(s-1),\pm(s-1)}$ and $\rho_{(s-1),\pm(s-1)}$. Then the dissipative term generates all components $\rho_{-s,\pm s}$ and $\rho_{s,\pm s}$. But components of the form $\rho_{-s,\pm(s-1)}$ or $\rho_{s,\pm(s-1)}$, etc. are never generated. Thus, the assumption $\rho_{-s,-s}(0) = 1$ implies that $\rho(t)$ has only eight non-vanishing components $\rho_{(mm')}$, with $(mm') = (-s,-s),\ (-s,s),\ (s,-s),\ (s,s),\ (-(s-1),-(s-1)),\ (-(s-1),s-1),\ (s-1,-(s-1)),\ (s-1,s-1)$ and equation (K.2) reads

$$\frac{d}{dt}\rho(t) = i\rho(t) \tag{K.39}$$

where the 8×8 matrix M is

$$
M =
\begin{bmatrix}
-\epsilon/\tau & 0 & 0 & 0 & 1/\tau & 0 & 0 & 0 \\
0 & -\epsilon/\tau & 0 & 0 & 0 & 0 & 0 & 0 \\
0 & 0 & -\epsilon/\tau & 0 & 0 & 0 & 0 & 0 \\
0 & 0 & 0 & -\epsilon/\tau & 0 & 0 & 0 & 1/\tau \\
\epsilon/\tau & 0 & 0 & 0 & -1/\tau & -i\omega_{\mathrm{T}} & i\omega_{\mathrm{T}} & 0 \\
0 & 0 & 0 & 0 & -i\omega_{\mathrm{T}} & -1/\tau & 0 & i\omega_{\mathrm{T}} \\
0 & 0 & 0 & 0 & i\omega_{\mathrm{T}} & 0 & -1/\tau & -i\omega_{\mathrm{T}} \\
0 & 0 & 0 & \epsilon/\tau & 0 & i\omega_{\mathrm{T}} & -i\omega_{\mathrm{T}} & -1/\tau
\end{bmatrix}
\tag{K.40}
$$

where $\epsilon = \exp(-\beta\Delta E)$ and ΔE is the energy of the excited doublet. The imaginary elements are not diagonal because the basis vectors are not exact eigenstates.

REFERENCES

Abbati, G.L., Brunel, L.C., Casalta, H., Cornia, A., Fabretti, A.C., Gatteschi, D., Hassan, A.K., Jansen, A.G.M., Maniero, A.L., Pardi, L.A., Paulsen, C. and Segre, U. (2001). *Chem. Eur. J.* **7**, 1796.

Abdi, A.N., Bucher, J.P., Rabu, P., Toulemonde, O., Drillon, M. and Gerbier, P. (2004). *J. Appl. Phys.* **95**, 7345.

Abragam, A. and Bleaney, B. (1986). *Electron paramagnetic resonance of transition ions.* Dover, New York.

Affronte, M., Lasjaunias, J.-C., Wernsdorfer, W., Sessoli, R., Gatteschi, D., Heath, S.L., Fort, A. and Rettori, A. (2002a). *Phys. Rev. B* **66**, art. no. 064408.

Affronte, M., Cornia, A., Lascialfari, A., Borsa, F., Gatteschi, D., Hinderer, J., Horvatic, M., Jansen, A.G.M. and Julien, M.H. (2002b). *Phys. Rev. Lett.* **88**, art. no.167201.

Affronte, A., Sessoli, R., Gatteschi, D., Wernsdorfer, W., Lasjaunias, J.C., Heath, S.L., Powell, A.K., Fort, A. and Rettori, A. (2004a). *J. Phys. Chem. Solids* **65**, 745.

Affronte, M., Ghirri, A., Carretta, S., Amoretti, G., Piligkos, S., Timco, G.A. and Winpenny, R.E.P. (2004b). *Appl. Phys. Lett.* **84**, 3468.

Amigó, R., Hernandez, J.M., García-Santiago, A. and Tejada, J. (2003a). *Phys. Rev. B* **67**, art. no. 220402.

Amigó, R., Hernandez, J.M., García-Santiago, A. and Tejada, J. (2003b). *Appl. Phys. Lett.* **82**, 4528.

Amoretti, G., Caciuffo, R., Combet, J., Murani, A. and Caneschi, A. (2000). *Phys. Rev. B* **62**, 3022.

Alvarez S. (2005). *Dalton Trans.*: 2209.

An, J., Chen, Z.-D., Bian, J., Chen, J.-T., Wang, S.-X., Gao, S. and Xu, G.-X. (2000). *Inorg. Chim. Acta* **299**, 28.

An, L., Owens, J.M., McNeil, L.E. and Liu, J. (2002). *J. Am. Chem. Soc.* **124**, 13688.

Anderson, P.W. and Hasegawa, H. (1955). *Phys. Rev.* **100**, 675.

Anderson, P.W. (1959). *Phys. Rev* **115**, 2.

Anderson, P.W. (1963). 'Magnetic Exchange', in: Rado, G.T. and Suhl, H., (eds.), *Magnetism.* Academic Press, New York, Vol. 1, pp. 25.

Andres, H., Basler, R., Güdel, H.U., Aromi, G., Christou, G., Buttner, H. and Ruffle, B. (2000). *J. Am. Chem. Soc.* **122**, 12469.

Anfuso, L., Rovai, D., Sessoli, R., Gaudin, G. and Villain, J. (2004). *J. Magn. Magn. Mater.* **272–276**, 1097.

Aromi, G., Aubin, S.M.J., Bolcar, M.A., Christou, G., Eppley, H.J., Folting, K., Hendrickson, D.N., Huffman, J.C., Squire, R.C., Tsai, H.L., Wang, S. and Wemple, M.W. (1998). *Polyhedron* **17**, 3005.

Artus, P., Boskovic, C., Yoo, J., Streib, W.E., Brunel, L.C., Hendrickson, D.N. and Christou, G. (2001). *Inorg. Chem.* **40**, 4199.

Ashcroft N. W. and Mermin, N.D. (1976). *Solid state physics.* Holt-Saunders, Philadelphia.

Aubin, S.M.J., Dilley, N.R., Pardi, L., Krzystek, J., Wemple, M.W., Brunel, L.C., Maple, M.B., Christou, G. and Hendrickson, D.N. (1998). *J. Am. Chem. Soc.* **120**, 4991.

Aubin, S.M.J., Sun, Z., Pardi, L., Krzystek, J., Folting, K., Brunel, L.-C., Rheingold, A.L., Christou, G. and Hendrickson, D.N. (1999). *Inorg. Chem.* **38**, 5329.

Aubin, S.M.J., Sun, Z.M., Eppley, H.J., Rumberger, E.M., Guzei, I.A., Folting, K., Gantzel, P.K., Rheingold, A.L., Christou, G. and Hendrickson, D.N. (2001). *Inorg. Chem.* **40**, 2127.

Awaga, K., Okuno, T., Yamaguchi, A., Hasegawa, M., Inabe, T., Maruyama, Y. and Wada, N. (1994). *Phys. Rev. B* **49**, 3975.

Axenovich, M. and Luban, M. (2001). *Phys. Rev. B* **63**, art. no. 100407.

Bachmann, R., Di Salvo, F.J., Geballe, T.H., Greene, R.L., Howard, R.E., King, C.N., Kirsch, H.C., Lee, K.N., Schwall, R.E., Thomas, H.U. and Zubeck, R.B. (1972). *Rev. Sci. Instr.* **43**, 205.

Bai, C. (2000). *Scanning tunneling microscopy and its application.* Springer-Verlag, New York.

Bal, M., Friedman, J.R., Suzuki, Y., Rumberger, E.M., Hendrickson, D.N., Avraham, N., Myasoedov, Y., Shtrikman, H. and Zeldov, E. (2005). *Europhys. Lett.* **71**, 110.

Ballhausen, C.J. (1962). *Introduction to ligand field theory.* McGraw-Hill, New York.

Barbara, B., Wernsdorfer, W., Sampaio, L.C., Park, J.G., Paulsen, C., Novak, M.A., Ferre, R., Mailly, D., Sessoli, R., Caneschi, A., Hasselbach, K., Benoit, A. and Thomas, L. (1995). *J. Magn. Magn. Mater.* **140**, 1825.

Barnett, S.J. (1935). *Rev. Mod. Phys.* **7**, 129.

Barra, A.L., Debrunner, P., Gatteschi, D., Schulz, Ch.E. and Sessoli, R. (1996). *Europhys. Lett.* **35**, 133.

Barra, A.L., Gatteschi, D., Sessoli, R., Abbati, G.L., Cornia, A., Fabretti, A.C. and Uytterhoeven, M.G. (1997a). *Angew. Chem. Int. Ed. Engl.* **36**, 2329.

Barra, A.L., Gatteschi, D. and Sessoli, R. (1997b). *Phys. Rev. B* **56**, 8192.

Barra, A.L., Brunel, L.C., Gatteschi, D., Pardi, L. and Sessoli, R. (1998). *Acc. Chem. Res.* **31**, 460.

Barra, A.L., Caneschi, A., Cornia, A., De Biani, F.F., Gatteschi, D., Sangregorio, C., Sessoli, R. and Sorace, L. (1999). *J. Am. Chem. Soc.* **121**, 5302.

Barra, A.L., Gatteschi, D. and Sessoli, R. (2000). *Chem. Eur. J.* **6**, 1608.

Barra, A.L. (2001). *Appl. Magn. Reson.* **21**, 619.

Barra, A.L., Bencini, F., Caneschi, A., Gatteschi, D., Paulsen, C., Sangregorio, C., Sessoli, R. and Sorace, L. (2001). *Chemphyschem* **2**, 523.

Bashkin, J.S., Chang, H.-R., Streib, W.E., Huffman, J.C., Hendrickson, D.N. and Christou, G. (1987). *J. Am. Chem. Soc.* **109**, 6502.

Basler, R., Boskovic, C., Chaboussant, G., Güdel, H.U., Murrie, M., Ochsenbein, S.T. and Sieber, A. (2003). *Chemphyschem* **4**, 910.

Baxter, P.N.W., Lehn, J.M., Fischer, J. and Youinou, M.T. (1994). *Angew. Chem. Int. Ed. Engl.* **33**, 2284.

Bellessa, G., Vernier, N., Barbara, B. and Gatteschi, D. (1999). *Phys. Rev. Lett.* **83**, 416.

Belorizky, E. and Fries, P.H. (1993). *J. Chim. Phys.* **90**, 1077.

Bencini, A. and Gatteschi, D. (1990). *EPR of exchange coupled systems.* Springer-Verlag, Berlin.

Bencini, A., Ciofini, I. and Uytterhoeven, M.G. (1998). *Inorg. Chim. Acta* **274**, 90.

Bencini, A. and Gatteschi, D. (1999). 'Electron Paramagnetic Resonance Spectroscopy', in: Lever, A.B.P. and Solomon, E.I., (eds.), *Inorganic electronic structure and spectroscopy, Vol I.* Wiley Interscience, New York, pp. 93-159.

Benelli, C. and Gatteschi, D. (2002). *Chem. Rev.* **102**, 2369.

Berliner, L.J. and Reuben, J. (1981). *Biological magnetic resonance.* Plenum Press, New York.

Berry, M.V. (1984). *Proc. R. Soc. London A* **392**, 45.

Bertini, I. and Luchinat, C. (1996). *Coord. Chem. Rev.* **150**, 1.

Bethe, H. (1929). *Ann. Physik* **3**, 133.

Bian, G.Q., Kuroda-Sowa, T., Konaka, H., Hatano, M., Maekawa, M., Munakata, M., Miyasaka, H. and Yamashita, M. (2004). *Inorg. Chem.* **43**, 4790.

Bimberg, D., Grundmann, M. and Ledentsov, N.N. (1999). *Quantum dots heterostructures.* John Wiley & Sons, New York.

Binder, K. and Young, A.P. (1986). *Rev. Mod. Phys.* **58**, 801.

Bircher, R., Chaboussant, G., Sieber, A., Gudel, H.U. and Mutka, H. (2004). *Phys. Rev. B* **70**, art. no. 212413.

Bleaney, B. Bowers, K. D. (1952). *Proc. R. Soc. London A* **68**, 57.

Blinc, R., Cevc, P., Arcon, D., Dalal, N.S. and Achey, R.M. (2001). *Phys. Rev. B* **63**, art. no. 212401.

Blume, M., (1963). *Phys. Rev* **130**, 1670.

Blundell, S.J. (2001). 'Muon-spin rotation studies of Molecule-based magnets', in: Miller, J.S. and Drillon, M., (eds.), *Magnetism: Molecules to materials*. Wiley-VCH, Weinheim, pp. 235–256.

Blundell, S.J. and Pratt, F.L. (2004). *J. Phys.: Condens. Matter* **16**, R771.

Bogani, L., Caneschi, A., Fedi, M., Gatteschi, D., Massi, M., Novak, M.A., Pini, M.G., Rettori, A., Sessoli, R. and Vindigni, A. (2004). *Phys. Rev. Lett.* **92**, art. no. 207204.

Bogoliubov, N.N. and Bogoliubov, N.N.Jr. (1982). *Introduction to quantum statistical mechanics*. World Scientific, Singapore.

Bokacheva, L., Kent, A.D. and Walters, M.A. (2000). *Phys. Rev. Lett.* **85**, 4803.

Bonadio, F., Gross, M., Stoeckli-Evans, H. and Decurtins, S. (2002). *Inorg. Chem.* **41**, 5891.

Borden W. T. (1999). 'Qualitative and quantitative predictions and measurements of Singlet-Triplet splittings in non-Kekulé hydrocarbon diradicals and heteroatom derivatives', in: Lahti, M.P., (ed.), *Magnetic properties of organic materials*. Marcel Dekker, New York Inc., pp. 61–102.

Borras-Almenar, J.J., Clemente-Juan, J.M., Coronado, E., Georges, R., Palii, A.V. and Tsukerblat, B.S. (1996). *J. Chem. Phys.* **105**, 6892.

Borras-Almenar, J.J., Coronado, E., Palii, A.V., Tsukerblat, B.S. and Georges, R. (1998a). *Chem. Phys.* **226**, 231.

Borras-Almenar, J.J., Clemente-Juan, J.M., Coronado, E., Palii, A.V. and Tsukerblat, B.S. (1998a). *J. Phys. Chem A* **102**, 200.

Borras-Almenar, J.J., Clemente-Juan, J.M., Coronado, E., Palii, A.V. and Tsukerblat, B.S. (1998c). *Phys. Lett. A* **238**, 164.

Borras-Almenar, J.J., Clemente-Juan, J.M., Coronado, E. and Tsukerblat, B.S. (1999). *Inorg. Chem.* **38**, 6081.

Borras-Almenar, J.J., Clemente-Juan, J.M., Coronado, E., Palii, A. and Tsukerblat, B.S. (2001a). 'Magnetic properties of mixed-valence clusters: theoretical approaches and applications', in: Miller, J.S. and Drillon, M., (eds.), *Magnetism: molecules to materials*. Wiley-VCH, Weinheim, pp. 155–210.

Borras-Almenar, J.J., Clemente-Juan, J.M., Coronado, E., Palii, A.V. and Tsukerblat, B.S. (2001b). *J. Chem. Phys.* **114**, 1148.

Boskovic, C., Pink, M., Huffman, J.C., Hendrickson, D.N. and Christou, G. (2001). *J. Am. Chem. Soc.* **123**, 9914.

Bouwen, A., Caneschi, A., Gatteschi, D., Goovaerts, E., Schoemaker, D., Sorace, L. and Stefan, M. (2001). *J. Phys. Chem B* **105**, 2658.

Boyd, P.D., Li, Q., Vincent, V.B., Folting, K., Chang, H.-R., Streib, W.E., Huffman, J.C., Christou, G. and Hendrickson, D.N. (1988). *J. Am. Chem. Soc.* **110**, 8537.

Braun, P.A. (1993). *Rev. Mod. Phys.* **65**, 115.

Brown, I.D. and Wu, K.K. (1976). *Acta Crystallogr.* **B32**, 1957.

Brune, M., Hagley, E., Dreyer, J., Maitre, X., Maali, A., Wunderlich, C., Raimond, J.M. and Haroche, S. (1996). *Phys. Rev. Lett.* **77**, 4887.

Caciuffo, R., Amoretti, G., Murani, A., Sessoli, R., Caneschi, A. and Gatteschi, D. (1998). *Phys. Rev. Lett.* **81**, 4744.

Cador, O., Gatteschi, D., Sessoli, R., Larsen, F.K., Overgaard, J., Barra, A.L., Teat, S.J., Timco, G.A. and Winpenny, R.E.P. (2004). *Angew. Chem. Int. Ed.* **43**, 5196.

Cage, B., Russek, S. E., Zipse, D. and Dalal, N. S. (2005). *J. Appl. Phys.* **97**, 10M507.

Cagnac, B. and Pébay-Pyroula (1983). *Physique atomique Vol. 1.* Dunod, Paris.

Caldeira, A. O., Leggett , A. J. (1983). *Ann. Phys.* (New York) **149**, 374.

Caneschi, A., Gatteschi, D., Laugier, J., Rey, P., Sessoli, R. and Zanchini, C. (1988). *J. Am. Chem. Soc.* **110**, 2795.

Caneschi, A., Gatteschi, D., Renard, J.P., Rey, P. and Sessoli, R. (1989). *Inorg. Chem.* **28**, 1976.

Caneschi, A., Gatteschi, D., Sessoli, R., Barra, A.-L., Brunel, L.C. and Guillot, M. (1991). *J. Am. Chem. Soc.* **113**, 5873.

Caneschi, A., Cornia, A. and Lippard, S.J. (1995). *Angew. Chem. Int. Ed. Engl.* **34**, 467.

Caneschi, A., Cornia, A., Fabretti, A.C., Foner, S., Gatteschi, D., Grandi, R. and Schenetti, L. (1996). *Chem. Eur. J.* **2**, 1379.

Caneschi, A., Gatteschi, D., Sessoli, R. and Schweizer, J. (1997). *Physica B* **241**, 600.

Caneschi, A., Ohm, T., Paulsen, C., Rovai, D., Sangregorio, C. and Sessoli, R. (1998). *J. Magn. Magn. Mater.* **177**, 1330.

Caneschi, A., Cornia, A., Fabretti, A.C. and Gatteschi, D. (1999). *Angew. Chem. Int. Ed. Engl.* **38**, 1295.

Caneschi, A., Gatteschi, D., Lalioti, N., Sangregorio, C., Sessoli, R., Venturi, G., Vindigni, A., Rettori, A., Pini, M.G. and Novak, M.A. (2001). *Angew. Chem. Int. Ed.* **40**, 1760.

Caneschi, A., Gatteschi, D., Lalioti, N., Sangregorio, C., Sessoli, R., Venturi, G., Vindigni, A., Rettori, A., Pini, M.G. and Novak, M.A. (2002). *Europhys. Lett.* **58**, 771.

Cannon, D.R. and White, R.P. (1988). *Progr. Inorg. Chem.* **36**, 195.

Carlin, R.L. (1986). *Magnetochemistry.* Springer-Verlag, Berlin.

Carretta, S., Van Slageren, J., Guidi, T., Liviotti, E., Mondelli, C., Rovai, D., Cornia, A., Dearden, A.L., Carsughi, F., Affronte, M., Frost, C.D., Winpenny, R.E.P., Gatteschi, D., Amoretti, G. and Caciuffo, R. (2003a). *Phys. Rev. B* **67**, art. no. 094405.

Carretta, S., Santini, P., Liviotti, E., Magnani, N., Guidi, T., Caciuffo, R. and Amoretti, G. (2003b). *Eur. Phys. J. B* **36**, 169.

Casimir, H.B.J. and Du Pré, F.K. (1938). *Physica* **V**, 507.

Chaboussant, G., Ochsenbein, S.T., Sieber, A., Güdel, H.U., Mutka, H., Müller, A. and Barbara, B. (2004). *Europhys. Lett.* **66**, 423.

Chakov, N.E., Wernsdorfer, W., Abboud, K.A., Hendrickson, D.N. and Christou, G. (2003). *Dalton Trans.* 2243.

Chakov, N.E., Wernsdorfer, W., Abboud, K.A. and Christou, G. (2004). *Inorg. Chem.* **43**, 5919.

Chapelier, C., El Khatib, M., Perrier, P., Benoit, A. and Mailly, D. (1993). Koch, H. and Lübbig, H. (eds.), *SQUID91, Superconducting devices and their applications.* Springer-Verlag, Berlin, pp. 286–291.

Cheesman, M.R., Oganesyan, V.S., Sessoli, R., Gatteschi, D. and Thomson, A.J. (1997). *Chem. Commun.*: 1677.

Chiolero, A. and Loss, D. (1998). *Phys. Rev. Lett.* **80**, 169.

Chiorescu, I., Wernsdorfer, W., Müller, A., Bogge, H. and Barbara, B. (2000a). *Phys. Rev. Lett.* **84**, 3454.

Chiorescu, I., Giraud, R., Jansen, A.G.M., Caneschi, A. and Barbara, B. (2000b). *Phys. Rev. Lett.* **85**, 4807.

Chouteau, G. and Veyret-Jeandey, C. (1981). *J. Phys.* **42**, 1441.

Christou, G. (1989). *Acc. Chem. Res.* **22**, 328.

Christou, G., Gatteschi, D., Hendrickson, D.N. and Sessoli, R. (2000). *Mater. Res. Bull.* **25**, 66.

Chudnovsky, E.M. (1994). *Phys. Rev. Lett.* **72**, 3433.

Chudnovsky, E.M. (2000). *Phys. Rev. Lett.* **84**, 5676.

Chudnovsky, E.M. and Garanin, D.A. (2001). *Phys. Rev. Lett.* **87**, art. no. 187203.

Chudnovsky, E.M. and Martinez-Hidalgo, X. (2002). *Phys. Rev. B* **66**, art. no. 054412.

Chudnovsky, E.M. and Garanin, D.A. (2002). *Phys. Rev. Lett.* **89**, art. no. 157201.

Cianchi, L., Del Giallo, F., Spina, G., Reiff, W. and Caneschi, A. (2002). *Phys. Rev. B* **65**, art. no. 064415.

Ciofini, I. and Daul, C.A. (2003). *Coord. Chem. Rev.* **238**, 187.

Clarke, J. (1990). 'SQUIDs: Principles, noise, and applications', in: Ruggiero, S.T. and Rudman, D.A., (eds.), *Superconducting devices*. Academic Press, San Diego.

Clemente-Leon, M., Soyer, H., Coronado, E., Mingotaud, C., Gomez-Garcia, C.J. and Delhaes, P. (1998). *Angew. Chem., Int. Ed. Engl.* **37**, 2842.

Clemente-Juan, J.M., Andres, H., Borras-Almenar, J.J., Coronado, E., Güdel, H.U., Aebersold, M., Kearly, G., Buttner, H. and Zolliker, M. (1999). *J. Am. Chem. Soc.* **121**, 10021.

Clerac, R., Miyasaka, H., Yamashita, M. and Coulon, C. (2002). *J. Am. Chem. Soc.* **124**, 12837.

Cohen, E. R., Taylor, B. N. (1987). *Rev. Mod. Phys.* **59**, 1121.

Cohen-Tannoudji, C., Diu, B. and Laloë, F. (1986). *Mécanique quantique.* Hermann, Paris.

Cohen-Tannoudji, C., Dupont-Roc, J. and Grynberg, G. (1987). *Photons et atomes.* Les Ulis: EDP Sciences.

Cole, K.S. and Cole R. H. (1941). *J. Chem. Phys.* **9**, 341.

Condorelli, G.G., Motta, A., Fragala, I.L., Giannazzo, F., Raineri, V., Caneschi, A. and Gatteschi, D. (2004). *Angew. Chem. Int. Ed.* **43**, 4081.

Corbino, O.M. (1911). *Phys. Z.* **12**, 292.

Cornia, A., Gatteschi, D. and Hegetschweiler, K. (1994). *Inorg. Chem.* **33**, 1559.

Cornia, A., Jansen, A.G.M. and Affronte, M. (1999). *Phys. Rev. B* **60**, 12177.

Cornia, A., Affronte, M., Jansen, A.G.M., Gatteschi, D., Caneschi, A. and Sessoli, R. (2000). *Chem. Phys. Lett.* **322**, 477.

Cornia, A., Fabretti, A.C., Sessoli, R., Sorace, L., Gatteschi, D., Barra, A.L., Daiguebonne, C. and Roisnel, T. (2002a). *Acta Crystallogr., Sect. C: Cryst. Struct. Commun.* **58**, m371.

Cornia, A., Sessoli, R., Sorace, L., Gatteschi, D., Barra, A.L. and Daiguebonne, C. (2002b). *Phys. Rev. Lett.* **89**, art. no. 257201.

Cornia, A., Fabretti, A.C., Pacchioni, M., Zobbi, L., Bonacchi, D., Caneschi, A., Gatteschi, D., Biagi, R., Del Pennino, U., De Renzi, V., Gurevich, L. and Van Der Zant, H.S.J. (2003). *Angew. Chem. Int. Ed.* **42**, 1645.

Cornia, A., Fabretti, A.C., Garrisi, P., Mortalo, C., Bonacchi, D., Gatteschi, D., Sessoli, R., Sorace, L., Wernsdorfer, W. and Barra, A.L. (2004). *Angew. Chem. Int. Ed.* **43**, 1136.

Coronado, E. and Gomez-García, C.J. (1998). *Chem. Rev.* **98**, 273.

Coronado, E., Forment-Aliaga, A., Gaita-Ariño, A., Giménez-Saiz, C, Romero, F.M., Wernsdorfer, W. (2004). *Angew. Chem. Int. Ed.* **43**, 6152.

Costes, J.P., Clemente-Juan, J.M., Dahan, F. and Milon, J. (2004). *Inorg. Chem.* **43**, 8200.

Coulon, C., Clerac, R., Lecren, L., Wernsdorfer, W. and Miyasaka, H. (2004). *Phys. Rev. B* **69**, art. no. 132408.

Coxall, R.A., Harris, S.G., Henderson, D.K., Parsons, S., Tasker, P.A. and Winpenny, R.E.P. (2000). *Dalton Trans.* 2349.

Cronin, L. (2004). 'High Nuclearity Clusters: Iso and Heteropolyanions and Relatives', in: Fujita, M., Powell, A. and Creutz, C. (eds.), *Comprehensive cordination chemistry II: from biology to nanotechnology.* Elsevier, Oxford, pp. 1–56.

Cuccoli, A., Fort, A., Rettori, A., Adam, E. and Villain, J. (1999). *Eur. Phys. J. B* **12**, 39.

Cullum, J.K. and Willoughby, R.A. (1985). *Lanczos algorithm for large symmetric eigenvalue computations.* Birkhauser, Boston.

De Loth, P., Daudey, J.P., Astheimer, H., Walz, L. and Haase W. (1985). *J. Chem. Phys.* **82**, 5048.

Del Barco, E., Kent, A.D., Rumberger, E.M., Hendrickson, D.N. and Christou, G. (2002). *Europhys. Lett.* **60**, 768.

Del Barco, E., Kent, A.D., Rumberger, E.M., Hendrickson, D.N. and Christou, G. (2003). *Phys. Rev. Lett.* **91**, art. no. 047203.

Del Barco, E., Kent, A.D., Yang, E.C. and Hendrickson, D.N. (2004). *Phys. Rev. Lett.* **93**, art. no. 157202.

Del Barco, E., Kent, A.D., Hill, S., North, J. M., Dalal, N. S., Rumberger, E.M., Hendrickson, D.N. Chakov, N. and Christou, G. (2005). *J. Low. Temp. Phys.* **140**, 119.

Dendrinou-Samara, C., Alexiou, M., Zaleski, C.M., Kampf, J.W., Kirk, M.L., Kessissoglou, D.P. and Pecoraro, V.L. (2003). *Angew. Chem. Int. Ed.* **42**, 3763.

De W. Horrocks, W.Jr. and De W. Hall, D. (1971). *Coord. Chem. Rev.* **6**, 147.

Dekker, C., Arts, A.F.M., Wijn, H.W., van Duyneveldt, A.J. and Mydosh, J.A. (1989). *Phys. Rev. B* **40**, 11243.

Dicke, R.H. (1954). *Phys. Rev.* **93**, 99.

Dirac, P.A.M. (1929). *Proc. R. Soc. London A* **123**, 714.

Diu, B., Guthmann, C., Lederer, D. and Roulet, B. (1995). *Physique statistique.* Hermann, Paris.

Domingo, N., Williamson, B.E., Gomez-Segura, J., Gerbier, Ph., Ruiz-Molina, D., Amabilino, D.B., Veciana, J. and Tejada, J. (2004). *Phys. Rev. B* **69**, art. no. 052405.

Eaton, G.R. and Eaton, S.S. (1999). *Appl. Magn. Reson.* **16**, 161.

Eisenstein, J. (1951). *Phys. Rev.* **84**, 548.

Ekert, A. and Jozsa, R. (1996). *Rev. Mod. Phys.* **68**, 733.

Eppley, H.J., Tsai, H.-L., Devries, N., Folting, K., Christou, G. and Hendrickson, D.N. (1995). *J. Am. Chem. Soc.* **117**, 301.

Eppley, H.J., Aubin, S.M.J., Wemple, M.W., Adams, D.M., Tsai, H.L., Grillo, V.A., Castro, S.L., Sun, Z.M., Folting, K., Huffman, J.C., Hendrickson, D.N. and Christou, G. (1997). *Mol. Cryst. Liq. Cryst.* **305**, 167.

Evangelisti, M. and Bartolomé, J. (2000). *J. Magn. Magn. Mater.* **221**, 99.

Evangelisti, M., Luis, F., Mettes, F.L., Aliaga, N., Aromi, G., Alonso, J.J., Christou, G. and de Jongh, L.J. (2004). *Phys. Rev. Lett.* **93**, art. no. 117202.

Ferbinteanu, M., Miyasaka, H., Wernsdorfer, W., Nakata, K., Sugiura, K., Yamashita, M., Coulon, C. and Clerac, R. (2005). *J. Am. Chem. Soc.* **127**, 3090.

Ferlay, S., Mallah, T., Ouahes, R., Veillet, P. and Verdaguer, M. (1995). *Nature* **378**, 701.

Fernández, J.F., Luis, F. and Bartolomé, J. (1998). *Phys. Rev. Lett.* **80**, 5659.

Fernández, J.F. and Alonso, J.J. (2000). *Phys. Rev. B* **62**, 53.

Fernández, J.F. and Alonso, J.J. (2002). *Phys. Rev. B* **65**, art. no. 189901.

Fernández, J.F. and Alonso, J.J. (2003). *Phys. Rev. Lett.* **91**, art. no. 047202.

Fernández, J.F. (2003). in: Garrido, P.L. and Marro, J., (eds.) , *Modelling of complex systems.* AIP-Melville, New York.

Fernández, J.F. and Alonso, J.J. (2004). *Phys. Rev. Lett.* **92**, art. no. 119702.

Feynman, R. P. (1948). *Rev. Mod. Phys.* **20**, 367.

Feynman, R.P. and Hibbs, A.R. (1965). *Quantum mechanics and path integrals.* McGraw-Hill, New York.

Finn, R.C. and Zubieta, J. (2000). *J. Cluster science* **11**, 461.

Fleury, B., Catala, L., Huc, V., David, C., Zhong, W.Z., Jegou, P., Baraton, L., Palacin, S., Albouy, P.A. and Mallah, T. (2005). *Chem. Commun.*: 2020.

Fominaya, F., Fournier, T., Gandit, P. and Chaussy, J. (1997a). *Rev. Sci. Instrum.* **68**, 4191.

Fominaya, F., Villain, J., Gandit, P., Chaussy, J. and Caneschi, A. (1997b). *Phys. Rev. Lett.* **79**, 1126.

Fominaya, F., Villain, J., Fournier, T., Gandit, P., Chaussy, J., Fort, A. and Caneschi, A. (1999). *Phys. Rev. B* **59**, 519.

Foner, S. (1959). *Rev. Sci. Instr.* **30**, 548.

Fort, A., Rettori, A., Villain, J., Gatteschi, D. and Sessoli, R. (1998). *Phys. Rev. Lett.* **80**, 612.

Fort, A. (2001) Thesis. Università di Firenze, Italy.

Friedman, J.R., Sarachik, M.P., Tejada, J. and Ziolo, R. (1996). *Phys. Rev. Lett.* **76**, 3830.

Furukawa, Y., Kumagai, K., Lascialfari, A., Aldrovandi, S., Borsa, F., Sessoli, R. and Gatteschi, D. (2001). *Phys. Rev. B* **64**, art. no. 094439.

Furukawa, Y., Watanabe, K., Kumagai, K., Borsa, F. and Gatteschi, D. (2001). *Phys. Rev. B* **64**, art. no. 104401.

Furukawa, Y., Kawakami, S., Kumagai, K., Baek, S.-H. and Borsa, F. (2003). *Phys. Rev. B* **68**, art. no. 180405.

Gaita, A. (2004), PhD Thesis, University of Valencia, Spain.

Garanin, D.A. (1991). *J. Phys. A: Math. Gen.* **24**, L61.

Garanin, D.A. and Chudnovsky, E.M. (1997). *Phys. Rev. B* **56**, 11102.

Garanin, D.A., Chudnovsky, E.M. and Schilling, R. (2000). *Phys. Rev. B* **61**, 12204.

Garanin, D.A. and Chudnovsky, E.M. (2002). *Phys. Rev. B* **65**, art. no. 094423.

Garg, A. (1993). *Europhys. Lett.* **22**, 205.

Garg, A. (1999). *Phys. Rev. Lett.* **83**, 4385.

Gatteschi, D., Pardi, L., Barra, A.-L., Müller, A. and Döring, J. (1991). *Nature* **354**, 463.

Gatteschi, D. and Pardi, L. (1993). 'Spin levels of high nuclearity spin clusters', in: O'Connor, C.J., (ed.), *Research frontiers in magnetochemistry*. World Scientific, Singapore, pp. 67–86.

Gatteschi, D., Caneschi, A., Pardi, L. and Sessoli, R. (1994). *Science* **265**, 1054.

Gatteschi, D., Sessoli, R., Plass, W., Müller, A., Krickemeyer, E., Meyer, J., Solter, D. and Adler, P. (1996). *Inorg. Chem.* **35**, 1926.

Gatteschi, D. and Sessoli, R. (2003). *Angew. Chem. Int. Ed.* **42**, 268.

Gaudin, G., Gandit, P., Chaussy, J. and Sessoli, R. (2002). *J. Magn. Magn. Mater.* **242**, 915.

Gerbier, P., Ruiz-Molina, D., Domingo, N., Amabilino, D.B., Vidal-Gancedo, J., Tejada, J., Hendrickson, D.N. and Veciana, J. (2003). *Monatsh. Chem.* **134**, 265.

Gerritsen, H.J.C. and Sabisky, E.S. (1963). *Phys. Rev.* **132**, 1507.

Gider, S., Awschalom, D.D., Douglas, T., Mann, S. and Chaparala, M. (1995). *Science* **268**, 77.

Gillon, B. (2001). 'Spin distribution in molecular systems with interacting transition metal ions', in: Miller, J.S. and Drillon, M., (eds.), *Magnetism: molecules to materials*. Wiley-VCH, Weinheim, pp. 357–378.

Giraud, R., Wernsdorfer, W., Tkachuk, A.M., Mailly, D. and Barbara, B. (2001). *Phys. Rev. Lett.* **87**, art. no. 057203.

Girerd, J.-J., Kahn, O. and Verdaguer, M. (1980). *Inorg. Chem.* **19**, 274.

Girerd, J.-J. (1983). *J. Chem. Phys.* **79**, 1766.

Glaser, T., Heidemeier, M. and Lugger, T. (2003). *Dalton Trans.* 2381.

Glauber, R.J. (1963). *J. Math. Physics* **4**, 294.

Goldberg, D.P., Caneschi, A., Delfs, C.D., Sessoli, R. and Lippard, S.J. (1995). *J. Am. Chem. Soc.* **117**, 5789.

Gomes, A.M., Novak, M.A., Sessoli, R., Caneschi, A. and Gatteschi, D. (1998). *Phys. Rev. B* **57**, 5021.

Gomes, A.M., Novak, M.A., Nunes, W.C. and Rapp, R.E. (2001). *J. Magn. Magn. Mater.* **226**, 2015.

Gomez-Segura, J., Lhotel, E., Paulsen, C., Luneau, D., Wurst, K., Veciana, J., Ruiz-Molina, D. and Gerbier, P. (2005). *New. J. Chem.* **29**, 499.

Goodenough, J.B. (1958). *J. Phys. Chem. Solids* **6**, 287.

Goodenough, J.B. (1963). *Magnetism and the chemical bond.* Interscience, New York.

Goodwin, J.C., Sessoli, R., Gatteschi, D., Wernsdorfer, W., Powell, A.K. and Heath, S.L. (2000). *Dalton Trans.* 1835.

Gorter, C.J. and Brons, F. (1937). *Physica* **4**, 579.

Gorter, C.J. (1947). *Paramagnetic relaxation.* Elsevier, Amsterdam.

Gorun, S.M. and Lippard, S.J. (1991). *Inorg. Chem.* **30**, 1625.

Goto, T., Koshiba, T., Kubo, T. and Awaga, K. (2003). *Phys. Rev. B* **67**, art. no.-104408.

Grahl, M., Kötzler, J. and Sessler, I. (1990). *J. Magn. Magn. Mat.* **90&91**, 187.

Griffith, J.S. (1961). *The theory of transition metal ions.* Cambridge University Press, Cambridge.

Grifoni, M. and Hanggi, P. (1998). *Phys. Rep.* **304**, 229.

Gross, M. and Haroche, S. (1982). *Phys. Rep.* **93**, 301.

Guidi, T., Carretta, S., Santini, P., Liviotti, E., Magnani, N., Mondelli, C., Waldmann, O., Thompson, L.K., Zhao, L., Frost, C.D., Amoretti, G. and Caciuffo, R. (2004). *Phys. Rev. B* **69**, art. no. 104432.

Gull, S.F. and Daniell, G.J. (1978). *Nature* **272**, 686.

Gunther, L. and Barbara, B. (1995). *Quantum tunneling of magnetization-QTM '94.* Kluwer, Dordrecht.

Haldane, F.D.M. (1983). *Phys. Rev. Lett.* **50**, 1153.

Harter, A. G., Chakov, N.E., Roberts, B., Achey, R., Reyes, A., Kuhns, P., Christou, G., Dalal, N. S. (2005). *Inorg. Chem.* **44**, 2122.

Hartmann-Boutron, F., Politi, P. and Villain, J. (1996). *Int. J. Mod. Phys. B* **10**, 2577.

Hartmann-Boutron, F. (1996). *J. Physique I* **5**, 1281.

Hatfield, W.A. (1983). *Inorg. Chem.* **22**, 833.

Hay, P.J., Thibeault, J.C. and Hoffmann, R. (1975). *J. Am. Chem. Soc.* **97**, 4884.

Heath, S.L. and Powell, A.K. (1992). *Angew. Chem. Int. Ed. Engl.* **31**, 191.

Hegetschweiler, K., Morgenstern, B., Zubieta, J., Hagrman, P.J., Lima, N., Sessoli, R. and Totti, F. (2004). *Ang. Chem. Int. Ed.* **43**, 3436.

Heisenberg, W. (1926). *Z. Phys.* **38**, 411.

Hennion, M., Pardi, L., Mirebeau, I., Suard, E., Sessoli, R. and Caneschi, A. (1997). *Phys. Rev. B* **56**, 8819.

Hernandéz, J.M., Zhang, X.X., Luis, F., Bartolomé, J., Tejada, J. and Ziolo, R. (1996). *Europhys. Lett.* **35**, 301.

Herpin, A. (1968). *Théorie du magnétisme.* Presses Univ. France, Paris.

Hibbs, W., Rittenberg, D.K., Sugiura, K., Burkhart, B.M., Morin, B.G., Arif, A.M., Liable-Sands, L., Rheingold, A.L., Sundaralingam, M., Epstein, A.J. and Miller, J.S. (2001). *Inorg. Chem.* **40**, 1915.

Hill, S., Perenboom, J.A.A.J., Dalal, N.S., Hathaway, T., Stalcup, T. and Brooks, J.S. (1998). *Phys. Rev. Lett.* **80**, 2453.

Hill, S., Edwards, R.S., Aliaga-Alcalde, N. and Christou, G. (2003). *Science* **302**, 1015.

Honecker, A., Meier, F., Loss, D. and Normand, B. (2002). *Eur. Phys. J. B* **27**, 487.

Huang, S.M., Fu, Q., An, L. and Liu, J. (2004). *Phys. Chem. Chem. Phys.* **6**, 1077.

Ibach, H. and Lüth, H. (1999). *Festkörperphysik.* Springer-Verlag, Berlin.

Ishikawa, N., Sugita, M., Ishikawa, T., Koshihara, S. and Kaizu, Y. (2003a). *J. Am. Chem. Soc.* **125**, 8694.

Ishikawa, N., Sugita, M., Okubo, T., Tanaka, N., Lino, T. and Kaizu, Y. (2003b). *Inorg. Chem.* **42**, 2440.

Ishikawa, N., Sugita, M., Ishikawa, T., Koshihara, S. and Kaizu, Y. (2004). *J. Phys. Chem. B* **108**, 11265.

Ishikawa, N., Sugita, M. and Wernsdorfer, W. (2005). *J. Am. Chem. Soc.* **127**, 3650.

Itoh, K. (1978). *Pure Appl. Chem.* **50**, 1251.

Jérome, D. and Schulz, H.J. (2002). *Adv. Phys.* **51**, 293.

Kagan, Yu. and Maksimov, L.A. (1980). *Sov. Phys. JETP* **52**, 688.

Kagan, Yu. and Legget, A.J. (1992). 'Quantum Tunneling in Condensed Media, in: *Modern problem in condenser matter.* Elsevier, Amsterdam.

Kahn, O. and Briat, B.J. (1976). *J. Chem. Soc. Faraday Trans. II* **72**, 268.

Kahn, O., Pei, Y., Verdaguer, M., Renard, J.-P. and Sletten, J. (1988). *J. Am. Chem. Soc.* **110**, 782.

Kahn, O. (1993). *Molecular magnetism.* VCH, Weinheim.

Kahn, O. and Martinez, C.J. (1998). *Science* **279**, 44.

Kambe, K. (1950). *J. Phys. Soc. Jpn.* **5**, 48.

Kanamori J. (1959). *J. Phys. Chem. Solids* **10**, 87.

Kanamori, J. (1963). Rado, G.T. and Suhl, H., (eds.), *Magnetism.* Academic Press, New York.

Katsnelson, M.I., Dobrovitski, V.V. and Harmon, B.N. (1999). *Phys. Rev. B* **59**, 6919.

Kayanuma, Y. (1984). *J. Phys. Soc. Jpn.* **53**, 108.

Keçecioğlu, E. and Garg, A. (2002). *Phys. Rev. Lett.* **88**, art. no. 237205.

Keçecioğlu, E. and Garg, A. (2003). *Phys. Rev. B* **67**, art. no. 054406.

Kent, A.D., von Molnàr, S., Gider, S. and Awschalom, D.D. (1994). *J. Appl. Phys.* **76**, 6656.

Kent, A.D., Zhong, Y.C., Bokacheva, L., Ruiz, D., Hendrickson, D.N. and Sarachik, M.P. (2000). *Europhys. Lett.* **49**, 521.

Khan, M. I. and Zubieta, J. (1995). *Prog. Inorg. Chem.* **43**, 1.

Khan, M.I., Chen, Q., Salta, J., O'Connor, C.J. and Zubieta, J. (1996). *Inorg. Chem.* **35**, 1880.

Korenblitt, I.Ya. and Shender, E.F. (1978). *Sov. Phys. JEPT* **48**, 937.

Kortus, J. and Pederson, M.R. (2000). *Phys. Rev. B* **62**, 5755.

Kortus, J., Hellberg, C.S. and Pederson, M.R. (2001). *Phys. Rev. Lett.* **86**, 3400.

Kortus, J., Baruah, T., Bernstein, N. and Pederson, M.R. (2002). *Phys. Rev. B* **66**, art. no. 092403.

Kozlov, G. V. , Volkov, A. A. (1998). In: Gruner G. (ed.), *Millimeter and submillimeter wave spectroscopy of solid.* Springer, Berlin, p. 51.

Kramers, H.A. (1930). *Proc. Acad. Sci. Amsterdam* **33**, 959.

Kubo, R. (1957) *J. Phys. Soc. Jpn.* **12**, 570.

Kubo, R., Toyabe, T. (1967). In: Blinc, R. (ed.) *Magnetic resonance and relaxation.* North-Holland, Amsterdam, p. 810.

Kubo, T., Goto, T., Koshiba, T., Takeda, K. and Awaga, K. (2002). *Phys. Rev. B* **65**, art. no. 224425.

Kuroda-Sowa, T., Lam, M., Rheingold, A.L., Frommen, C., Reiff, W.M., Nakano, M., Yoo, J., Maniero, A.L., Brunel, L.C., Christou, G. and Hendrickson, D.N. (2001). *Inorg. Chem.* **40**, 6469.

Kuroda-Sowa, T., Handa, T., Kotera, T., Maekawa, M., Munakata, M., Miyasaka, H. and Yamashita, M. (2004). *Chem. Lett.* **33**, 540.

Köhler, F.H. (2001). 'Probing spin density by use of NMR spectroscopy ', in: Miller, J.S. and Drillon, M., (eds.), *Magnetism: molecules to materials.* Wiley-VCH, Weinheim, pp. 379–340.

Lahti, P.M. (1999). *Magnetic properties of organic materials.* Marcel Dekker Inc., New York.

Lancaster, T., Blundell, S., Pratt, F., Brooks, M.L., Manson, J.L., Brechin, E.K., Cadiou, C., Low, D., McInnes, E.J.L. and Winpenny, R. E. P (2004). *J. Phys. Condens. Mat.* **16**, S4563.

Landau, L. (1932). *Phys. Z. Sowjetunion* **2**, 46.

Landau, L. and Lifshitz E. M. (1960). *Mécanique*. Moscow: MIR.

Landau, L. and Lifshitz E. M. (1969). *Electrodynamique des Milieux Continus*. Moscow: MIR.

Langan, P., Robinson, R., Brown, P.J., Argyriou, D., Hendrickson, D. and Christou, G. (2001). *Acta Cryst.* **C57**, 909.

Larionova, J., Gross, M., Pilkington, M., Andres, H., Stoeckli-Evans, H., Güdel, H.U. and Decurtins, S. (2000). *Ang. Chem. Int. Ed. Engl.* **39**, 1605.

Larsen, F.K., McInnes, E.J.L., El Mkami, H., Rajaraman, G., Rentschler, E., Smith, A.A., Smith, G.M., Boote, V., Jennings, M., Timco, G.A. and Winpenny, R.E.P. (2003). *Ang. Chem. Int. Ed. Engl.* **42**, 101.

Lascialfari, A., Jang, Z.H., Borsa, F., Carretta, P. and Gatteschi, D. (1998). *Phys. Rev. B* **57**, 514.

Lascialfari, A., Ullu, R., Affronte, M., Cinti, F., Caneschi, A., Gatteschi, D., Rovai, D., Pini, M.G. and Rettori, A. (2003). *Phys. Rev. B* **67**, art. no. 224408.

Laye, R.H. and McInnes, E.J.L. (2004). *Eur. J. Inorg. Chem.* 2811.

Le Gall, F., Fabrizi de Biani, F., Caneschi, A., Cinelli, P., Cornia, A., Fabretti, A.C. and Gatteschi, D. (1997). *Inorg. Chim. Acta* **262**, 123.

Leggett, A.J. (1995). 'Macroscopic quantum effects in magnetic systems: an overview', in: Gunther, L. and Barbara, B., (eds.), *Quantum tunneling of magnetization-QTM '94*. Kluwer, Dordrecht, pp. 1–18.

Lehn, J.-M. (1995). *Supramolecular chemistry. Concepts and perspectives*. VCH, Weinheim.

Lescouezec, R., Vaissermann, J., Ruiz-Perez, C., Lloret, F., Carrasco, R., Julve, M., Verdaguer, M., Dromzee, Y., Gatteschi, D. and Wernsdorfer, W. (2003). *Angew. Chem. Int. Ed.* **42**, 1483.

Leuenberger, M.N. and Loss, D. (1999). *Europhys. Lett.* **46**, 692.

Leuenberger, M.N. and Loss, D. (2001). *Nature* **410**, 789.

Lever, A.B.P. and Solomon, E.I. (1999). 'Ligand Field Theory and the Properties of Transition Metal Complexes', in: Lever, A.B.P. and Solomon, E.I., (eds.), *Inorganic electronic structure and spectroscopy, Vol. I*. Wiley Interscience, New York, pp. 1–91.

Lis, T. (1980). *Acta Crystallog. B* **36**, 2042.

Liu, T.F., Fu, D., Gao, S., Zhang, Y.Z., Sun, H.L., Su, G. and Liu, Y.J. (2003). *J. Am. Chem. Soc.* **125**, 13976.

Liviotti, E., Carretta, S. and Amoretti, G. (2002). *J. Chem. Phys.* **117**, 3361.

Lovesey, S.W. (1986). *Theory of neutron scattering from condensed matter.* Oxford Univ. Press, Oxford.

Low, D.M., Jones, L.F., Bell, A., Brechin, E.K., Mallah, T., Riviere, E., Teat, S.J. and Mcinnes, E.J.L. (2003). *Angew. Chem. Int. Ed. Engl.* **42**, 3781.

Luis, F., Bartolomé, J. and Fernàndez, J.F. (1998). *Phys. Rev. B* **57**, 505.

Lüthi, B. (1980). 'Interactions of Magnetic Ions with Phonons', in: Horton, G.K. and Maradudin, A.A., (eds.), *Dynamical properties of solids.* North-Holland, Amsterdam.

Mandel, A., Schmitt, W., Womack, T.G., Bhalla, R., Henderson, R.K., Heath, S.L. and Powell, A.K. (1999). *Coord. Chem. Rev.* **192**, 1067.

Manriquez, J.M., Yee G. T. , McLean, R.S., Epstein, A.J. and Miller, J.S. (1991). *Science* **252**, 1415.

Martinez-Hidalgo, X., Chudnovsky, E.M. and Aharony, A.A. (2001). *Europhys. Lett.* **55**, 273.

Marvaud, V., Decroix, C., Scuiller, A., Tuyeras, F., Guyard-Duhayon, C., Vaissermann, J., Marrot, M., Gonnet, F. and Verdaguer, M. (2003a). *Chem. Eur. J.* **9**, 1692.

Marvaud, V., Decroix, C., Scuiller, A., Guyard-Duhayon, C., Vaissermann, J., Gonnet, F. and Verdaguer, M. (2003b). *Chem. Eur. J.* **9**, 1677.

McConnell, J. (1980). *Rotational brownian motion and dielectric theory.* Academic Press, New York.

McCusker, J.K., Vincent, J.B., Schmitt, E.A., Mino, M.L., Shin, K., Coggin, D.K., Hagen, P.M., Huffman, J.C., Christou, G. and Hendrickson, D.N. (1991). *J. Am. Chem. Soc.* **113**, 3012.

McGarvey, B.R. (1966). *Transition Met. Chem.* **3**, 89.

McInnes, E.J.L., Anson, C., Powell, A.K., Thomson, A.J., Poussereau, S. and Sessoli, R. (2001). *Chem. Commun.* 89.

McInnes, E.J.L., Pidcock, E., Oganesyan, V.S., Cheesman, M.R., Powell, A.K. and Thomson, A.J. (2002). *J. Am. Chem. Soc.* **124**, 9219.

Meier, F. and Loss, D. (2001). *Phys. Rev. Lett.* **86**, 5373.

Meier, F., Levy, J. and Loss, D. (2003). *Phys. Rev. Lett.* **90**, art. no. 047901.

Mertes, K.M., Suzuki, Y., Sarachik, M.P., Paltiel, Y., Shtrikman, H., Zeldov, E., Rumberger, E., Hendrickson, D.N. and Christou, G. (2001). *Phys. Rev. Lett.* **87**, art. no. 227205.

Mertes, K.M., Suzuki, Y., Sarachik, M.P., Myasoedov, Y., Shtrikman, H., Zeldov, E., Rumberger, E.M., Hendrickson, D.N. and Christou, G. (2003). *J. Appl. Phys.* **93**, 7095.

Messiah, A. (1965). *Quantum mechanics.* North Holland, Amsterdam.

Miller , J.S., Calabrese, J.C., Rommelmann H., Chittapeddi S.R., Zhang, J.H., Reiff, W.M. and Epstein, A.J. (1987). *J. Am. Chem. Soc.* **109**, 769.

Miller, J.S., Drillon, M. (eds). 2001–2005 *Magnetism: Molecules to materials.* Vol. I–V. Wiley-VCH, Weinheim.

Mirebeau, I., Hennion, M., Casalta, H., Andres, H., Güdel, H.U., Irodova, A.V. and Caneschi, A. (1999). *Phys. Rev. Lett.* **83**, 628.

Mitra, S. (1977). *Prog. Inorg. Chem.* **22**, 309.

Miyasaka, H., Clerac, R., Mizushima, K., Sugiura, K., Yamashita, M., Wernsdorfer, W. and Coulon, C. (2003). *Inorg. Chem.* **42**, 8203.

Miyasaka, H., Nezu, T., Sugimoto, K., Sugiura, K.C., Yamashita, M. and Clerac, R. (2004). *Inorg. Chem.* **43**, 5486.

Miyashita, S. and Saito, K. (2001). *J Phys. Soc. Jpn.* **70**, 3238.

Mohr, P. J. and Taylor, B.N. (2003). *Phys. Today* **56**, BG6.

Morello, A., Mettes, F.L., Luis, F., Fernández, J.F., Krzystek, J., Aromi, G., Christou, G. and De Jongh, L.J. (2003). *Phys. Rev. Lett.* **90**, art. no. 017206.

Morello, A., Bakharev, O.N., Brom, H.B., Sessoli, R. and de Jongh, L.J. (2004). *Phys. Rev. Lett.* **93**, art. no. 197202.

Moriya, T. (1960). *Phys. Rev.* **117**, 635.

Moriya, T. (1963). Rado G.T. and Suhl, H., (eds.), *Magnetism.* Academic Press, New York.

Morrish, A.H. (1966). *The physical principles of magnetism.* Wiley, New York.

Mossin, S., Weihe, H., Sorensen, H.O., Lima, N. and Sessoli, R. (2004). *Dalton Trans.* 632.

Mukhin, A.A., Travkin, V.D., Zvezdin, A.K., Lebedev, S.P., Caneschi, A. and Gatteschi, D. (1998). *Europhys. Lett.* **44**, 778.

Mukhin, A., Gorshunov, B., Dressel, M., Sangregorio, C. and Gatteschi, D. (2001). *Phys. Rev. B* **63**, art. no. 214411.

Müller, A., Sessoli, R., Krickemeyer, E., Bogge, H., Meyer, J., Gatteschi, D., Pardi, L., Westphal, J., Hovemeier, K., Rohlfing, R., Doring, J., Hellweg, F., Beugholt, C. and Schmidtmann, M. (1997). *Inorg. Chem.* **36**, 5239.

Müller, A., Peters, F., Pope, M.T. and Gatteschi, D. (1998). *Chem. Rev.* **98**, 239.

Müller, A., Sarkar, S., Shah, S.Q.N., Bogge, H., Schmidtmann, M., Sarkar, Sh., Kögerler, P., Hauptfleisch, B., Trautwein, A.X. and Schunemann, V. (1999). *Angew. Chem. Int. Ed. Engl.* **38**, 3238.

Müller, A., Kögerler, P. and Dress, A.W.M. (2001a). *Coord. Chem. Rev.* **222**, 193.

Müller, A., Luban, M., Schröder, C., Modler, R., Kögerler, P., Axenovich, M., Schnack, J., Canfield, P., Bud'ko, S. and Harrison, N. (2001b). *Chemphyschem* **2**, 517.

Müller, A., Beckmann, E., Bogge, H., Schmidtmann, M. and Dress, A. (2002). *Angew. Chem. Int. Ed. Engl.* **41**, 1162.

Müller, A. and Roy, S. (2003). *Coord. Chem. Rev.* **245**, 153.

Murugesu, M., Habrych, M., Wernsdorfer, W., Abboud, K.A. and Christou, G. (2004a). *J. Am. Chem. Soc.* **126**, 4766.

Murugesu, M., Clerac, R., Anson, C.E. and Powell, A.K. (2004b). *Chem. Commun.* 1598.

Mydosh, J.A. (1993). *Spin glasses: an experimental introduction.* Taylor & Francis Ltd, London.

Nakamura, N., Inoue, K. and Iwamura, H. (1993). *Angew. Chem. Int. Ed. Engl.* **32**, 872.

Nakano, H. and Miyashita, S. (2001). *J. Phys. Soc. Jpn.* **70**, 2151.

Néel L. (1949). *Ann. Geophys.* **5**, 99.

Noodleman, L. (1981). *J. Chem. Phys* **74**, 5737.

Noodleman, L., Post, D. and Baerends, E.J. (1982). *Chem. Phys.* **64**, 159.

Noodleman, L. and Davidson, E.R. (1986). *Chem. Phys.* **109**, 131.

Noodleman, L., Peng, C.Y., Case, D.A. and Mouesca, J.-M. (1995). *Coord. Chem. Rev.* **144**, 199.

Novak, M.A. and Sessoli, R. (1995). 'AC Suseptibility Relaxation Studies on a Manganese Organic Cluster Compound: Mn12Ac', in: Gunther, L. and Barbara, B., (eds.), *Quantum tunneling of magnetization - QTM'94.* Kluwer, Dordrecht, pp. 171–188.

Novak, M. A., Folly, W. S. D., Sinnecker, J. P, and Soriano, S. (2005). *J. Magn. Magn. Mater.* **294**, 133.

Ohm, T., Sangregorio, C. and Paulsen, C. (1998). *J. Low Temp. Phys.* **113**, 1141.

Osa, S., Kido, T., Matsumoto, N., Re, N., Pochaba, A. and Mrozinski, J. (2004). *J. Am. Chem. Soc.* **126**, 420.

Oshio, H., Hoshino, N., Ito, T., Nakano, M., Renz, F. and Gutlich, P. (2003). *Angew. Chem. Int. Ed. Engl.* **42**, 223.

Pacchioni, M., Cornia, A., Fabretti, A.C., Zobbi, L., Bonacchi, D., Caneschi, A., Chastanet, G., Gatteschi, D. and Sessoli, R. (2004). *Chem. Commun.* 2604.

Palacio, F., Antorrena, G., Castro, M., Burriel, R., Rawson, J.M., Smith, J.N.B., Bricklebank, N., Novoa, J. and Ritter, C. (1997). *Phys. Rev. Lett.* **79**, 2336.

Pankhurst, Q.A., Connolly, J., Jones, S.K. and Dobson, J. (2003). *J. Phys. D: Appl. Phys.* **36**, R167.

Papoular, R.J. and Gillon, B. (1990). *Europhys. Lett.* **13**, 429.

Pardo, E., Ruiz-Garcia, R., Lloret, F., Faus, J., Julve, M., Journaux, Y., Delgado, F. and Ruiz-Perez, C. (2004). *Adv. Mater.* **16**, 1597.

Park, H., Park, J., Lim, A.K.L., Anderson, E.H., Alivisatos, A.P. and McEuen, P.L. (2000). *Nature* **407**, 57.

Park, K., Baruah, T., Bernstein, N. and Pederson, M.R. (2004). *Phys. Rev. B* **69**, art. no. 144426.

Paulsen, C. and Park, K.J.G. (1995). Gunther, L. and Barbara, B., (eds.), *Quantum tunneling of magnetization-QTM'94*. Kluwer, Dordrecht, pp. 19–58.

Paulsen, C., Park, J.G., Barbara, B., Sessoli, R. and Caneschi, A. (1995). *J. Magn. Magn. Mater.* **140**, 379.

Pederson, M.R. and Khanna S. N. (1999). *Phys. Rev. B* **60**, 9566.

Pederson, M.R., Porezag, D.V., Kortus, J. and Khanna, S.N. (2000). *J. Appl. Phys.* **87**, 5487.

Pederson, M.R., Bernstein, N. and Kortus, J. (2002). *Phys. Rev. Lett.* **89**, art. no. 097202.

Perenboom, J.A.A.J., Brooks, J.S., Hill, S., Hathaway, T. and Dalal, N.S. (1998a). *Physica B* **246**, 294.

Perenboom, J.A.A.J., Brooks, J.S., Hill, S., Hathaway, T. and Dalal, N.S. (1998b). *Phys. Rev. B* **58**, 330.

Petukhov, K., Wernsdorfer, W., Barra, A.-L. and Mosser, V. (2005). *Phys. Rev. B* **72** art. no. 052401,

Pilbrow, J.R. (1990). *Transition ion electron paramagnetic resonance.* Oxford University Press, Oxford.

Pohl, I.A.M., Westin, L.G. and Kritikos, M. (2001). *Chem. Eur. J.* **7**, 3438.

Pontillon, Y., Caneschi, A., Gatteschi, D., Sessoli, R., Ressouche, E., Schweizer, J. and Lelievre-Berna, E. (1999). *J. Am. Chem. Soc.* **121**, 5342.

Press, W.H., Flannery, B.P., Teukolsky, S.A. and Vetterling, W.T. (1989). *Numerical recipes.* Cambridge University Press, Cambridge.

Prokofev, N.V. and Stamp, P.C.E. (1996). *J. Low Temp. Phys.* **104**, 143.

Prokofev, N.V. and Stamp, P.C.E. (1998a). *Phys. Rev. Lett.* **80**, 5794.

Prokofev, N.V. and Stamp, P.C.E. (1998b). *J. Low Temp. Phys.* **113**, 1147.

Prokofev, N.V. and Stamp, P.C.E. (2000). *Phys. Rev. Lett.* **84**, 5677.

Rabenau, A. (1985). *Angew. Chem. Int. Ed. Engl.* **24**, 1026.

Raghu, C., Rudra, I., Sen, D. and Ramasesha, S. (2001). *Phys. Rev. B* **64**, art. no. 064419.

Razavy, M. (2003). *Quantum theory of tunneling.* World Scientific, Singapore.

Redl, F.X., Cho, K.S., Murray, C.B. and O'brien, S. (2003). *Nature* **423**, 968.

Regnault, N., Jolicoeur, T., Gatteschi, D., Sessoli, R. and Verdaguer, M. (2002). *Phys. Rev. B.* **66**, art. no. 054409.

Robin, M.B. and Day, P. (1967). *Adv. Inorg. Chem. Radiochem.* **10**, 247.

Robinson, R.A., Brown, P.J., Argyriou, D.N., Hendrickson, D.N. and Aubin, S.M.J. (2000). *J. Phys.: Condens. Matter* **12**, 2805.

Rosemberg, B., Van Camp, L., Trosko, J.E. and Mansour, V.H. (1969). *Nature* **222**, 385.

Rudra, L., Ramasesha, S. and Sen, D. (2001). *Phys. Rev. B* **64**, art. no. 014408.

Ruiz, E., Alvarez S., Rodriguez-Fortea, A., Alemany, P., Pouillon, Y. and Massobrio, C. (2001). 'Electronic Structure and Magnetic Behaviour in Poly-nucelar Transition Metal Complexes', in: Miller, J.S. and Drillon, M., (eds.), *Magnetism: Molecules to materials* II Wiley-VCH, Weinheim.

Ruiz, E., Rajaraman, G., Alvarez, S., Gillon, B., Stride, J., Clerac, R., Larionova, J. and Decurtins, S. (2005). *Angew. Chem. Int. Ed.* **44**, 2711.

Ruiz, D., Sun, Z., Albela, B., Folting, K., Ribas, J., Christou, G., and Hendrickson, D.N. (1998). *Angew. Chem. Int. Ed. Engl.* **37**, 300.

Ruiz-Molina, D., Gerbier, P., Rumberger, E., Amabilino, D. B., Guzei, I. A., Folt-ing, K., Huffman, J.C., Rheingold, A., Christou, G., Veciana, J. and Hendrickson, D. N. (2002). *J. Mat. Chem.* **12**, 1152.

Saalfrank, R.W., Bernt, I., Uller, E. and Hampel, F. (1997). *Angew. Chem. Int. Ed. Engl.* **36**, 2482.

Saalfrank, R.W., Bernt, I., Chowdhry, M.M., Hampel, F. and Vaughan, G.B.M. (2001). *Chemistry Eur. J.* **7**, 2765.

Saha, D.K., Padhye, S., Anson, C.E. and Powell, A.K. (2002). *Inorg. Chem. Commun.* **5**, 1022.

Saint Paul, M. and Veyret, C. (1973). *Phys. Lett. A* **45**, 362.

Saint Pierre, T.G., Webb, J. and Mann, S. (1989). 'Ferritin and Hemosiderin: Structural and Magnetic Studies of the Iron Core', in: Mann, S., Webb, J. and Williams, J.P., (eds.), *Biomineralization.* VCH, Weinheim, pp. 295–344.

Salman, Z. (2002). Preprint cond-mat/0209497 and PhD Thesis, Technion-Israel Institute of Technology, Haifa, Israel.

Salta, J., Chen, Q., Chang, Y.D. and Zubieta, J. (1994). *Angew. Chem. Int. Ed. Engl.* **33**, 757.

Sangregorio, C., Ohm, T., Paulsen, C., Sessoli, R. and Gatteschi, D. (1997). *Phys. Rev. Lett.* **78**, 4645.

Santini, P., Carretta, S., Amoretti, G., Guidi, T., Caciuffo, R., Caneschi, A., Rovai, D., Qiu, Y. and Copley, J.R.D. (2005). *Phys. Rev. B* **71**, art. no.184405.

Sato, O., Iyoda, A., Fujishima, K. and Hashimoto, K. (1996). *Science* **272**, 704.

Schäffer, C.E. (1968). *Struct. Bonding (Berlin)* **5**, 68.

Schäffer, C.E. (1973). *Struct. Bonding (Berlin)* **12**, 50.

Schelter, E.J., Prosvirin, A.V. and Dunbar, K.R. (2004). *J. Am. Chem. Soc.* **126**, 15004.

Schenck, A., Gygax, F.N. (1995). In: Buschow, K. H. J. (ed.), *Handbook of magnetic materials.* Elsevier, Amsterdam, vol. 9, p. 57.

Schiff, L.I. (1946). *Quantum mechanics*. McGraw-Hill, New York.

Schilling, R. (1995). 'Quantum spin tunneling and path integral approach', in: Gunther, L. and Barbara, B., (eds.), *Quantum tunneling of magnetization-QTM '94*. Kluwer, Dordrecht, pp. 59–76.

Schmitt, W., Anson, C.E., Pilawa, B. and Powell, A.K. (2002). *Z. Anorg. Allg. Chem.* **628**, 2443.

Schnack, J. and Luban, M. (2001). *Phys. Rev. B* **63**, art. n° 014418.

Schwartz, A., Dressel, M., Blank, A. Csiba, T., Gruner, G., Volkov, A.A., Gorshunov, B.P., Kozlov, G.V. (1995). *Rev. Sci. Instrum.* **66**, 2943.

Shor, P. (1994). In Goldwasser, S. (ed.), *Proc. 35th Annu. Symp. on the Foundations of Computer Science*. IEEE Computer Society Press, Los Alamitos, California, pp. 124–134.

Schrödinger, E. (1960). *Statistical thermodynamics*. Cambridge Univ. Press, Cambridge.

Schweizer, J. and Ressouche, E. (2001). 'Neutron scattering and spin densities in free radicals', in: Miller, J.S. and Drillon, M., (eds.), *Magnetism: molecules to materials*. Wiley-VCH, Weinheim, pp. 325–355.

Sen, K.D. (2002). *J. Chem. Phys.* **116**, 9570.

Sessoli, R., Gatteschi, D., Caneschi, A. and Novak, M.A. (1993a). *Nature (London)* **365**, 141.

Sessoli, R., Tsai, H.L., Schake, A.R., Wang, S.Y., Vincent, J.B., Folting, K., Gatteschi, D., Christou, G. and Hendrickson, D.N. (1993b). *J. Am. Chem. Soc.* **115**, 1804.

Sessoli, R. (1995). *Mol. Cryst. Liq. Cryst.* **273**, 145.

Sessoli, R., Caneschi, A., Gatteschi, D., Sorace, L., Cornia, A. and Wernsdorfer, W. (2001). *J. Magn. Magn. Mater.* **226**, 1954.

Shaikh, N., Parja, A., Goswami, S., Banerjee, P., Vojtisek, P., Zhang, Y.Z., Su, G. and Gao, S. (2004). *Inorg. Chem.* **43**, 849.

Shapira, Y., Liu, M.T., Foner, S., Dube, C.E. and Bonitatebus, P.J. (1999). *Phys. Rev. B* **59**, 1046.

Shapira, Y., Liu, M.T., Foner, S., Howard, R.J. and Armstrong, W.H. (2001). *Phys. Rev. B* **63**, art. no. 094422.

Sharpe, A. G. (1976). *The chemistry of cyano complexes of transition metals*. Academic Press, New York.

Silver, B.L. (1976). *Irreducible tensor methods, an introduction for chemists*. Academic Press, New York.

Slater, J.C. (1968). *Quantum theory of matter*. McGraw-Hill, New York.

Slichter C. P. (1963). *Principles of magnetic resonance*. Harper, New York.

Sokol, J.J., Shores, M.P. and Long, J.R. (2002a). *Inorg. Chem.* **41**, 3052.

Sokol, J.J., Hee, A.G. and Long, J.R. (2002b). *J. Am. Chem. Soc.* **124**, 7656.

Soler, M., Chandra, S.K., Ruiz, D., Davidson, E.R., Hendrickson, D.N. and Christou, G. (2000). *Chem. Commun.* 2417.

Soler, M., Artus, P., Folting, K., Huffman, J. C., Hendrickson, D. N. and Christou, G. (2001a). *Inorg. Chem.* **40**, 4902.

Soler, M., Chandra, S.K., Ruiz, D., Huffman, J.C., Hendrickson, D.N. and Christou, G. (2001b). *Polyhedron* **20**, 1279.

Soler, M., Wernsdorfer, W., Sun, Z., Ruiz, D., Huffman, J.C., Hendrickson, D.N. and Christou, G. (2003). *Polyhedron* **22**, 1783.

Soos, Z.G. and Ramasesha, S. (1990). In: Klein, D.J. and Trinajstic, N., (eds.), *Valence bond theory and chemical structure.* Elsevier, New York, p. 81.

Sorace, L., Wernsdorfer, W., Thirion, C., Barra, A.L., Pacchioni, M., Mailly, D. and Barbara, B. (2003). *Phys. Rev. B* **68**, art. no. 220407.

Stemmler, A. J. , Kampf, J. W. and Pecoraro, V. L. (1996). *Angew. Chem. Int. Ed.* **35**, 23.

Steiner, M., Villain, J. and Windsor, C.G. (1976). *Adv. Phys.* **25**, 87.

Stoll, S.L., Steckel, J.S., Persky, N.S., Martinez, C.R., Barnes, C.L., Fry, E.A., Kulkarni, J., Burgess, J.D. and Pacheco, J.D. (2004). *Nano Letters* **4**, 1167.

Stückelberg, E.C.G. (1932). *Helv. Phys. Acta* **5**, 369.

Sugimoto, T. (ed.) (2000). *Fine Particles: Synthesis, Characterization, and Mechanisms of Growth.* Marcel Dekker, New York.

Sullivan, P.F. and Seidel, G. (1968a). *Ann. Acad. Sci Fennicae* **210**, 58.

Sullivan, P.F. and Seidel, G. (1968b). *Phys. Rev.* **173**, 679.

Sun, S.H. and Murray, C.B. (1999). *J. Appl. Phys.* **85**, 4325.

Sun, Z.M., Ruiz, D., Rumberger, E., Incarvito, C.D., Folting, K., Rheingold, A.L., Christou, G. and Hendrickson, D.N. (1998). *Inorg. Chem.* **37**, 4758.

Sun, Z.M., Ruiz, D., Dilley, N. R., Soler, M., Ribas, J., Folting, K., Maple, M.B., Christou, G. and Hendrickson, D.N. (1999). *Chem. Commun.*: 1973.

Taft, K.L., Delfs, C.D., Papaefthymiou, G.C., Foner, S., Gatteschi, D. and Lippard, S.J. (1994). *J. Am. Chem. Soc.* **116**, 823.

Takeda, K. and Awaga, K. (1997). *Phys. Rev. B* **56**, 14560.

Takeda, K., Awaga, K. and Inabe, T. (1998). *Phys. Rev. B* **57**, R11062.

Tamura M., Nakazawa, Y., Shiomi, D., Nozawa, Y., Hosokoshi, M., Ishikawa, M., Takahashi, M. and Kinoshita, M. (1991). *Chem. Phys. Lett.* **186**, 401.

Tari, A. (2003). *The Specific heat of matter at low temperatures.* Imperial College Press, London.

Tasiopoulos, A.J., Wernsdorfer, W., Moulton, B., Zaworotko, M.J. and Christou, G. (2003). *J. Am. Chem. Soc.* **125**, 15274.

Tasiopoulos, A.J., Vinslava, A., Wernsdorfer, W. , Abboud, K.A. and Christou, G. (2004). *Angew. Chem. Int. Ed. Engl.* **43**, 2117.

Tasset, F. (2001). *J. Phys. IV*, **11**, 159.

Tejada, J., Zhang, X.X., Del Barco, E., Hernandéz, J.M. and Chudnovsky, E.M. (1997). *Phys. Rev. Lett.* **79**, 1754.

Tejada, J., Amigó, R., Hernandéz, J.M. and Chudnovsky, E.M. (2003). *Phys. Rev. B* **68**, art. no. 014431.

Thomas, L., Lionti, F., Ballou, R., Gatteschi, D., Sessoli, R. and Barbara, B. (1996). *Nature* **383**, 145.

Thomas, L., Caneschi, A. and Barbara, B. (1999). *Phys. Rev. Lett.* **83**, 2398.

Thompson, L.K. (2002). *Coord. Chem. Rev.* **233**, 193.

Thompson, L.K., Waldmann, O. and Xu, Z. (2003). 'Magnetic Properties of Self-assembled [2 × 2] and [3 × 3] Grids.', in: Miller, J.S. and Drillon, M. (eds.), *Magnetism: Molecules to materials IV*. Wiley-VCH, Weinheim, pp. 173.

Tiron, R., Wernsdorfer, W., Foguet-Albiol, D., Aliaga-Alcalde, N. and Christou, G. (2003). *Phys. Rev. Lett.* **91**, art. no. 227203.

Toraldo di Francia, G. and Bruscaglioni, P. (1988). *Onde elettromagnetiche.* Zanichelli, Bologna.

Troiani, F., Affronte, M., Carretta, S., Santini, P. and Amoretti, G. (2005a). *Phys. Rev. Lett.* **94**, art. no. 190501.

Troiani, F., Ghiri, A., Affronte, M., Carretta, S., Santini, P., Amoretti, G., Piligkos, S., Timco, G. and Winpenny, R.E.P. (2005b). *Phys. Rev. Lett.* **94**, art. no. 207208.

Tsai H.L.; Chen, D.M., Yang, C.I., Jwo, T.Y., Wur, C.S., Lee, G.H. and Wang, Y. (2001). *Inorg. Chem. Commun.* **4**, 511.

Tsukerblat, B.S., Belinski, M.I. and Fainzil'berg, V.E. (1987). *Sov. Sci.Rev. B. Chem. (Engl. Transl.)* **9**, 339.

Tupitsyn, I.S. and Barbara, B. (2002). 'Quantum Tunneling of magnetization in molecular complexes with large spins- effects of the environment', in: Miller, J.S. and Drillon, M., (eds.), *Magnetism: Molecules to materials* III Wiley-VCH, Weinheim, pp. 109.

Tupitsyn, I.S., Prokofev, N.V. and Stamp, P.C.E. (1997). *Intern. J. Mod. Phys. B* **11**, 2901.

Tupitsyn, I.S. and Stamp, P.C.E. (2004). *Phys. Rev. Lett.* **92** art. no. 119701.

Ullmann, A. (1991). *Introduction to ultrathin organic films: from Langmuir Blodgedtt to self-assembly.* Academic Press, San Diego.

Vandersypen, L.M.K, Steffen, M., Breyta, G., Yannoni, C.S., Sherwood, M.H., and Chuang, I.L. (2001). *Nature* **414**, 883.

Van Hemmen, J.L. and Sütö, A. (1986). *Europhys. Lett.* **1**, 481.

Van Hemmen, J.L. and Sütö, A. (1995). 'Theory of mesoscopic quantum tunneling in magnetism: a WKB approach', in: Gunther, L. and Barbara, B., (eds.), *Quantum tunneling of the magnetization- QTM '94*. Kluwer, Dordrecht, pp. 19–57.

Van Niekerk. and J. N., Schoening, F.R.L. (1953). *Acta Crystallogr.* **6**, 227.

Van Slageren, J., Sessoli, R., Gatteschi, D., Smith, A.A., Helliwell, M., Winpenny, R.E.P., Cornia, A., Barra, A.L., Jansen, A.G.M., Rentschler, E. and Timco, G.A. (2002). *Chem. Eur. J.* **8**, 277.

Van Slageren, J., Vongtragool, S., Gorshunov, B., Mukhin, A.A., Karl, N., Krzystek, J., Telser, J., Müller, A., Sangregorio, C., Gatteschi, D. and Dressel, M. (2003). *Phys. Chem. Chem. Phys.* **5**, 3837.

Van Vleck, J.H. (1932). *The theory of electric and magnetic susceptibility.* Oxford Univ. Press, Oxford.

Verdaguer, M., Bleuzen, A., Marvaud, V., Vaissermann, J., Seuileman, M., Desplanches, C., Scuiller, A., Train, C., Garde, R., Gelly, G., Lomenech, C., Rosenman, I.V.P., Cartier, C. and Villain, F. (1999a). *Coord. Chem. Rev.* **190–192**, 1023.

Veyret, C. and Blaise, A. (1973). *Mol. Phys.* **25**, 873.

Villain, J., Hartmann-Boutron, F., Sessoli, R. and Rettori, A. (1994). *Europhys. Lett.* **27**, 159.

Villain, J., Würger, A., Fort, A. and Rettori, A. (1997). *J. Phys. I* **7**, 1583.

Villain, J. and Fort, A. (2000). *Eur. Phys. J.* **17**, 69.

Villain, J. (2003). *Ann. Phys.Paris* **28**, 1.

Vincent, J.B., Chang, H.-R., Folting, K., Huffman, J.C., Christou, G. and Hendrickson, D.N. (1987). *J. Am. Chem. Soc.* **109**, 5703.

von Neumann, J. and Wigner, E.P. (1929). *Phys. Z.* **30**, 467.

Wada, N., Kobayashi, T., Yano, H., Okuno, T., Yamaguchi, A. and Awaga, K. (1997). *J. Phys. Soc. Jpn.* **66**, 961.

Waldmann, O. (2002a). *Phys. Rev. B* **65**, art. no. 024424.

Waldmann, O. (2002b). *Europhys. Lett.* **60**, 302.

Waldmann, O., Guidi, T., Carretta, S., Mondelli, C. and Dearden, A.L. (2003). *Phys. Rev. Lett.* **91**, art. no. 237202.

Waldmann, O., Carretta, S., Santini, P., Koch, R., Jansen, A.G.M., Amoretti, G., Caciuffo, R., Zhao, L. and Thompson, L.K. (2004). *Phys. Rev. Lett.* **92**, art. no. 096403.

Watton, S.P., Fuhrmann, P., Pence, L.E., Caneschi, A., Cornia, A., Abbati, G.L. and Lippard, S.J. (1997). *Angew. Chem. Int. Ed. Engl.* **36**, 2774.

Wei, Y.-G., Zhang, S.-W., Shao, M.-C. and Tang, Y.-Q. (1997). *Polyhedron* **16**, 1471.

Weigert, S. (1994). *Europhys. Lett.* **26**, 561.

Weihe, H. and Güdel, H.U. (1997a). *Inorg. Chem.* **36**, 3632.

Weihe, H. and Güdel, H.U. (1997b). *J. Am. Chem. Soc.* **119**, 6539.

Weinland, R.F. and Fischer, G. (1921). *Z. Anorg. Allg. Chem.* **120**, 161.

Wemple, M.W., Tsai, H.L., Folting, K., Hendrickson, D.N. and Christou, G. (1993). *Inorg. Chem.* **32**, 2025.

Wernsdorfer, W., Hasselbach, K., Benoit, A., Barbara, B., Mailly, D., Tuaillon, J., Perez, J.P., Dupuis, V., Dupin, J.P., Giraud, G. and Perex A. (1995). *J. Appl. Phys.* **78**, 7192.

Wernsdorfer, W. and Sessoli, R. (1999). *Science* **284**, 133.

Wernsdorfer, W., Ohm, T., Sangregorio, C., Sessoli, R., Mailly, D. and Paulsen, C. (1999). *Phys. Rev. Lett.* **82**, 3903.

Wernsdorfer, W., Sessoli, R., Caneschi, A., Gatteschi, D., Cornia, A. and Mailly, D. (2000a). *J. Appl. Phys.* **87**, 5481.

Wernsdorfer, W., Sessoli, R., Caneschi, A., Gatteschi, D. and Cornia, A. (2000b). *Europhys. Lett.* **50**, 552.

Wernsdorfer, W., Caneschi, A., Sessoli, R., Gatteschi, D., Cornia, A., Villar, V. and Paulsen, C. (2000c). *Phys. Rev. Lett.* **84**, 2965.

Wernsdorfer, W., Sessoli, R., Caneschi, A., Gatteschi, D. and Cornia, A. (2000d). *J. Phys. Soc. Jpn.* **69**, 375.

Wernsdorfer, W. (2001). *Adv. Chem. Phys.* **118**, 99.

Wernsdorfer, W., Aliaga-Alcalde, N., Hendrickson, D.N. and Christou, G. (2002). *Nature* **416**, 406.

White, R. (1983). *Quantum theory of magnetism.* Springer, Berlin.

Wieghardt, K., Pohl, K., Jibril, I. and Huttner, G. (1984). *Angew. Chem. Int. Ed. Engl.* **23**, 77.

Willet, R.D., Gatteschi, D. and Kahn, O. (1983). *Magneto-structural correlations in exchange coupled systems.* Reidel Publishing, Dordrecht.

Williams, W.G. (1988). *Polarized neutrons.* Oxford Univ. Press, Oxford.

Winpenny, R.E.P. (2002). *Dalton Trans.* 1.

Winpenny, R.E.P. (2004). 'High Nuclearity Clusters: Clusters and Aggregates with Paramagnetic Centers:Oxygen and Nitrogen Bridged Systems', in: Fujita, M., Powell, A. and Creutz, C., (eds.), *Comprehensive coordination chemistry II: From biology to nanotechnology.* Elsevier, Oxford, pp. 125.

Wong, K.K.W., Douglas, T., Gider, S., Awschalom, D.D. and Mann, S. (1998). *Chem. Mater.* **10**, 279.

Würger, A. (1998). *J. Phys.: Condens. Matter* **10**, 10075.

Xia, Y.N., and Whitesides, G.M. (1998). *Angew. Chem. Int. Ed.* **37**, 551.

Yang, J.Y., Shores, M.P., Sokol, J.J. and Long, J.R. (2003). *Inorg. Chem.* **42**, 1403.

Zaleski, C.M., Depperman, E.C., Kampf, J.W., Kirk, M.L. and Pecoraro, V.L. (2004). *Angew. Chem. Int. Ed.* **43**, 3912.

Zener, C. (1932). *Proc. R. Soc. London A* **137**, 696.

Zener, C. (1951). *Phys. Rev.* **82**, 403.

Zhao, H.H., Berlinguette, C.P., Bacsa, J., Prosvirin, A.V., Bera, J.K., Tichy, S.E., Schelter, E.J. and Dunbar, K.R. (2004). *Inorg. Chem.* **43**, 1359.

Zhao, L., Xu, Z.Q., Thompson, L.K., Heath, S.L., Miller, D.O. and Ohba, M. (2000a). *Angew. Chem. Int. Ed. Engl.* **39**, 3114.

Zhao, L.A., Matthews, C.J., Thompson, L.K. and Heath, S.L. (2000b). *Chem. Commun.* 265.

Zhong, Y.C., Sarachik, M.P., Friedman, J.R., Robinson, R.A., Kelley, T.M., Nakotte, H., Christianson, A.C., Trouw, F., Aubin, S.M.J. and Hendrickson, D.N. (1999). *J. Appl. Phys.* **85**, 5636.

Zhong, Z.J., Seino, H., Mizobe, Y., Hidai, M., Fujishima, A., Ohkoshi, S. and Hashimoto, K. (2000). *J. Am. Chem. Soc.* **122**, 2952.

Zobbi, L., Mannini, M., Pacchioni, M., Chastanet, G., Bonacchi, D., Zanardi, C., Biagi, R., Del Pennino, U., Gatteschi, D., Cornia, A. and Sessoli, R. (2005). *Chem. Commun.:* 1640.

Zurek, W.H. (1991). *Phys. Today* **44**, 36.

Zurek, W.H. (2003). *Rev. Mod. Phys.* **75**, 715.

INDEX